Monitoring the Comprehensive Nuclear-Test-Ban Treaty: Source Processes and Explosion Yield Estimation

Edited by
Göran Ekström
Marvin Denny
John R. Murphy

Springer Basel AG

Reprint from Pure and Applied Geophysics
(PAGEOPH), Volume 158 (2001), No. 11

Editors:

Göran Ekström
Harvard University
Department of Earth & Planetary Sciences
20 Oxford Street
Cambridge, Massachusetts 02138
USA
e-mail: ekstrom@seismology.harvard.edu

Marvin Denny
Lawrence Livermore National Laboratory
P.O. Box 808, MS L-205
Livermore, CA 94550-0808
USA

John R. Murphy
Maxwell Technologies, Inc.
11800 Sunrise Valley Dr., Suite 1212
Reston, VA 20191
USA
e-mail: jrm@maxwell.com

A CIP catalogue record for this book is available from the Library of Congress, Washington D.C., USA

Deutsche Bibliothek Cataloging-in-Publication Data

Monitoring the comprehensive nuclear test ban treaty. - Springer Basel AG
(Pageoph topical volumes)

Source processes and explosion yield estimation / ed. by Göran Ekström .... - 2001
ISBN 978-3-7643-6552-3 ISBN 978-3-0348-8310-8 (eBook)
DOI 10.1007/978-3-0348-8310-8

Originally published by Birkhäuser Verlag, Basel - Boston - Berlin 2001
Member of the BertelsmannSpringer Publishing Group
Printed on acid-free paper produced from chlorine-free pulp. TCF ∞

9 8 7 6 5 4 3 2 1

Contents

Pure appl. geophys. 158 (2001) 1861–1862
0033–4553/01/111861–02 $ 1.50 + 0.20/0

Pure and Applied Geophysics

Monitoring the Comprehensive Nuclear-Test-Ban Treaty

Preface

The first nuclear bomb was detonated in 1945, thus ushering in the nuclear age. A few political leaders quickly saw a need to limit nuclear weapons through international cooperation and first proposals to do so were made later that same year. The issue of nuclear testing, however, was not formally addressed until 1958 when the United States, the United Kingdom, and the Soviet Union, initiated talks intended to establish a total ban on that testing (a Comprehensive Test-Ban Treaty or CTBT). Those talks ended unsuccessfully, ostensibly because the participants could not agree on the issue of on-site verification.

Less comprehensive treaties did, however, place constraints on nuclear testing. The United States, the United Kingdom, and the Soviet Union, in 1963, negotiated the Limited Test-Ban Treaty (LTBT) which prohibited nuclear explosions in the atmosphere, outer space and under water. The Threshold Test-Ban Treaty (TTBT), signed by the United States and the Soviet Union in 1974, limited the size, or yield, of explosions permitted in nuclear tests to 150 kilotons.

Seismological observations played an important role in monitoring compliance with those treaties. Many of the world's seismologists set aside other research projects and contributed to that effort. They devised new techniques and made important discoveries about the earth's properties that have enhanced our ability to detect nuclear events, to determine their yield, and to distinguish them from earthquakes. Seismologists are rightfully proud of their success in developing methods for monitoring compliance with the LTBT and TTBT.

Although seismologists have also worked for many years on research related to CTBT monitoring, events of recent years have caused them to redouble their efforts in that area. Between 1992 and 1996 Russia, France and the United States all placed moratoria on their nuclear testing, though France did carry out a few tests at the end of that period. In addition, the United States decided to use means other than testing to ensure the safety and reliability of its nuclear arsenal, and all three countries, joined by the United Kingdom, agreed to continue moratoria as long as no other country tested. Those developments, as well as diplomatic efforts by many nations, led to the renewal of multilateral talks on a CTBT that began in January 1994. The talks led to the Comprehensive Nuclear-Test-Ban Treaty. It was adopted by the United Nations General Assembly on 10 September 1996 and has since been signed by 161 nations. Entry of the treaty into force, however, is still uncertain since it requires ratification by all 44 nations that have some nuclear capability and, as of 15 June 2001, only 31 of those nations have done so.

Although entry of the CTBT into force is still uncertain, seismologists and scientists in related fields, such as radionuclides, have proceeded with new research on issues relevant to monitoring compliance with it. Results of much of that research may be used by the International Monitoring System, headquartered in Vienna, and by several national centers and individual institutions, to monitor compliance with the CTBT.

New issues associated with CTBT monitoring in the 21st century have presented scientists with many new challenges. They must be able to effectively monitor compliance by several countries that have not previously been nuclear powers. Effective monitoring requires that we be able to detect and locate much smaller nuclear events than ever before and to distinguish them from small earthquakes and other types of explosions. We must have those capabilities in regions that are seismically active and geologically complex, and where seismic waves might not propagate efficiently.

Major research issues that have emerged for monitoring the CTBT are the precise location of events, and discrimination between nuclear explosions, earthquakes, and chemical explosions, even when those events are relatively small. These issues further require that we understand how seismic waves propagate in the solid earth, the oceans and atmosphere, especially in regions that are structurally complex, where waves undergo scattering and, perhaps, a high degree of absorption. In addition, we must understand how processes occurring at the sources of explosions and earthquakes manifest themselves in recordings of ground motion.

Monitoring a CTBT has required, and will continue to require, the best efforts of some of the world's best seismologists. They, with few exceptions, believe that methods and facilities that are currently in place will provide an effective means for monitoring a CTBT. Moreover, they expect that continuing improvements in those methods and facilities will make verification even more effective in the future. This topical series on several aspects of CTBT monitoring is intended to inform readers of the breadth of the CTBT research program, and of the significant progress that has been made toward effectively monitoring compliance with the CTBT.

The following set of papers, edited by Drs. Gorän Ekstrm, Marvin Denny and John Murphy, presents research results on Source Processes and Explosion Yield Estimation that are applicable for monitoring a CTBT. It is the fifth of eight topics addressed by this important series on *Monitoring the Comprehensive Nuclear-Test-Ban Treaty*. Previously published topics are Source Location, Hydroacoustics, Regional Wave Propagation and Crustal Structure, and Surface Waves. Topics to appear in ensuing issues are Source Discrimination, Data Processing, and Infrasound.

Brian J. Mitchell
Saint Louis University
Series Editor

Pure appl. geophys. 158 (2001) 1863–1865
0033–4553/01/111863–03 $ 1.50 + 0.20/0

Pure and Applied Geophysics

Introduction

GÖRAN EKSTRÖM[1], MARVIN DENNY[2] and JOHN R. MURPHY[3]

An important goal for nuclear testing monitoring research is the development of an understanding of the processes that make up the explosion, and how these processes give rise to seismic and other signals that can be used to characterize the explosions and differentiate them from natural seismic events. The current compilation of eight papers on Source Processes and six papers on Explosion Yield Estimation in the context of nuclear test monitoring addresses a variety of issues related to the general topic of seismic source characterization, and provides a good sampling of current research activities in this area.

Research in the area of Source Processes is conducted in many different ways, for example through theoretical modeling, laboratory experiments, and analysis of actual nuclear explosions and tests. Related to this goal, and crucial for discrimination, is an understanding of how non-nuclear explosions are different from nuclear tests. A significant effort has consequently also been devoted to the study of conventional large explosions such as those conducted in mines. The papers on Source Processes in this volume address a number of these issues. The first contribution, by Johnson and Sammis, discusses the generation of shear waves by explosions. A model of the growth of pre-existing cracks is proposed as a likely source of shear-wave generation and is described quantitatively as an increase in rock damage. The paper by Liu and Ahrens describes experimental results of measuring P- and S- wave generation for explosions in a cavity in the laboratory. In the paper by Patton, the thorny issues of magnitude scaling and transportability between different geophysical regions are taken up in an extensive review of Ms, mb(Pn), mb(Lg), Mw and Mo. The relationships between these quantities are developed and biases between regions are tested as well. In the contribution by Bonner et al., the structure at the former Soviet test site in Kazakhstan is interpreted from close-in data acquired during the depth-of-burial

[1] Harvard University, Department of Earth & Planetary Sciences, 20 Oxford Street, Cambridge, Massuchusetts 02138, USA. E-mail: ekstrom@seismology.harvard. edu
[2] Lawrence Livermore National Laboratory, P.O. Box 808, MS L-205 Livermore, California 94550-0808, USA. E-mail: dennyl@llnl.gov
[3] Maxwell Technologies, Inc., 11800 Sunrise Valley Dr., Suite 1212, Reston, VA 20191, USA. E-mail: jrm@maxwell.com

experiment conducted by the Defense Threat Reduction Agency in 1997 and 1998. In the paper by Baumgardt and Leith, the seismological data from a large, routine mining blast is studied from the point of view of seismological discrimination. The contribution of Der and Baumgardt addresses particular aspects of source characteristics of multiple explosions observed in seismic data, such as signal decorrelation and spectral modulation. The source characteristics of numerous peaceful nuclear explosions in the former Soviet Union are determined and discussed in the contribution by Murphy, Kitov, Barker and Sultanov. In the final contribution, Murphy, Sultanov, Rimer and Barker discuss the source characteristics of some unique nuclear explosion tests conducted in water-filled cavities.

Seismic estimation of the yields of underground nuclear explosions has a long and controversial history. Uncertainties associated with the determination of seismic coupling efficiency as a function of source medium and depth of burial, and with the estimation of the effects due to differences in seismic propagation path characteristics from the various test areas of interest have made it difficult to formulate routine, quantitative inversion procedures of sufficient accuracy to support stringent treaty monitoring requirements. These uncertainties led to significant international political problems related to verification of the 1974 Threshold Test-Ban Treaty. Although explosion yield estimation is not an explicit element of the CTBT, it underlies all assessments of monitoring capability and, therefore, continues to be a subject of research interest. The six papers treating Explosion Yield Estimation included in this volume address a number of different aspects of this complex problem.

The first contribution, by Murphy and Barker, presents a summary of the development of a set of procedures for estimating network-averaged teleseismic P-wave spectra for underground nuclear explosions and for analytically inverting these spectra to obtain estimates of explosion yield. The effectiveness of these procedures is then illustrated through applications to seismic data recorded from explosions at various Soviet, French and Chinese nuclear test sites. The effects of different approximate source models and near-source propagation on seismic yield estimation capability are addressed in a paper by Saikia, Helmberger, Stead and Woods. In this contribution they review some of the issues involved in the application of Nevada Test Site scaling laws to other test sites and summarize some of the results of their deterministic modeling analyses of seismic data recorded from a wide range of nuclear tests. The use of surface wave data in the estimation of explosion yield is addressed in papers by Adushkin and by Stevens and Murphy. The Adushkin paper describes the development of a statistical procedure for estimating explosion yields based on the peak amplitudes of short-period surface waves recorded at near-regional distances, and documents the results of applying this algorithm to a large sample of Semipalatinsk explosions. The contribution by Stevens and Murphy focuses on long-period surface waves and addresses the issues of how complications associated with tectonic strain release and source region geometry can affect the accuracy of yield estimates inferred from such data. The relationship between yield estimates based on regional and teleseismic Lg observations is investigated in a paper by Schlittenhardt in which he compares results obtained using Lg data recorded at the Borovoye station in Kazakhstan with those obtain using corresponding Lg data recorded at the Gräfenberg array in Germany. In the final contribution, Shumway analyzes some of the statistical issues associated with estimating explosion yields using classical and Bayesian methods, and illustrates his derived methodology through applications to the estimation of the yields of the recent underground nuclear tests conducted by India and Pakistan.

Source Processes

Pure appl. geophys. 158 (2001) 1869–1908
0033–4553/01/111869–40 $ 1.50 + 0.20/0

Pure and Applied Geophysics

Effects of Rock Damage on Seismic Waves Generated by Explosions

LANE R. JOHNSON[1] and CHARLES G. SAMMIS[2]

Abstract—In studying the physical processes involved in the generation of seismic waves by explosions, it is important to understand what happens in the region of high stresses immediately surrounding the explosion. This paper examines one of the processes that takes place in this region, the growth of pre-existing cracks, which is described quantitatively as an increase in rock damage. An equivalent elastic method is used to approximate the stress field surrounding the explosion and a micro-mechanical model of damage is used to calculate the increase in damage. Simulations for a 1 kt explosion reveal that the region of increased damage can be quite large, up to ten times the cavity radius. The damage is initiated on a damage front that propagates outward behind the explosive stress wave with a velocity intermediate between that of *P* and *S* waves. Calculations suggest that the amount of increased damage is controlled primarily by the initial damage and the extent of the region of increased damage is controlled primarily by the initial crack radius. The motions that occur on individual cracks when damage increases are converted to seismic moment tensors which are then used to calculate secondary elastic waves which radiate into the far field. It is found that, while the contribution from an individual crack is small, the combined effect of many cracks in a large region of increased damage can generate secondary waves that are comparable in amplitude to the primary waves generated by the explosion. Provided that there is asymmetry in the damage pattern, this process is quite effective in generating *S* waves, thus providing a quantitative explanation of how *S* waves can be generated by an explosion. Two types of asymmetry are investigated, a shear pre-stress and a preferred orientation of cracks, and it is found that both produce similar effects.

Key words: Elastic waves, explosions, cracks, rock damage.

Introduction

In the case of a tamped underground explosion there are a number of different processes that occur in the rocks immediately surrounding the explosion. These processes are illustrated schematically in Figure 1 (see RODEAN, 1971, for a more complete discussion). The explosion is initially contained within a cavity of radius r_c which has been excavated from the surrounding rock. At the time of detonation a hot

[1] Seismological Laboratory, University of California, Berkeley, CA 94720, USA. Fax: 510-643-9980, E-mail: lane@seismo.berkeley.edu

[2] Department of Earth Sciences, University of Southern California, Los Angeles, CA 90089-0740, USA.

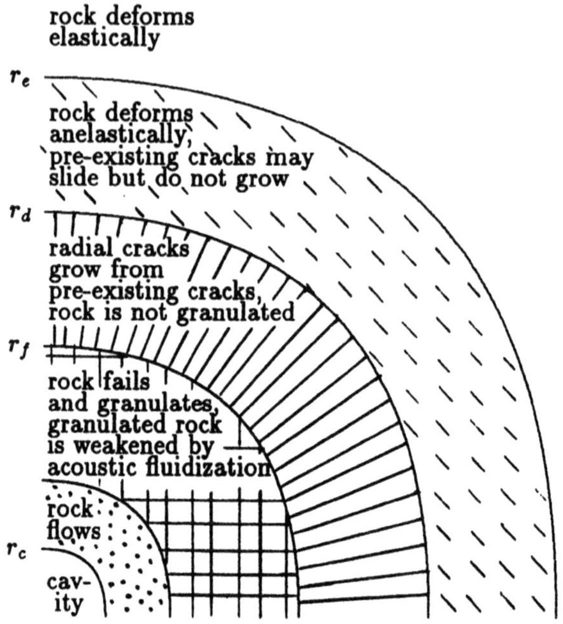

Figure 1
Schematic diagram of the region surrounding a contained explosion in rock.

pressurized gas is created within the cavity, which causes it to expand. Some of the surrounding rock may be vaporized and added to the cavity at this time also. The sudden expansion of the cavity generates a shock wave which propagates outward and causes major damage to the surrounding rock, resulting in a series of zones where the material is: First reduced to one having the properties of a fluid; then is stressed beyond the point of failure and becomes granulated; then enters a range where cracks grow but failure is not reached; then is deformed anelastically; and finally behaves as a normal elastic material. The energy density of this shock wave decreases with distance from the explosion, partly due to the fact that it is spreading in three dimensions and partly due to the fact that energy is being used to deform and fracture the rock. Thus the shock wave gradually decays into an inelastic wave involving nonlinear motions, which further decays with distance until eventually a radius is reached where the motions are small enough to be described by the ordinary elastodynamic equations of linear elasticity. Beyond this distance r_e, the elastic radius, the disturbance caused by the explosion can be modeled as linear elastic waves that propagate throughout the rest of the earth.

These processes that occur around an explosion can have strong effects upon the elastic waves that are radiated beyond the elastic radius r_e. In this paper we concentrate upon the growth of pre-existing cracks, which can affect the radiated elastic waves in at least three different ways. First, intense cracking will significantly lower the bulk modulus and shear modulus near the source (O'CONNELL and BUDIANSKY, 1974; RIMER *et al.*, 1998). Second, it has been speculated that when cracking extends into the regime of failure, acoustic fluidization can lower the basic strength of the rock (SAMMIS, 1998). Third, motions on the cracks serve as secondary sources of elastic waves that can contribute to the net seismic radiation field. The primary objective of this paper is to provide a quantitative assessment of this third aspect. The intent is to determine how these secondary sources modify the P waves radiated by explosions and to test the hypothesis that they may generate significant S waves in the far field.

We approach this problem by using the formulation of ASHBY and SAMMIS (1990) to model the nucleation and growth of cracks in the source region, which is here referred to as rock damage. Calculation of this damage requires knowledge of the stress field in the region surrounding the source, which is approximated in this study by using an equivalent elastic method. The micro-mechanical model of damage provides the parameters necessary to represent each crack as a seismic moment tensor that is then used to calculate the elastic waves radiated by the damage. Finally, we consider the cumulative effect of all the cracks in the damage zone in order to estimate seismic P and S waves in the far field.

To provide a reality check for these calculations, simulations have been performed for the chemical explosion detonated as part of the Non-Proliferation Experiment (NPE) at the Nevada Test Site in September, 1993 (DENNY, 1994). This 1 kt explosion provides a convenient test case because abundant information is available for this event, including waveform data recorded near the explosion.

Damage Mechanics

The concept of damage used in this study is that described by ASHBY and SAMMIS (1990). In that paper the conditions under which an initial crack can nucleate additional cracking are derived. Only the basic definitions and equations from that paper that are needed for the damage calculations will be repeated here. In three dimensions the initial damage is defined as

$$D_0 = \tfrac{4}{3}\pi(a\cos\chi)^3 N_v \ , \tag{1}$$

where a is the radius of penny-shaped cracks, χ is the angle describing the orientation of the cracks (see Fig. 2), and N_v is the number of cracks per unit volume. During deformation wing cracks of length l grow off opposite edges of the initial crack. Then the damage is

$$D = \tfrac{4}{3}\pi(l + a\cos\chi)^3 N_v \ . \tag{2}$$

Note that both D_0 and D are dimensionless numbers.

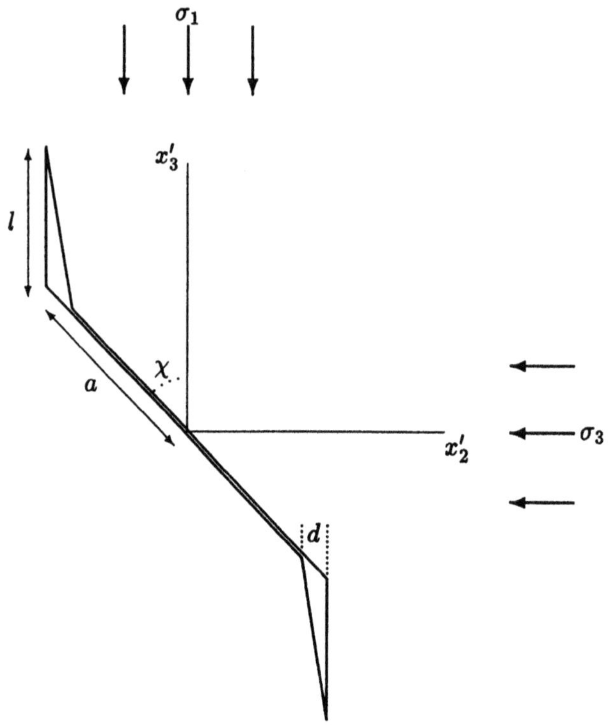

Figure 2
Geometry of a penny-shaped crack of radius a which is extended by wing cracks of length l.

The value of l is determined by letting the crack grow until the stress intensity factor at the tip decreases to the fracture toughness of the medium. The equation that must be solved in order to determine the final amount of damage is (ASHBY and SAMMIS, 1990)

$$S_1 = -\frac{C_2\left(\left(\frac{D}{D_0}\right)^{1/3} - 1 + \frac{\beta}{\cos\chi}\right)^{2/3} - S_3\left[C_1\left(1 + \left(\frac{C_3 D_0^{2/3}}{1-D^{2/3}}\right)\left(\left(\frac{D}{D_0}\right)^{1/3} - 1\right)^2\right) + C_4\left(\left(\frac{D}{D_0}\right)^{1/3} - 1\right)^2\right]}{1 + C_3\frac{D_0^{2/3}}{1-D^{2/3}}\left(\left(\frac{D}{D_0}\right)^{1/3} - 1\right)^2} . \tag{3}$$

Here S_1 and S_3 are the maximum and minimum normalized principal compressive stresses

$$S_1 = \frac{\sigma_1}{K_{Ic}/\sqrt{\pi a}}, \quad S_3 = \frac{\sigma_3}{K_{Ic}/\sqrt{\pi a}} , \tag{4}$$

and K_{Ic} is the critical stress intensity factor for mode I cracks, a material property. The constants in this equation are

$$C_1 = \frac{(1+\eta^2)^{1/2}+\eta}{(1+\eta^2)^{1/2}-\eta},$$
$$C_2 = \pi(\cos\chi)^{3/2}\sqrt{3/\beta}\frac{1}{(1+\eta^2)^{1/2}-\eta},$$
$$C_3 = 2, \tag{5}$$
$$C_4 = 2\pi(\cos\chi)^2\sqrt{3/\beta}\frac{1}{(1+\eta^2)^{1/2}-\eta}.$$

In these last equations η is the coefficient of sliding friction and β is a correction factor for the effective length of the crack, typically 0.45, introduced by ASHBY and SAMMIS (1990) to bring their approximate analytic model into agreement with numerical simulations in the limit of small l. In addition, the angle χ is generally assumed to be 45°.

Given the initial damage D_0 and the major and minor principal stresses σ_1 and σ_3, equation (3) can be used to calculate the final state of the damage. This amounts to solving a fairly complicated cubic equation, but the solution can be obtained by numerical methods. When the maximum compressive stress σ_1 is only slightly greater than the minimum compressive stress σ_3 this equation has no real positive roots, which means that there is no increase in damage, i.e., no crack growth. Friction on the pre-existing cracks requires a threshold of resolved shear stress before motion will occur. However, as the difference between the maximum and minimum compressive stresses increases, solutions that indicate an increase in damage begin to emerge from the equation. With a further increase in stress the damage will continue to increase until the maximum value of σ_1 allowing a solution is reached. This point on the damage surface is interpreted by ASHBY and SAMMIS (1990) as the onset of failure, as beyond this point the solution for damage continues to increase with decreasing stress. In this study we have not allowed the damage to increase beyond this point of failure, here labeled D_f, because we do not know of a suitable method for estimating the maximum damage that will occur. Thus, all of the calculations of damage in this study should be considered as minimum estimates because they are stopped at the inception of failure. It should be emphasized that a fairly simple model of damage has been used in this study in which new shear cracks are not created and the interaction between cracks is only on a first order global basis. Based on studies such as those of LOCKNER and MADDEN (1991) and RECHES and LOCKNER (1994), it appears that the addition of more complicated features to the model might change some of the fine details of the damage calculations but would not alter any of the basic conclusions of this study.

Stress Field Surrounding an Explosion

In the approach outlined in the previous section, the primary ingredient necessary for the investigation of rock damage surrounding an explosion is knowledge of the stress field generated by the explosion. The estimation of this stress field is made difficult by the existence of a zone dominated by nonlinear processes between the cavity radius of the explosion and the effective elastic radius, beyond which the assumptions of ordinary linear elasticity are valid. A number of computer codes has been developed for modeling this nonlinear zone, which include hydrodynamic effects, shock waves, and nonlinear equations of state (see, for example, RODEAN, 1971; KING *et al.*, 1989; GLENN, 1993; GLENN and GOLDSTEIN, 1994, for discussion and further references). Because they are fairly complicated and numerical in nature, these codes do not readily provide insight into questions about which parameters are playing critical roles in determining the radiated elastic waves. In this study we have followed a different approach by using an equivalent elastic method for the calculation of the stresses surrounding the explosion. While this is an approximate approach that does not incorporate all of the physics contained in complete equation-of-state approaches, it does provide a very efficient method of estimating the basic characteristics of elastic waves radiated by an explosion. To the extent that it provides a reasonably accurate estimate of the radial and transverse stresses as a function of distance from the explosion, which can be checked with experimental data, this approach should suffice for the purposes of the damage calculations.

The concept of an equivalent elastic medium has been used quite successfully by earthquake engineers to model the nonlinear behavior of soils that occurs during strong ground motion. An advantage of this approach is that it makes use of the analytical solutions developed for linear problems, thus providing a relatively simple and versatile algorithm. This is important in the present problem, particularly as regards calculating secondary elastic waves radiated from the damage. The central idea of the method is to make the material properties a function of the stress in the outward propagating pressure pulse, and then adjust these material properties in an iterative process until the appropriate values are present at all distances from the source (JOHNSON, 1993). In effect, a nonlinear stress–strain relationship is modeled by a series of linear relationships which change with the level of the stress so as to approximate the nonlinear behavior. The present formulation relates density and bulk elastic properties to the peak pressure and shear and anelastic properties to the maximum shear strain. Note that the process is different for the bulk and shear properties.

Consider first the bulk properties. Let ρ_u be the ultimate or maximum density and P be the peak overpressure. Then, following MAJTENYI and FOSTER (1992), we approximate the density dependence upon pressure as

$$\rho(P) = \rho_u - (\rho_u - \rho_0)e^{-AP} , \tag{6}$$

where

$$A = \frac{1}{V_{p0}^2} \frac{1}{\rho_u - \rho_0} , \tag{7}$$

and ρ_0 and V_{p0} are zero pressure values. The compressional velocity as a function of pressure is

$$V_p(P) = \left(V_{p0}^2 + \frac{2}{\rho_u}\left[P + \frac{1}{A}\ln\left\{\frac{\rho(P)}{\rho_0}\right\}\right]\right)^{1/2} . \tag{8}$$

Next consider the shear properties. The small strain shear modulus μ_0 can be obtained from the small strain shear velocity V_{s0} by

$$\mu_0 = \rho_0 V_{s0}^2 \ . \tag{9}$$

If τ_{max} is the maximum shear stress that the material will sustain, a reference strain can be defined by

$$e_r = \frac{\tau_{max}}{\mu_0} \ . \tag{10}$$

The apparent strain e associated with the shear stress τ is given by

$$e = \frac{\tau}{\mu_0} \ . \tag{11}$$

Then a hyperbolic stress–strain relationship (HARDIN and DRNEVICH, 1972) yields the effective shear modulus

$$\mu(e) = \mu_0 \frac{1}{1 + e/e_r} \ , \tag{12}$$

and the shear velocity as a function of pressure and shear strain follows from

$$V_s(P, e) = \left(\frac{\mu(e)}{\rho(P)}\right)^{1/2} . \tag{13}$$

Similarly, the effective damping ratio is (HARDIN and DRNEVICH, 1972)

$$\delta(e) = \delta_{max} \frac{e/e_r}{1 + e/e_r} \ , \tag{14}$$

where the maximum damping δ_{max} is a material property. The damping ratio is related to the attenuation quality factor by

$$Q_\mu^{-1} = \frac{2\delta}{\sqrt{1 - 4\delta^2}} \ . \tag{15}$$

In the present implementation of the method it has been assumed that the quality factor for the bulk modulus Q_κ does not change.

Application of the equivalent elastic method to the case of explosive sources requires solutions of the elasto-dynamic equations of motion for spherical waves in the case where the material properties vary as a function of radius. Even in the case where the material properties are initially homogeneous, the equivalent elastic adjustments will temporarily alter the properties so as to make them a function of radius. A procedure for calculating such a solution is outlined in Appendix A.

The basic steps of the computational scheme for the equivalent elastic method can now be described. The first step is to initialize the model of the material properties surrounding the explosion. This consists of dividing the region into an appropriate number of spherical shells and specifying the initial material properties for these shells. For the NPE simulations described below, the shell boundaries were at radial distances of 5.9, 10, 20, 40, 60, 80, 100, 200, 300, 400, 500, and 600 m. The material properties are assumed to be constant within each shell. The parameters associated with the equivalent elastic treatment, such as the ultimate density, the reference shear strain, and the maximum damping must also be specified at this time. The outside radius of the last shell should be large enough so that the material properties at this and greater distances can be safely assumed to be linear visco-elastic.

The second step is to initialize the properties of the explosive source. The size of the explosion is specified by giving its chemical energy E_0. The initial radius of the cavity r_c is also specified, which leads to the initial volume

$$V_0 = \frac{4}{3}\pi r_c^3 \ . \tag{16}$$

The explosion is assumed to take place instantaneously so that at $t = 0$ the initial pressure within the cavity increases by an amount

$$P_0 = \frac{E_0(g_c - 1)}{V_0} \ , \tag{17}$$

where g_c is the gas constant for the explosive gases. This initial pressure is the radial stress at $r = r_c$, which is one of the boundary conditions for the elastic wave solution.

The third step is to obtain the elastic wave solution for the entire region surrounding the source using the propagator solution method of Appendix A. Given this solution, the stresses and strains at all distances from the source can be calculated. In general the stresses associated with this solution will be sufficiently large so that the initial material properties are not consistent with the relationships given in equations (6) through (14). Thus the material properties are adjusted on the basis of the calculated stresses and the equivalent elastic treatment outlined above. For the purposes of this adjustment the increase in pressure is

$$P = -\frac{\tau_{rr} + \tau_{\theta\theta} + \tau_{\phi\phi}}{3} \ , \tag{18}$$

and the increase in shear stress is

$$\tau = \frac{\tau_{\theta\theta} - \tau_{rr}}{2} \ . \tag{19}$$

Having adjusted the material properties in all of the shells, the elastic wave solution can be obtained for the adjusted model and the entire process repeated. This iterative process of solving for the stresses and adjusting the material properties is repeated until a solution is obtained in which the stresses are compatible with the material properties at all distances. Generally, only three or four iterations are required.

After convergence of the process of adjusting the material properties, the source radius is also adjusted to allow for the inelastic growth of the cavity. The cavity radius r_c is increased by successively eliminating shells until the static pressure is less than some factor times the lithostatic pressure

$$P(r_c, t \to \infty) \leq fP_l \ . \tag{20}$$

Here the lithostatic pressure P_l is just that due to the overburden and is given by

$$P_l = \rho_0 g z_s \ , \tag{21}$$

where g is the acceleration of gravity and z_s is the source depth. Recall that the calculated pressure is in excess of the lithostatic pressure, therefore this condition actually corresponds to the static cavity pressure being $1 + f$ times the lithostatic pressure. For the calculations in tuff the factor f was assumed to have a value of 0.5 (MUELLER and MURPHY, 1971).

After the adjustment of the cavity radius, a final iteration is performed on the adjustment of the material properties to insure that the stresses and material properties at all distances from the source are compatible. At the completion of this step the final solution is available, which consists of displacements and stresses as a function of time at all points surrounding the source. Note that in the equivalent elastic method all stresses are deviations from any initial static stress field. However, the stresses that enter into the damage calculations are total values that include any static stress fields that might be present. Because of this, hydrostatic stress such as P_l or any pre-existing shear stress in the vicinity of the explosion may have an effect upon the damage calculations.

Results for the NPE Explosion

In order to test the potential effect of damage on realistic explosive sources, calculations were performed for the chemical explosion detonated as part of the Non-Proliferation Experiment (NPE) in September, 1993, at the Nevada Test Site. This 1 kt explosion provides a convenient test case because of the large amount of information available for this experiment. The detonation consisted of 1.29 million kg of an ammonium nitrate and fuel oil mixture which filled a cylindrical cavity 15.2 m in diameter and 5.5 m high, located 389 m below the surface of Rainier Mesa. The energy release of the explosion was $1.07\,\text{kt} = 4.48 \cdot 10^{12}$ J (DENNY, 1994).

The first matter to be checked is the ability of the equivalent elastic method to simulate the stress field around the explosion. The NPE was simulated as an explosion in saturated tuff having material properties $V_p = 2.6\,\mathrm{km/sec}$, $V_s = 1.9\,\mathrm{km/sec}$, $\rho = 1.9\,\mathrm{gm/cm^3}$ (EMERICK, 1966; BALDWIN *et al.*, 1994). The initial cavity radius was taken as 5.9 m and the adiabatic gas constant as $g_c = 1.75$. The final cavity radius turned out to be 15 m. The calculations were carried out to a radial distance of 600 m, with 17 spherical shells intervening between this distance and the cavity radius. Beyond this distance the material was treated as being elastic. The parameters for the equivalent elastic treatment were $\rho_u = 2.58\,\mathrm{gm/cm^3}$, $e_r = 10^{-3}$, $\delta_{\max} = 0.40$.

Free-field ground motion data from the NPE were recorded in a suite of tunnels at approximately the same depth as the explosion. These data were acquired from the LANL NTS Ground Motion Data Base (APP and TUNNELL, 1994). The most common type of measurement is particle acceleration, which is typically integrated and presented as particle velocity (SMITH, 1994; OLSEN and PERATT, 1994). Figure 3

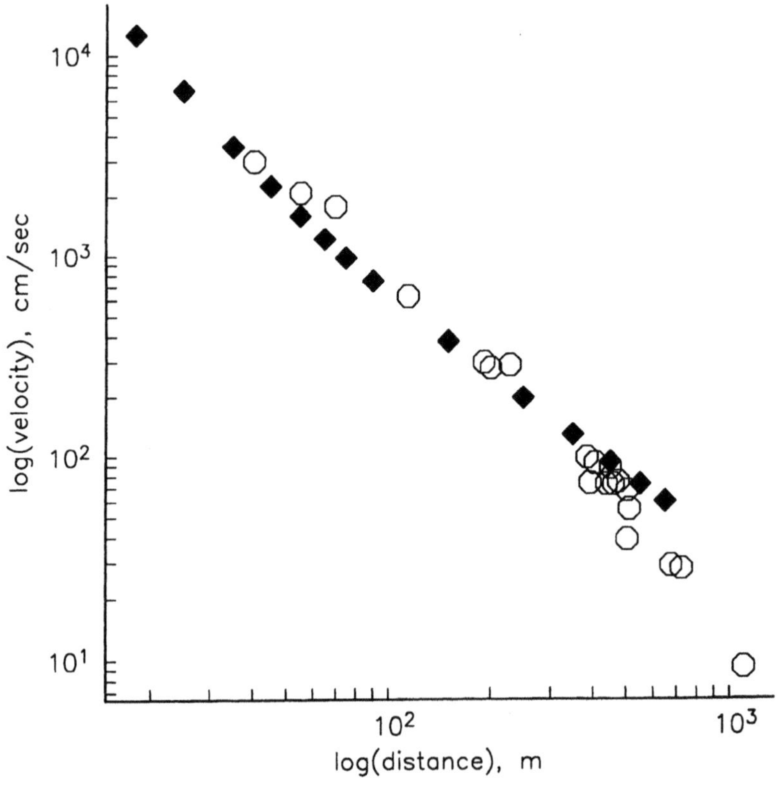

Figure 3
Measured and calculated peak velocities for the NPE explosion. The open symbols are observations taken from SMITH (1994) and OLSEN and PERATT (1994), while the filled symbols were calculated with the equivalent elastic method.

is a comparison between such peak radial velocity measurements and peak velocity calculated with the equivalent elastic approach. There is general agreement between the observations and the simulations in terms of both the levels of velocity and the trend with distance, although beyond a distance of about 500 m the observed velocities decay more rapidly than those of the simulations. Figure 4 is a comparison of waveforms, with the traces labeled *tm* being selected observations from OLSEN and PERATT (1994) and the remaining traces coming from the simulations at various distances. The main difference is that the simulated waveforms are more oscillatory and do not achieve the *saw-tooth* shape that is usually associated with nonlinear wave propagation. This is expected, as the equivalent elastic method only approximates the nonlinear effects. On the basis of these comparisons it was assumed that, at least for the NPE event, the equivalent elastic method could be used to approximate the stress field surrounding the explosion.

It should be emphasized that the equivalent elastic method of calculating the stress field surrounding an explosion was used in this study primarily because it could

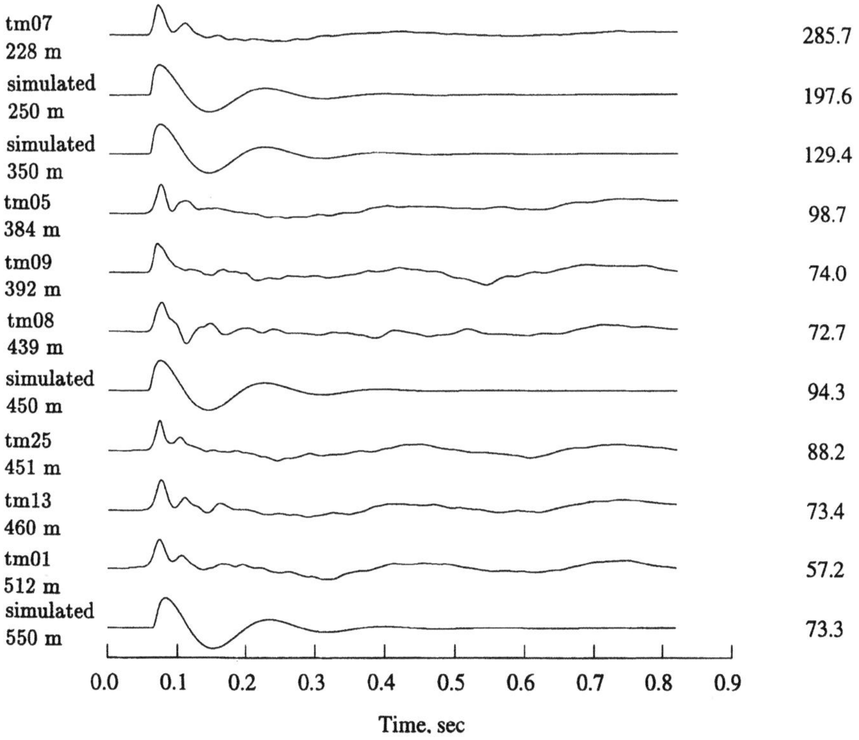

Figure 4

Comparison between measured and calculated waveforms for the NPE explosion. The traces labeled with *tm* are the radial velocities measured by OLSEN and PERATT (1994) at the radial distances shown. The traces labeled with *simulated* are the radial velocities calculated with the equivalent elastic method at the radial distances shown. The numbers on the right are maximum velocities on the traces in units of cm/sec.

be combined easily with the damage calculations. There are other methods of calculating the stress field which yield more accurate results. For instance, KAMM and BOS (1995) used a two-dimensional Lagrangian stress-wave propagation code that included the effect of geologic layering to model the NPE explosion, and obtained considerably better matches to the observed waveforms than those shown in Figure 4. However, it is not clear whether this numerical solution of the problem could incorporate motions on cracks of dimension 0.10 cm or less, which is one of the objectives of the present study.

The next step was to determine whether the outward propagating elastic waves could cause an increase in damage. The damage parameters used for this calculation were $K_{Ic} = 1$ MPa $m^{1/2}$, $\eta = 0.6$, $a = 0.10$ cm, and $\chi = 45°$ (ASHBY and SAMMIS, 1990). The initial damage was taken as $D_0 = 0.10$, which is a representative value for rocks (ASHBY and SAMMIS, 1990). Subsequently, given the equivalent elastic estimates of the radial and transverse stress as a function of time at all points surrounding the explosion, equation (3) was evaluated at each time point to determine if there was a growth in damage. As mentioned earlier, the stresses used in the damage calculation included both the dynamic stresses due to the explosion and any pre-existing static stress field. In the present case the static field consisted of a hydrostatic pressure of 7.3 MPa, a value estimated to be appropriate for the 389 m depth of the explosion.

Figure 5 shows the results of such a calculation at a radial distance of 35 m from the center of the explosion, slightly over two times the final cavity radius. As can be seen in this figure, there is an interval of about 0.05 sec beginning 0.05 sec after the arrival of the first pressure pulse when the stress conditions are such that damage can increase, which is interpreted to mean that cracks will grow. In this calculation the damage grew from an initial value of 0.10 to a maximum value of 0.21. The delay in the onset of damage indicates that large stresses by themselves are not sufficient to cause damage, rather it is the difference between the maximum and minimum stresses that is important in causing the damage. In this case the damage occurs because the tangential stress relaxes more rapidly than the radial stress.

Calculations of results such as those shown in Figure 5 were performed at the center of each spherical shell surrounding the explosion, which corresponded to distances of 25, 35, 45, 55, 65, 75, 90, 150, 250, 350, 450, and 550 m from the NPE explosion. Damage was found to increase in a zone that extended from the final cavity radius (15 m) out to a radius that varied between 25 m and 150 m, depending upon the material properties that were assumed. At closer distances the damage was always driven into the post-failure stage, whereas at greater distances it was possible to have an increase in damage without failure. The limits to these zones of damage with failure and damage without failure will be referred to as r_f and r_d, respectively (see Fig. 1). The delay in the onset of damage increased with distance and it was found that the damage front propagated with a velocity intermediate between that of P and S waves.

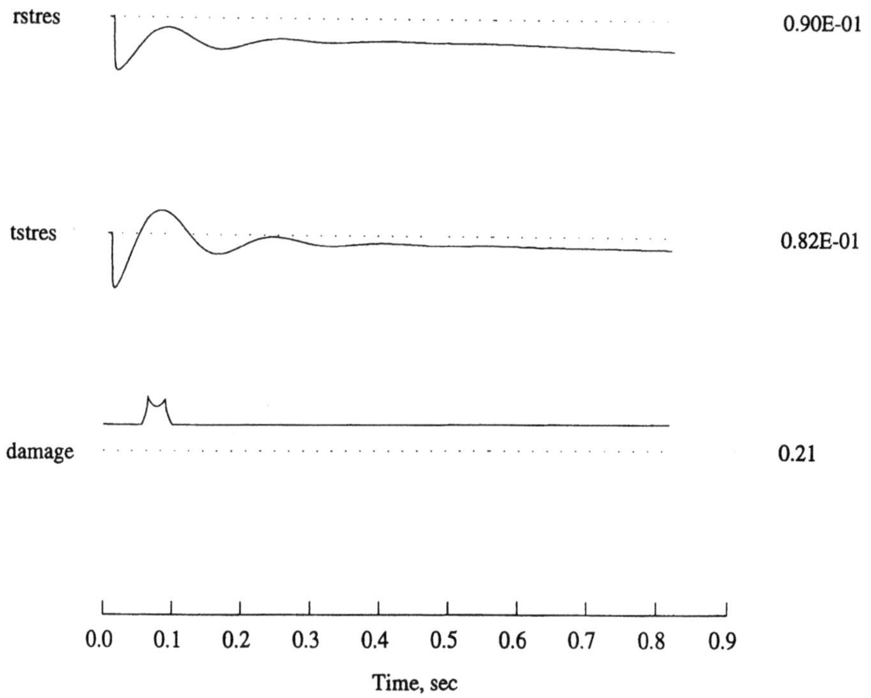

Figure 5
Calculated radial stress, tangential stress, and damage for a distance of 35 m from the NPE explosion. The numbers on the right are peak stress in units of GPa for the stresses. An initial damage of $D_0 = 0.10$ and an initial crack radius of $a = 0.10$ cm were used in these calculations.

Damage as a Source of Secondary Elastic Waves

Having shown that additional damage can be generated by the elastic waves propagating outward from an explosion, it will now be considered whether this damage can have appreciable effects upon the general explosion process. One obvious effect is the fact that a crack on which motion occurs as part of the damage will act as a secondary source of elastic waves. In this section we derive a moment tensor representation of the damage and then calculate the radiated elastic waves using a Green function approach.

The damage of ASHBY and SAMMIS (1990) is modeled as an initial crack which is extended by wing cracks. This can be represented as the combination of two separate source moment tensors, a shear dislocation on the initial crack and a tensile opening on the wing crack. In Appendix B it is shown that the scalar moment tensor per unit volume for the tensile crack is

$$m_t = \frac{3}{2}\frac{(\lambda + 2\mu)^2}{\mu(\lambda + \mu)}\frac{K_{Ic}}{(\pi a \cos\chi)^{1/2}}D_0\left[\left(\frac{D}{D_0}\right)^{1/3} - 1\right]^{5/2} , \tag{22}$$

while that for the shear crack is

$$m_s = \frac{9}{2}\frac{\lambda + 2\mu}{\lambda + \mu}\frac{K_{Ic}}{(\pi a \cos\chi)^{1/2}}\frac{D_0}{\sin\chi\cos^2\chi}\left[\left(\frac{D}{D_0}\right)^{1/3} - 1\right]^{1/2} . \tag{23}$$

Note that

$$\frac{m_t}{m_s} = \frac{\lambda + 2\mu}{3\mu}\sin\chi\cos^2\chi\left[\left(\frac{D}{D_0}\right)^{1/3} - 1\right]^2 , \tag{24}$$

which shows that the relative effect of the tensile crack tends to increase as the damage increases.

A complete specification of the moment tensor for the damage requires that the orientation of the cracks be considered. Figure 6 illustrates how this was done. Let $\mathbf{x}$ be a fixed coordinate system with its origin at the center of the explosion and the x_1 axis positive north, the x_2 axis positive east, and the x_3 axis positive down. The crack is described in a local coordinate system $\mathbf{x}'$ which has its origin at a radial distance r from the center of the explosion, with the x_3' axis along the radial direction. The orientation of the $\mathbf{x}'$ system with respect to the $\mathbf{x}$ system is described in terms of the three Euler angles ϕ, θ, and ψ (Goldstein, 1959) with the transformation matrix being

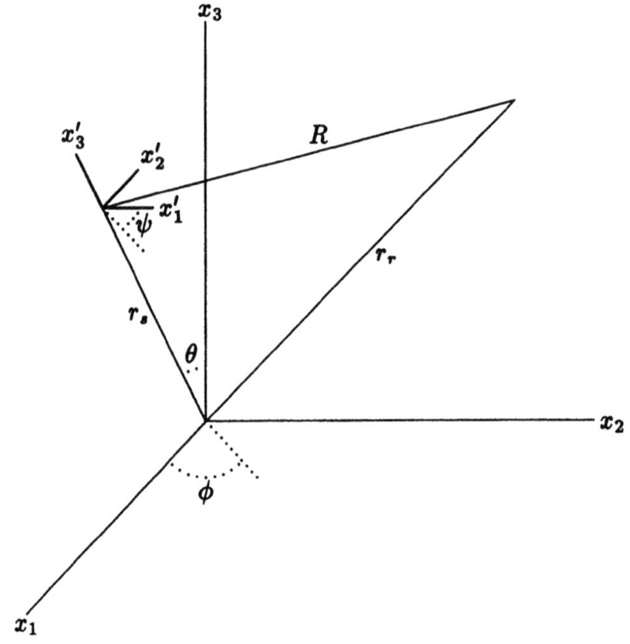

Figure 6
Relationship between the local coordinate system of the damage calculation $\mathbf{x}'$ and a fixed coordinate system for the wave propagation calculation $\mathbf{x}$.

$$\mathbf{T} = \begin{bmatrix} \cos\psi\cos\phi - \sin\psi\sin\phi\cos\theta & \cos\psi\sin\phi + \sin\psi\cos\phi\cos\theta & \sin\psi\sin\theta \\ -\sin\psi\cos\phi - \cos\psi\sin\phi\cos\theta & -\sin\psi\sin\phi + \cos\psi\cos\phi\cos\theta & \cos\psi\sin\theta \\ \sin\phi\sin\theta & -\cos\phi\sin\theta & \cos\theta \end{bmatrix} . \tag{25}$$

Then if $\mathbf{m}'$ is the moment density tensor of the damage in the local coordinate system, in the fixed coordinate system it is given by

$$\mathbf{m} = \mathbf{T}^T\mathbf{m}'\mathbf{T} . \tag{26}$$

Now consider the part of the damage which is a tensile crack with an opening in the x_2' direction (see Fig. 2). Its moment density tensor in the local coordinate system is (AKI and RICHARDS, 1980)

$$\mathbf{m}' = m_t \begin{bmatrix} \zeta & 0 & 0 \\ 0 & 1 & 0 \\ 0 & 0 & \zeta \end{bmatrix} , \tag{27}$$

where m_t is given by equation (22) and

$$\zeta = \frac{\lambda}{\lambda + 2\mu} = 1 - 2\frac{V_s^2}{V_p^2} . \tag{28}$$

In the fixed coordinate system the components of $\mathbf{m}$ are

$$\begin{aligned} m_{11} &= m_t\left\{\zeta + (1-\zeta)[\sin\psi\cos\phi + \cos\psi\sin\phi\cos\theta]^2\right\}, \\ m_{21} &= m_{12} = m_t(1-\zeta)[\sin\psi\cos\psi(\sin^2\phi - \cos^2\phi)\cos\theta \\ &\qquad + (\sin^2\psi - \cos^2\psi\cos^2\theta)\sin\phi\cos\phi], \\ m_{31} &= m_{13} = -m_t(1-\zeta)[\sin\psi\cos\phi + \cos\psi\sin\phi\cos\theta]\cos\psi\sin\theta, \\ m_{22} &= m_t\left\{\zeta + (1-\zeta)[\sin\psi\sin\phi - \cos\psi\cos\phi\cos\theta]^2\right\}, \\ m_{32} &= m_{23} = -m_t(1-\zeta)[\sin\psi\sin\phi - \cos\psi\cos\phi\cos\theta]\cos\psi\sin\theta, \\ m_{33} &= m_t\{\zeta + (1-\zeta)\cos^2\psi\sin^2\theta\} . \end{aligned} \tag{29}$$

Note that

$$\operatorname{trace}(\mathbf{m}) = \operatorname{trace}(\mathbf{m}') = m_t(1 + 2\zeta) , \tag{30}$$

which shows that there is a net change in volume connected with the tensile crack.

Next consider the part of the damage which is the original shear crack with its normal lying in the $x_2' - x_3'$ plane (again consult Fig. 2). Its moment density tensor in the local coordinate system is (AKI and RICHARDS, 1980)

$$\mathbf{m}' = m_s \begin{bmatrix} 0 & 0 & 0 \\ 0 & 1 & 0 \\ 0 & 0 & -1 \end{bmatrix} , \tag{31}$$

where m_s is given by equation (23) and it has been assumed that $\chi = 45°$. In the fixed coordinate system the components of $\mathbf{m}$ are

$$
\begin{aligned}
m_{11} &= m_s\left\{(\sin\psi\cos\phi + \cos\psi\sin\phi\cos\theta)^2 - (\sin\phi\sin\theta)^2\right\},\\
m_{21} &= m_{12} = m_s\{(\sin^2\psi - \cos^2\psi\cos^2\theta + \sin^2\theta)\sin\phi\cos\phi\\
&\qquad + \sin\psi\cos\psi(\sin^2\phi - \cos^2\phi)\cos\theta\},\\
m_{31} &= m_{13} = -m_s[\sin\psi\cos\psi\cos\phi + \sin\phi(1+\cos^2\psi)\cos\theta]\sin\theta,\\
m_{22} &= m_s\left\{(\sin\psi\sin\phi - \cos\psi\cos\phi\cos\theta)^2 - (\cos\phi\sin\theta)^2\right\},\\
m_{32} &= m_{23} = -m_s[\sin\psi\cos\psi\sin\phi - \cos\phi(1+\cos^2\psi)\cos\theta]\sin\theta,\\
m_{33} &= m_s\{\cos^2\psi\sin^2\theta - \cos^2\theta\}\ .
\end{aligned}
\tag{32}
$$

Note that

$$\mathrm{trace}(\mathbf{m}) = \mathrm{trace}(\mathbf{m}') = 0\ , \tag{33}$$

which shows that there is no change in volume connected with the shear crack.

The expression for the displacement field from a moment density tensor source is conveniently expressed in the form (AKI and RICHARDS, 1980)

$$u_n = \iiint G_{np,q} \otimes m_{pq}\,\mathrm{d}x^3\ , \tag{34}$$

where u_n is the displacement in the n direction at the receiver location, m_{pq} is the moment density tensor at the source location, G_{np} is the Green function that connects the two locations, and the integral is taken over the volume where m_{pq} is non-zero. Here $\otimes$ represents convolution in time. When the time dependence is a step function and the material is homogeneous, the integrand of equation (34) (the contribution of the moment density from a unit volume) is

$$
\begin{aligned}
u_n = {} & \{15\gamma_n\gamma_p\gamma_q - 3\gamma_n\delta_{pq} - 3\gamma_p\delta_{nq} - 3\gamma_q\delta_{np}\}\frac{m_{pq}}{4\pi\rho R^4}\\
& \cdot\frac{1}{2}\left(\left[t^2 - \left(\frac{R}{\alpha}\right)^2\right]H(t-R/\alpha) - \left[t^2 - \left(\frac{R}{\beta}\right)^2\right]H(t-R/\beta)\right)\\
& + \{6\gamma_n\gamma_p\gamma_q - \gamma_n\delta_{pq} - \gamma_p\delta_{nq} - \gamma_q\delta_{np}\}\frac{m_{pq}}{4\pi\rho\alpha^2R^2}H(t-R/\alpha)\\
& - \{6\gamma_n\gamma_p\gamma_q - \gamma_n\delta_{pq} - \gamma_p\delta_{nq} - 2\gamma_q\delta_{np}\}\frac{m_{pq}}{4\pi\rho\beta^2R^2}H(t-R/\beta)\\
& + \{\gamma_n\gamma_p\gamma_q\}\frac{m_{pq}}{4\pi\rho\alpha^3R}\delta(t-R/\alpha)\\
& - \{\gamma_n\gamma_p\gamma_q - \gamma_q\delta_{np}\}\frac{m_{pq}}{4\pi\rho\beta^3R}\delta(t-R/\beta)\ .
\end{aligned}
\tag{35}
$$

Here R is the distance between the source and receiver locations and the γ_i are direction cosines. Let the location of the damage crack be r_s, θ_s, ϕ_s and let the location of the receiver be r_r, θ_r, ϕ_r (Fig. 6). Consequently the distance between the two is

$$R = \sqrt{r_r^2 + r_s^2 - 2r_r r_s \cos\Theta} \ , \tag{36}$$

where

$$\cos\Theta = \cos\theta_r \cos\theta_s + \sin\theta_r \sin\theta_s \cos(\phi_r - \phi_s) \ . \tag{37}$$

The direction cosines are given by

$$\begin{aligned}
\gamma_1 &= \frac{r_r \sin\theta_r \cos\phi_r - r_s \sin\theta_s \cos\phi_s}{R} , \\
\gamma_2 &= \frac{r_r \sin\theta_r \sin\phi_r + r_s \sin\theta_s \sin\phi_s}{R} , \\
\gamma_3 &= \frac{r_r \cos\theta_r - r_s \cos\theta_s}{R} \ .
\end{aligned} \tag{38}$$

The relationships between the spherical coordinates of the damage crack and its Euler angles are simply $\theta_s = \theta$ and $\phi_s = \phi - \pi/2$.

With the scalar moments of equations (22) and (23), the rotated moment components of equations (29) and (32), and the Green function of equation (35), it is possible to calculate the elastic waves radiated by a volume element of damage cracks at any specified receiver location. What remains is to sum up the effects of all the damage that occurs in the region surrounding the source. This is accomplished by evaluating an integral over the entire damage zone, taking into account the amount of damage, the orientation of the cracks, and the time history of the motion on the cracks at each point in this zone. Such integrals can be obtained by numerical methods using the following procedure:

- The region surrounding the source is divided into a large number of small volume elements.
- The equivalent elastic method is used to calculate the principal stresses within each volume element as a function of time.
- Equation (3) is used to decide if an increase in damage occurs within the volume element and, if so, the amount of increase and the time it begins are determined.
- The increase in damage is converted to a moment tensor using equations (22), (23), (29), and (32).
- Equation (35) is used to calculate the displacements at the receiver location contributed by every volume element where there is an increase in damage.
- The vector displacements generated at the receiver location by the secondary waves from all volume elements are summed.

In calculating the secondary elastic waves generated by damage according to the procedure just outlined, a couple of assumptions have to be made. First it is assumed that the movement on an individual damage crack takes place in a time interval that is short with respect to the time scale of interest, and thus it can be approximated by a step function. Note that it is only the first increase in damage that is considered, and any later change in damage at a given location is ignored. Second, a Green function for a homogeneous material is used to calculate the waves radiated by the damage (equation 35), however the material properties are taken as the mean values along the straight ray path between source and receiver, using the final values of the equivalent elastic calculations. This means that waves radiated by damage on one side of the source must pass back through the source region to reach a receiver on the other side, consequently they experience the reduced S velocities and increased attenuation in the vicinity of the source. This adds a fundamental asymmetry to the problem.

To illustrate the effects of the secondary elastic waves caused by an increase in damage, consider first the case of cracks which have completely random orientations. The moment tensor expressions of equations (29) and (32) can then be averaged assuming that the Euler angle ψ is uniformly distributed between 0 and 2π, leading to a considerable simplification. The results of such a simulation for the NPE event are shown in Figure 7 where the observation point is taken to be at a radius of 450 m from the center of the explosion. The top two traces in this figure show observed velocity waveforms at approximately this distance and the third trace shows the simulation obtained with the equivalent elastic method. The stresses associated with this simulation resulted in an increase in damage, such as that shown in Figure 5, at distances between the cavity radius and 45 m. The result of summing up the radiated waves from all of the motions associated with the increase in damage is shown in the fourth trace. It consists of a relatively high frequency P wave arriving shortly after the direct arrival from the explosion with an amplitude that is about 10% of the direct arrival in this case. Arriving at a time of about 0.5 sec is a larger low frequency arrival that is the near-field contribution from the damage, which travels with the S velocity.

All of the traces shown in Figure 7 are for the radial component of motion. The simulations for this case of cracks with completely random orientations does not produce any S waves and thus the two components of motion orthogonal to the radial component are not shown. This is not to say that S waves are not present, as the individual cracks associated with an increase in damage produce S waves that can be greater in amplitude than the P waves. However, the net effect outside the source region is the vector sum of the waves generated by all of the cracks, and in this case of spherical symmetry the S waves all cancel out to yield a zero net sum. This does not happen in the case of the P waves because there is a net dilatancy associated with the damage, which produces an isotropic P wave.

It helps to understand the type of waves radiated by the damage if the composite moment tensor for the entire damage region is considered. At distances from the source that are large compared to its dimensions and for wavelengths large compared

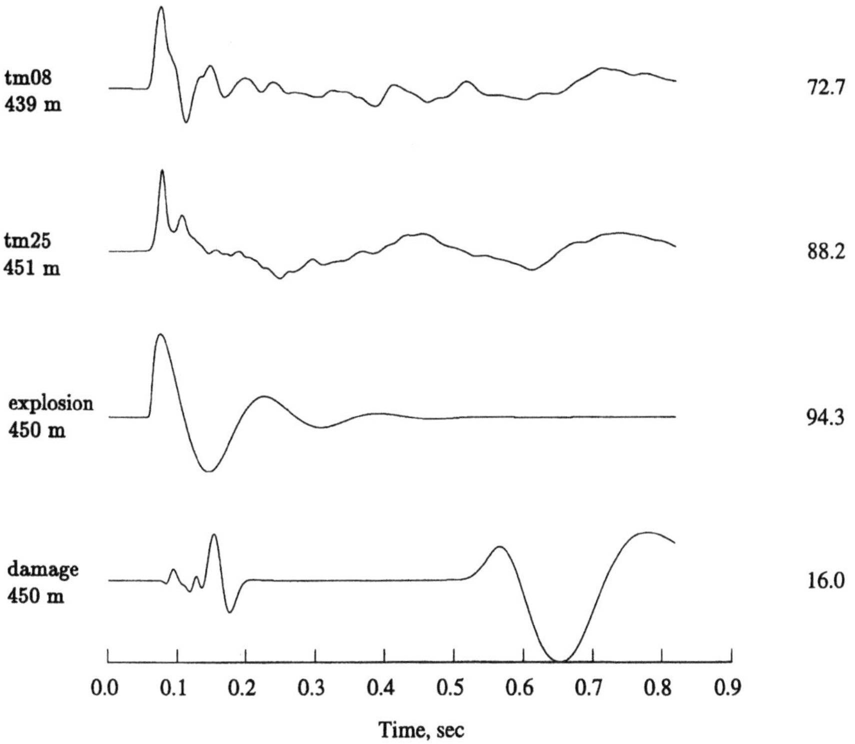

Figure 7
Comparison of measured and calculated waveforms for the NPE explosion with waves radiated from damage in the case of initial cracks that have completely random orientations. The top two traces are radial velocities measured by OLSEN and PERATT (1994) at the radial distances shown. The third trace is a simulation of the NPE explosion. The bottom trace is a simulation of the secondary waves radiated by the increase in damage associated with the third trace. The numbers on the right are maximum velocities on the traces in units of cm/sec. In simulating the damage the initial damage was $D_0 = 0.10$ and the initial crack radius was $a = 0.10$ cm.

to the source dimensions, the source can be approximated by a point source. This means that equation (34) can be written

$$u_n = G_{np,q} \otimes M_{pq} \; , \tag{39}$$

where M_{pq} is the volume integral of the moment tensor density

$$M_{pq} = \iiint m_{pq} \, \mathrm{d}x^3 \; . \tag{40}$$

Assuming the increase in damage is uniform in a spherical region between the cavity radius r_c and the limit of the damage r_d, the integral in equation (40) can be evaluated. For the case of the tension crack the expressions of equation (29) produce

$$M_{11} = M_{22} = M_{33} = m_t \frac{4\pi}{3}(r_d^3 - r_c^3)\frac{1+2\zeta}{3}, \quad \text{all other} \quad M_{ij} = 0 \ . \tag{41}$$

On the other hand, integrating the expressions of equation (32) for the shear crack yields

$$M_{ij} = 0 \ . \tag{42}$$

This result says that only P waves should be radiated by this type of source, which is consistent with the fact that it has complete spherical symmetry. Furthermore, comparing equation (41) with equations (B12) and (B8), it is easy to show that the amplitude of the P waves is proportional to the dilatancy of the tension cracks.

In the calculations that produced Figure 7 it was necessary to select specific values for the initial damage D_0 and the crack radius a. In the earth these parameters can take on various values, so it is worth exploring the effect of these parameters upon the increase in damage and, in particular, the secondary waves radiated by the damage. This is done in Tables 1 and 2 and in Figures 8 and 9 for waves simulated at a radial distance of 450 m. Table 1 and Figure 8 display the effect of the initial

Table 1

Effect of initial damage

Initial damage D_0	Failure damage D_f	Failure radius r_f m	Damage radius r_d m	Maximum velocity V_{max} (450 m) cm/sec
0.01	0.09	25	25	0.9
0.05	0.26	35	45	8.4
0.10	0.40	35	45	16.1
0.20	0.53	35	45	28.2
0.40	0.73	35	45	44.8
0.80	1.13	35	45	88.9

Crack radius $a = 0.1$ cm in all cases.

Table 2

Effect of initial crack radius

Crack radius a	Failure damage D_f	Failure radius r_f m	Damage radius r_d m	Maximum velocity V_{max} (450 m) cm/sec
0.01	0.22	18	18	2.6
0.10	0.40	35	45	16.1
1.00	0.39	75	90	21.1
10.0	0.40	90	150	13.1

Initial damage $D_0 = 0.1$ in all cases.

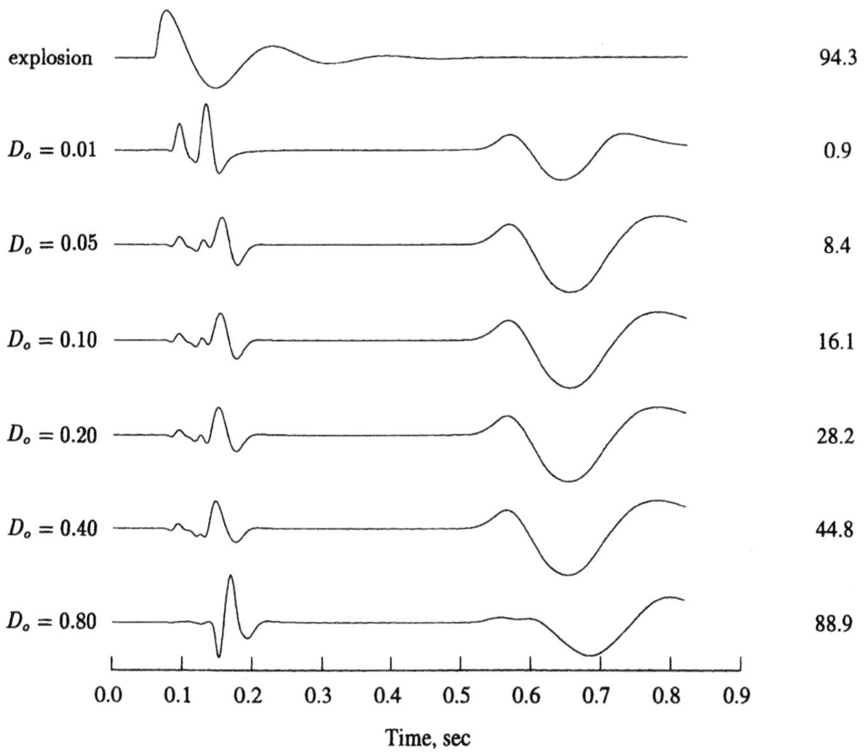

Figure 8
Effect of initial damage upon the radiated secondary waves. The top trace shows the simulated radial velocity at a radial distance of 450 m from a 1 kt explosion. The traces below show the simulated radial velocity at the same distance caused by an increase in damage when the initial damage has the value shown on the left and the initial crack radius was held constant at $a = 0.10$ cm. The numbers on the right are maximum velocities on the traces in units of cm/sec.

damage while the crack radius is held constant at $a = 0.10$ cm. There is a strong effect here as the damage at failure D_f increases along with the initial damage D_0, which is consistent with the fact that the solution for damage (equation 3) is mainly a function of the ratio D/D_0. Also obvious in Table 1 and Figure 8 is the strong dependence of the secondary waves upon the initial damage, which is consistent with the linear dependence of the moment tensors upon D_0 (equations 22 and 23). Some caution must be exercised in interpreting these results, however, as the same equivalent elastic parameters were used in all of the calculations. This is obviously an approximation, as one would expect the material to be weaker for larger values of initial damage, however, this level of sophistication is as yet unincorporated into our formulation of the problem.

Table 2 and Figure 9 examine the effect of the crack radius while the initial damage is held constant at $D_0 = 0.10$. An increase in the crack radius has minimal effect on the maximum damage, but it causes a major increase in the size of the

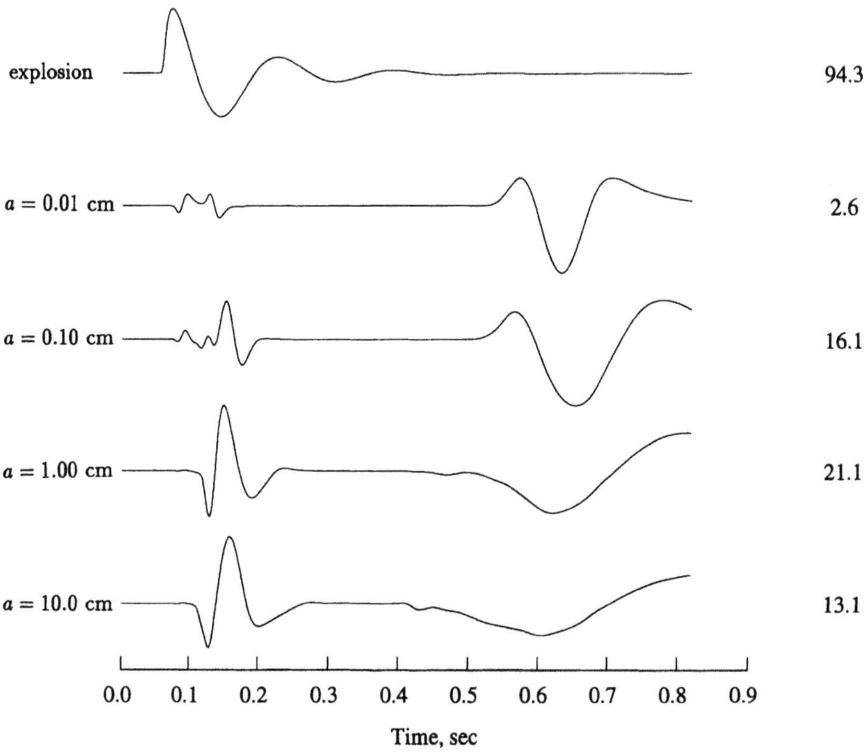

Figure 9
Effect of initial crack radius upon the radiated secondary waves. The top trace shows the simulated radial velocity at a radial distance of 450 m from a 1 kt explosion. The traces below show the simulated radial velocity at the same distance caused by an increase in damage when the initial crack radius has the value shown on the left in cm and the initial damage was held constant at $D_0 = 0.10$. The numbers on the right are maximum velocities on the traces in units of cm/sec.

damaged zone. This is mainly because the stress needed to extend cracks decreases with their size, which in this problem enters through the scaling of the principal stresses (equation 4). With respect to the amplitude of the secondary waves, the increase in the volume of the damaged zone caused by an increase in crack radius is partly compensated by an inverse dependence of the moment tensors on the crack radius (equations 22 and 23). Also, an increase in the size of the effective source region causes a decrease in the efficiency of radiating high frequency waves into the far field, which appears in Figure 9 as longer period waves for the larger crack radii. Also note in Figure 9 that the near field contribution begins to arrive earlier as the crack radius increases, another consequence of the increase in the effective source size. Therefore the main effect of a larger crack radius is an increase in the volume of the damaged zone, thus enlarging the effective source size for the secondary waves and producing an increase in their low-frequency components and a relative decrease in their high-frequency components.

The results in Table 2 and Figure 9 all assume a single crack size for the individual calculations. A more realistic situation would be a distribution of crack sizes, but the method could easily be modified to handle this case. Calculations could be performed for a range of crack radii, the moment tensors generated by the damage for a particular crack radius weighted by the proportion of the cracks having that radius, and the results summed. The results are predictable. Near the cavity radius the entire range of crack sizes would be growing and generating secondary elastic waves, while as the distance from the source increased there would be a gradual transition to the case where only the largest cracks would be growing and generating secondary elastic waves. However, this scenario assumes that cracks with different sizes grow independently and this may not be valid, as the large cracks may grow first and relieve the stress field so that small cracks do not grow.

Tables 1 and 2 reveal rather different dependencies for the limits of the region where damage is increased. The initial damage has essentially no effect upon these limits, with the radius of failure extending out to about 35 m and the limit of increased damage extending out to about 45 m. An exception is the case of very small initial damage, in which the extent of damage and the amplitude of secondary waves is much smaller than the other cases, suggesting that there is a threshold of crack density that must exist in order to nucleate significant crack growth. On the other hand, the initial crack radius has a strong effect upon the limits of the damaged region, with the largest cracks of 10 cm radius manifesting failure out to 90 m and increased damage out to 150 m. There is also an exception to this result, with very small cracks showing a decidedly smaller damaged region and amplitude of secondary waves. It should be pointed out that the numbers listed for the limits of damage in Tables 1 and 2 are only approximate, as the selection of radial layers in the calculations limited their resolution to about ± 5 m.

All of the calculations in this study were performed assuming a critical stress intensity factor of $K_{Ic} = 1\,\text{MPa m}^{1/2}$, which is a reasonable value for hard rocks (ASHBY and SAMMIS, 1990). However, this parameter only enters the calculations in the ratio $K_{Ic}/\sqrt{a}$ (see equations 4, 22, and 23), therefore all of the effects described above for the crack radius a can be converted into equivalent effects for the critical stress intensity factor K_{Ic}. Thus decreasing K_{Ic} to a value less than $1\,\text{MPa m}^{1/2}$, possibly to make it more appropriate for the tuff emplacement material of the NPE, is equivalent to increasing the crack radius a, which would have the effect of increasing the limits of the damaged region but would not appreciably increase the amplitude of the radiated secondary waves.

Effects of Pre-stress

In the previous section it was assumed that the initial cracks were uniformly distributed and randomly oriented in the medium surrounding the explosion. This leads to a considerable simplification in the integration over the damage zone. It also results in a problem with spherical symmetry so that no S waves are radiated into the far field by the motion on the cracks. In the real earth there are a number of different ways that this symmetry is likely to be broken, thus presenting the possibility of generating secondary S waves. It is common to observe SH waves generated by explosions, even though such waves should not exist for a symmetric explosion in a horizontally layered, homogeneous, isotropic medium. One of the earliest and still most viable explanations for this phenomenon is the existence of shear stress in the vicinity of the explosion prior to its detonation (PRESS and ARCHAMBEAU, 1962; TOKSÖZ *et al.*, 1965). A quantitative estimate of the magnitude of such a mechanism can be obtained by calculating the energy content of the pre-stress in some volume surrounding the source, and assuming that all of this energy is released at the time of the explosion (ARCHAMBEAU and SAMMIS, 1970). The method described in this paper permits a more direct approach to this problem by calculating the effect of the pre-stress upon the damage mechanics.

As mentioned earlier, the stresses used in the estimate of damage increase include both the dynamic effects of the explosion and any pre-existing static stress field. Whereas the previous calculations assumed a hydrostatic pre-stress due only to the effects of the overburden, it is relatively straightforward to add an additional static stress field due to tectonic effects. In the fixed coordinate system let this tectonic pre-stress be

$$\tau = \tau_t \begin{bmatrix} 0 & 0 & 0 \\ 0 & 1 & 0 \\ 0 & 0 & -1 \end{bmatrix} , \tag{43}$$

which corresponds to compression in the x_3 direction and extension in the x_2 direction. The maximum shear stress, which is half the difference in the maximum and minimum principal stresses, has a value of τ_t in this case.

It is necessary to express this pre-stress in a rotated coordinate system so that it can be added to the stresses generated by the elastic waves from the explosion. Thus we need

$$\tau' = \mathbf{R}\tau\mathbf{R}^T , \tag{44}$$

where $\mathbf{R}$ is the rotation matrix given earlier in equation (25) with the condition that $\psi = 0$. The result is

$$\tau' = \tau_t \begin{bmatrix} \sin^2\phi & \sin\phi\cos\phi\cos\theta & -\sin\phi\cos\phi\sin\theta \\ \sin\phi\cos\phi\cos\theta & \cos^2\phi\cos^2\theta - \sin^2\theta & -(1+\cos^2\phi)\sin\theta\cos\theta \\ -\sin\phi\cos\phi\sin\theta & -(1+\cos^2\phi)\sin\theta\cos\theta & \cos^2\phi\sin^2\theta - \cos^2\theta \end{bmatrix} . \tag{45}$$

At each point in time and at each location surrounding the explosion, this pre-stress is added to the dynamic stresses due to the explosion, new principal stresses are determined, and the damage calculation is performed. Because the pre-stress can take on both positive and negative values, its presence can either increase or decrease the likelihood of additional damage.

A representative calculation that includes the effect of a tectonic pre-stress is shown in Figures 10 and 11. The pre-stress was taken to have a value of $\tau_t = 7$ MPa, which is a value estimated by ARCHAMBEAU and SAMMIS (1970) for the nuclear explosion Bilby at NTS and one slightly less than the lithostatic pressure at the depth of the NPE. The orientation of the pre-stress is an east-west tensional regime that is consistent with the normal faulting in the Basin and Range province near NTS. Other parameters for the damage calculation were $D_0 = 0.10$ and $a = 0.10$ cm. The initial damage is still assumed to consist of randomly oriented cracks. The addition of the pre-stress means that the zone of additional damage is no longer spherical, extending out to only 35 m in some directions and more than 55 m in other directions (recall that the resolution in these numbers is

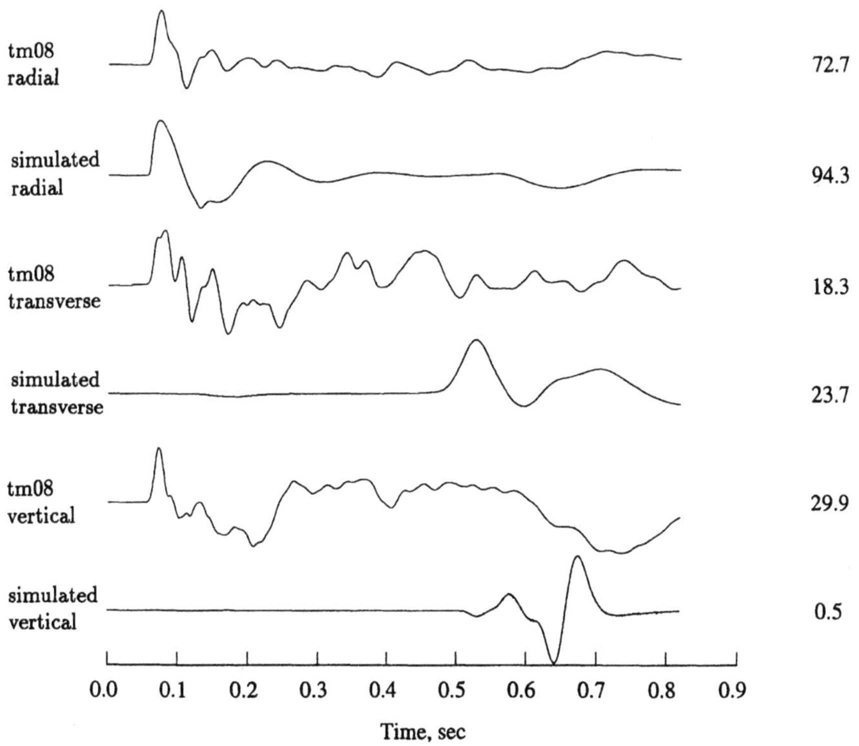

Figure 10
Comparison of measured waveforms for the NPE explosion with waves simulated for an explosion including the effects of pre-stress. The initial damage was $D_0 = 0.10$ and the initial crack radius was $a = 0.10$ cm for the simulations. Gauge *tm*08 was at a radial distance of 439 m (OLSEN and PERATT, 1994) and the simulations were for a distance of 450 m. The numbers on the right are maximum velocities on the traces in units of cm/sec.

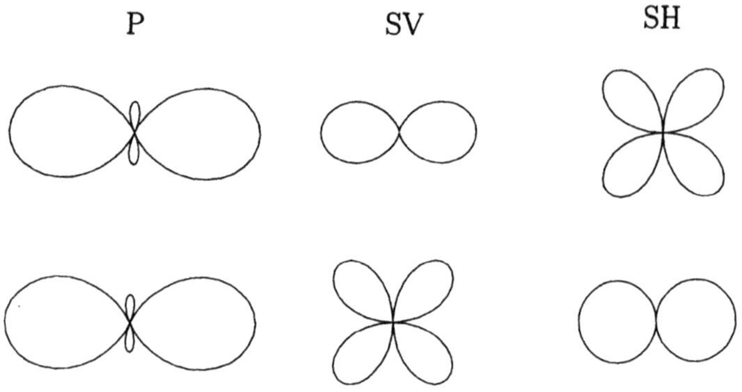

Figure 11
Radiation patterns of far-field *P*, *SV*, and *SH* waves for the case of a pre-stress. The upper figures show the patterns cut by a horizontal plane with north being up, while the lower figures show the patterns cut by a vertical plane along an azimuth where the pattern has a maximum.

±5 m). Because of this asymmetry, the increase in damage is now an effective source of secondary *S* waves.

In Figure 10 the results of the calculations including pre-stress are compared with the three components of velocity measured at gauge *tm*08 by OLSEN and PERATT (1994), which was at a radial distance of 439 m at the same depth as the NPE and at an azimuth of N58E. The simulations were for a distance of 450 m and the same depth and azimuth as *tm*08. The simulated radial component contains the contributions of both the explosion and the secondary waves from increased damage, with the damage contribution, attaining a maximum amplitude of 21.7, being somewhat difficult to identify (compare with the third trace of Fig. 7). On the simulated transverse component there is now a clear *SH* wave, arriving at a time of about 0.5 sec and with an amplitude greater than the *P* wave due to the damage alone. Because of the symmetry chosen for the pre-stress, the vertical component at this location has a very small amplitude in the simulations. It should be emphasized that no attempt has been made here to fit the observed data at gauge *tm*08, and the primary point of Figure 10 is to show that, with a pre-stress present, the increased damage is capable of generating significant *S* waves that are compatible with the general nature of the observational data in terms of both the amplitude and frequency content of the waveforms. There are of course many other processes that could also explain the observed waveforms on the transverse and vertical components, such as heterogeneity or anisotropy of the medium, converted waves at discontinuities, or the scattering effects of the tunnels that were used to emplace the gauges.

Just as in the case of a uniform distribution of initial cracks with no pre-stress, it is possible to obtain a composite moment tensor for the effects of increased damage when a pre-stress is present. In this case it was obtained numerically, summing up the contributions from all of the individual moment tensor densities in the region of increased damage while ignoring any differences in onset time of damage. The far-field radiation patterns appropriate for this composite moment tensor are shown in Figure 11 for *P*, *SV*, and *SH* waves. The composite moment tensor has relative values of approximately $M_{11} = -0.2$, $M_{22} = 1.0$, $M_{33} = -0.2$, which is equivalent to an isotropic part of $M_{iso} = 0.6$ and a deviatoric part of $M'_{11} = -0.4$, $M'_{22} = 0.8$, $M'_{33} = -0.4$. This deviatoric part can be interpreted as a compensated linear vector dipole, and the radiation pattern of the *P* waves in Figure 11 is characteristic of such a source. The radiation patterns of the *SV* and *SH* waves are more typical of earthquakes modeled as a simple shear dislocation. The scalar part of the composite moment tensor has a value of $2.85 \cdot 10^{13}$ Nm in this simulation, which is about 15% of the value determined for the NPE explosion from surface observations (JOHNSON, 1994).

The simulations in Figure 10 that include the effect of pre-stress are for one particular choice of magnitude and orientation of the pre-stress and for one particular choice of initial damage and initial crack radius. It is obvious that different results would be obtained for different choices of these parameters (see for example Figs. 8 and 9). To illustrate just one of these choices, the calculations that led to Figure 10 were repeated with the initial crack radius increased from 0.10 cm to 1.00 cm, and the results are shown in Figure 12. With larger initial cracks present, the zone of increased damage is now considerably extended, having a minimum radius of 75 m in some directions and a maximum radius of 150 m in other directions. The secondary waves radiated into the far field now have amplitudes that are approximately doubled for both *P* and *S* waves, have slightly earlier arrival times, and have longer periods when compared to the case of an 0.10 cm initial crack radius (Fig. 10). The secondary *P* waves are now large enough to have an observable effect upon the waveform of the primary *P* wave from the explosion, making it more similar to the observed data. The longer periods of the simulated *S* waves also seem to be more in agreement with the observed data. The radiation patterns do not change appreciably from those shown in Figure 11 for the smaller initial crack radius. The scalar part of the composite moment tensor now has a value of $8.42 \cdot 10^{13}$ Nm, which is about 40% of that estimated for the NPE explosion.

Preferred Orientation of Cracks

Another way of breaking the symmetry of the waves radiated by increased damage is to allow the initial cracks to have a preferred orientation. This condition is not as easily simulated as the case of pre-stress, nonetheless we attempted to do so by allowing the initial damage to vary as a function of position around the explosion. This can be interpreted to mean that the density of cracks that are favorably oriented

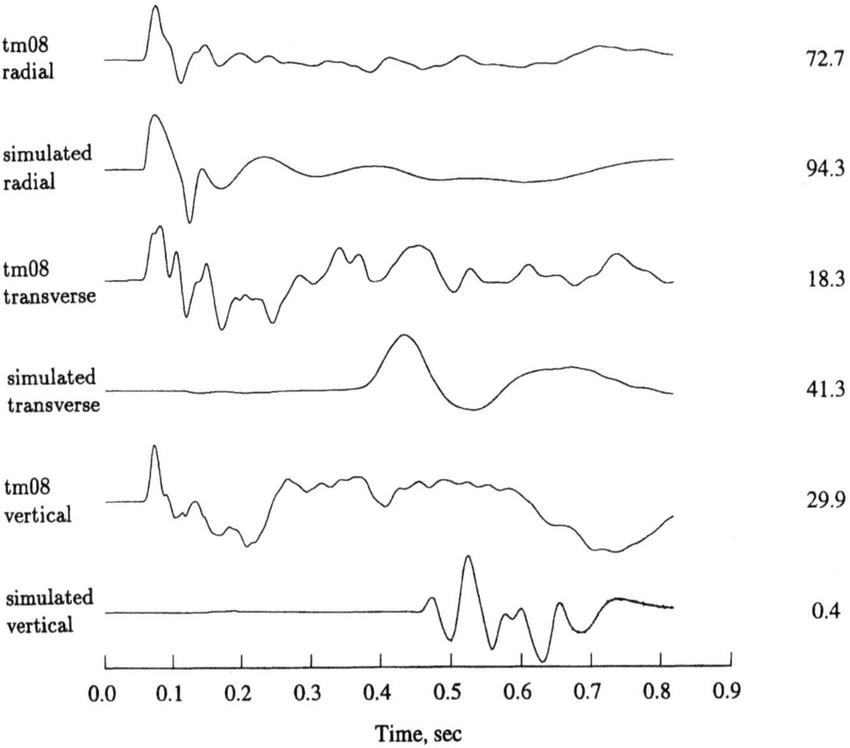

Figure 12
Similar to Figure 10 except the initial crack radius has been increased from 0.10 cm to 1.00 cm.

to allow slip is not the same at all locations. From Table 1 and Figure 8 it is already clear that the initial damage strongly affects the final damage, therefore this representation of preferred orientation of cracks should lead to asymmetric radiation of secondary waves from the damaged zone.

Consider the case in which the initial cracks have a preferred orientation with their normal parallel to the x_2 axis (see Fig. 6). Assuming that the cracks will slip when the principal stress along the radial direction is at an angle of approximately 45° with the normal to the crack, the initial damage was taken to be

$$D_0 = D_{om}[\sin(2\phi)\sin(2\theta)]^2 \;, \tag{46}$$

where D_{om} is the maximum initial damage. Unlike the case of randomly oriented cracks, it is also necessary to determine the orientation of the crack that is most likely to slip at each location. This is achieved by choosing the angle χ so that the x_1' axis lies in the plane of the crack (see Fig. 2), which is

$$\tan\chi = -\frac{\sin\phi}{\cos\phi\cos\theta} \;. \tag{47}$$

Inserting this result into equations (29) and (32) provides the moment tensors for the increase in damage at each location, which then can be used to calculate the secondary elastic waves.

A representative calculation that includes the effects of cracks that have a preferred orientation as described in the previous paragraph is shown in Figures 13 and 14. The maximum initial damage was $D_{om} = 0.10$ and the crack radius was $a = 0.10$ cm. As in the case of a pre-stress, the zone of additional damage is no longer spherical, being as small as the cavity radius in some directions and as great as 45 m in other directions. Figure 13 is similar to Figure 10 in that the simulated waveforms are compared with the observations at gauge *tm*08 for the NPE explosion. The results are actually quite similar to those of Figure 10, with about a 50% increase in the amplitude of the *S* waves caused by increased damage and about a 50% decrease in the amplitude of the *P* waves. Figure 14 shows the far-field radiation patterns for this case of oriented cracks, which is again quite similar to the case of pre-stress shown in Figure 11. The composite

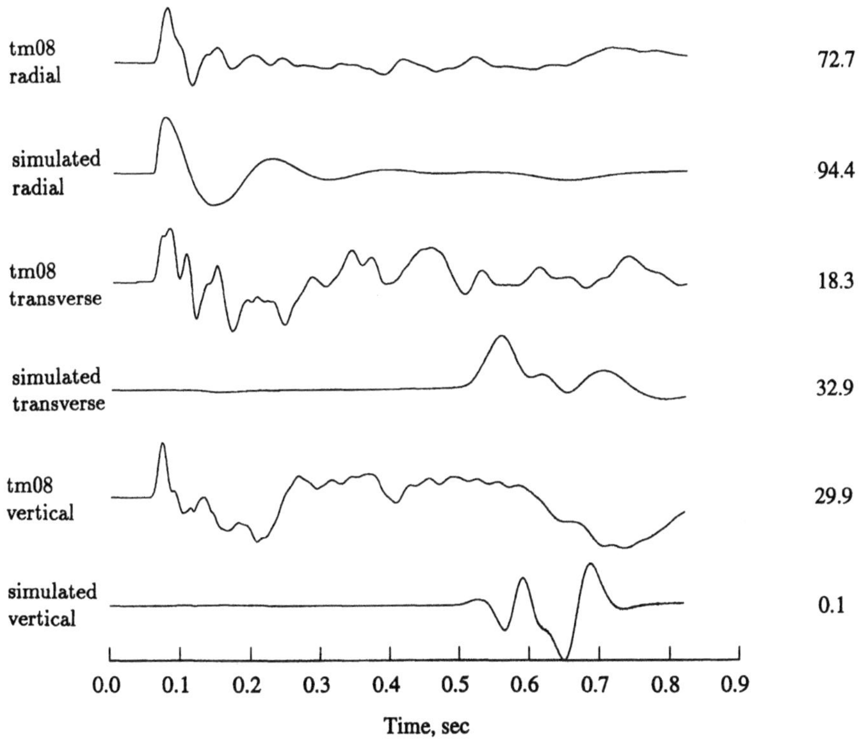

Figure 13

Comparison of measured waveforms for the NPE explosion with waves simulated for an explosion including the effects of oriented cracks. The initial damage was $D_0 = 0.10$ and the initial crack radius was $a = 0.10$ cm for the simulations. Gauge *tm*08 was at a radial distance of 439 m (OLSEN and PERATT, 1994) and the simulations were for a distance of 450 m. The numbers on the right are maximum velocities on the traces in units of cm/sec.

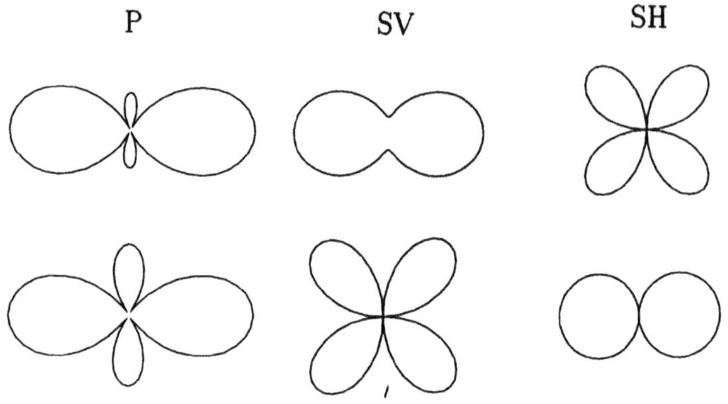

Figure 14
Radiation patterns of far-field *P*, *SV*, and *SH* waves for the case of oriented cracks. The upper figures show the patterns cut by a horizontal plane with north being up, while the lower figures show the patterns cut by a vertical plane along an azimuth where the pattern has a maximum.

moment tensor for Figure 14 has relative values of approximately $M_{11} = -0.30$, $M_{22} = 1.0$, $M_{33} = -0.55$, which is equivalent to an isotropic part of $M_{iso} = 0.15$ and a deviatoric part of $M'_{11} = -0.35$, $M'_{22} = 0.95$, $M'_{33} = -0.60$. In this case the scalar part of the composite moment tensor has a value of $3.04 \cdot 10^{13}$ Nm .

A comparison of the results for a pre-stress (Figs. 10 and 11) with those for oriented cracks (Figs. 13 and 14) shows general similarities and also subtle differences. The general similarity is not surprising, as the orientation of the cracks was chosed to be consistent with the pre-stress in that the normal to the initial cracks was parallel to the maximum tensional pre-stress. Given only the far-field elastic waves shown in Figures 10 and 12, it would be difficult to distinguish between a cause due to pre-stress and one due to oriented cracks. However, there do exist some real differences in the asymmetry of the composite moment tensor and slight differences in the radiation patterns (Figs. 11 and 14) that might possibly be diagnostic. Of course, in realistic situations it is highly likely that both a pre-stress and oriented cracks will be present, and the results of this study would suggest that sorting out their relative contributions in such an instance would be difficult. It is also worth noting that a medium with oriented cracks will be anisotropic on a macroscopic scale and this anisotropy will contribute its own effects, including the generation of *S* waves by an explosive source (BEN-MENAHEM and SENA, 1990; BEN-MENAHEM and GIBSON, 1995) and more complicated radiation patterns and particle motions (TSVANKIN and CHESNOKOV, 1990).

Discussion and Conclusions

This study has considered one of the processes which may occur in the region of high stresses immediately surrounding an explosion detonated in rock, the extension of pre-existing cracks. Using a combination of analytical and numerical methods, the physics of this crack extension process has been modeled. The results appear to have interesting and significant implications with regard to the elastic waves that are radiated by explosive sources.

Combining an equivalent elastic method of calculating stresses in the near-source region of an explosion with the micro-mechanical damage model of ASHBY and SAMMIS (1990) leads to an approximate but efficient method of estimating the growth of pre-existing cracks. Sample calculations for the NPE chemical explosion revealed that the stress waves from the explosion were capable of generating significant amounts of crack growth in an extensive region surrounding the source, with the extent and shape of this region depending upon such parameters as the initial damage, the size of the cracks, the amount of pre-stress and the orientation of the cracks. These calculations suggest that the extent of the damaged zone is most sensitive to the initial crack radius, varying from a radial distance of about 45 m for a crack radius of 0.10 cm to about 150 m for a crack radius of 10 cm for this 1 kt explosion. These numbers are quite consistent with the results reported by RIMER *et al.* (1998) for Russian explosions in hard rocks, which underwent intensive fracturing extending out to a scaled radius of $40\,\mathrm{m/kt}^{1/3}$ and occasional additional fracturing observed out to about $200\,\mathrm{m/kt}^{1/3}$.

These calculations also suggest that the growth of the cracks does not occur at the time of peak stress, but slightly later when the tangential stress is relaxing more rapidly than the radial stress. This leads to a damage front that follows behind the outward propagating stress wave with a velocity between those of P and S waves. In addition to calculating the time when the cracks begin to grow, it is also possible to calculate the amount of growth and the orientation of the cracks that are growing. This information is sufficient to calculate a seismic moment tensor for the crack growth, which in turn can be used to calculate the secondary elastic waves generated by the crack growth. Because the crack growth is accomplished by two different processes, sliding on the pre-existing cracks and the extension of wing cracks, the moment tensor also consists of two different parts. Thus one of the contributions of this study is a procedure for estimating the secondary elastic waves that are generated when there is growth of cracks in a zone surrounding an explosion.

While the contribution from a single crack is very small, the integration over the many cracks in a large zone of increased damage surrounding an explosion may lead to the generation of seismic waves that are comparable in amplitude to those of the explosion itself. For an isotropic and homogeneous distribution of cracks, the composite moment tensor is isotropic and thus radiates only P waves into the far field. However, any departure from spherical symmetry in the problem leads to the generation of S waves. In this paper we investigated two types of asymmetry; a pre-stress in the vicinity of the explosion and a preferred orientation of the cracks in the vicinity of the explosion. Both types of asymmetry have similar effects if the pre-stress and crack orientations are chosen in a consistent manner. Included in these effects are significant S waves that emerge from the region surrounding the explosion and thus appear to be generated by the explosion itself. This is entirely consistent with the strong SH waves that McLaughlin *et al.* (1983) observed from a nuclear explosion at NTS. The radiation patterns of these secondary waves also show a general similarity with the radiation patterns of nuclear explosions that generated strong SH waves (Toksöz *et al.*, 1965; Archambeau and Sammis, 1970; Lambert *et al.*, 1972).

The primary purposes of this paper were to demonstrate a method for calculating the growth of cracks in the region surrounding an explosive source and to determine if the secondary waves generated by this crack growth were significant. The results clearly indicate that the secondary elastic waves caused by crack growth can be comparable in size to the primary waves generated by the explosion. Furthermore, provided some type of asymmetry is present in the distribution or orientation of the crack growth, large S waves can be generated by this process. What remains to be done is to further test this approach by comparing its predictions with several different sets of observational data. The ideal data set should contain information about the explosion parameters, the physical properties of the emplacement material, the initial and final crack densities, and waveforms recorded beyond the elastic radius. Tests could then be performed that would compare calculations and observations as a function of such parameters as initial crack density, initial crack radius, type of emplacement material, depth of burial, pre-stress, and crack orientation. Such comparisons would help evaluate the general utility of the approach outlined in this paper for modeling elastic waves generated by explosive sources in realistic environments.

The effects of crack growth calculated in this paper should be regarded as minima because the increase in damage was stopped at the inception of failure. Damage is likely to continue beyond this point, but we are not aware of a suitable method of calculating damage, along with stress and strain, in the unstable post-failure regime. This is one aspect of the problem that definitely needs further research. Sammis (1998) has suggested acoustic fluidization (Melosh, 1979) as a possible method of modeling this type of behavior, and this aspect of the problem is currently being pursued.

Another area in which further research would be useful is a more complete connection between the equivalent elastic treatment and the damage mechanics. In the present study the shear modulus surrounding an explosion is modified by using a hyperbolic relationship employing an empirical reference strain, and the resulting stresses are used to calculate an increase in damage. In principle, it should be possible to calculate a change in the elastic moduli caused by a change in damage using a method such as that of O'Connell and Budiansky (1974), and this could be used in place of the hyperbolic relationship in the equivalent elastic method, thus completing the circle between stresses, elastic moduli, and damage. Unfortunately, there are likely to be other physical processes besides crack growth that affect the moduli, such as crushing of pores and other nonlinear effects, and these other processes would also have to be modeled in order to completely avoid the empirical aspects of the equivalent elastic treatment.

Appendix A

Elastic Wave Propagator Solution

The use of the equivalent elastic method requires that the elastodynamic equations of motion be solved for an explosive source in a radially inhomogeneous medium, which is possible with a propagator method. Consider an explosive source centered at the origin and assume complete spherical symmetry so that all displacements will be in the radial direction and only compressional waves will be generated. Adopting a spherical coordinate system and assuming a harmonic time dependence of the form $e^{i\omega t}$, the displacement can be written as

$$\mathbf{u} = u(r,\omega)\hat{\mathbf{r}} \ , \tag{A1}$$

where $\hat{\mathbf{r}}$ is the unit vector in the radial direction. In a region where the equations of isotropic linear elasticity are obeyed the displacement must satisfy

$$(\lambda + 2\mu)\nabla^2\mathbf{u} - \nabla \times \nabla\mathbf{u} + \rho\omega^2\mathbf{u} = 0 \ , \tag{A2}$$

where ρ is the density and λ, μ are the Lamé elastic constants. In a region where these elastic parameters are independent of the spatial coordinates, a general solution to the above equation is

$$u(r,\omega) = a^{(1)}h_1^{(1)}(k_p r) + a^{(2)}h_1^{(2)}(k_p r) \ . \tag{A3}$$

Here $h_l^{(1)}$ and $h_l^{(2)}$ are spherical Hankel functions of the first and second kind, respectively, and

$$k_p = \frac{\omega}{V_p} \ , \tag{A4}$$

where

$$V_p = \left[\frac{\lambda + 2\mu}{\rho}\right]^{1/2} \ . \tag{A5}$$

It is easily demonstrated that the function $h_l^{(1)}$ represents inward traveling compressional waves, while $h_l^{(2)}$ represents outward traveling waves. The terms $a^{(1)}$ and $a^{(2)}$ are arbitrary constants at this point. Corresponding to the solution for $u(r,\omega)$ given above, the radial stress is

$$\begin{aligned}\tau_{rr}(r,\omega) = a^{(1)}&\left[(\lambda + 2\mu)k_p h_0^{(1)}(k_p r) - \frac{4\mu}{r}h_1^{(1)}(k_p r)\right] \\ &+ a^{(2)}\left[(\lambda + 2\mu)k_p h_0^{(2)}(k_p r) - \frac{4\mu}{r}h_1^{(2)}(k_p r)\right] \ . \end{aligned}\tag{A6}$$

The tangential stresses are

$$\tau_{\theta\theta}(r,\omega) = \tau_{\phi\phi}(r,\omega) = a^{(1)}\left[\lambda k_p h_0^{(1)}(k_p r) + \frac{2\mu}{r} h_1^{(1)}(k_p r)\right] + a^{(2)}\left[\lambda k_p h_0^{(2)}(k_p r) + \frac{2\mu}{r} h_1^{(2)}(k_p r)\right] . \tag{A7}$$

Now consider the region around the source to be divided into a number of spherical shells, with the material properties constant within each shell but possibly changing at the boundaries between shells. Shell n will be bounded by radii r_n and r_{n+1} and will have material properties ρ_n, λ_n, μ_n. We also have $k_n = \omega/V_{pn}$. Using the results developed above, the displacement and radial stress within the n-th shell can be written in matrix form as

$$\begin{bmatrix} u \\ t_{rr} \end{bmatrix} = \mathbf{A}_n \begin{bmatrix} a_n^{(1)} \\ a_n^{(2)} \end{bmatrix} , \tag{A8}$$

where the shell matrix $\mathbf{A}_n$ is defined as

$$\mathbf{A}_n(r,\omega) = \begin{bmatrix} h_1^{(1)}(k_n r) & h_1^{(2)}(k_n r) \\ (\lambda_n + 2\mu_n) k_n h_0^{(1)}(k_n r) - \frac{4\mu_n}{r} h_1^{(1)}(k_n r) & (\lambda_n + 2\mu_n) k_n h_0^{(2)}(k_n r) - \frac{4\mu_n}{r} h_1^{(2)}(k_n r) \end{bmatrix} . \tag{A9}$$

It is easily shown that the inverse of this matrix exists and is given by

$$\mathbf{A}_n^{-1}(r,\omega) = \frac{i k_n r^2}{2(\lambda_n + 2\mu_n)} \begin{bmatrix} [(\lambda_n + 2\mu_n) k_n h_0^{(2)}(k_n r) - \frac{4\mu_n}{r} h_1^{(2)}(k_n r)] & -h_1^{(2)}(k_n r) \\ -[(\lambda_n + 2\mu_n) k_n h_0^{(1)}(k_n r) - \frac{4\mu_n}{r} h_1^{(1)}(k_n r)] & h_1^{(1)}(k_n r) \end{bmatrix} . \tag{A10}$$

The boundary conditions for this problem are that u and t_{rr} should be continuous everywhere. This means that at the radius r_n, where shells n and $n-1$ are in contact, it is required that

$$\begin{bmatrix} u \\ t_{rr} \end{bmatrix} = \mathbf{A}_{n-1}(r_n,\omega) \begin{bmatrix} a_{n-1}^{(1)} \\ a_{n-1}^{(2)} \end{bmatrix} = \mathbf{A}_n(r_n,\omega) \begin{bmatrix} a_n^{(1)} \\ a_n^{(2)} \end{bmatrix} . \tag{A11}$$

It follows that

$$\begin{bmatrix} a_n^{(1)} \\ a_n^{(2)} \end{bmatrix} = \mathbf{A}_n^{-1}(r_n,\omega) \mathbf{A}_{n-1}(r_n,\omega) \begin{bmatrix} a_{n-1}^{(1)} \\ a_{n-1}^{(2)} \end{bmatrix} . \tag{A12}$$

It is convenient to define an interface matrix as

$$\mathbf{B}_n(\omega) \equiv \mathbf{A}_n^{-1}(r_n,\omega) \mathbf{A}_{n-1}(r_n,\omega) . \tag{A13}$$

Letting $z_n = k_n r_n$ and $z_{n-1} = k_{n-1} r_n$, the individual elements of $\mathbf{B}_n$ are

$$\begin{aligned}
B_{11} &= \frac{z_n e^{-i(z_n - z_{n-1})}}{2(\lambda_n + 2\mu_n)} \left[(\lambda_n + 2\mu_n) \frac{i + z_{n-1}}{z_{n-1}^2} - (\lambda_{n-1} + 2\mu_{n-1}) \frac{i - z_n}{z_n^2} \right. \\
&\qquad \left. + 4i(\mu_n - \mu_{n-1}) \frac{(i - z_n)(i + z_{n-1})}{z_n^2 z_{n-1}^2} \right], \\
B_{12} &= \frac{-z_n e^{-i(z_n + z_{n-1})}}{2(\lambda_n + 2\mu_n)} \left[(\lambda_n + 2\mu_n) \frac{i - z_{n-1}}{z_{n-1}^2} - (\lambda_{n-1} + 2\mu_{n-1}) \frac{i - z_n}{z_n^2} \right. \\
&\qquad \left. + 4i(\mu_n - \mu_{n-1}) \frac{(i - z_n)(i - z_{n-1})}{z_n^2 z_{n-1}^2} \right], \\
B_{21} &= \frac{z_n e^{i(z_n + z_{n-1})}}{2(\lambda_n + 2\mu_n)} \left[(\lambda_n + 2\mu_n) \frac{i + z_{n-1}}{z_{n-1}^2} - (\lambda_{n-1} + 2\mu_{n-1}) \frac{i + z_n}{z_n^2} \right. \\
&\qquad \left. + 4i(\mu_n - \mu_{n-1}) \frac{(i + z_n)(i + z_{n-1})}{z_n^2 z_{n-1}^2} \right], \\
B_{22} &= \frac{-z_n e^{i(z_n - z_{n-1})}}{2(\lambda_n + 2\mu_n)} \left[(\lambda_n + 2\mu_n) \frac{i - z_{n-1}}{z_{n-1}^2} - (\lambda_{n-1} + 2\mu_{n-1}) \frac{i + z_n}{z_n^2} \right. \\
&\qquad \left. + 4i(\mu_n - \mu_{n-1}) \frac{(i + z_n)(i - z_{n-1})}{z_n^2 z_{n-1}^2} \right].
\end{aligned} \tag{A14}$$

This process of matching boundary conditions at an interface can be repeated at successive boundaries, and thus the solution can be propagated through an arbitrary number of shells. Assuming the process to have started in shell 1, the solution in shell n is

$$\begin{bmatrix} a_n^{(1)} \\ a_n^{(2)} \end{bmatrix} = \mathbf{B}_n(\omega)\mathbf{B}_{n-1}(\omega) \cdots \mathbf{B}_2(\omega) \begin{bmatrix} a_1^{(1)} \\ a_1^{(2)} \end{bmatrix}. \tag{A15}$$

The next step is to assume that shell 0 is a cavity containing a gas with pressure P and adiabatic gas constant g_c. Then, allowing approximately for the change in pressure caused by the expansion of the cavity, the displacement–stress vector at the radius r_1 will be

$$\begin{bmatrix} u \\ t_{rr} \end{bmatrix} = \begin{bmatrix} u \\ -P(1 - 3g_c u / r_1) \end{bmatrix} = \mathbf{A}_1(r_1, \omega) \begin{bmatrix} a_1^{(1)} \\ a_1^{(2)} \end{bmatrix}. \tag{A16}$$

Two independent trial solutions can be obtained for this system. First assume that $a_1^{(2)} = 0$ and then it is possible to solve for

$$a_1^{(1)} = \frac{-P}{(\lambda_1 + 2\mu_1) k_1 h_0^{(1)}(k_1 r_1) - \frac{4}{r_1}\left(\mu_1 + \frac{3g_c P}{4} h_1^{(1)}(k_1 r_1)\right)}. \tag{A17}$$

The second solution is obtained by assuming that $a_1^{(1)} = 0$ and then

$$a_1^{(2)} = \frac{-P}{(\lambda_1 + 2\mu_1)k_1 h_0^{(2)}(k_1 r_1) - \frac{4}{r_1}\left(\mu_1 + \frac{3g_c P}{4} h_1^{(2)}(k_1 r_1)\right)} \ . \tag{A18}$$

The general solution is obtained by introducing the parameter ϵ, and then the sum of ϵ times the first trial solution and $(1 - \epsilon)$ times the second trial solution will satisfy the boundary conditions at r_1.

The parameter ϵ is determined by the radiation condition. Assume that shell N extends to infinity so that there are only outward propagating waves in this region, and thus $a_N^{(1)} = 0$. This means that

$$\epsilon a_{N1}^{(1)} + (1 - \epsilon) a_{N2}^{(1)} = 0 \ , \tag{A19}$$

where a_{N1} and a_{N2} are the two solutions that are obtained in shell N when the two different trial solutions from r_1 are propagated independently to r_N. It follows that

$$\epsilon = \frac{a_{N2}^{(1)}}{a_{N2}^{(1)} - a_{N1}^{(1)}} \ . \tag{A20}$$

Having determined ϵ, the two trial solutions can be summed with the proper weighting factors at any radius to yield the total solution.

The possibility that elastic waves will be attenuated as they propagate is included by introducing anelasticity into the problem. This is achieved by using the correspondence principle to generalize the previous results. This consists of introducing the complex elastic constants

$$\lambda + 2\mu \ \rightarrow \ (\lambda + 2\mu)(1 + iQ_p^{-1}) \tag{A21}$$

and

$$\mu \ \rightarrow \ \mu(1 + iQ_\mu^{-1}) \ , \tag{A22}$$

where

$$Q_p^{-1} = Q_\kappa^{-1} + \frac{4V_s^2}{3V_p^2} Q_\mu^{-1} \ . \tag{A23}$$

With attenuation present, the velocities will in general be dispersive. Assuming that Q_κ and Q_μ are not functions of frequency in the band of interest, causality is maintained by requiring that (LIU *et al.*, 1976)

$$V_p(\omega) = V_p(\omega_r)\left(1 + \frac{Q_p^{-1}}{\pi} \ln\left[\frac{\omega}{\omega_r}\right]\right) \ , \tag{A24}$$

$$V_s(\omega) = V_s(\omega_r)\left(1 + \frac{Q_\mu^{-1}}{\pi}\ln\left[\frac{\omega}{\omega_r}\right]\right) , \tag{A25}$$

where ω_r is some reference frequency.

Appendix B

Scalar Moment Tensors for the Damage

A convenient method of calculating the elastic waves generated by an increase in damage is to first represent the damage in terms of a seismic moment tensor. In this appendix, the scalar part of this moment tensor is derived. The damage is best represented as a combination of two separate moment tensors, one due to a shear dislocation on the initial crack and another due to the tensile opening of the wing cracks.

In order to estimate the opening of the wing cracks one can use the results for a force applied to the center of a crack (TADA *et al.*, 1973). The critical stress factor is then

$$K_{1c} = \frac{F_3}{(\pi l)^{3/2}} \tag{B1}$$

and the volume of the opened crack is

$$\delta V = \frac{8}{\pi}\frac{(1 - \nu^2)}{E}F_3 l , \tag{B2}$$

where F_3 is opening force, ν is Poisson's ratio, and E is Young's modulus. Note that in ASHBY and SAMMIS (1990, equation 20) the expression for the critical stress factor contains $l + \beta a$ in the denominator where the βa term has been added to prevent the stress from blowing up as l goes to zero. This βa term is not included in the present examination of the geometry of the cracks, as it leads to a condition in which the opening of the crack does not go to zero for negligible damage.

Eliminating the opening force and using

$$\nu = \frac{\lambda}{2(\lambda + \mu)} , \tag{B3}$$

$$E = \mu\frac{3\lambda + 2\mu}{\lambda + \mu} , \tag{B4}$$

the volume becomes

$$\delta V = \frac{2\pi^{1/2}}{\mu}\frac{\lambda + 2\mu}{\lambda + \mu}K_{1c}l^{5/2} . \tag{B5}$$

This is the opening of a single crack. From the damage equation (1) the number of cracks per unit volume is

$$N_v = \frac{1}{\frac{4}{3}\pi(a\cos\chi)^3} D_0 \ , \tag{B6}$$

and also

$$l = a\cos\chi\left[\left(\frac{D}{D_0}\right)^{1/3} - 1\right] . \tag{B7}$$

Thus the total change in volume per unit volume becomes

$$\frac{\delta V}{V} = \frac{3}{2\pi^{1/2}} \frac{\lambda + 2\mu}{\lambda + \mu} \frac{K_{1c}}{\mu(a\cos\chi)^{1/2}} D_0 \left[\left(\frac{D}{D_0}\right)^{1/3} - 1\right]^{5/2} . \tag{B8}$$

This is the dilatancy associated with the wing cracks. One can also write

$$\delta V = \frac{\pi}{3} l^2 d \ , \tag{B9}$$

where d is the opening of the wing crack. This has the solution

$$d = \frac{6}{\pi^{1/2}} \frac{\lambda + 2\mu}{\mu(\lambda + \mu)} K_{1c} l^{1/2} . \tag{B10}$$

For a tensile crack the moment per unit volume is (AKI and RICHARDS, 1980)

$$m_t = \frac{(\lambda + 2\mu)\delta V}{V} \ , \tag{B11}$$

which now becomes

$$m_t = \frac{3}{2\pi^{1/2}} \frac{(\lambda + 2\mu)^2}{\mu(\lambda + \mu)} \frac{K_{1c}}{(a\cos\chi)^{1/2}} D_0 \left[\left(\frac{D}{D_0}\right)^{1/3} - 1\right]^{5/2} . \tag{B12}$$

For the shear crack the moment per unit volume is (AKI and RICHARDS, 1980)

$$m_s = \frac{\mu\pi a^2 d}{V\sin\chi} \ , \tag{B13}$$

and this becomes

$$m_s = \frac{9}{2\pi^{1/2}} \frac{\lambda + 2\mu}{\lambda + \mu} \frac{K_{1c}}{(a\cos\chi)^{1/2}} \frac{D_0}{\sin\chi(\cos\chi)^2} \left[\left(\frac{D}{D_0}\right)^{1/3} - 1\right]^{1/2} . \tag{B14}$$

Acknowledgments

We gratefully acknowledge the use of waveform data for the NPE explosion which were obtained from the NTS Ground Motion Data Base maintained by Los Alamos National Laboratory. We also thank an anonymous reviewer for numerous helpful suggestions. Partial support for this research was provided by the Defense Special Weapons Agency under grants DSWA-01-97-1-0026 (LRJ) and DSWA-01-97-1-0005 (CGS). All computations were done at the Center for Computational Seismology (CCS) at Lawrence Berkeley National Laboratory (LBNL), which is operated by the University of California for the US Department of Energy under contract No. DE-AC03-76SF00098. CCS is supported by the DOE Office of Basic Energy Sciences.

References

AKI, K., and RICHARDS, P. G., *Quantitative Seismology: Theory and Methods*, vol. I (W. H. Freeman and Co., New York, 1980) 557 pp.

APP, R. N., and TUNNELL, T. W. (1994), *User Manual for the NTS Ground Motion Data Base Retrieval Program ntsgm*, Los Alamos National Laboratory Report LA-UR-94-1538, 40 pp.

ARCHAMBEAU, C., and SAMMIS, C. (1970), *Seismic Radiation from Explosions in Pre-stressed Media and the Measurement of Tectonic Stress in the Earth*, Rev. Geophys. Space Phys. *8*, 473–499.

ASHBY, M. F., and SAMMIS, C. G. (1990), *The Damage Mechanics of Brittle Solids in Compression*, Pure appl. geophys. *133*, 489–521.

BALDWIN, M. J., BRADFORD, R. P., HOPKINS, S. P., TOWNSEND, D. R., and HARRIS-WEST, B. L., *Geology, geophysics, and physical properties of the U12n.25 non-proliferation experiment site*. In *Proc. Symposium on the Non-Proliferation Experiment (NPE): Results and Implications for Test Ban Treaties* (ed. M. D. Denny), CONF-9404100 (Department of Energy, 1994).

BEN-MENAHEM, A., and SENA, A. G. (1990), *Seismic Source Theory in Stratified Anisotropic Media*, J. Geophys. Res. *95*, 15,395–15,427.

BEN-MENAHEM, A., and GIBSON R. L., Jr. (1995), *Radiation of Elastic Waves from Sources Embedded in Anisotropic Inclusions*, Geophys. J. Int. *122*, 249–265.

DENNY, M. D., *Introduction and highlights*. In *Proc. Symposium on the Non-Proliferation Experiment (NPE): Results and Implications for Test Ban Treaties* (ed. M. D. Denny), CONF-9404100 (Department of Energy, 1994).

EMERICK, W. L. (1966), *Physical Properties of Volcanic Rocks, Rainier Mesa and Vicinity, Nevada Test Site, Tech. Let. Area 12-18*, U.S. Geological Survey, 21 pp.

GLENN, L. A. (1993), *Energy-density Effects on Seismic Decoupling*, J. Geophys. Res. *98*, 1933–1942.

GLENN, L. A., and GOLDSTEIN, P. (1994), *Seismic Decoupling with Chemical and Nuclear Explosions in Salt*, J. Geophys. Res. *99*, 11,723–11,730.

GOLDSTEIN, H., *Classical Mechanics* (Addison-Wesley, Reading, 1959) 300 pp.

HARDIN, B. O., and DRNEVICH, V. P. (1972), *Shear Modulus and Damping in Soils: Design Equations and Curves*, J. Soil Mech. and Found. Div., Proc. Am. Soc. Civil Eng. *98*, 667–692.

JOHNSON, L. R. (1993), *Wave Propagation near Explosive Sources*, PL-TR-93-2118, Phillips Laboratory, 22 pp.

JOHNSON, L. R., *Recording experiment on Rainier Mesa in conjunction with a reflection survey*. In *Proc. Symposium on the Non-Proliferation Experiment (NPE): Results and Implications for Test Ban Treaties* (ed. M. D. Denny), CONF-9404100 (Department of Energy, 1994).

KAMM, J. R., and BOS, R. J. (1995), *Comparison of Chemical and Nuclear Explosions: Numerical Simulations of the Non-Proliferation Experiment*, Los Alamos National Laboratory Report LA-12942-MS, 98 pp.

King, D. S., Freeman, B. E., Eilers, D. D., and Johnson, J. D. (1989), *The Effective Yield of a Nuclear Explosion in a Small Cavity in Geologic Material: Enhanced Coupling Revisited*, J. Geophys. Res. *94*, 12,375–12,385.

Lambert, D. G., Flinn, E. A., and Archambeau, C. B. (1972), *A Comparative Study of the Elastic Wave Radiation from Earthquakes and Underground Explosions*, Geophys. J. R. Astr. Soc. *29*, 403–432.

Liu, H.-P., Anderson, D. L., and Kanamori, H. (1976), *Velocity Dispersion due to Anelasticity; Implications for Seismology and Mantle Composition*, Geophys. J. R. Astr. Soc. *47*, 41–58.

Lockner, D. A., and Madden, T. R. (1991), *A Multiple-crack Model of Brittle Fracture*, 1. *Non-time Dependent Simulations*, J. Geophys. Res. *96*, 19,623–19,642.

Majtenyi, S. I., and Foster, E. L. (1992), *Propagation Velocity of Shock Waves and Pressure Waves in Soil and Rock*, Geophys. Res. Lett. *19*, 865–868.

McLaughlin, K. L., Johnson, L. R., and McEvilly, T. V. (1983), *Two-dimensional Array Measurements of Near-source Ground Accelerations*, Bull. Seismol. Soc. Am. *73*, 349–375.

Melosh, H. J. (1979), *Acoustic Fluidization: A New Geologic Process*, J. Geophys. Res. *84*, 7513–7520.

Mueller, R. A., and Murphy, J. R. (1971), *Seismic Characteristics of Underground Nuclear Detonations, Part I. Seismic Spectrum Scaling*, Bull. Seismol. Soc. Am. *61*, 1675–1692.

O'Connell, R. J., and Budiansky, B. (1974), *Seismic Velocities in Dry and Saturated Cracked Solids*, J. Geophys. Res. *79*, 5412–5426.

Olsen, K. H., and Peratt, A. L. *Free-field ground motion for the non-proliferation experiment: Preliminary comparisons with nearby nuclear events.* In *Proc. Symposium on the Non-Proliferation Experiment (NPE): Results and Implications for Test Ban Treaties* (ed. M. D. Denny), CONF-9404100 (Department of Energy, 1994).

Press, F., and Archambeau, C. (1962), *Release of Tectonic Strain by Underground Nuclear Explosions*, J. Geophys. Res. *67*, 337–343.

Reches, Z., and Lockner, D. A. (1994), *Nucleation and Growth of Faults in Brittle Rocks*, J. Geophys. Res. *99*, 18,159–18,173.

Rimer, N., Stevens, J. L., Murphy, J. R., and Kocharyan, G. G. *Estimating seismic characteristics of explosions in hard rock using a micro-mechanical damage model.* In *Proc. 20th Annual Seismic Research Symposium on Monitoring a Comprehensive Test Ban Treaty* (eds. J. Fantroy, D. Heatley, J. Warren, F. Chavez, and C. Meade) (Department of Defense and U.S. Department of Energy, 1998), pp. 392–400.

Rodean, H. C., *Nuclear-Explosion Seismology*, TID-25572, (U. S. Atomic Energy Commission, 1971), 156 pp.

Sammis, C. G., *Acoustic fluidization in the damage regime of explosions in crystalline rock.* In *Proc. 20th Annual Seismic Research Symposium on Monitoring a Comprehensive Test Ban Treaty* (eds. J. Fantroy, D. Heatley, J. Warren, F. Chavez, and C. Meade) (Department of Defense and U.S. Department of Energy, 1998), pp. 417–421.

Smith, C. W., *NPE – close-in stress and motion measurements.* In *Proc. Symposium on the Non-Proliferation Experiment (NPE): Results and Implications for Test Ban Treaties* (ed. M. D. Denny), CONF-9404100 (Department of Energy, 1994).

Tada, H., Paris, P., and Irwin, G., *The Stress Analysis of Cracks Handbook* (Del Research Corp., Hellertown, Pa., 1973).

Toksöz, M. N., Harkrider, D. G., and Ben-Menahem, A. (1965), *Determination of Source Parameters by Amplitude Equalization of Seismic Surface Waves*, 2. *Release of Tectonic Strain by Underground Explosions and Mechanisms of Earthquakes*, J. Geophys. Res. *70*, 907–922.

Tsvankin, I. D., and Chesnokov, E. M. (1990), *Synthesis of Body-wave Seismograms from Point Sources in Anisotropic Media*, J. Geophys. Res. *95*, 11,317–11,331.

(Received June 28, 1999, accepted October 20, 1999)

Pure appl. geophys. 158 (2001) 1909–1949
0033–4553/01/111909–41 $ 1.50 + 0.20/0

Pure and Applied Geophysics

Wave Generation from Explosions in Rock Cavities

CANGLI LIU[1] and THOMAS J. AHRENS[1]

Abstract—We have developed a measurement method to monitor *P*- and *S*-waves generated from laboratory-scale explosions in meter-sized rock samples at a series of stations, as well as invented a device to drill spherical cavities in rock, with diameters up to 10 centimeters. We applied these to experiments in Bedford limestone in which spherical/cylindrical explosives (0.2 to 1.9 g) were centrally placed in 1.2- to 3-cm diameter cavities. Stress waves generated by the explosions were recorded within a radius of 25 cm. The radial stress wave records and post-explosion studies demonstrate that *S*-waves are generated from explosions in cavities as a result of both wave mode-conversion from the cavity wall and crack propagation in rocks. The experimental results of wave generation from the explosions in spherical and cylindrical cavities demonstrate the cavity geometrical effect on the resulting wave pattern. The *P*- and *S*-waves generated by explosions and crack propagation in rocks are analyzed. A simple analytic model for *P*-wave generation is proposed to explain the differences of *P*-wave-induced displacement histories between the observed waveforms and those predicted by a step-pressure source. Generally, the qualitative predictions of this model fit the observations. The present results demonstrate the importance of rock cracking and cavities in *P*- and *S*-wave generation.

Key words: Explosion, cavities, *P*-wave, *S*-wave, cracks.

1. Introduction

In the past 30 years, extensive research has been conducted to discriminate underground explosions from earthquakes. As a result, MURPHY (1996) concluded that the seismology of earthquakes and underground explosions is at a state in which it is possible to identify tamped explosions with a yield over 1 kiloton ($m_b \geq 4$) on a worldwide basis. Explosions can be distinguished from earthquakes by global seismic stations using initial *P*-wave motion and M_s to m_b ratio. In fact, the discrimination of tamped explosions from earthquakes depends mainly on empirical correlations of seismic data from previous underground nuclear explosions, explosions from mining and quarrying operations and earthquakes. Although many basic problems related to tamped explosions still remain unresolved (MASSE, 1981 and MURPHY, 1996), the discrimination problem, that is, how to distinguish earthquakes from decoupled

[1] Seismological Laboratory, California Institute of Technology, Pasadena, CA 91125, USA. E-mail: tja@caltech.edu

nuclear explosions with complex geometries and chemical explosions, remains the most challenging problem.

The characteristics of decoupled explosions (MURPHY, 1996) are: (1) Resultant values of m_b and M_s are too low to use the global monitoring system (for a decoupled 1 kt explosion, $m_b \leq 3$); (2) the amplitudes of surface waves generated from decoupled explosions are much lower than from tamped explosions; (3) the excitation of direct S waves complicates the discrimination problem. Although there have been many studies related to decoupled explosions (GLENN *et al.*, 1985; GLENN, 1993; GLENN and GOLDSTEIN, 1994, 1996; MURPHY *et al.*, 1996, 1997; SYKES, 1996; LANGSTON, 1983), it is still unclear as to what really controls the generation of S waves from decoupled explosions (MURPHY, 1996). Therefore, the mechanism of S-wave generation from explosions, especially from decoupled explosions in cavities, appears to be one of the most important discrimination problems.

Sources for S waves from explosions include: (1) Asymmetric brittle fracture of rock near explosions; (2) asymmetric plastic deformation near the explosion due to asymmetric decoupling geometry and the inherent anisotropy of rocks; (3) tectonic release of shear energy induced by the explosion; and (4) mode conversion of P- to S-waves at boundaries such as cavity wall and the earth's free surface.

The objective of this work is to experimentally investigate P- and S-wave generation from explosions in cavities, dynamic fracture in rock and the conversion of shock waves at the surface of the cavities. If shock-wave pressure in a cavity is not too high, failure via compressive comminution and shear failure is not expected. However, tensional brittle failure of the cavity rock may occur. Radial fractures may be initiated from the inner surface of a cavity because tangential (hoop) stresses always achieve the maximum value of pressure at cavity-rock interfaces. Radial fractures generally can be further driven by the abundant high-pressure explosive products (COURSEN, 1985). The propagation of radial cracks may result in P- and S-wave generation. The reflection of the shock waves at cavity surfaces will also induce S-waves.

In small-scale laboratory experiments, wave profiles have decidedly higher frequencies than the usual seismic waves. Therefore, conventional seismic recording systems cannot be used to measure these wave profiles. One method we considered was that developed for superhigh strain rate experiments (KIM *et al.*, 1977). It is better suited for metallic samples than for the rocks of the present experiments because surfaces must remain coherent on the scales of a wavelength of light. Magnetic velocity gauges have long been used to study shock-wave decay in rocks (FLORENCE *et al.*, 1993). However this method requires embedding gauges in rock and thus adds artificial internal interfaces that will generate S waves and make the S-wave train more complicated. Based on the analysis of the interaction between P- and S-waves and free-surfaces, we developed a method to measure P- and S-wave profiles at the free-surfaces of rock. The main feature of the method is to use the different characteristics of the interactions between P and S waves and the free-surface as discussed in Section 2. In order to produce a spherical cavity in rock, we invented a spherical drilling tool (Section 3). Using the measurement method and this drilling tool, we conducted wave generation experiments by setting off explosions in cavities inside Bedford limestone.

In order to study the far-field seismic radiation, we chose the explosive charge mass to be sufficiently small such that the source dimension of the cavity was much less than the resulting wave propagation path.

Future work might address the question of the time history of tensile failure around gas-filled cavities and the generation of regional and teleseismic *S* waves.

In this report, the details of the methods are described, including the measurement method and the spherical cavity drilling method. Experimental data for seismic wave generation from spherical and cylindrical cavities are reported. In addition, a simple analytic model to explain the observed *P*- and *S*-wave profiles is discussed.

2. *Measurement Method*

When *P*- and *S*-waves arrive at a free surface, their interactions with the free surface are very different. The characteristics of these interactions provide an opportunity for measuring the incident *P*- and *S*-wave amplitudes from strain measurements on the free surfaces of rocks. The method developed is to use two strain gauges to measure the strains along two perpendicular directions at one point on the free surface. The incident amplitude of *P* and *S* waves can be obtained from these strain measurements.

The basic assumptions made below are as follows: (1) Waves in rocks near free surfaces are treated locally as plane waves because the strain gauges used in the experiments have dimensions of 3×3 mm and are located at least 150 mm away from the explosive source; and (2) the deformation of rock near the free surface is essentially elastic and is in the far-field radiation pattern. In experiments (Fig. 8) that induce cracking, these cracks eventually propagated to the free surface, however, during most of the recording intervals of the present data sets (60 μs) the crack is enclosed by rock and is expected to only have propagated a few cm, and thus to good approximation, our data are largely in the far-field.

In the following, we review the relationship between surface strains and the displacement associated with *P*- and *S*-waves that would exist at the location of the strain gauges for an infinite medium.

2.1 Sensors

A group of strain gauges are employed to monitor wave profiles at different locations along two directions on the free surface of rock samples. All gauges are located along a line defined by a plane perpendicular to the "rock joint" (Fig. 1),

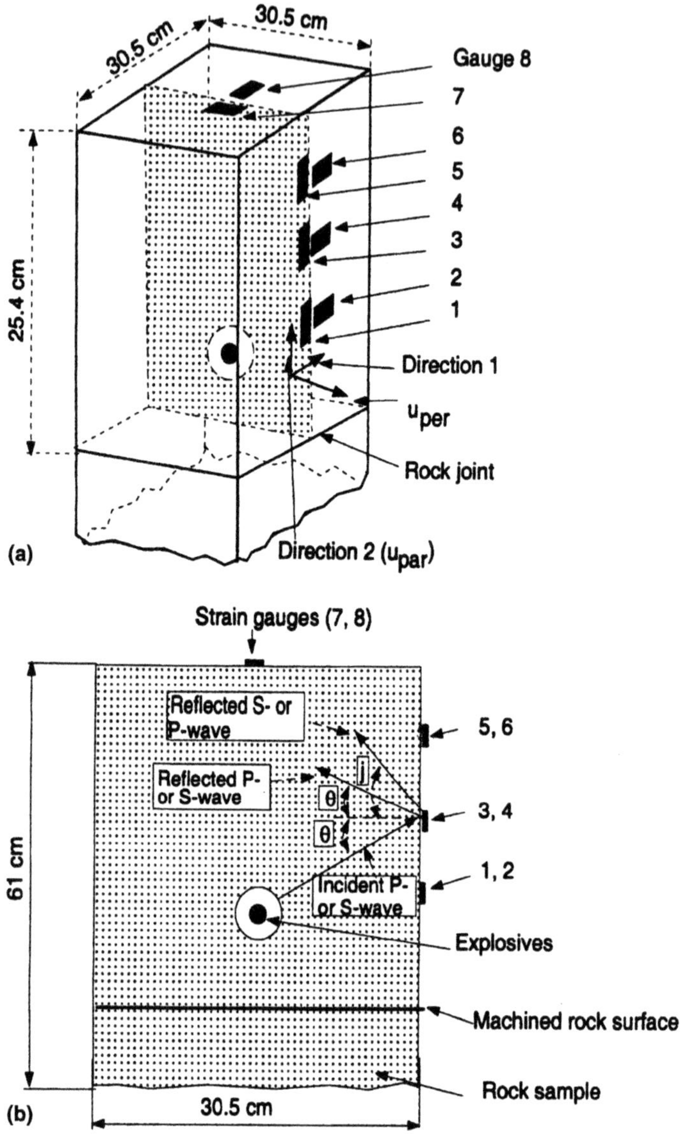

Figure 1
Strain gauge (A) and sample (B) layout and circuit diagram (C) for spherical cavity experiments. The relation of the spherical cavity to the joint is shown in Figure 5.

bisecting the explosives and the 30.5 cm width of the rock sample. This surface is expected to allow *P*- to *S*-mode conversion. The gauges and amplifiers are powered by batteries in order to suppress AC ripple (60 Hz) and achieve a high signal-to-noise ratio. A mercury relay is used to calibrate the system.

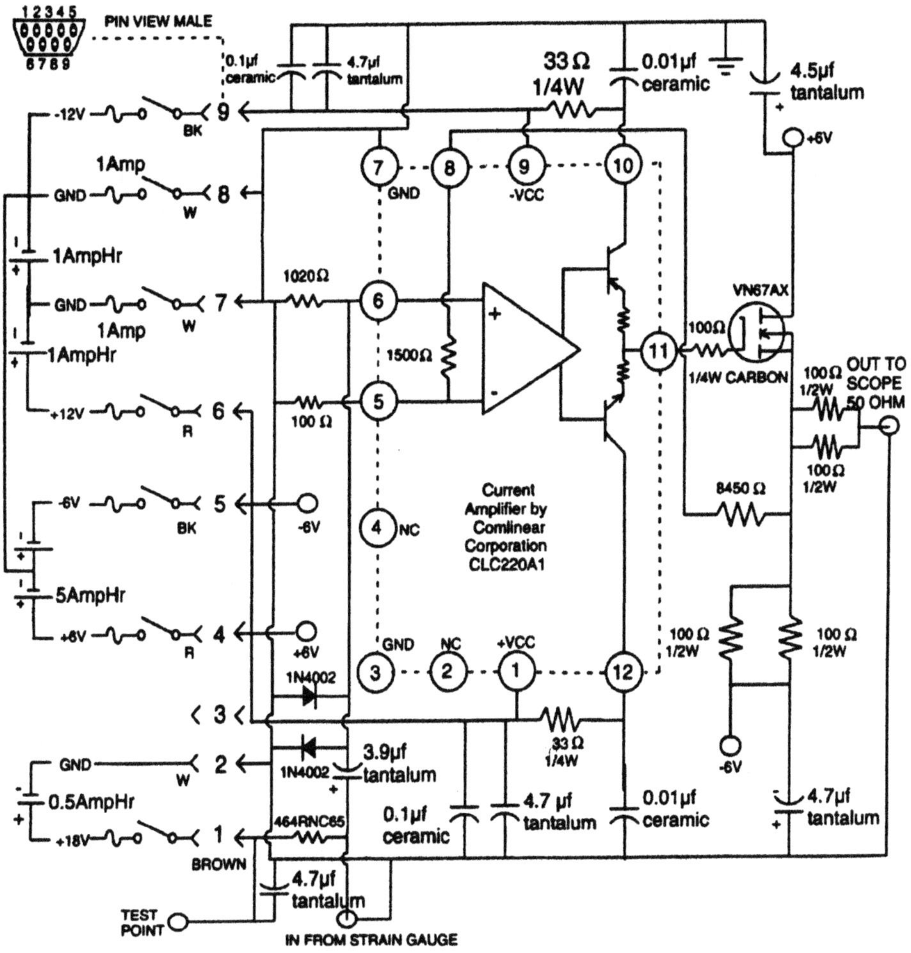

Figure 1C

The relation between strain, ε, and voltage change, ΔV, is

$$\varepsilon = \frac{\Delta V}{fgV_0} , \tag{1}$$

where V_0 is initial voltage supplied to the gauge, and f and g are gauge factor and amplifier gain, respectively.

A simplified circuit diagram is shown in Figure 1C.

2.2 Data Reduction Method

A strain gauge yields elongation along the gauge direction. When a strain gauge is bonded to a free surface, elongation of the free surface during reflections can be obtained. Elongation includes the contributions from incident and reflected *P*- and

S-waves. The relationship between the strains recorded by the gauges and the incident P- and S-wave amplitudes is derived below.

2.2.1 Planar P- and S-wave reflection at free surfaces

When a P-wave arrives at a free surface with incident angle θ, the resultant displacements of the particles on the free surfaces after wave reflection are

$$u^p_{\text{par}} = H_{\text{par}} u^I_p \ , \tag{2}$$

$$u^p_{\text{per}} = H_{\text{per}} u^I_p \ , \tag{3}$$

where u^I_p is the particle displacement of the incident P-wave, and u^p_{par} and u^p_{per} indicate, respectively, the resultant particle displacements along direction 2 and the direction that is perpendicular to the free surface after reflection, as shown in Figure 1. The values of H_{par} and H_{per} are

$$H_{\text{par}} = \frac{2\cos\theta\sin(2j)}{A+B} \ , \tag{4}$$

$$H_{\text{per}} = \frac{2\cos\theta\cos(2j)}{A+B} \ , \tag{5}$$

where j is the reflected S-wave angle. The derivation of Eqs. (4) and (5) and the definitions of A and B are given in Appendix 1.

When an S-wave arrives at a free surface with incident angle θ, the resultant displacements of the particles on the free surfaces are

$$u^{sv}_{\text{par}} = G_{\text{par}} u^I_{sv} \ , \tag{6}$$

$$u^{sv}_{\text{per}} = G_{\text{per}} u^I_{sv} \ , \tag{7}$$

where u^I_{sv} is the particle displacement of the incident SV wave and u^{sv}_{par} and u^{sv}_{per} are, respectively, the resultant particle displacements on the free surface after reflection along direction 2 and the direction that is perpendicular to the free surface, as shown in Figure 1. The values of G_{per} and G_{par} are

$$G_{\text{par}} = \frac{2\cos(2\theta)\cos\theta}{A_s + B_s} \ , \tag{8}$$

$$G_{\text{per}} = -2\frac{\frac{\beta}{\alpha}\cos j\sin(2\theta)}{A_s + B_s} \ , \tag{9}$$

where j is the reflected P-wave angle and α and β are P- and S-wave velocities, respectively. The derivations of Eqs. (8) and (9) and definitions of A_s and B_s are given in Appendix 2.

2.2.2 Strains induced by incident plane P-waves

1. Strains measured along direction 1

Because u^P_{per} is perpendicular to the free surface and is along the symmetrical axis direction of the waves, the strain due to u^P_{per} along direction 1 is

$$\varepsilon_1^{\text{per}} = \frac{H_{\text{per}}}{r_0} u_p^I \ , \tag{10}$$

where $\varepsilon_1^{\text{per}}$ is the strain along direction 1 induced by u^P_{per}, and r_0 is the distance from the center of the cavity to the free surface at $\theta = 0$.

Because u^P_{per} does not at any time result in any strains in the gauges along direction 1, the total strain induced by the incident P waves is

$$\varepsilon_1^P = H_1 \frac{u_p^I}{r_0} \ , \tag{11}$$

where $H_1 = H_{\text{per}}$.

2. Strains measured along direction 2

Because u^P_{per} and u^P_{par} both contribute to the strains measured by the gauges along direction 2, we must consider the resultant displacements. The length of the gauge after the reflection l_s^r, is

$$l_s^r = \left(r_n^2 + \left(\frac{\partial r_n}{\partial \theta} \right)^2 \right)^{\frac{1}{2}} \delta\theta \ , \tag{12}$$

where $\delta\theta = l_s/r_n$ in which l_s is the initial length of the strain gauge, and r_n the distance from the center of the cavity to the position of the gauge upon wave reflection is,

$$r_n = r + u\cos(\eta - \theta) \ , \tag{13}$$

where r is the distance from the center of the cavity to the gauge before P-wave reflection, u the resultant displacement of the point at θ on the free surface, is

$$u = u_p^I \frac{2\cos(\theta)}{A + B} \ , \tag{14}$$

and η, the angle between u and u^P_{per},

$$\eta = 2j \ . \tag{15}$$

To a first-order approximation, l_s^r is

$$l_s^r \approx \left(\left(\frac{r}{\cos\theta} \right)^2 + 2ru_p^I \left(W(\theta) + \tan(\theta) \frac{dW}{d\theta} \right) \right)^{\frac{1}{2}} \frac{l_s \cos\theta}{r} \ , \tag{16}$$

where $W = 2\cos(\theta)\cos(\eta - \theta)/(A + B)$.

If $x \ll 1$, we have $(1+x)^{1/2} \simeq 1 + x/2$. Because $u_p^I/r \ll 1$, l_s^r is

$$l_s^r \approx \left(r\left(1+\frac{\tan^2\theta}{2}\right) + u_p^I\left(W(\theta)\left(1-\frac{\tan^2\theta}{2}\right) + \tan\theta\frac{dW}{d\theta}\right)\right)\delta\theta\ . \tag{17}$$

Then, the strain is

$$\varepsilon_2^p = \frac{l_s^r - l_s}{l_s}\ , \tag{18}$$

and therefore

$$\varepsilon_2^p = H_2\frac{u_p^I}{r_0}\ , \tag{19}$$

where

$$H_2 = \frac{\left(\cos\theta\left(W\left(1-\frac{\tan^2\theta}{2}\right)\right) + \tan\theta\frac{dW}{d\theta}\right)}{\left(1+\frac{\tan^2\theta}{2}\right)}\ , \tag{20}$$

and $r_0 = r\cos\theta$.

2.2.3 Strains induced by incident plane *S*-waves

Using the same method, all the strains induced by incident *SV*-waves can be obtained. The final expressions are listed as following:

1. Strains measured by gauges along direction 1 are

$$\varepsilon_1^{sv} = G_1\frac{u_{sv}^I}{r_0}\ , \tag{21}$$

where $G_1 = G_{\text{per}}$.

2. The strains measured by the gauges along direction 2 can be found using the same method as for the incident *P*-wave. The resultant strains induced by the incident *SV*-waves along direction 2 are

$$\varepsilon_2^{sv} = G_2\frac{u_{sv}^I}{r_0}\ , \tag{22}$$

where

$$G_2 = \frac{\left(\cos\theta\left(W_s\left(1-\frac{\tan^2\theta}{2}\right)\right) + \tan\theta\frac{dW_s}{d\theta}\right)}{\left(1+\frac{\tan^2\theta}{2}\right)}\ , \tag{23}$$

$$W_s = \frac{2\cos(\eta-\theta)\cos\theta}{A_s+B_s}\left(\cos^2(2\theta) + 4\left(\frac{\beta}{\alpha}\right)^2\cos^2 j\sin^2\theta\right)^{\frac{1}{2}}\ , \tag{24}$$

$$\tan\eta = -\frac{\alpha\cos\theta}{\beta\cos j\tan(2\theta)}\ . \tag{25}$$

2.2.4 Amplitudes of incident *P*- and *S*-waves

From the expressions given above, the displacements of incident *P*- and *SV*-waves can be obtained through the strains recorded by the gauges along the two directions.

1. The displacements from an incident *P*-wave is

$$u_p^I = r_0 \frac{\varepsilon_1^p}{H_1} = r_0 \frac{\varepsilon_2^p}{H_2} \quad . \tag{26}$$

2. The displacement from an incident *SV*-wave is

$$u_{sv}^I = r_0 \frac{\varepsilon_{sv}^1}{G_1} = r_0 \frac{\varepsilon_{sv}^2}{G_2} \quad . \tag{27}$$

2.3 Characteristics of the Strains along the Two Directions

The variations of H_1, H_2, G_1 and G_2 with wave incident angle are calculated using Eqs. (5), (20), (9) and (23) for Bedford limestone in which $\beta = 2.8$ km/s and $\alpha = 4.45$ km/s (Fig. 2).

2.3.1 Strains induced by incident P-waves

1. The dimensionless constant, H_1, varies from approximately 2.0 and 1.1, and H_2 varies from 2 to -0.8 when the incident angle varies from 0 to 55° for Bedford limestone, as shown in Figure 2.

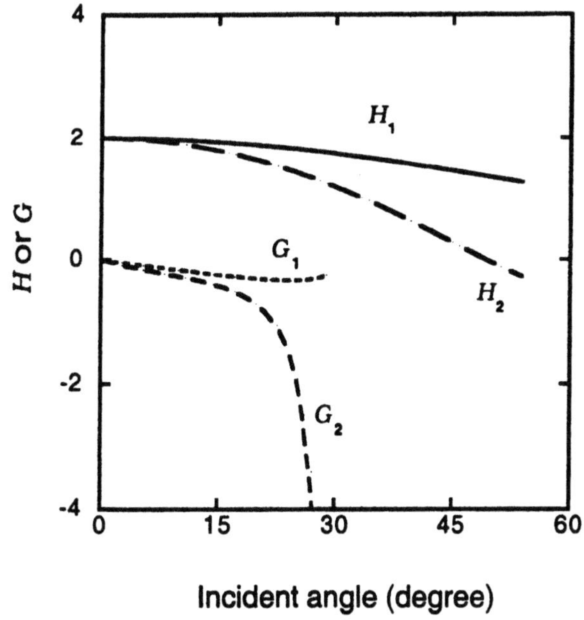

Figure 2
H and *G* versus incident angle.

2. From Figure 2, the strains induced by *P*-waves along direction 1 are always positive, but the strains along direction 2 are positive when θ is less than 47° and negative when θ is larger than 47°.

3. The characteristics of the strains induced by incident *P*-waves have been verified experimentally, as shown in Figures 6, 7, 18 and 19. From the experimental results, it is seen that all the strains along direction 1 (labeled with even gauge numbers) are positive at all the stations and the strains along direction 2 (labeled with odd gauge numbers) are positive when the incident angle is smaller than the critical angle and negative when the incident angle is larger than the critical angle.

2.3.2 Strains induced by incident *S*-waves

1. From the calculated results shown in Figure 2, the gauges along direction 1 are not sensitive to incident *SV*-waves, the gauges along direction 2 are very sensitive to incident *SV*-waves.

2. The measurement of *SV*-waves is limited by the start of Rayleigh surface wave generation. For Bedford limestone, the Rayleigh wave is generated when the incident angle θ is larger than 38° (determined from Snell's law).

3. The polarities of the strains along direction 2 are negative, and the polarities of the strains along direction 1 are determined by the direction of the particle motion (the calculation is made assuming that the motion is toward the increasing direction of the *SV*-wave incident angle).

4. The characteristics of the strains induced by incident *S*-waves have been verified experimentally, as shown in Figures 6, 7, 18 and 19. From the experimental results, it is seen that all the strains along direction 1 (labeled with even gauge numbers) are generally very small compared to the corresponding strain amplitudes along direction 2 at all the stations, and the strains along direction 2 (labeled with odd gauge numbers) are negative at all the stations.

3. Wave Generation From Explosions in Spherical Cavities in Rock

3.1 Experiments and Results

3.1.1 Spherical cavity drill

In order to drill spherical cavities in rock, a special tool has been developed. The device is schematically shown in Figure 3. The working principle of the drill is to use one of the two shanks (Shank A) to transfer the force from the drilling machine to the other shank (Shank B), which rotates the blade and at the same time pushes the blade to cut the rock. During the drilling process, the shank positioning plate is used to protect the drill from vibrations induced by asymmetrical forces acting on the shank from the blade. The device can be used to drill spherical cavities with diameters up to 10 cm in rocks that have the strength of Bedford limestone. The procedure to

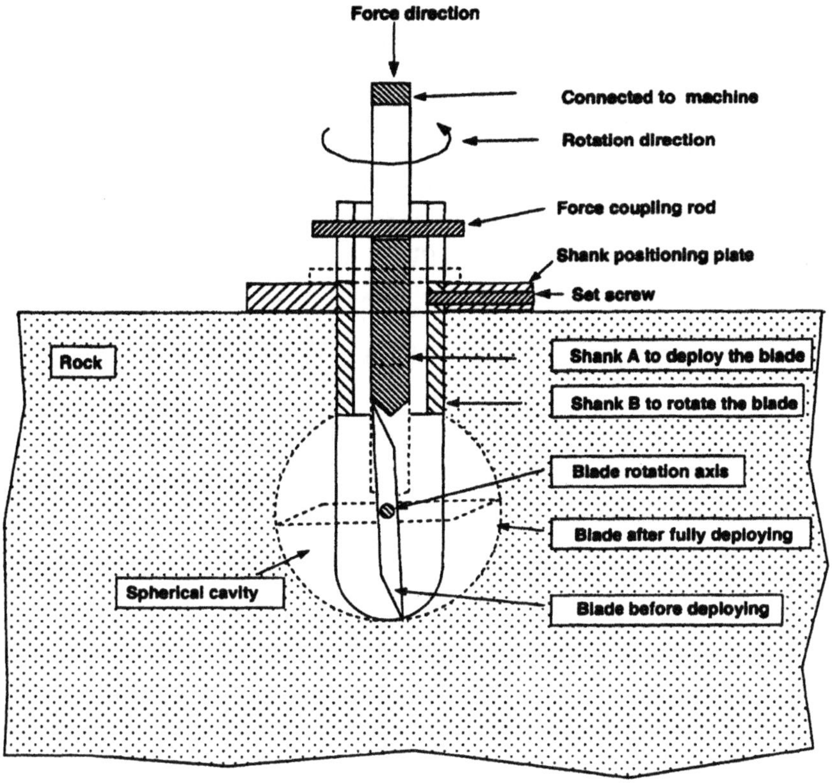

Figure 3
Schematic of spherical cavity drill.

produce a spherical cavity inside rock is as follows: (1) A cylindrical hole of 1 cm diameter is drilled. The hole depth is the deepest point of the spherical cavity; (2) a blade with a 1.5 cm diameter is put into the cylindrical hole, and the position plate is screwed onto Shank B in order to fix the depth of the spherical cavity. During drilling, pressurized air is used to blow out rock dust and cool the blade. The air flow rate is roughly 2 1/minute. The rotation speed is about 240 rpm, and the feed rate is about 1 mm/minute; (3) a large blade with a 3.0 cm diameter is used to enlarge the cavity. This device and procedure have been successfully used to make spherical cavities in Bedford limestone.

3.1.2 Experimental details

The rock sample (Bedford limestone) is assembled from two blocks as shown in Figure 1. The rock sample with strain gauges is placed inside a tank pressurized to 10 bar. The type of strain gauge used in this work is Measurements Group's CEA-00-062UT-120. The dimensions of the gauges are 3×3 mm. The initial resistance of the gauges is 120 $\Omega \pm 0.4\%$ and the gauge factor is $2.09 \pm 0.5\%$. Before the gauges are bonded to the surface. the surface is polished with #240 sandpaper and cleaned with acetone. The gauges are attached to the surface using 910 adhesive (Permabond International). After the adhesive dries, a thin layer of epoxy (5-minute epoxy, ITWDevcon) is applied to the gauge surface in order to protect it. The voltage on the gauge is typically 6 V, measured before each experiment.

The explosives used in this work are Detasheet PETN (DuPont) and commercially pressed pellets of PETN (Reynolds Industries System Incorporated). The initial density is 1.5 g/cm^3 for Detasheet PETN and 1.4 g/cm^3 of pressed PETN. Detonators (Reynolds Industries System Incorporated) have a diameter of 2 mm and length of 8 mm.

We fired several decoupled explosion tests in spherical cavities. Due to the inherent difficulty of properly detonating a small charge, we put considerable effort into completing a very few, highly successful experiments. The experimental setup is sketched in Figures 1 and 4. Figure 5 schematically shows the spherical explosive source. Table 1 lists the experimental parameters and Table 2 gives the location of the gauges.

3.1.3 Experimental records

Using the experimental system given above and the measurement method described in Section 2, the waves generated from the explosions in the spherical cavity were recorded. Experimental records for the two directions are shown in Figure 6 for SW4 and Figure 7 for SW9.

3.1.4 Crack pattern from the recovered samples

(a) SW4: The recovered experimental sample shows that four large cracks (~10 cm) and many small cracks (~5 mm) were induced by the explosion (Fig. 8). The lengths of the cracks with the explosion product trace range from 50 to 80 mm.

(b) SW9: From the recovered sample, cracks were not seen on the free surfaces. After the sample was cut along the axis, many small cracks were found near the cavity, as well as two very long cracks about 50 mm in length.

3.2 Analysis of the Recorded Wave Series

1. *P*-wave velocity

From the experimental records, the *P*-wave velocity is determined to be 4.22 km/s under the experimental conditions, as shown in Figure 9 (the data from SW9 were not used because of the noise at the beginning of the signals). The *P*-wave velocity determined here is lower than the 4.45 km/s found using the ultrasonic method. Possible reasons for the low *P*-wave velocity in the experiments may be the nonlinear elastic behavior of the rock and/or the non-ideal spherical wave (radial *P*-wave velocity is the compressive wave velocity only when the wave is an ideal spherical wave) MURPHY *et al.* (1997).

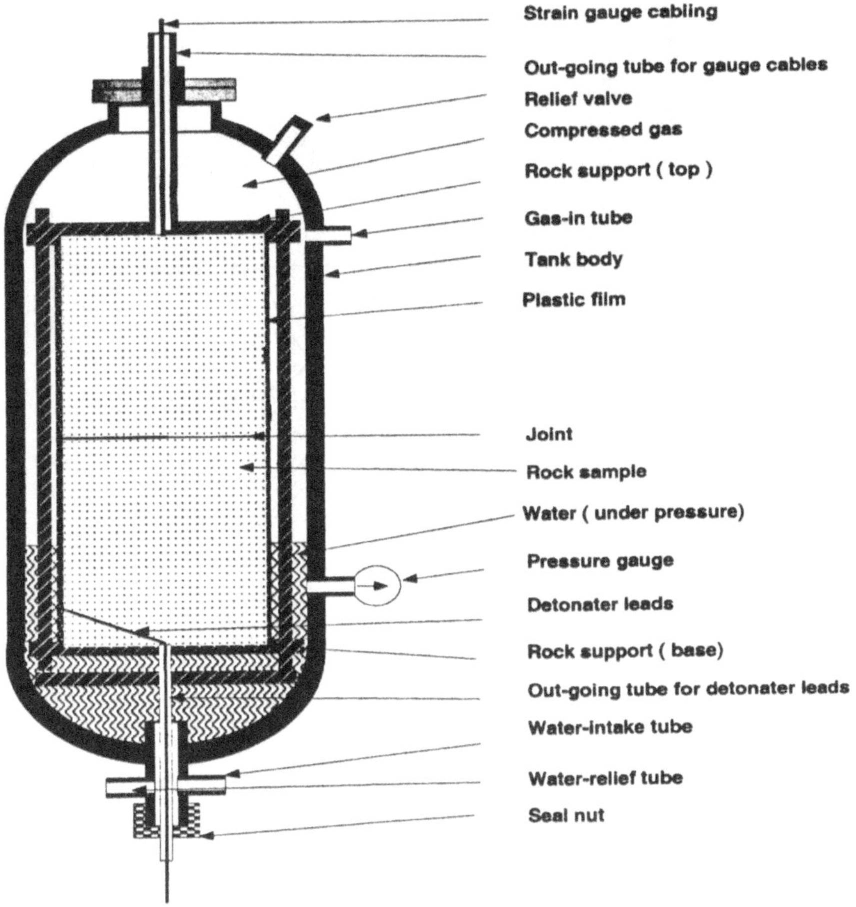

Figure 4
Mechanical configuration of the test chamber.

2. Measurement time window

Because of the finite dimensions of the rock sample, the waves reflected from the other free surfaces will arrive at a gauge after the arrival of the direct waves from the source.

For the cubic samples used in the experiments, the first reflected waves arriving at gauges 1 to 6 must be from the midpoints of the lines that connect the projection points of the source center to the gauge locations on the two adjacent vertical free surfaces; the first reflected waves arriving at gauges 7 and 8 are from the four mid-lines on the four vertical free surfaces. Then the earliest arrival of the reflected waves, Tr, is

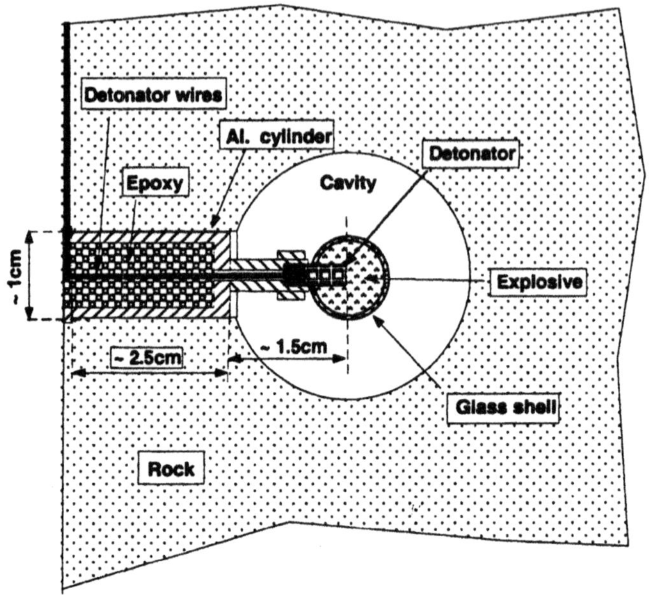

Figure 5
Spherical explosive source arrangement.

Table 1

Explosive source used in the experiments

Shot #		Dimensions of explosive	Mass of explosive*	Type of explosive	Dimensions of cavities
SW1	decoupled	ϕ 5 × 6 mm	0.24 g	Pellet pressed PETN	ϕ 30 × 30 mm
SW2	tamped	ϕ 10 × 10 mm	1.2 g	Detasheet PETN	ϕ 10 × 10 mm
SW4	decoupled	10 mm sphere	1.9 g	Detasheet PETN	30 mm sphere
SW9	decoupled	7 mm sphere	0.8 g	Detasheet PETN	30 mm sphere

* Explosive mass excludes 29 mg of PETN explosive in detonator (RP-3, Reynolds Industries System Inc.).

Table 2

Gauge locations

Shot #		Gauges 1 & 2	Gauges 3 & 4	Gauges 5 & 6	Gauges 7 & 8
SW1	r (mm)	152.5	157.6	189.1	243.5
	θ (°)	2.1	15.3	36.4	9.6
SW2	r (mm)	153.0	162.2	198.4	243.2
	θ (°)	6.4	20.9	39.8	9.9
SW4	r (mm)	152.5	158.9	174.5	209.0
	θ (°)	3.1	16.5	29.2	0
SW9	r (mm)	152.5	158.9	174.5	214.0
	θ (°)	3.0	16.5	39.2	0

The distance uncertainty is ± 0.1 mm.

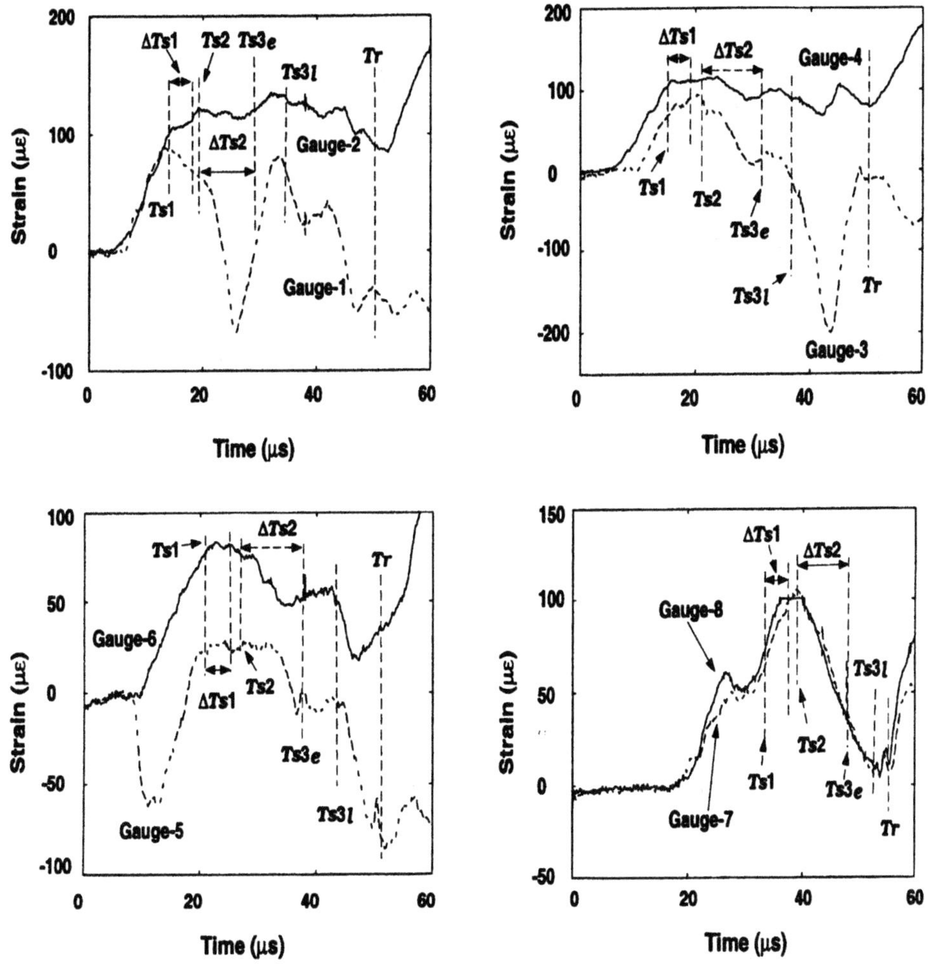

Figure 6
Strain signals of the gauges in SW4. $Ts1$ and $\Delta Ts1$: the arrival time and duration of the shear waves induced by microcracking (Eqs. (30) and (31)). $Ts2$ and $\Delta Ts2$: the arrival time and duration of the shear waves induced by shock wave-to-S conversion at the cavity wall (Eq. (29)). $Ts3_e$ and $Ts3_l$: the earliest and latest possible arrival of the shear waves induced by macrocracking (Eqs. (33) and (34)). Tr is the earliest of the reflected waves from the other free surfaces of the sample.

$$Tr = \frac{\left(r^2 + 4r_0^2\right)^{1/2}}{\alpha} , \tag{28}$$

where r is the distance between the source and gauge and r_0 is half the width of the horizontal dimension.

The strains induced by the reflected waves are tensional at gauges 2, 4, and 6 because the reflected waves are coming from both directions along direction 1 and are also releasing waves. The strains from the reflected waves are compressive at

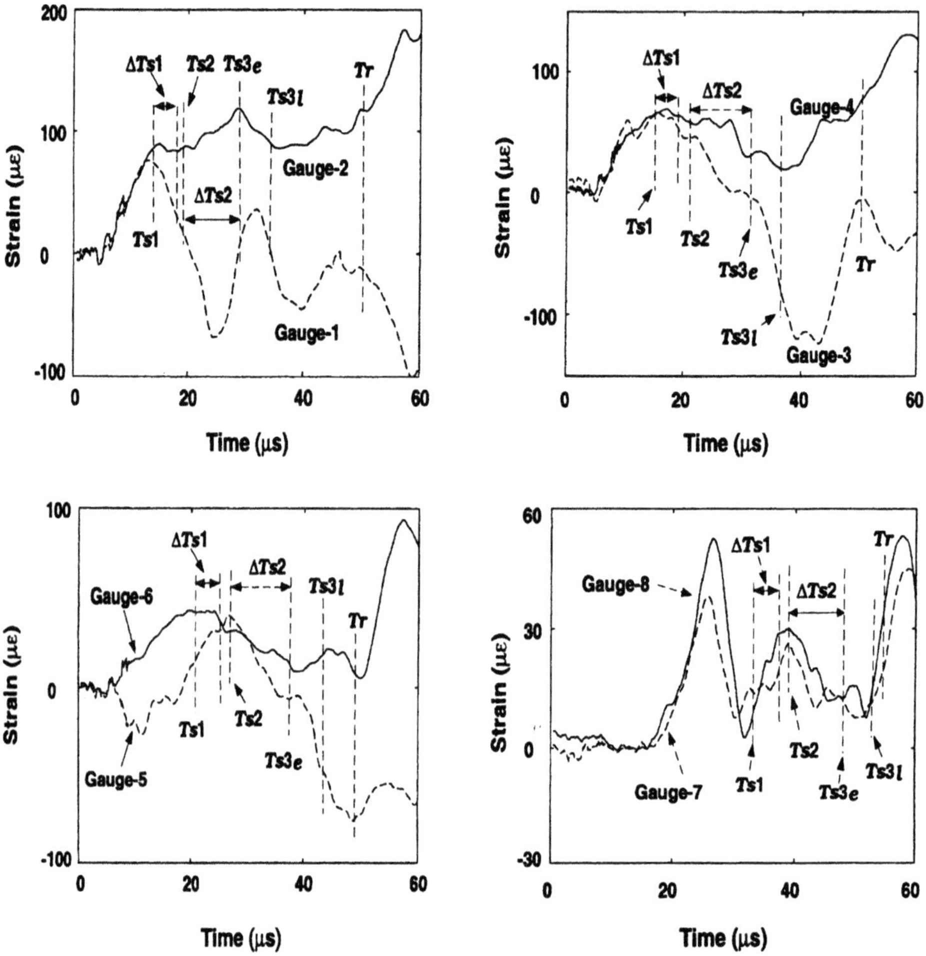

Figure 7

Strain signals of the gauges in SW9. *Ts*1 and Δ*Ts*1 : the arrival time and duration of the shear waves induced by microcracking (Eqs. (30) and (31)). *Ts*2 and Δ*Ts*2 : the arrival time and duration of the shear waves induced by shock wave-to-*S* conversion at the cavity wall (Eq. (29)). $Ts3_e$ and $Ts3_l$: the earliest and latest possible arrival of the shear waves induced by macrocracking (Eqs. (33) and (34)). *Tr* is the earliest of the reflected waves from the other free surfaces of the sample.

gauges 1, 3, and 5 because no waves are reflected along direction 2 and the gauge is compressed due to the tension along direction 1. Because the reflected waves arrive at gauges 7 and 8 from all directions at the same time, the strains induced are all tensional.

3. *S*-wave arrival

S-waves are generated from at least three different sources: (1) Microcracking; (2) shock wave-to-*S* conversion at the cavity wall; and (3) macrocracking.

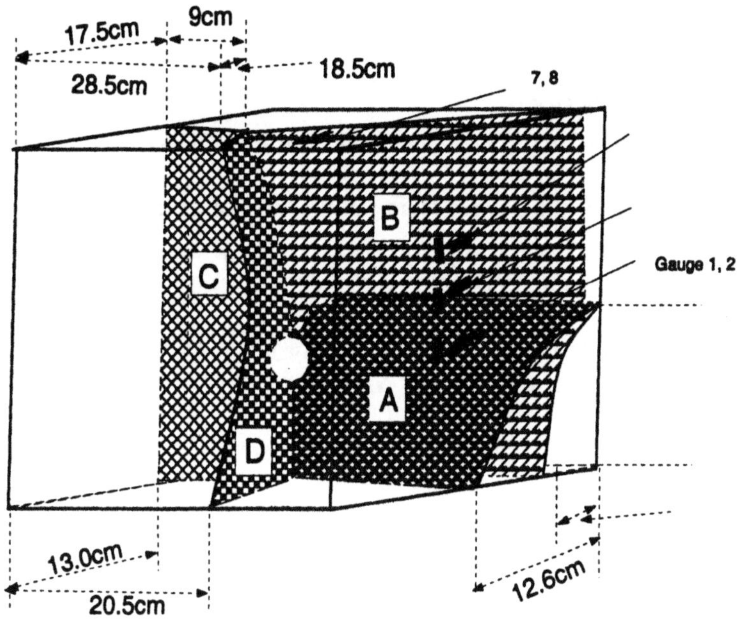

Figure 8
Schematic of the large cracks in relation to the spherical cavity.

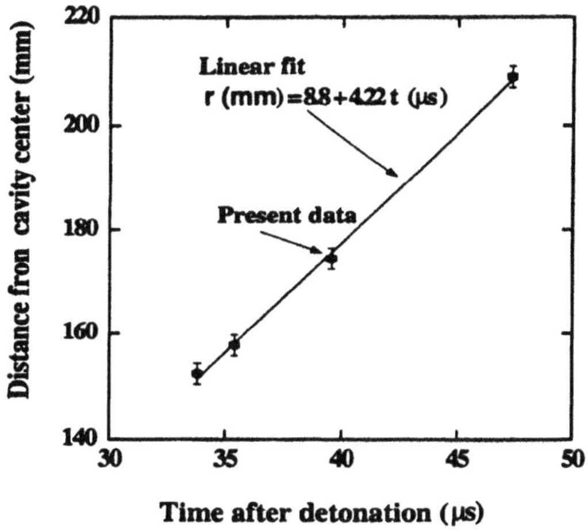

Figure 9
P-wave arrival versus radius.

For each of these sources, the arrival time of the S-wave from the second source can be estimated with relatively high accuracy because we know the P-wave arrival time at the cavity wall. The arrival time of the S-wave from the second source, $Ts2$, at a radial distance, r, is

$$Ts2 = \frac{\alpha - \beta}{\alpha\beta}(r - a) \ , \tag{29}$$

where β is S-wave velocity.

The duration of the S-wave $\Delta Ts2$, should be approximately equal to the rise time of the P-waves.

The arrival time of the S-wave from the first source at radial distance, r, is

$$Ts1 = \frac{\alpha - \beta}{\alpha\beta}(r - R_c) + \frac{r - a}{\alpha} \ , \tag{30}$$

where R_c is the radius beyond which there are no more microcracks induced by P-waves. The duration of this S-wave, $\Delta Ts1$, is

$$\Delta Ts1 = \frac{\alpha - \beta}{\alpha\beta}(R_c - a) \ . \tag{31}$$

The accuracy of this arrival time is dependent upon how well R_c is determined. Here based on the experimental data, R_c is chosen to be 50 mm.

The arrival time of the S-wave from the third source, $Ts3$, is

$$Ts3 = \frac{\alpha - \beta}{\alpha\beta}(r - a) + t_0 + \frac{(a^2 + r^2 - 2ar\cos(\theta_c))^{1/2} - (r - a)}{\beta} \ , \tag{32}$$

where θ_c is the angle between the two radial lines, one of which points to the nearest macrocrack generated and the other points to the gauge location. The last term on the right in Eq. (32) accounts for the S-wave that arrives later, from the cracks that are not along the line connecting the gauge with the center of the cavity. The earliest and latest arrival times, $Ts3_e$ and $Ts3_l$, correspond to $\theta_c = 0$ and $\theta_c = 90°$, respectively. Then, we have

$$Ts3_e = \frac{\alpha - \beta}{\alpha\beta}(r - a) + t_0 \ , \tag{33}$$

$$Ts3_l = \frac{\alpha - \beta}{\alpha\beta}(r - a) + t_0 + \frac{(a^2 + r^2)^{1/2} - (r - a)}{\beta} \ . \tag{34}$$

Here, t_0 is calculated to be ~10 μsec. This is not really resolvable on account of variations in the effective values of V_P and V_S in rock samples.

4. Comparisons between the predicted and measured arrivals of the waves

Using Eq. (28), the arrival times of the reflected waves are labeled in Figures 6 and 7. There is very good correlation between the calculated arrival and the waves recorded. Also, the strains are tensile at gauges 2, 4, 6, 7 and 8 and compressive at gauges 1, 3 and 5. Therefore, the waves arriving before Tr are all direct waves from the source. Using Eqs. (29), (30), (31), (33) and (34), the three S-wave arrivals are marked on the records shown in Figures 6 and 7. These figures show very good agreement between the experimental data and predictions for $Ts2$ and $Ts3$. The data clearly demonstrate the *S*-waves generated from shock wave-to-*S* conversion on the cavity wall and from cracking. In addition, the characteristics of the strains seen by the gauges along the two directions are the same as predicted in Section 2.

For the strains reported by gauges 1 and 2 in both of the experiments, those induced by the first arriving *P*-waves are identical before time *Ts*1 because the *P*-wave incident angle is small. After *Ts*1, the strain along gauge 2 is approximately constant however the strain along gauge 1 starts to decrease. This characteristic of the strains fits the strains induced by incident *S*-waves. After time *Ts*2, the strain along gauge 2 increases slowly but the strain along gauge 1 drops sharply. This means that another *S*-wave has arrived at the gauge location. Between $Ts3_e$ and $Ts3_l$, another *S*-wave arrives at the gauge location. Consequently, the reflected wave from other surfaces arrives at the gauge location. The strains from both gauges show that *P*-waves are continuously generated after the direct *P*-wave because neither returns to zero.

In Section 2, the strain induced by the incident *P* wave along gauge 4 is predicted to be larger than the strain along gauge 1. This is true for SW4 but not for SW9. This may be because the signals for SW9 were very noisy around the arrival time. For both experiments, the strain signals for gauges 3 and 4 show no wave arrival at predicted time *Ts*1, but do show very good correlation between the experimental records and predicted times for *Ts*2, *Ts*3 and *Tr*. Again, both of the gauges indicate that *P*-waves are generated continuously.

The strains are negative along gauge 5 but positive along gauge 6 for both of the experiments. This feature reflects the fact that the *P*- wave incident angle is larger than the critical angle that is discussed in Section 2, calculated to be 36°. The calculated *P*-wave incident angle at gauges 5 and 6 is 29.1° for a perfect spherical wave generated from the source, thus the wave generated from the source is not a perfect spherical wave.

For the strains along gauges 7 and 8, those induced by the incident *P*-wave are approximately the same for SW4 but have large differences for SW9, for unknown reasons. Because the two gauges are located at the symmetrical axis of the source (the axis of the detonator), they are not sensitive to any *S*-waves from the source. Therefore, the strains from gauges 7 and 8 are mainly induced by incident *P*-waves. The experimental results clearly show that a *P*-wave arrives at the gauges after the direct *P*-waves from the source. The arrival of this second *P*-wave for SW4 occurs earlier than that for SW9, as shown in Figure 10.

5. Particle displacement induced by *P*-waves

The maximum *P*-wave particle displacements, $u_{\max}$, at different distances from the cavity center, are calculated using the relations given in Section 2, and are plotted in Figure 11 together with the Cowboy data (MURPHEY 1961). These data can be fit by:

$$u_{\max} = 4.35 \times 10^4 r^{-1.23} \tag{35}$$

for SW4, and

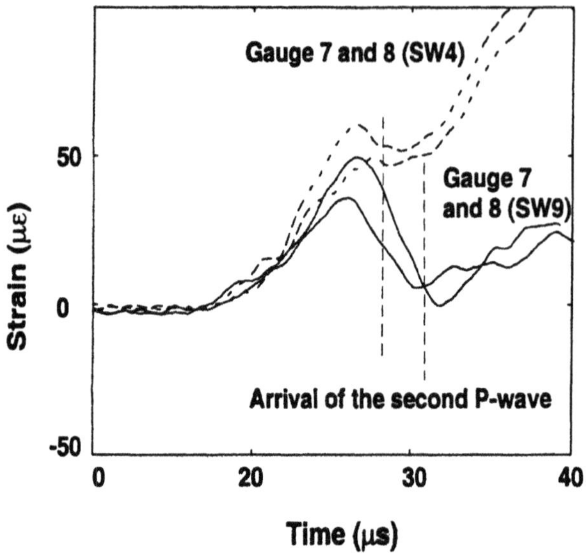

Figure 10
Second *P*-wave arrival times for SW4 and SW9. The signals are aligned at the arrival of the *P*-wave from the source.

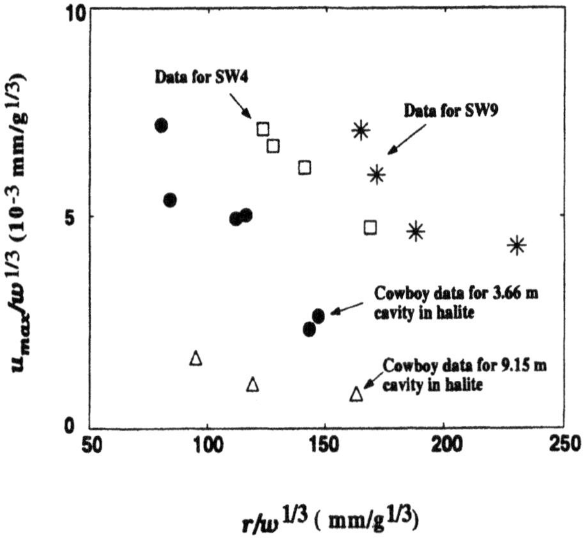

Figure 11
Scaled maximum *P*-wave particle displacement versus propagation distance. Attenuation of *P*-wave varies as $r^{-1.23}$ to $r^{-1.6}$ for two radii in halite.

$$u_{max} = 1.82 \times 10^4 r^{-1.59} \tag{36}$$

for SW9, where u_{max} is in μm, and r in mm, is the distance from the cavity center. However, Eqs. (35) and (36) are not plotted in Figure 11. MURPHEY (1961) found that the field results show the peak P-wave particle displacement to be dependent on the distance from the cavity center, proportional to $r^{-1.5}$. Thus the previous study supports the results of this work. However, the experimental data reveal that, in overdriven cavities, the coupling coefficient of explosive energy into elastic energy decreases with increasing explosive mass. These results suggest that $r/w^{1/3}$ may not be the optimum scaling parameter for overdriven cavities because the rate of the elastic energy propagating away from the source is limited by the rock's properties. However, the duration of the elastic P-wave remains scalable via this approach in overdriven cavities.

6. S-wave amplitudes.

Figure 12 shows the maximum strains from the gauges along direction 2 (Fig. 1) for shear waves generated from shock wave-to-S conversion and macrocracking, at different locations for shots SW4 and SW9. Due to the difficulty of determining the S-wave incident angle at each gauge directly, the two gauges at the same location are used to approximatey determine the incident angle; the peak displacement of the S waves is then determined. The procedure used is to first find the ratio of the strains from the gauges along direction 2 to those along direction 1, and then calculate the incident angle using this ratio and Eqs. (23) and (9). This method is not particularly precise because P waves interfere with S waves, nonetheless it gives at least a first-order estimate of S-wave amplitudes induced by shock wave-to-S conversion at the

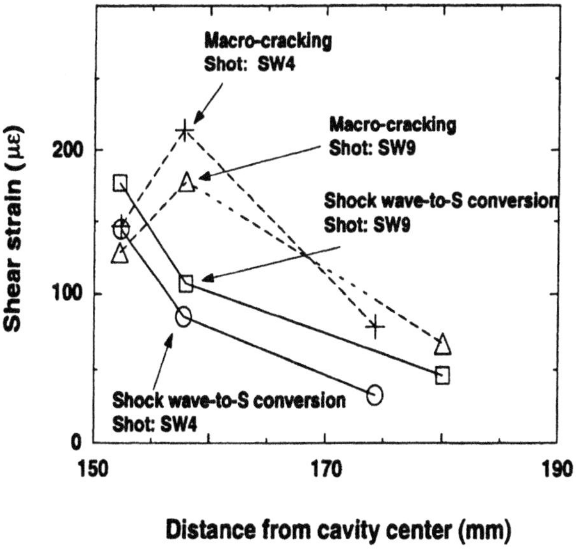

Figure 12
Maximum shear-wave strain versus propagation distance.

cavity wall and from crack propagation in rock. Figure 13 depicts the maximum displacement of S waves at different distances from the source. From this figure, it seems that the shear wave generated from shock wave-to-S conversion has higher regularity than the shear wave generated from macrocracking. The amplitude decay with distance clearly shows the complexity of shear waves due to crack propagation. The increase of the S-wave amplitude generated by shock wave-to-S conversion at the cavity wall is believed to originate from the source.

3.3 Comparison between Experimental Data and Theoretical Predictions

3.3.1 Displacement induced by pulse-spherical waves

The analytic solution for a pulsed-spherical wave generated from an explosion in a spherical cavity (TIMOSHENKO and GOODIER, 1970) is obtained by solving the wave equation

$$\frac{\partial^2 u}{\partial r^2} + \frac{2}{r}\frac{\partial u}{\partial r} - \frac{2u}{r^2} = \frac{1}{\alpha^2}\frac{\partial^2 u}{\partial t^2} , \tag{37}$$

where u is the particle displacement and α is P-wave velocity.

If pressure on the cavity boundary at $r = a$ is $p(t)$, the displacement can be calculated using (TIMOSHENKO and GOODIER, 1970)

$$u = \frac{f'}{\alpha r} + \frac{f}{r^2} , \tag{38}$$

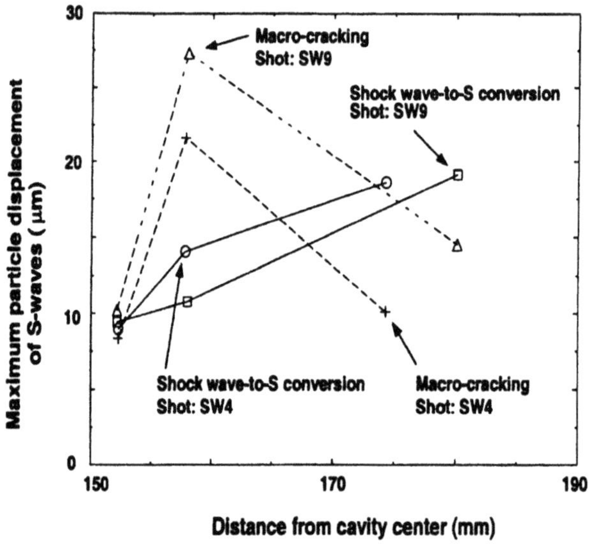

Figure 13
Maximum shear-wave displacement versus propagation distance.

where f is the solution to

$$f''(t) + 2\gamma f'(t) + \frac{2\gamma\alpha}{a} f(t) = \frac{ap(t)}{\rho} \ , \tag{39}$$

where

$$\gamma = \frac{1-2v}{1-v}\frac{\alpha}{a} \ , \tag{40}$$

v is Poisson's ratio and ρ is density.

If a step-like pressure history is assumed, i.e., $p(t) = p_0 H(t)$, the particular solution is (TIMOSHENKO and GOODIER 1970)

$$u(t,r) = \frac{A}{r^2} + \frac{A}{\alpha}\frac{e^{-\gamma\tau}}{r}\left(\left(\gamma s + \frac{\gamma}{s} - \frac{\alpha}{rs}\right)\sin(\gamma s\tau) - \frac{\alpha\cos(\gamma s\tau)}{r}\right) \ , \tag{41}$$

where $A = p_0 a^2/(2\rho\gamma\alpha)$.

The first term in the solution corresponds to the static solution, and the second is the dynamic term. The typical displacement history from Eq. (41) is shown in Figure 14 together with the experimental and numerical results. It can be concluded from this figure that the solution from pure pulse-spherical wave propagation alone does not explain the observed results.

3.3.2 Numerical simulation and experimental data

Figure 14 displays the comparison between the experimental results from SW4 and the numerical simulation by JOHNSON (1998). The numerical simulation did not include any cracking processes. The figure illustrates that the numerical simulation results are generally in good agreement with experimental data on the rise times and amplitudes, but have major differences in profile after the peak displacement.

The discussion above demonstrates that both the analytic solution for spherical waves and the first-principal numerical simulation results cannot explain the experimental data. Therefore, neither method includes important processes that are related to the long rise time of the direct P wave from the source and the generated second P waves. These processes are:

1. The pressure history in the cavity is not a single-step-like function. This arises from the large mismatch of impedance between the cavity, gas, and rock. Shock waves generated by explosives in the cavity must reverberate several times before the pressure in the cavity reaches its final value. This means that the pressure history in the cavity may have a multi-step history. This can be seen from the experimental data shown in Figure 15, in which the multiple wave arrivals are obvious.

2. The cracking processes may play an important role in the P-wave-induced displacement history, because radial cracking generated by the pressure in the cavity and P-wave propagation changes the boundary conditions. When pressure in a cavity

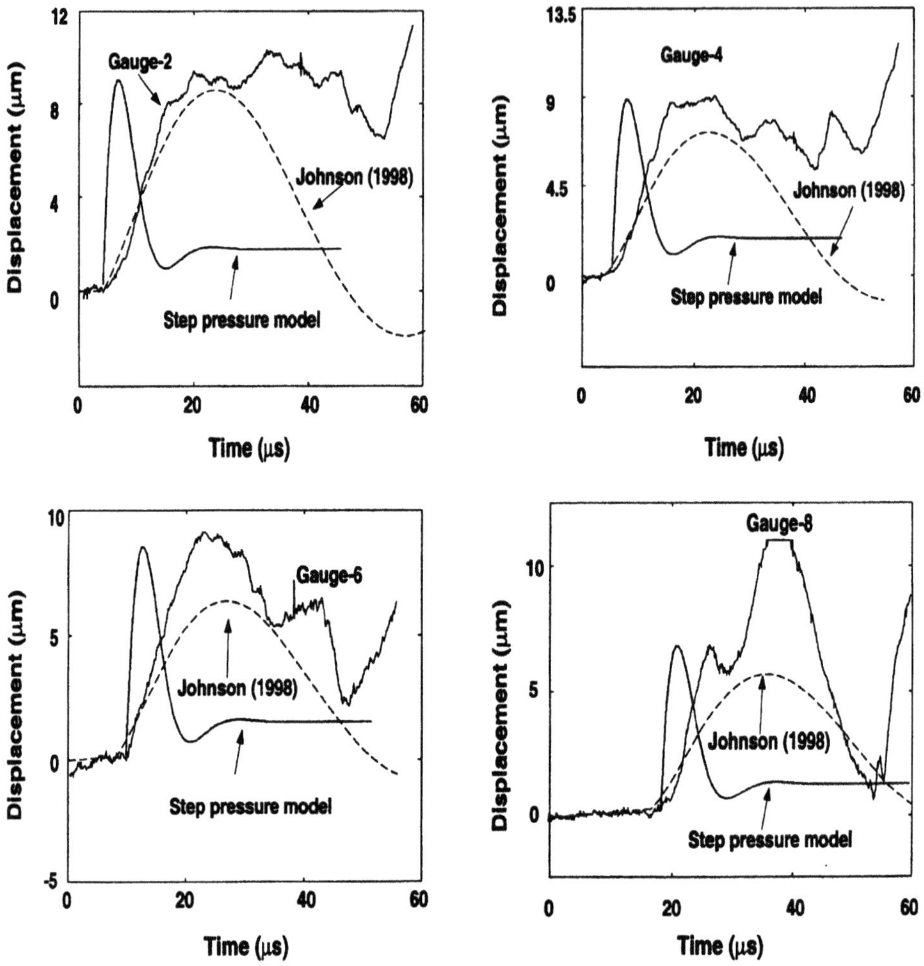

Figure 14
Comparison between numerical simulation and experimental results (SW4). The amplitude for the step-pressure model is arbitrary.

is larger than some critical value, radial cracks are generated from the inner cavity wall. The radial cracks will either cause the tangential stresses to become zero, if the explosion products do not penetrate into cracks, or make them the same pressure as in the cavity. Whether or not the explosion products penetrate into cracks, the tangential stress in rock has a large jump during cracking because the tangential stress is tensional before cracking and compressive after cracking. These rapid changes give rise to shear waves. Radial cracking also generates P waves because cracking is equivalent to an increase in cavity radius (a crack opening does not, however, change the pressure in the cavity).

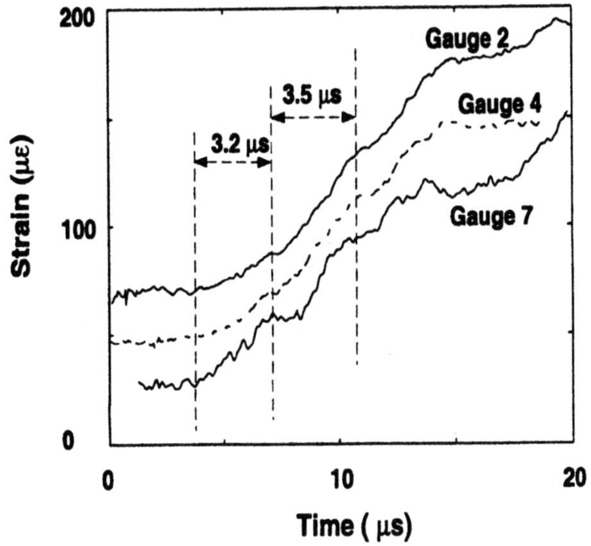

Figure 15
Wave reverberations in the cavity. The vertical amplitude is arbitrary. All signals are aligned at arrival.

3.3.3 Displacement induced by the P waves generated from explosions and cracking

Based on the above analysis, we constructed a model considering the reverberation of shock waves in a cavity and the P-waves generated form the radial cracking process.

1. Pressure history

In this model, we assume that the pressure history is divided into three time periods, i.e., a reverberation-dominated period, a cracking-dominated period and a steady-state period. The pressure history on the cavity wall can be given by

$$p(t) = p_I(t)(H(t) - H(t - t_0)) + p_{II}(t)(H(t - t_0) - H(t - t_{00})) + p_{III}H(t - t_{00}) \ , \quad (42)$$

where $p_I(t), p_{II}(t)$ and $p_{III}(t)$ are the pressure histories that correspond to the three periods, respectively. H(t) is the Heaviside step function, and t_0 and t_{00} are the start and stop times of crack propagation, respectively. The pressure history for each period in the cavity is approximated.

(I) Reverberation-dominated period:

In this period, shock waves induced by the detonation of the explosive reverberate in a cavity, and the pressure acting on the cavity wall is multi-step; radial cracks have not yet propagated in the rock.

During this period, it is assumed that the pressure on the wall of the cavity between two adjacent reverberations is constant,

$$p_I(t) = \sum_{i=1}^{N} p_i H(t - t_i) \ , \tag{43}$$

where p_i is the pressure increment after each reflection, t_i is the reflection time for the i-th step, and N is the reverberation time.

(II) Cracking-dominated period:

In general, there are two types of cracks induced by an explosion in a cavity. The first type is the nucleation of new cracks and growth of existing cracks (microcracks) related to P-wave propagation. The second type is the macrocrack that starts from the cavity wall and propagates into rock. Microcracks appear just after the P-wave if stress caused by the P-wave is higher than the local tensile strength. Macrocrack propagation is controlled by fracture dynamics. In this part, only P-waves generated from macrocracking are considered because P-waves generated from microcracks have small amplitude and cannot be separated from the initial P-waves.

When the pressure in a cavity is larger than some critical value, radial cracks are generated from the inner cavity wall. The tangential stresses in rock change significantly during cracking because the tangential stress is tensile before cracking and compressive after. Radial cracking generates P-waves because cracking in a rock next to a cavity is equivalent to an increase in pressure in a cavity without adjacent cracking.

Two methods give an approximate estimate of the effect of cracking on displacements.

(a) The first is to solve the differential equation with moving boundary

$$f''(t) + 2\gamma(t)f'(t) + \frac{2\gamma(t)\alpha}{a(t)} f(t) = \frac{a^r(t)p(t)}{\rho} \ , \tag{44}$$

where

$$\gamma(t) = \frac{1-2\nu}{1-\nu} \frac{\alpha}{a^r(t)} \ , \tag{45}$$

and

$$a^r(t) = a + \int_{t_0}^{t} C^c(\tau)\, d\tau \ , \tag{46}$$

where $a^r(t)$ is the radius of a crack tip from the centre of the cavity, t_0 is the time at which cracks start to propagate, and C^c is the crack propagation velocity. Numerical solutions may be applicable to Eq. (44) but the moving boundary of the fracture is difficult to model.

(b) To obtain an approximate pressure profile, we assume that the pressure in a cavity changes, based on the fact that the total force acting on a spherical surface with radius, $a = p_0 + \Delta a$ is equal to the force acting on a surface with radius a_0 and pressure p_0.

$$p = p_0 \frac{a_0^2}{(a_0 + \Delta a)^2} , \tag{47}$$

where p_0 is the pressure in the cavity just before cracks start to propagate. We assume that the radial stress at a spherical surface with radius $a(t)$ is equal to p equivalent pressure in the cavity where p decreases as Δa increases until the value of p corresponds to the critical tensile strength p_{II}.

(III) Steady period:

After cracking stops, the pressure in the cavity is assumed to be constant:

$$p_{III} = p_{II}(t_{00}) . \tag{48}$$

2. Solutions for the three periods

(a) Solution for period I:

From Eq. (41), the displacement induced by the i-th reverberation at radius r is

$$u_i(\tau_i, r) = \frac{A_i}{r^2} + A_i \frac{e^{-\gamma\tau_i}}{r\alpha} \left(\left(\gamma s + \frac{\gamma}{s} - \frac{\alpha}{rs} \right) \sin(\gamma s \tau_i) - \frac{\alpha \cos(\gamma s \tau_i)}{r} \right) , \tag{49}$$

where $A_i = p_i a^2/(2\rho\gamma\alpha), \tau_i = t - t_i - (r - a)/\alpha$ and t_i is the i-th reverberation time.

The resultant solution is just the superposition of the solutions for all N reverberations. The displacement field induced by reverberations at r is

$$u^I = \sum_{i=1}^{N} u_i H(t - t_i) . \tag{50}$$

(b) Solution for period II:

In order to obtain an analytic expression, the crack velocity is assumed to be one of two simple cases:

- Case A: crack velocity remains constant during crack propagation, or
- Case B: crack velocity decreases linearly with time, i.e.,

$$C^c(t) = C_0^c \left(1 - \frac{t - t_0}{t_d} \right) , \tag{51}$$

where t_d is the total cracking duration, C_0^c is the initial crack propagation velocity and t_0 is the time for the cracks to initiate propagation.

For Case A, the pressure history in period II is

$$p(t) = p_0 \left(1 + \frac{C_0^c}{a} (t - t_0) \right)^2 . \tag{52}$$

From Appendix 3, the displacement history in period II is

$$u^{II}(\tau) = \left(\frac{X_1}{\alpha r} + \frac{X_0}{r^2}\right) + \left(\frac{2X_2}{\alpha r} + \frac{X_1}{r^2}\right)\tau + \frac{X_2}{r^2}\tau^2 + \left(\frac{\gamma s X_6 - \gamma X_5}{\alpha r} + \frac{X_5}{r^2}\right)K_1 + \left(\frac{X_6}{r^2} - \frac{\gamma X_6 + \gamma s X_5}{\alpha r}\right)K_2 \ . \qquad (53)$$

For Case B, the pressure history is calculable.

$$p = p_0\left(1 + \frac{C_0^c}{a}\left((t - t_0) - \frac{(t - t_0)^2}{2t_d}\right)\right)^2 \ . \qquad (54)$$

From Appendix 4, the displacement history in period II for Case B is

$$u^{II}(\tau) = \left(\frac{a_1}{\alpha r} + \frac{a_0}{r^2}\right) + \left(\frac{2a_2}{\alpha r} + \frac{a_1}{r^2}\right)\tau + \left(\frac{a_2}{r^2} + \frac{3a_3}{\alpha r}\right)\tau^2 + \left(\frac{4a_4}{\alpha r} + \frac{a_3}{r^2}\right)\tau^3 + \frac{a_4}{r^2}\tau^4 + \left(\frac{\gamma s a_6 - \gamma a_5}{\alpha r} + \frac{a_5}{r^2}\right)K_1 + \left(\frac{a_6}{r^2} - \frac{\gamma a_6 + \gamma s a_5}{\alpha r}\right)K_2 \ . \qquad (55)$$

(c) Solution for period III:

Substituting Eq. (48) into Eq. (39), the solution is

$$f(t) = \frac{p_1 a^2}{2\rho\gamma\alpha} 1 - e^{-\gamma t}\left(\cos(\gamma s t) - \frac{\sin(\gamma s t)}{s}\right) , \qquad (56)$$

then, substituting Eq. (56) into Eq. (38), the displacement is

$$u^{III}(\tau) = \frac{p_1 a^2}{2\rho\gamma\alpha r}\left(\frac{\alpha}{r} + e^{-\gamma\tau'}\left(\left(\frac{\gamma}{s} + \gamma s - \frac{\alpha}{rs}\right)\sin(\gamma s\tau') - \frac{\alpha\cos(\gamma s\tau')}{r}\right)\right) , \qquad (57)$$

where

$$\tau' = \tau + t_{III} \ . \qquad (58)$$

3. Displacement history induced by P-waves

The displacement history induced by P-waves is

$$u = u^I(H(\tau) - H(\tau - t_0)) + u^{II}(H(\tau - t_0) - H(\tau - t_{00})) + u^{III}H(\tau - t_{00}) \ . \qquad (59)$$

3.3.4 Comparison between the observation and theoretical predictions of displacement history induced by P waves

Parameters estimation

Because the ratio of cavity radius to explosive radius is about 3, there is no separation between the shock wave in the air in the cavity and the adiabatic release wave of the explosion products. The shock wave velocity in the air can be estimated using the Chapman-Joquet (CJ) theory. The particle velocity after CJ detonation is

$$U_p = \frac{D}{\gamma + 1} \ , \tag{60}$$

where D is the CJ-wave velocity and γ is the gas polytropic exponent.

For PETN, the typical CJ-wave velocity is 8 km/s (DOBRATZ, 1981). The particle velocity is then about 3.3 km/s. The vibration period in the cavity is

$$t_v \approx a/(2 * 3.3) \approx 2.3 \ \mu s \ . \tag{61}$$

From the experimental data, the risetime of the particle displacement induced by the P wave is ~10 μs, or about three reverberations. This can also be verified from the profiles of the displacement rise-fronts from the experiments. For this case, we have $N = 3$, $t_0 = 10.0$ μs, $t_1 = 0.0$, $t_2 = 4.6$ μs, $t_3 = 9.2$ μs, and $p'_1 = p'_2 = p'_3 = p_0/3$.

The maximum velocity of a mode-I crack is theoretically deduced to be the Rayleigh wave velocity (FREUND, 1990). Compared to the crack velocity in rocks from seismology (KANAMORI, 1994), the fault propagation velocity is ~70% of the theoretical maximum velocity of a mode-II crack, i.e., the shear-wave velocity. The crack propagation velocity, C^c, is assumed to be 70% of the Rayleigh wave velocity, C_R, because the cracks generated in the experiments are mode-I. Then, t_{00} is

$$t_{00} = \frac{L_c}{0.7C_R} + t_0, \quad \text{for Case A} \ , \tag{62}$$

$$t_{00} = \frac{2 * L_c}{0.7C_R} + t_0, \quad \text{for Case B} \ , \tag{63}$$

where L_c is the crack length.

For shot SW4, the crack length with the explosion product traces on it is roughly 50 mm, so $t_{00} \approx 38$ μs for Case A and 65 μs for Case B.

3.3.5 Displacement history

Figure 16 shows the calculated results of displacement time history, based on the pressure histories of Case A and Case B. From this figure, it seems that the displacement history recorded from experiments is between the two extreme conditions; this may reflect the reality that cracks propagate neither at constant velocity nor at constant deceleration. From the comparison between the theoretical estimation based on Cases A and B and the experimental data, the cracking process is seen to have a significant effect on the history of displacement or strain. First, the cracking process increases the amplitude of the low frequency component because cracking is equivalent to increasing the pressure duration in a cavity. Second, the cracking process increases the complexity of the directional dependence of the displacement history because of the nonuniform distribution of cracks.

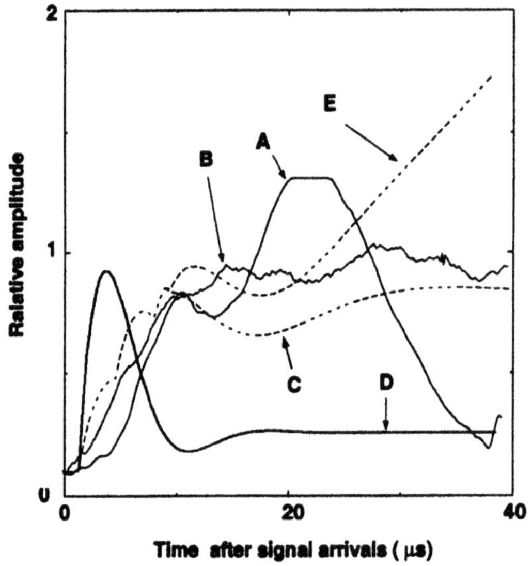

Figure 16
Comparison between the experimental data and results from the equivalent models. A: gauge 8 of SW4. B: gauge 2 of SW4. C: constant deceleration model (Eqs. (50) and (55)). D: constant step pressure model (Eq. (41)). E: constant crack velocity model (Eqs. (50) and (53)).

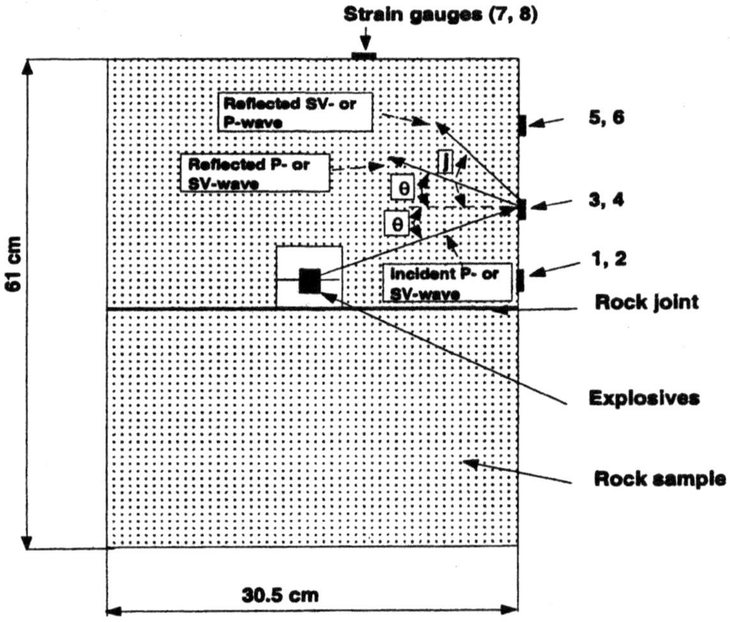

Figure 17
Sample and strain gauge layout for cylindrical cavity experiments.

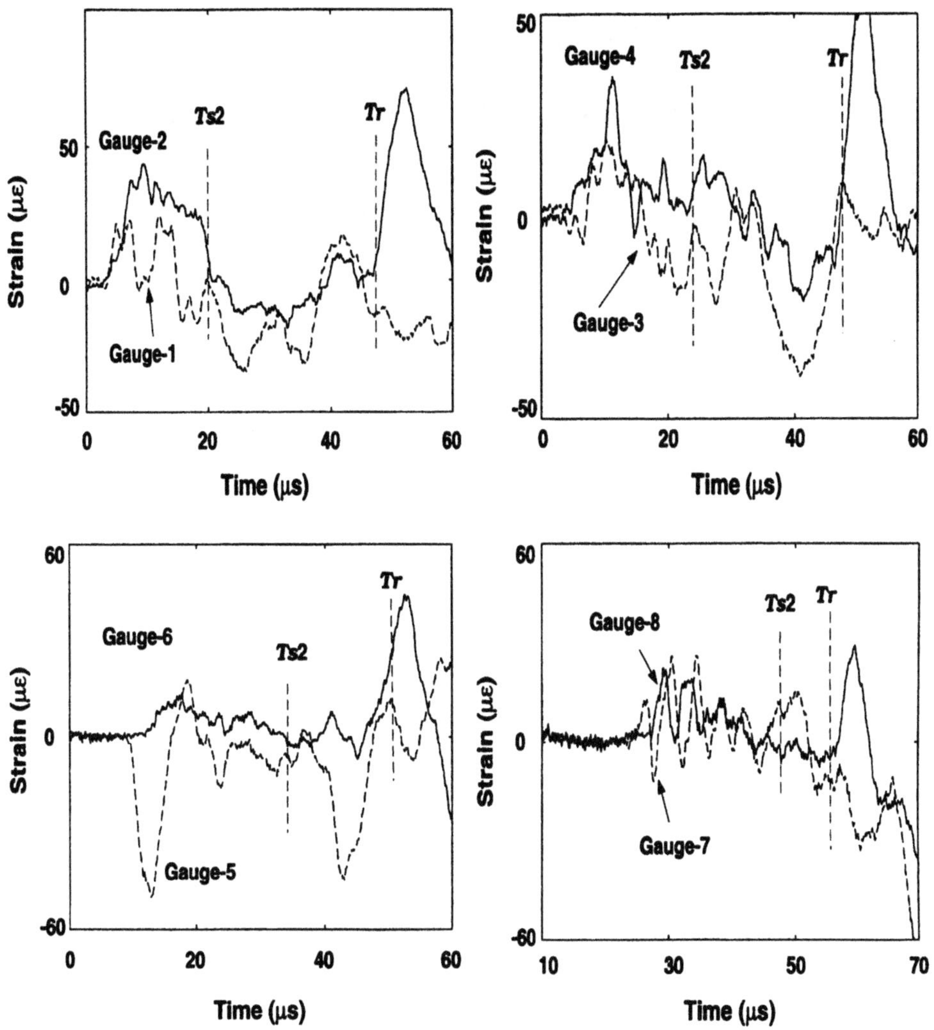

Figure 18
Strain signals at different locations for SW1.

4. *Wave Generation From Explosions in Cylindrical Cavities in Rock*

4.1 Experiments and Results

4.1.1 Experimental details

Experiments were conducted using a simple cylindrical charge detonated in a cylindrical hole centered in the rock sample depicted in Figures 4 and 17. The purpose was to improve our understanding of the cavity geometry effects on *P*- and *S*-waves generated from explosions, using the experimental data on spherical and cylindrical cavities. Two experiments were conducted, one with a decoupled

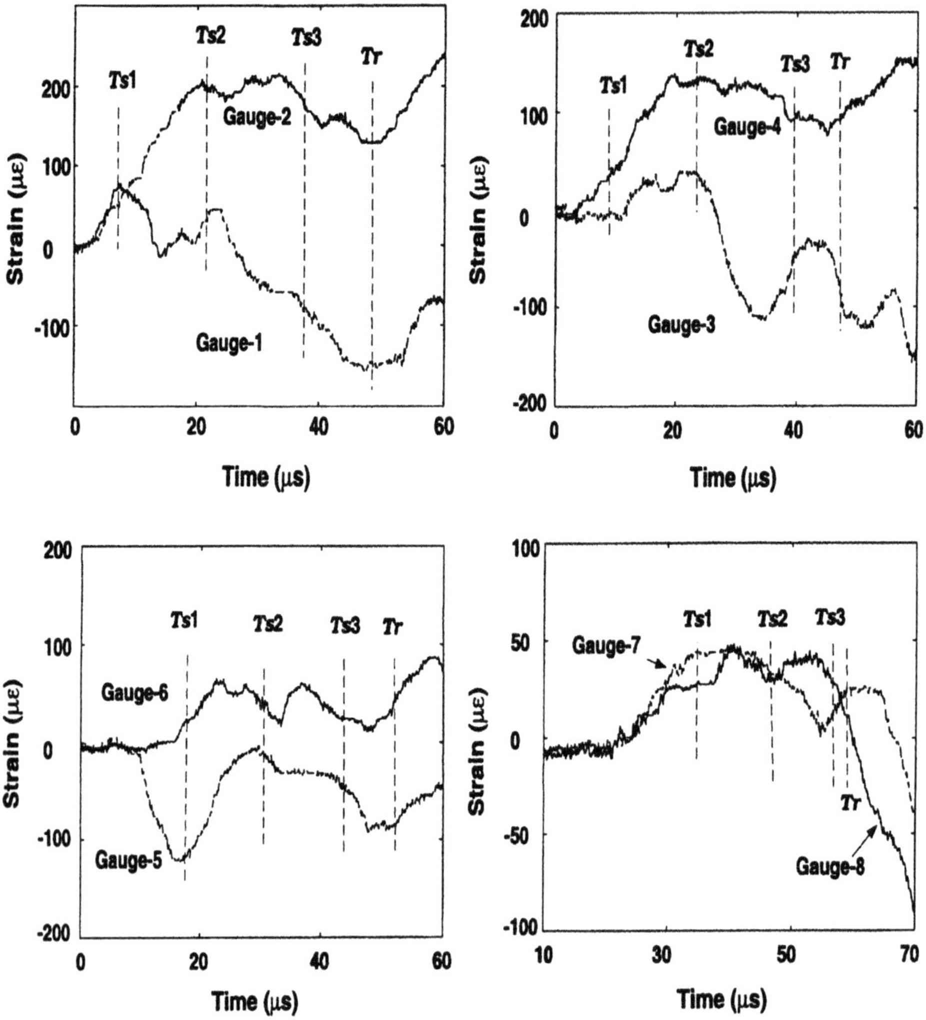

Figure 19
Strain signals at the different locations for SW2. *Ts*1: the arrival and duration of the shear waves induced by microcracking (Eq. (30)). *Ts*2: the arrival and duration of the shear waves induced by shock wave-to-*S* conversion at the cavity wall (Eq. (29)). *Ts*3: the arrival of the shear waves induced by macrocracking (Eq. (33)).

explosion, SW1, and the other a tamped explosion, SW2. The parameters used in the experiments are listed in Table 1. The gauge locations are given in Table 2. The rock sample is assembled using two blocks as shown in Figure 17.

4.1.2 Experimental results

The recorded strains for the two experiments are shown in Figure 18 for shot SW1 and Figure 19 for shot SW2. From the recovered rock samples, minimal

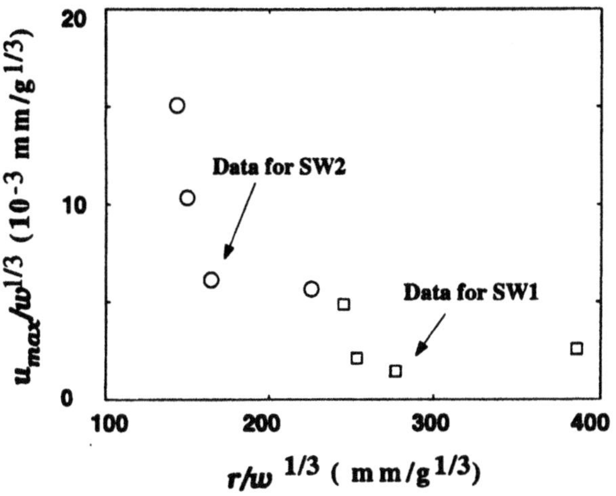

Figure 20
Scaled *P*-wave-induced displacement versus distance.

damage is observed near the cavity wall in the target used in the decoupled experiment, and many cracks of varying lengths up to 122 mm and of different orientations were observed in the target used in the tamped experiment.

From Figure 18, it can be seen that the wave profiles from the decoupled experiment, SW1, are very similar to the results from analytic solution shown in Figure 14, although the cavity is not spherical. From Figure 19, the wave profiles from the tamped experiment, SW2, have the typical characteristics of the wave profiles from the spherical experiments. The difference between the decoupled and tamped experiments once again demonstrates the role of cracking in the generation of *P* and *S* waves in that there is minimal damage in the decoupled experiment, and the rock sample is heavily cracked in the tamped experiment.

4.2 P Waves Generated from Explosions

4.2.1 *P* waves generated from the decoupled experiment

The peak particle displacements from the *P*-waves are shown in Figure 20. As mentioned above, although the *P*-wave profile shape can be explained quantitatively using the analytic solution based on the spherical source assumptions as shown in Figure 14, the experimental results show the direction-dependency of the wave shape.

4.2.2 *P* waves generated from the tamped experiment

Figure 20 gives the peak particle displacement induced by *P*-waves generated from the tamped explosions. This figure depicts more clearly the *P*-wave amplitude direction-dependence due to the geometry of the cavity. This is very different from the results in the spherical experiments as shown in Figure 14, in which the peak

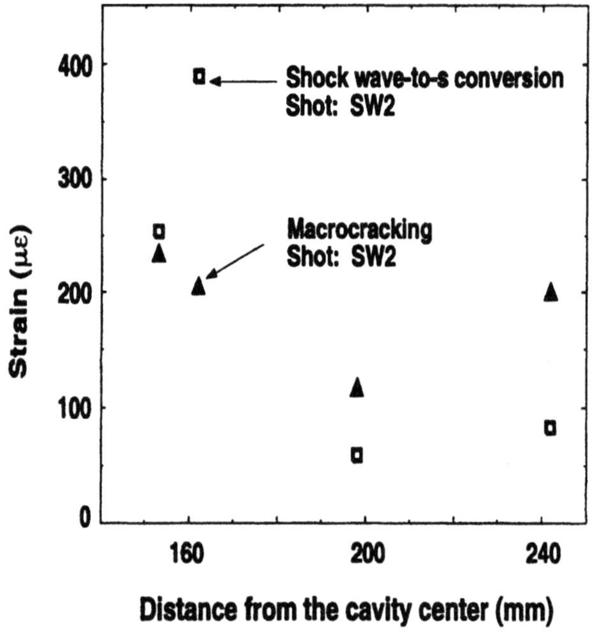

Figure 21
S-wave amplitude versus propagation distance for SW1 and SW2.

particle displacement can be fitted using the spherical solution. The difference between the *P*-wave profiles and peak displacements of the cylindrical and spherical cavity experiments clearly demonstrates the importance of geometry's effect on the source function.

4.3 S-waves Generated from Explosion

4.3.1 *S*-waves generated from the decoupled experiment

Figure 18 gives the estimated arrival time of the *S*-waves from the conversion of the *P*-wave at the cavity wall using Eq. (29). The arrival of the *S*-waves from micro- and macrocracking are not labeled because there are no cracks. It is very difficult to distinguish between vibrations and the *S*-waves converted from the *P*-wave.

4.3.2 *S*-waves generated from the tamped experiment

Figure 19 shows the arrival of the three *S*-waves from the three different sources as discussed in Section 3.4. As in the spherical experiments, we assume that the macrocracks start to propagate when the *P*-wave reaches its peak value. The microcrack radius is found to be 122 mm based on the cracks in the recovered target. Although the source is cylindrical, there is still good agreement between the predicted arrivals of the *S*-waves and the experimental results. We can see the *S*-waves from the shock wave-to-*S* conversion and the macrocracks on all the records. Figure 21 gives the maximum strain related to the shear waves generated from shock wave-to-*S* conversion and macrocracking along direction 2.

5. *Conclusions*

Based on the methods developed and the experimental results and the analyses given above, we conclude:

1. A measurement method to monitor *P*- and *S*-waves from explosions in newly ideal cylindrical and spherical cavities in rock in the laboratory has been developed. The experimental results agree with the predicted strain characteristics of *P*- and *S*-waves generated from explosions and cracking.

2. A device has been developed to drill spherical cavities with diameters up to 10 cm in rock such as limestone.

3. The records from the experiments clearly demonstrate that *P*-waves are generated from both the pressure exerted on the cavity by the explosive products from explosions and crack propagation (radially) outward from the cavity. The latter are driven by the gas from the explosive source. The peak particle displacement induced by *P*-waves ranges from 9 to 5 μm at 150 to 210 mm from the source with 1.9 g of explosives, and 8 to 4 μm at 150 to 220 mm from the source with 0.8 g of explosives for spherical experiments. The peak particle displacement induced by *P*-waves ranges from 16 to 5 μm at 150 to 240 mm from the source for the tamped cylindrical cavity experiment with 1.2 g of explosives, and 3 to 1 μm at 150 to 190 mm from the source for the decoupled cylindrical cavity experiment with 0.24 g explosives. The step-like radially symmetric pressure model applies to the experimental data for the peak *P*-wave-induced particle displacement versus propagation distance from the explosions in spherical cavities. This model fails for cylindrical cavities geometry. This result demonstrates the cavity's geometrical effects on wave pattern. The experimental results from both spherical and cylindrical cavities demonstrate that the *P*-wave profiles are similar to each other.

4. The experimental results in both spherical and cylindrical cavities verify the *S*-waves generated from both shock wave-to-*S* conversion at cavity walls and from crack propagation in rocks. The amplitude of the *S*-waves from shock wave-to-*S* conversion is from 10 to 27 μm at 150 to 180 mm from the source for the spherical experiment with 1.9 g of explosives, and 8 to 21 μm at 150 to 180 mm from the source for the spherical experiment with 0.8 g of explosives. The amplitude of *S*-waves generated by crack propagation is from 9 to 19 μm at 150 to 180 mm from the source for the spherical experiment with 1.9 g of explosives, and 8 to 19 μm at 150 to 180 mm from the source for the spherical experiment with 0.8 g of explosives.

5. A theoretical model incorporating multi-reverberation of sound waves in cavity gas followed by crack propagation into the rock is proposed to explain the differences in *P*-wave-induced displacement history between the observed data and the step-like pressure model. The *P*-wave profiles predicted from this model fit the observation qualitatively as shown in Figure 16.

From the results presented above, we believe that further study of the characteristics of *P*- and *S*-wave profiles generated from explosions, shock wave-to-*S* conversion and crack propagation driven by explosive products may lead to the development of criteria to discriminate underground chemical explosions from underground nuclear explosions because:

1. Cracks generated in chemical explosions are expected to be markedly longer and have larger openings than those from nuclear explosions, because of the increased quantity of detonation product goes from chemical explosives. Longer crack lengths will result in both the *P*- and *S*-waves generating more energy at low frequencies.

2. The efficient conversion of *P*-wave shock energy to *S*-wave energy at the cavity wall may only occur for chemical explosions. A detailed study of shear-wave generation for chemical explosives close to the explosion sources has not yet been carried out. We suggest this may prove useful for seismic discrimination if close in shear-wave data were measured.

3. Because the reverberation pressure are $\sim 10^{-1}$ GPa and not $\sim 10^{2}$ GPa as in the nuclear case, *P*-wave profiles may also contain information concerning the cavity dimensions in which they have been fired.

Acknowledgments

We thank Dr. R. Blandford, The Air Force Technical Applications Center, for his critical and beneficial comments, and Jeff Batten, E. Gelle, M. Long, and V. Nenow for their technical support. We also thank Prof. G. Ravichandran for his advice on crack theory. We are grateful to reveiwers for their helpful comments. The project was sponsored by The Air Force Material Command. Contribution number 8627, Division of Geological and Planetary Sciences, California Institute of Technology.

References

Aki, K., and Richards, P., *Quantitative Seismology Theory and Methods* (W. H. Freeman and Company 1980) pp. 151–161.

Coursen, L., *A gas penetration model of fragmentation*. In *Fragmentation by Blasting* (eds. Fourney, W., Boade, R., and Costin, O.) (Soc. Experim. Mechan. Connecti-cut 1985) pp. 1–10.

Dobratz, B. M. (1981), LLNL *Explosives Handbook – Properties of Chemical Explosives and Explosive Simulants*, Lawrence Livermore National Laboratory Report, UCRL-52997.

FLORENCE, A., MILLER, S., and KELLER, C. (1993), *Decoupling of Underground Explosions by Rubble-filled Cavities*, J. Geophys. Res. *98*, 14,197–14,209.

FREUND, L. B., *Dynamic Fracture Mechanics* (Cambridge University Press 1990) pp. 336–340.

GLENN, L., and GOLDSTEIN, P. (1994), *Seismic Decoupling with Chemical and Nuclear Explosions in Salt*, J. Geophys. Res. *99*, 11,732–11,730.

GLENN, L., and GOLDSTEIN, P. (1996), *Reply*, J. Geophys. Res. *101*, 851–854.

GLENN, L., LADD, A., MORAN, B., and WILSON, K. (1985), *Elastic Radiation from Explosively Loaded Ellipsoidal Cavities in an Unbounded Medium*, Geophys. J. R. astr. Soc. *81*, 231–241.

JOHNSON, L. (1998) Unpublished data.

KIM, S., CLIFTON, R., and KUMAR, P. (1977), *A Combined Normal and Transverse Displacement Interferometer*, J. Appl. Phys. *48*, 4132–4139.

KANAMORI, H. (1994), *Mechanics of Earthquakes*, Annu. Rev. Earth Planet. Sci. *22*, 207–237.

KLOPP, R. W., FLORENCE, A. L., and GRAN, J. K. (1994), *Spherical wave interaction with cylindrical holes in large limestone specimens*. In *International Conference on Mechanical and Physical Behavior of Materials under Dynamic Loading* (ed. Harding, J.) (Les Editions de Physique, Les Ulis 1994) pp. 735–740.

LANGSTON, C. (1983), *Kinematic Analysis of Strong Motion P and SV Waves from the Sterling Event*, J. Geophys. Res. *88*, 2486–3497.

MASSE, R. (1981), *Review of Seismic Source Models for Underground Nuclear Explosions*, Bull. Seismol. Soc. Am. *71*, 1249–1268.

MURPHEY, B. (1961), *Particle Motions near Explosions in Halite*, J. Geophys. Res. *66*, 947–957.

MURPHY, J., *Types of seismic events and their source descriptions*. In *Monitoring a Comprehensive Test-Ban Treaty* (eds. Husebye, and Dainty) (Kluwer Academic Publishers, The Netherlands 1996) pp. 247–293.

MURPHY, J., RIMER, N., and STEVENS, J. (1996), *Comment on Seismic Decoupling with Chemical and Nuclear Explosions in Salt by L. Glenn and P. Goldstein*, J. Geophys. Res. *101*, 845–850.

MURPHY, J. R., KITOV, I. O., RIMER, N., and ADUSHKIN, V. V. (1997), *Seismic Characteristics of Cavity Decoupled Explosions in Limestone: An Analysis of Soviet High Explosive Test Data*, J. Geophys. Res. *102*, 27,393–27,405.

SYKES, L., *Dealing with decoupled nuclear explosions under a Comprehensive Test-Ban Treaty*. In *Monitoring a Comprehensive Test-Ban Treaty* (eds. Husebye, and Dainty) (Kluwer Academic Publishers, The Netherlands 1996) pp. 225–245.

TIMOSHENKO, S. P., and GOODIER, J. N., *Theory of Elasticity* McGraw-Hill Book Company 1970) pp. 510–513.

Appendix 1

The displacement reflection coefficients for incident P-waves at free surfaces (AKI and RICHARDS, 1980) are

$$PP = \frac{B - A}{B + A} \ , \tag{64}$$

$$PS = \frac{2\frac{\beta}{\alpha}\sin(2\theta)\cos(2j)}{A + B} \ , \tag{65}$$

where PP and PS are the reflection coefficients for P- and SV-wave displacements due to the incident P-waves, respectively. α and β are P- and S-wave velocities, respectively. A and B are

$$A = \cos^2(2j) \ , \tag{66}$$

$$B = \left(\frac{\beta}{\alpha}\right)^2 \sin(2j)\sin(2\theta) \ , \tag{67}$$

where θ and j (see Fig. 1) are the *P*-wave incident angle and *S*-wave reflection angle, respectively.

$$\sin j = \frac{\beta}{\alpha}\sin\theta \ , \tag{68}$$

$$\cos j = \left(1 - \left(\frac{\beta}{\alpha}\right)^2 \sin^2 j\right)^{\frac{1}{2}} \ . \tag{69}$$

The resultant displacements of the particles on the free surfaces after the reflection are

$$u^P_{\text{par}} = u^I_p[(1 + PP)\sin\theta + PS\cos j] = H_{\text{par}}u^I_p \ , \tag{70}$$

$$u^P_{\text{per}} = u^I_p[(1 - PP)\cos\theta + PS\sin j] = H_{\text{per}}u^I_p \ , \tag{71}$$

Where u^I_p is the particle displacement of the incident *P*-wave, and u^P_{par} and u^P_{per} indicate, respectively, the resultant particle displacements along direction 2 and the direction that is perpendicular to the free surface after reflection, as shown in Figure 1. H_{par} and H_{per} are

$$H_{\text{par}} = (1 + PP)\sin\theta + PS\cos j \ , \tag{72}$$

$$H_{\text{per}} = (1 - PP)\cos\theta + PS\sin j \ . \tag{73}$$

Substituting *PP* with Eq. (64) and *PS* with Eq. (65), Eqs. (72) and (73) can be rewritten as

$$H_{\text{par}} = \frac{2\cos\theta\sin(2j)}{A + B} \ , \tag{74}$$

$$H_{\text{per}} = \frac{2\cos\theta\cos(2j)}{A + B} \ . \tag{75}$$

Appendix 2

For incident *SV* waves, the displacement reflection coefficients for *P* and *SV* waves (AKI and RICHARDS, 1980) are

$$SP = \frac{\frac{\beta}{\alpha}\sin(4\theta)}{A_s + B_s} \ , \tag{76}$$

$$SS = \frac{A_s - B_s}{A_s + B_s} \ , \tag{77}$$

where *SS* and *SP* are the reflection coefficients of *SV*- and *P*-wave displacements due to the incident *SV* waves.

$$A_s = \cos^2(2\theta) \ , \tag{78}$$

$$B_s = \left(\frac{\beta}{\alpha}\right)^2 \sin(2\theta)\sin(2j) \ , \tag{79}$$

where θ and j are the SV-wave incident angle and P-wave reflected angle, respectively, and

$$\cos j = \left(1 - \left(\frac{\alpha}{\beta}\right)^2 \sin^2\theta\right)^{\frac{1}{2}} \ . \tag{80}$$

The resultant displacements of the particles on the free surfaces after the reflection are

$$u^{sv}_{\text{par}} = u^I_{sv}[(1+SS)\cos\theta + SP\sin j] = G_{\text{par}}u^I_{sv} \ , \tag{81}$$

$$u^{sv}_{\text{per}} = u^I_{sv}[(SS-1)\sin\theta - SP\cos j] = G_{\text{per}}u^I_{sv} \ , \tag{82}$$

where u^{sv}_{par} and u^{sv}_{per} are the resultant particle displacements on the free surface after reflection along direction 2 as shown in Figure 1 and the direction that is perpendicular to the free surface, respectively. u^I_{sv} is the particle displacement of the incident SV-wave. The G_{per} and G_{par} are

$$G_{\text{par}} = (1+SS)\cos\theta + SP\sin j \ , \tag{83}$$

$$G_{\text{per}} = (SS-1)\sin\theta - SP\cos j \ . \tag{84}$$

Substituting SP with Eq. (76) and SS with Eq. (77), Eqs. (83) and (84) are rewritten as

$$G_{\text{par}} = \frac{2\cos(2\theta)\cos\theta}{A_s + B_s} \ , \tag{85}$$

$$G_{\text{per}} = -2\frac{\frac{\beta}{\alpha}\cos j\sin(2\theta)}{A_s + B_s} \ . \tag{86}$$

Appendix 3

Substituting Eq. (52) into Eq. (39), the solution is found to be

$$f(\tau) = X_0 + X_1\tau + X_2\tau^2 + X_5K_1(\tau) + X_6K_2(\tau) \ , \tag{87}$$

where

$$X_0 = A_0\left(\frac{p_0}{\gamma^2} + \frac{2p_1}{\gamma^3(1+s^2)} + \frac{p_2(3-s^2)}{\gamma^4(1+s^2)^2}\right) \ , \tag{88}$$

$$X_1 = A_0\left(\frac{p_1}{\gamma^2} + \frac{4p_2}{\gamma^2(1+s^2)}\right) \ , \tag{89}$$

$$X_2 = \frac{A_0 p_2}{\gamma^2} , \tag{90}$$

$$X_5 = A_0 \left(\frac{p_0}{\gamma^2} - \frac{2p_1}{\gamma^3(1+s^2)} + \frac{2p_2(s^2-3)}{\gamma^4(1+s^2)^2} \right) , \tag{91}$$

$$X_6 = A_0 \left(-\frac{p_0}{\gamma^2 s} + \frac{p_1(s^1-1)}{\gamma^3 s(1+s^2)} + \frac{2p_2(3_s - s^{-1})}{\gamma^4(1+s^2)^2} \right) , \tag{92}$$

$$A_0 = \frac{a}{\rho(1+s^2)} , \tag{93}$$

$$K_1 = e^{-\gamma\tau} \cos(\gamma s \tau) , \tag{94}$$

$$K_2 = e^{-\gamma\tau} \sin(\gamma s \tau) , \tag{95}$$

$$p_1 = 2p_0 \frac{C_0^c}{a} , \tag{96}$$

$$p_2 = p_0 \left(\frac{C_0^c}{a} \right)^2 \tag{97}$$

and

$$\tau = t - \frac{r-a}{\alpha} - t_0 . \tag{98}$$

Substituting Eq. (87) into Eq. (38), the displacement history in period II is

$$\begin{aligned} u^{II}(\tau) = & \left(\frac{X_1}{\alpha r} + \frac{X_0}{r^2} \right) + \left(\frac{2X_2}{\alpha r} + \frac{X_1}{r^2} \right) \tau + \frac{X_2}{r^2} \tau^2 \\ & + \left(\frac{\gamma s X_6 - \gamma X_5}{\alpha r} + \frac{X_5}{r^2} \right) K_1 + \left(\frac{X_6}{r^2} - \frac{\gamma X_6 + \gamma s X_5}{\alpha r} \right) K_2 . \end{aligned} \tag{99}$$

Appendix 4

For Case B, the pressure history is

$$p = p_0 \left(1 + \frac{C_0^c}{a} \left((t - t_0) - \frac{(t-t_0)^2}{2t_d} \right) \right)^2 . \tag{100}$$

Substituting Eq. (54) into Eq. (39), the solution is

$$f(\tau) = a_0 + a_1\tau + a_2\tau^2 + a_3\tau^3 + a_4\tau^4 + a_5 K_1 + a_6 K_2 , \tag{101}$$

where

$$a_0 = A_0(b_1p_0 + 2p_1b_2 + p_2b_3(3 - s^2) + 24p_3b_4(1 - s^2) + 24p_4b_5(1 - 6s^2 + s^4 + 4s - 4s^3)s^{-1}) \ , \quad (102)$$

$$a_1 = A_0(p_1b_1 + 4p_2b_2 + 6b_3(3 - s^2) + 96p_4b_4(1 - s^2)) \ , \quad (103)$$

$$a_2 = A_0(p_2b_1 + 6p_3b_2 + 12p_4b_3(3 - s^2)) \ , \quad (104)$$

$$a_3 = A_0(p_3b_1 + 8p_4b_2) \ , \quad (105)$$

$$a_4 = A_0p_4b_1 \ , \quad (106)$$

$$a_5 = A_0(-b_1p_0 - 2p_1b_2 + 2(s^2 - 3)p_2b_3 - 24p_3(1 - s^2)b_4 - 24p_4(1 - 6s^2 - 4s^3 + s^4 + 4s)b_5s^{-1}) \ , \quad (107)$$

and

$$a_6 = A_0(-b_1p_0s^{-1}b_2p_1(s^2 - 1)s^{-1} + 2b_3p_2(3s - s^{-1}) - 6b_4p_3(s^{-1} - 6s + s^3) - 24b_5s^{-1}p_4((1 + s^2)^5 - (1 + 4s - 6s^2 - 4s^3 + s^4)^2)^{0.5}) \ , \quad (108)$$

where $b_1 = 1/\gamma^2$, $b_2 = 1/(\gamma^3(1 + s^2))$, $b_3 = 1/(\gamma^4(1 + s^2)^2)$, $b_4 = 1/(\gamma^5(1 + s^2)^3)$, $b_5 = 1/(\gamma^6(1 + s^2)^4)$, $p_1 = 2C_0^c p_0/a$, $p_2 = p_0((C_0^c/a)^2 - C_0^c/at_d)$, $p_3 = -C_0^2 p_0/(a^2t_d)$, $p_4 = p_0/4(C_0^c/(at_d))^2$.

Substituting Eq. (101) into Eq. (38), the displacement history in period II for Case B is

$$u^{II}(\tau) = \left(\frac{a_1}{\alpha r} + \frac{a_0}{r^2}\right) + \left(\frac{2a_2}{\alpha r} + \frac{a_1}{r^2}\right)\tau + \left(\frac{a_2}{r^2} + \frac{3a_3}{\alpha r}\right)\tau^2 + \left(\frac{4a_4}{\alpha r} + \frac{a_3}{r^2}\right)\tau^3 + \frac{a_4}{r^2}\tau^4 + \left(\frac{\gamma s a_6 - \gamma a_5}{\alpha r} + \frac{a_5}{r^2}\right)K_1 + \left(\frac{a_6}{r^2} - \frac{\gamma a_6 + \gamma s a_5}{\alpha r}\right)K_2 \ . \quad (109)$$

(Received May 21, 1999, accepted October 31, 2000)

Pure appl. geophys. 158 (2001) 1951–2015
0033–4553/01/111951–65 \$ 1.50 + 0.20/0

Pure and Applied Geophysics

Regional Magnitude Scaling, Transportability, and M_s:m_b Discrimination at Small Magnitudes

HOWARD J. PATTON[1]

Abstract—Data sets of $m_b(Pn)$ and $m_b(Lg)$ measurements are presented for three continental regions in order to investigate scaling relationships with moment magnitude M_w and event discrimination at small magnitudes. Compilations of published measurements are provided for eastern North American and central Asian earthquakes, and new measurements are reported for earthquakes located in western United States. Statistical tests on M_w:m_b relationships show that the $m_b(Lg)$ scale of NUTTLI (1973) is transportable between tectonic regions, and a single, unified M_w:$m_b(Lg)$ relationship satisfies observations for M_w ~4.2–6.5 in all regions. A unified relationship is also developed for nuclear explosions detonated at the Nevada Test Site and test sites of the former Soviet Union. Regional m_b for explosions scale at higher rates than for earthquakes, and of significance is the finding that $m_b(Pn)$ for explosions scales at a higher rate than $m_b(Lg)$. A model is proposed where differences in scaling rates are related to effects of spectral overshoot and near-field *Rg* scattering on the generation of *Pn* and *Lg* waves by explosions. For earthquakes, $m_b(Pn)$ and $m_b(Lg)$ scale similarly, showing rates near 1.0 or $2/3 \cdot \log_{10}M_o$ (seismic moment).

M_w:$m_b(Lg)$ scaling results are converted to unified M_s:$m_b(Lg)$ relationships using scaling laws between $\log M_o$ and M_s. For earthquakes with M_s greater than 3.0, the scaling rate is $0.69 \cdot M_s$, which is the same as it is for nuclear explosions if M_s is proportional to $1.12 \cdot \log M_o$, as determined by NTS observations. Thus, earthquake and explosion populations are parallel and separated by 0.68 m_b units for large events. For small events ($M_s < 3.0$), populations may converge or diverge depending on the tectonic region in which earthquakes occur and the scaling rate of explosions at small yields. Earthquakes scale as 0.64 and 0.75 on M_s:$m_b(Lg)$ plots for stable and tectonic regions, respectively. While the scaling rate for explosions is ~0.69, this value is uncertain due to paucity of M_o observations at small yields. Measurements of $[m_b(P) - m_b(Lg)]$ for earthquakes in the western United States have an average value of $-0.33 \pm .03$ m_b units, in good agreement with Nuttli's estimate of m_b bias for NTS. This result suggests that Nuttli's method for estimating test site bias can be extended to earthquakes to make estimates of bias on regional scales. In addition, a new approach for quick assessments of regional bias is proposed where M_s:$m_b(P)$ observations are compared with M_s:$m_b(Lg)$ relationships. Catalog M_s:$m_b(P)$ data suggest that m_b bias is significant for tectonic regions of southern Asia, averaging about -0.4 m_b units.

Key words: Seismic magnitudes, regional phases, transportability, event discrimination.

1. Introduction

Effective monitoring of a Comprehensive Test-Ban Treaty (CTBT) depends in large part on regional broadband seismic data and robust techniques for accurate

[1] Los Alamos Seismic Research Center, Los Alamos National Laboratory, Los Alamos, NM, USA. E-mail: patton@geophysics.lanl.gov

location, characterization, and discrimination of small events. For years, seismologists focused on teleseismic techniques because teleseismic data is all that was available for many remote areas of the world. Major advances were made on all technologies with teleseismic data. For example, M_s:m_b observations proved to be a very successful teleseismic discriminant. However, it was recognized that regional observations held the key to better monitoring at low magnitudes, and interest is currently high to extend M_s:m_b applicability to regional data.

Surface-wave technologies for M_s determination are rapidly being developed, tested, and evaluated using regional data for many key monitoring regions. A host of short-period magnitude scales, applicable to near-regional and regional *P*-waves phases (e.g., *Pg* and *Pn*), local and regional *S*-wave phases (e.g., *Lg*), and local and regional coda waves, have been devised and applied over the years. Many scales have proven to be very effective, but better understanding of their scaling with source size and physical basis is needed.

For CTBT monitoring, performance of regional magnitude scales must satisfy a number of desirable qualities. In addition to the usual demands for stability and low variance, regional magnitudes must work well at small magnitudes, be effective over the regional distance range (200–2000 km), and perform well in a variety of geological environments. Crustal and upper mantle structures vary significantly on regional scales, and transportability of seismic measurements between radically different geologic environments is important for regional monitoring.

The importance of transportability can be illustrated from experience monitoring the Threshold Test-Ban Treaty (TTBT). The TTBT stipulated an upper bound limit of 150 kilotons (kt) on all underground nuclear explosions. The principal monitoring technique relied upon estimates of seismic yield based on measurements of teleseismic body-wave magnitude, $m_b(P)$. For the United States (US), calibration made use of explosions at the Nevada Test Site (NTS), where nuclear yields were known and correlated against observations of $m_b(P)$. Seismic yields of foreign nuclear tests were estimated from measurements of $m_b(P)$ using the NTS-calibrated m_b-yield relationship. Unfortunately, this technique did not work well because $m_b(P)$ is susceptible to regional biases caused by upper mantle attenuation of short-period *P* waves. NTS lies squarely over highly attenuating upper mantle associated with the anomalous Basin and Range province. Consequently, seismic yield estimates of foreign tests were often biased high, and protocols to the TTBT had to be renegotiated for measures that would enhance the US ability to verify. Non-transportability of $m_b(P)$ was a factor in the US government losing confidence in seismic verification of the TTBT.

The focus of this paper is regional magnitude scaling, transportability of regional magnitude scales, and extensions of M_s:m_b using regional data. I present data sets of magnitudes based on ~1-Hz amplitudes of *Lg* waves, *Pn* waves, and teleseismic *P* waves for three different tectonic regions. I investigate their scaling properties using measurements of seismic moment M_o for earthquakes and explosions, and evaluated transportability by comparing scaling relationships determined for different regions. Regional M_s:m_b relationships are derived for continental earthquakes and nuclear explosions, and discrimination is investigated at small magnitudes.

I will show that $m_b(Lg)$, as devised by NUTTLI (1973), is transportable between regions investigated in this study, which are the eastern North America, western US and central Asia. Because the range of crust and upper mantle structures is so great in these regions, this result is very promising for $m_b(Lg)$ transportability to all continental regions. Unified (or global) moment-magnitude versus $m_b(Lg)$ relationships are derived for earthquakes and explosions using earthquakes for all three tectonic regions and explosions conducted at NTS and test sites of the former Soviet Union. I use these relationships in investigate M_s:$m_b(Lg)$ scaling and discrimination behavior at low magnitudes ($M_s \sim 1.5$). The global relationships for $m_b(Lg)$ are also used to estimate $m_b(P)$ bias, and the results suggest significant $m_b(P)$ bias for regions of monitoring interest in southern Asia.

Throughout this paper, I adopt moment magnitude, which I designate M_w, as the independent variable upon which scaling behavior of 1-Hz magnitudes in investigated. I used the equation, $M_w = 2/3 \cdot \log_{10} M_o - 10.66$, where M_o has units of dyn-cm (HANKS and KANAMORI, 1979). The scaling model is linear, i.e., $m_b = a \cdot M_w + b$, where a and b are unknowns. In a few instances, this model is tested against a higher-order model (quadratic), but all of the results reported in this paper are for the linear model. Error estimates are one standard deviation ($1-\sigma$) unless noted otherwise. I use the abbreviation 'mu', magnitude units, for all magnitudes, and $E[\cdot]$ is used to designate the mean, usually mean difference between two magnitudes, e.g., $E[m_b(P) - m_b(Lg)]$.

2. *Earthquake Data for Three Tectonic Regions*

2.1. *Eastern North America (ENA) Data Set*

Data. In developing a regional magnitude scale based on Lg, NUTTLI (1973) recognized the robust character (large, stable amplitudes at 1 Hz) exhibited by Lg wherever it propagates in the North American continent east of the Rocky Mountains. This character is due in part to low anelastic attenuation and the existence of fairly uniform crustal waveguides, which reduces amplitude loss due to scattering. Nuttli's magnitude scale rapidly gained acceptance throughout eastern United States and Canada by operators of regional networks and researchers investigating seismicity and seismic hazards of sites of nuclear power plants. As a result, many research papers have been written on ENA seismicity where waveform data from old recordings and recordings made by newer digital systems were processed to determine magnitudes using Lg. Furthermore, network operators in the east routinely publish $m_b(Lg)$ in either seismicity catalogs they produce or supply them to organizations with responsibility for generating catalogs on a global scale.

Because of wide availability of $m_b(Lg)$ measurements in ENA, I undertook a literature compilation of events located east of 105°W in continental areas of the United States and southern Canada, with reported $m_b(Lg)$ and estimates of M_o. Figure 2.1 is a map of epicenters for 103 earthquakes which satisfied these criteria. I limited the search to events with $m_b(Lg)$ greater than about 3.0. Events tend to cluster in two regions, the New Madrid seismic zone and the seismic zones of the St. Lawrence seaway and New England. Table 2.1 is a list of these ENA events. I make no claims this is a comprehensive list for this region, as the body of literature is substantial. However, I believe I have captured a large percentage of events satisfying the selection criteria I followed (also see subsection entitled "*Other event selection issues*" below).

NUTTLI (1973) calibrated $m_b(Lg)$ against teleseismic $m_b(P)$ for events located in the New Madrid seismic zones. The calibration was established as follows: For events with $m_b(P) \sim 5.0$, Nuttli determined that the 1-Hz amplitude of "hypothetical" Lg waves at a distance of 10 km is 110 microns. This hypothetical Lg amplitude for the i-th event $A_i(10)$ is computed with the formula

$$A_i(10) = A(\Delta_i) \cdot (\Delta_i/10)^{1/3} \cdot [\sin(\Delta_i/111.1)/\sin(10/111.1)]^{1/2} \cdot \exp[\gamma \cdot (\Delta_i - 10)] \ ,$$

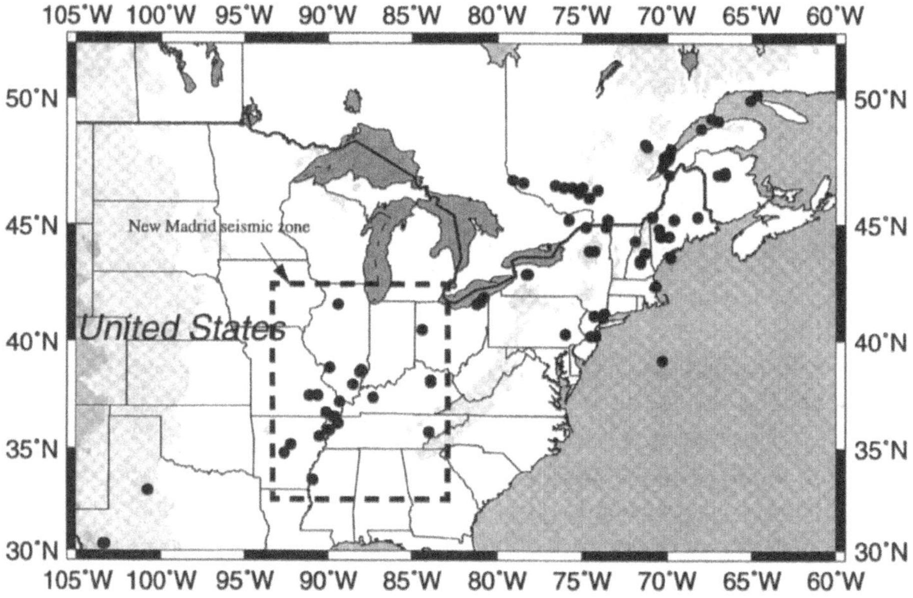

Figure 2.1
Map of eastern north America between 30°N and 52°N, showing locations of events used in this study. Box encloses events in the New Madrid seismic zone referred to in the text, which includes areas in the central US outside the seismic zone proper.

Table 2.1

Magnitudes for Eastern North American earthquakes

yrmodahrmnsc	m_b P	m_b Lg	m_b (Lg) #1	m_b (Lg) #2	M_w Avg	M_w #1	M_w #2	Region
250301021920[a]	6.7[b]	**6.6**	6.6[c]	(7.0[d])	**6.28**	6.25[c]	6.3[e]	La Malbaie Quebec
290812112448[f]	–	**5.2**	5.2[c]	(5.4[d])	**4.75**	4.75[c]	–	Attica NY
351101060340[a]	6.1[b]	**6.2**	6.2[c]	(6.3[d])	**6.08**	5.65[c]	6.5[b]	Timiskaming Quebec
380823033631[f]	–	**3.9**	3.9[c]	–	**3.59**	3.59[c]	–	central NJ
380823050453[f]	–	**4.0**	4.0[c]	–	**3.62**	3.62[c]	–	southern NY
380823070328[f]	–	**3.7**	3.7[c]	–	**3.64**	3.64[c]	–	central NJ
391019115358[a]	5.4[b]	**5.6**	5.6[c]	–	**5.23**	5.16[c]	5.3[b]	Charlevoix Quebec
401220072726[f]	5.5[b]	**5.5**	5.5[c]	(5.7[d])	**5.40**	5.31[c]	5.5[b]	Ossipee NH
401225050343[f]	–	**3.7**	3.7[c]	–	**3.44**	3.44[c]	–	Ossipee NH
401227195609[f]	–	**3.8**	3.8[c]	–	**3.48**	3.48[c]	–	Ossipee NH
430114213237[f]	–	**4.3**	4.3[c]	–	**3.99**	3.99[c]	–	Dover ME
440905043845[a]	4.6[b]	**5.8**	5.8[c]	(5.9[d])	**5.69**	5.58[c]	5.8[b]	Quebec/NY border
491005023347[f]	–	**4.4**	4.4[c]	–	**4.23**	4.23[c]	–	west-central ME
510903212624[f]	–	**3.8**	3.8[c]	–	**3.44**	3.44[c]	–	northern NJ
521014220342[a]	5.0[a]	**4.9**	4.9[c]	–	**4.08**	4.08[c]	–	southern Quebec
570426114008[f]	–	**4.7**	4.7[c]	–	**4.42**	4.42[c]	–	off coast of ME
630303173011	4.8[g]	**4.7**	4.7[g]	–	**4.64**	4.70[h]	4.59[i]	southeastern MO
631016143100	–	**3.9**	3.9[c]	–	**3.46**	3.46[c]	–	northeastern MA
650814131356	3.8[g]	**3.8**	3.8[g]	–	**3.57**	3.65[h]	3.49[i]	southern IL
651021020438	4.9[g]	**5.0**	5.0[g]	–	**4.60**	4.64[h]	4.57[i]	southern MO
660101132338	4.7	**4.6**	4.6[i]	–	**4.31**	4.31[h]	–	western NY
670604161413	4.5[g]	**4.4**	4.4[g]	–	**4.33**	4.33[h]	–	west-central MS
670613190854	3.9	**4.4**	4.4[j]	–	**4.12**	4.12[h]	–	western NY
670701161933	–	**3.4**	3.4[c]	–	**3.34**	3.34[c]	–	south-central ME
670721091448	4.2[g]	**4.4**	4.4[g]	–	**4.03**	4.08[h]	3.98[i]	southeastern MO
681109170142	5.5[g]	**5.4**	5.4[g]	–	**5.33**	5.33[h]	–	southern IL
690101233536	4.3[g]	**4.5**	4.5[g]	–	**4.38**	4.38[h]	4.39[i]	central AR
701117021355	4.4[g]	**4.4**	4.4[g]	–	**4.10**	4.14[h]	4.06[i]	AR/MO/TN border
720915052215	4.4[g]	**4.4**	4.4[g]	–	**4.12**	4.16[h]	4.08[i]	northern IL
730107225606	3.2[i]	**3.2**	3.2[k]	–	**3.01**	3.01[i]	–	western KY
730615010904	4.8	**5.1**	5.0[j]	5.2[k]	**4.53**	4.53[h]	–	southern Quebec
731003035020	3.4[i]	**3.4**	3.4[k]	–	**3.04**	3.04[i]	–	AR/TN border
731009201526	3.7[i]	**3.8**	3.8[k]	–	**3.31**	3.31[i]	–	New Madrid MO
731130074841	5.6	**4.6**	4.6[l]	4.6[m]	**4.09**	4.12[h]	4.06[i]	eastern TN
740108011237	4.1	**4.3**	4.3[k]	–	**3.66**	3.66[i]	–	MO/TN border
740403230502	4.5	**4.7**	4.7[k]	–	**4.40**	4.39[h]	4.42[i]	southern IN
740605080611	4.0	**3.6**	3.6[k]	–	**3.19**	3.19[i]	–	southern IL
750609183923	–	**3.8**	3.5[c]	4.2[n]	**3.34**	3.34[c]	–	northern NY
750613224027	4.3	**4.3**	4.3[k]	–	**3.78**	3.78[h]	–	MO/TN border
750712123715	4.2	**(4.1)**	(4.1[o])	–	**4.16**	4.16[o]	–	Maniwaki Quebec
751006222141	5.2	**4.8**	4.8[c]	–	**4.10**	4.10[c]	–	off coast of Nova Scotia
751103205455	4.0	**4.0**	4.0[c]	3.9[n]	**3.74**	3.74[c]	–	near Goodnow NY
760325004120	4.9	**5.0**	5.0[k]	–	**4.67**	4.67[h]	–	northeastern AR
760325010011	4.1	**4.5**	4.5[k]	–	**4.27**	4.27[h]	–	northeastern AR
780218144825	4.1	**3.8**	3.8[p]	4.1[q]	**3.94**	3.94[r]	–	southern Quebec
780616114654	4.4	**4.6**	4.6[k]	4.7[m]	**4.55**	4.55[s]	–	western Texas
790819224931	4.6	**(5.0)**	(5.0[t])	5.4[q]	**4.79**	4.79[u]	–	Charlevoix Quebec

Table 2.1

Continued

yrmodahrmnsc	m_b P	m_b Lg	$m_b(Lg)$		M_w			Region
			#1	#2	Avg	#1	#2	
800403165724	–	**(4.0)**	(4.0[v])	4.1[q]	**3.6**	3.6[v]	–	southern Quebec
800727185221	5.1	**5.1**	5.2[m]	5.0[n]	**5.08**	5.08[w]	–	Sharpsburg, KY
811028195610	–	**(3.9)**	3.9[q]	–	**3.5**	3.5[v]	–	southern Quebec
820109125351	5.7	**5.8**	5.8[l]	5.7[q]	**5.59**	5.53[x]	5.51[y] &	Miramichi N Brunswick
820109163642	5.1	**5.0**	5.0[l]	–	**4.38**	4.6[z]	4.16[y]	Miramichi N Brunswick
820111214108	5.4	**5.6**	5.5[l]	5.7[aa]	**5.11**	5.2[d]	4.84[y] &	Miramichi N Brunswick
820113175641	–	**3.9**	3.9	4.0[q]	**3.2**	3.2[v]	–	Miramichi N Brunswick
820117133356	–	**3.7**	3.7[aa]	3.6[q]	**3.19**	3.19[aa]	–	Miramichi N Brunswick
820119001442	4.5	**4.5**	4.5[l]	4.8[q]	**4.2**	4.0[v]	4.4[d]	Gaza NH
820121003354	4.5	**4.6**	4.5[ab]	4.7[m]	**3.96**	3.96[y]	–	Enola AR
820122235422	–	**3.8**	3.7[l]	3.9[m]	**3.43**	3.43[y]	–	Enola AR
820124032244	–	**4.2**	4.3[ab]	4.0[l]	**3.68**	3.68[y]	–	Enola AR
820331210220	5.0	**4.8**	4.5[p]	5.0[aa]	**4.33**	4.4[d]	4.21[y] &	Miramichi N Brunswick
820402135011	4.6[ac]	**4.3**	4.1[ad]	4.5[aa]	**3.75**	3.7[v]	3.80[aa]	Miramichi N Brunswick
820411180051	–	**4.1**	4.0	4.2[aa]	**3.50**	3.5[v]	3.49[aa]	Miramichi N Brunswick
820418224721[q]	–	**4.2**	4.2[aa]	4.1[q]	**3.50**	3.5[v]	3.51[aa]	Miramichi N Brunswick
820506162807	–	**4.2**	4.2[aa]	4.0[q]	**3.50**	3.5[v]	3.51[aa]	Miramichi N Brunswick
820616114330	4.7	**4.6**	4.6[ad]	4.6[q]	**3.96**	3.89[y]	4.0[v] &	Miramichi N Brunswick
820813010641	3.7[ac]	**(4.3)**	4.3[q]	–	**3.6**	3.6[v]	–	southern Quebec
821204160832	–	**(3.9)**	3.9[q]	–	**3.4**	3.4[v]	–	southern Quebec
830117193551	4.5	**(4.1)**	4.1[q]	–	**3.6**	3.6[v]	–	southern Quebec
830513234057	–	**4.1**	4.1[ad]	3.9[q]	**3.7**	3.7[v]	–	Miramichi N Brunswick
830529054549	4.2	**3.9**	3.9[p]	4.1[q]	**3.7**	3.7[v]	–	west-central ME
831007101846	5.1	**5.3**	5.3[l]	–	**4.98**	4.99[x]	4.86[ae] &	Goodnow NY
831011041055	4.2	**(4.1)**	4.1[q]	–	**3.60**	3.6[v]	3.6[d]	Ottawa Ontario
851005053414	–	**(3.9)**	3.9[q]	–	**3.5**	3.5[v]	–	Miramichi N Brunswick
851019100740	3.6	**4.2**	4.2[af]	4.1[q]	**3.6**	3.6[v]	–	Ardsley NY
860111133028	–	**(4.0)**	4.0[q]	–	**3.4**	3.4[v]	–	southern Quebec
860131164643	5.0	**5.4**	5.4[af]	5.3[q]	**4.85**	4.67[ag]	5.03[x]	Painesville, OH
860712081937	4.5	**(4.7)**	4.5[z]	4.9[q]	**4.40**	4.3[v]	4.5[z]	Lima OH
860919155301	4.9[ac]	**(4.2)**	4.2[q]	–	**3.6**	3.6[v]	–	southern Quebec
861025171638	3.9	**3.9**	3.9	–	**3.9**	3.9[v]	–	Northfield NH
861109195719[q]	–	**(4.2)**	4.2[q]	–	**3.7**	3.7[v]	–	southern Quebec
870610234854	4.9	**5.1**	5.1[k]	–	**5.00**	5.00[ah]	–	southeast IN
870713054917	3.5	**3.8**	3.8	4.1[q]	**3.6**	3.6[v]	–	near Painesville OH
880907022809	4.5	**4.6**	4.6[l]	–	**4.21**	4.21[ai]	–	northeast KY
881123091127	4.3	**(4.6)**	4.6[q]	–	**4.10**	4.1[v]	4.1[z]	Saguenay Quebec
881125234604	5.9	**6.5**	6.5[af]	6.5[q]	**5.89**	5.81[aj]	5.90[x] &	Saguenay Quebec
881126033808	–	**4.1**	4.1	4.1[q]	**3.5**	3.5[v]	–	Saguenay Quebec
890210010424	–	**(4.3)**	4.3[q]	–	**3.8**	3.8[v]	–	eastern Quebec
890309094132	3.9	**(4.3)**	4.3[q]	–	**3.82**	3.8[v]	3.83[ak]	southern Quebec
890311083152	–	**(4.4)**	4.4[q]	–	**3.86**	3.86[ak]	–	southern Quebec
891116092452	–	**(4.0)**	4.0[q]	–	**3.6**	3.6[v]	–	southern Quebec
901007084730	–	**(3.9)**	3.9[q]	–	**3.6**	3.6[v]	–	southern Quebec
901019070157	4.6	**5.1**	5.1[af]	5.0[q]	**4.65**	4.6[v]	4.7[z]	Mt Laurier Quebec
920822122032	4.8	**4.7**	4.7	–	**4.42**	4.42[al]	–	off coast of NJ
930830051528	–	**3.3**	3.3	3.7[q]	**3.16**	3.16[al]	–	southern Quebec

Table 2.1

Continued

yrmodahrmnsc	m_b P	m_b Lg	$m_b(Lg)$ #1	$m_b(Lg)$ #2	M_w Avg	M_w #1	M_w #2	Region
930923064528	–	**3.5**	3.5	3.8[q]	**3.28**	3.28[al]	–	southern Quebec
931016063005	–	**3.4**	3.4	3.7[q]	**3.26**	3.26[al]	–	Painesville OH
931116093144	3.8	**4.1**	4.1	4.3[q]	**3.78**	3.78[al]	–	Napierville Quebec
940116014916	4.6	**4.6**	4.6	–	**4.31**	4.31[al]	–	Reading PA
940916074011	–	**3.3**	3.3	3.3[q]	**3.29**	3.29[al]	–	Springfield ME
940916074406	–	**3.3**	3.3	3.3[q]	**3.31**	3.31[al]	–	Springfield ME
950415143329	–	**4.0**	4.0	–	**3.70**	3.70[am]	–	western Texas
950601010615	–	**3.5**	3.5	–	**3.37**	3.37[am]	–	western Texas
950616121311	–	**3.8**	3.8	3.8[q]	**3.57**	3.57[al]	–	VT/NH border

[a] BASHAM *et al.* (1979); [b] EBEL *et al.* (1986); [c] STREET and TURCOTTE (1977); [d] BOORE and ATKINSON (1987); [e] BENT (1922); [f] NEIC Catalog for Eastern, Central, and Mountain States of US, 1534–1986; [g] NUTTLI and ZOLLEWEG (1974); [h] HERRMANN (1979); [i] STREET *et al.* (1975); [j] STREET (1976); [k] St. Louis University, MO; [l] Virginia Polytechnical Institute and State University, Blackburg, VA; [m] Oklahoma Geological Survey, Tulsa, OK; [n] Columbia University, Lamont–Doherty Earth Observatory, Palisades, NY; [o] HORNER *et al.* (1978); [p] Boston College, Weston Observatory; [q] Geophysics Division, Geological Survey of Canada, Ottawa, CANADA; [r] HORNER *et al.* (1979),[s] VOSS and HERMANN (1980); [t] HASEGAWA (1983); [u] HASEGAWA and WETMILLER (1980); [v] BOATWRIGHT (1994); [w] HERRMANN *et al.* (1982); [x] Harvard Centroid Moment Tensor (CMT); [y] STREET (1984); [z] ATKINSON (1993); [aa] SHIN and HERRMANN (1987); [ab] Tennessee Earthquake Information Center; [ac] International Seismic Center (ISC); [ad] CHAEL (1987); [ae] NABELEK and SUAREZ (1989); [af] KIM (1998); [ag] GLASSMOYER and BORCHERDT (1990); [ah] TAYLOR *et al.* (1989); [ai] STREET *et al.* (1993); [aj] SOMERVILLE *et al.* (1990); [ak] LI *et al.* (1995); [al] SHI *et al.* (1998); [am] XIE (1998).

Notes. (·) ≡ High frequency $m_b(Lg)$; underscore ≡ an average of $m_b(Lg)$ values; & ≡ additional estimates of M_w were used to compute average M_w: 820109 – CHOY *et al.* (1983), WETMILLER *et al.* (1984), NABELEK (1984), BOATWRIGHT and CHOY (1992) 820111 – SHIN and HERRMANN (1987), BOATWRIGHT and CHOY (1992) 820331 – SHIN and HERRMANN (1987), BOATWRIGHT (1994) 820616 – BOATWRIGHT (1994) 831007 – WU (1984) 881125 – BOATWRIGHT and CHOY (1992); no footnote on $m_b(P)$ or $m_b(Lg)$ means NEIC value.

where Δ is epicentral distance in km, γ is attenuation coefficient, and "exp" is the exponential function. The 1-Hz Lg amplitude $A(\Delta_i)$ is measured off regional recordings made on short-period sensors, which, at the time, were World Wide Standard Seismographic Network (WWSSN) instrumentation. This amplitude is corrected for Airy phase propagation, geometrical spreading, and attenuation from the receiver back to a distance of 10 km. The calibration formula is

$$m_b(Lg) = 5.0 + \log_{10}[A_i(10)/110] \ .$$

While Nuttli recognized the need for region-specific γ estimates, he boldly asserted that the calibration formula is transportable to any continental structure. This is a remarkable claim considering the large variations of crustal structures around the world, which might be expected to affect the source amplitude of Lg waves. Thus the transportability of $m_b(Lg)$ relies on the ability to make accurate path corrections on a regional basis and on the invariance or insensitivity of 1-Hz Lg Green's function response for seismogenic depths in the crust.

Other event selection issues. In making event selections, I placed no particular requirements on the way in which M_o was measured, other than assurance that the amplitude measurement was valid for the flat portion of the long-period moment rate source spectrum. Thus, studies that I have drawn upon (see Table 2.1) utilize a variety of time- and frequency-domain methods applied to teleseismic, regional and local data. Some of the data may have come from fairly narrow-band seismographic systems, but in all studies where this may be true, care was taken by investigators to make allowances for this limitation in the data.

On the other hand, I was particularly selective about $m_b(Lg)$, making sure there was adequate information provided about its source (organizations and/or investigators) and the way in which measurements were taken and magnitude computed. Here the issue concerns mainly the frequency of *Lg* waves. Nuttli's original $m_b(Lg)$ formula is predicated on the sustained 1-Hz amplitude within a specified time window. In practice, the frequency at which a measurement is taken may vary, and the farther it is from 1 Hz, the less valid are propagation corrections which Nuttli developed for ENA (NUTTLI, 1973) and the calibration equation with $m_b(P)$. Adaptations of Nuttli's formula to other frequencies have been developed (e.g., HERRMANN and KIJKO, 1983), but these developments must utilize frequency-dependent propagation corrections and source theory to "equalize" for frequency differences so that the calibration still applies. To insure uniformity within and between data sets for ENA, western United States and central Asia, I preferred $m_b(Lg)$ values determined by the original formulas.

EBEL (1982) and KIM (1998) point out systematic differences in the practice followed by the Geological Survey of Canada (GSC) to compute *Lg* magnitude using an extension of Nuttli's original method. The GSC magnitude, which is designated m_N or M_N, uses amplitudes of *Lg* waves for higher frequencies, typically between 2 to 10 Hz. Comparisons between m_N, 1-Hz $m_b(Lg)$, and Richter's local magnitude M_L show that GSC's m_N is larger on average than $m_b(Lg)$ or M_L by as much as 0.37 to 0.47 mu for magnitudes between 2.5 to 4.0. This is a significant difference, and I am careful to point out in Table 2.1 those magnitudes that were likely at frequencies above 2 Hz. Unfortunately, it is sometimes difficult to ascertain the nature of magnitudes from studies in the literature because of loose nomenclature. Studies refer to M_N, m_{Lg}, or sometimes just m_b, and it may be unclear whether or not the frequency of *Lg* is 1 Hz or not. For this reason, there are ambiguities for magnitudes I designated as "high frequency" in Table 2.1; for 1 Hz $m_b(Lg)$, I am fairly confident that measurements were made within one-octave of 1 Hz.

Magnitude relationships. Figure 2.2 is a graph of $m_b(Lg)$ scaling for events in ENA. This graph has all 103 events plotted, but is differentiated by symbol type between $m_b(Lg)$ values based on high frequency and 1-Hz amplitudes. Regression line in the figure was obtained by fitting just the 1-Hz data. For comparison purposes, the quadratic regression line is shown from the study of JOHNSTON (1996) based on 1-Hz $m_b(Lg)$ observations for the entire North American craton. Johnston's

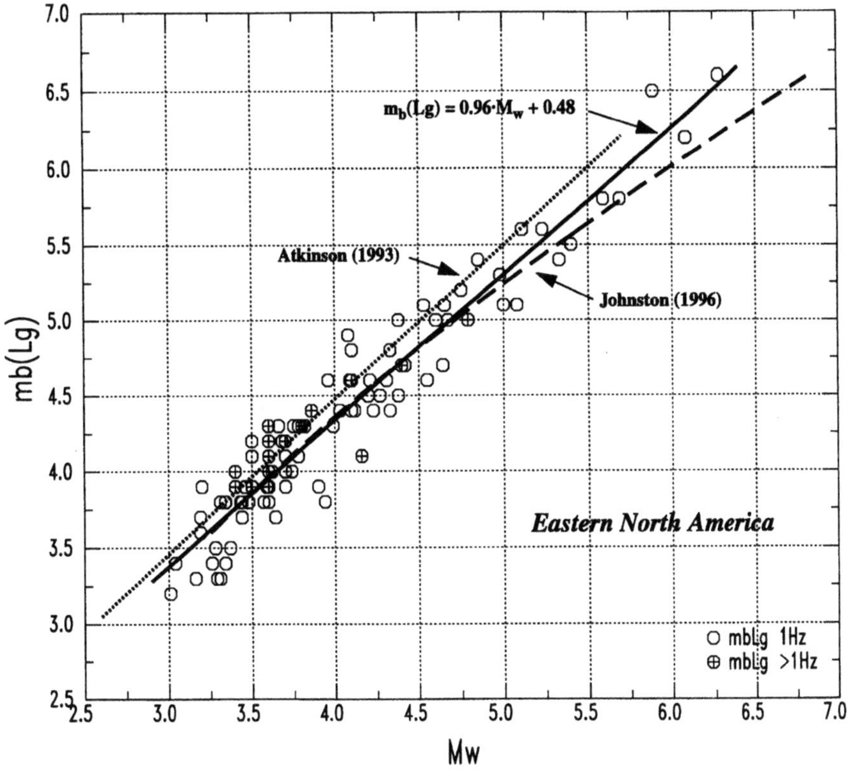

Figure 2.2
Scaling of $m_b(Lg)$ with M_w for events in ENA. All 103 events in the ENA data set are plotted, and events which only had high frequency $m_b(Lg)$ values are designated with a circled + sign. Solid line is regression result using only 1-Hz $m_b(Lg)$ data. The quadratic regression line of JOHNSTON (1996) is dash line and dotted line is the scaling result from ATKINSON (1993) based on m_N measurements reported by the Geological Survey of Canada.

study includes northwest territories and northern Superior Province of Canada, in addition to the areas of ENA. Johnston's results virtually overlay for M_w less than 4.75. At larger magnitudes, where observations are fewer in both data sets, the quadratic line rolls over, suggesting effects of saturation. However, statistical tests indicate that a roll-over effect is not required by the observations (JOHNSTON, 1996), and there is no convincing evidence in either data set for saturation of the $m_b(Lg)$ scale in ENA (see DISCUSSION section). Overall, the agreement between Johnston's $m_b(Lg)$ scaling results and the linear result of this study is very good.

Also shown in Figure 2.2 is the regression line of ATKINSON (1993), which relates M_w to m_N for about 100 events with m_N ranging between 3.0 and 6.0 in ENA. Comparing this line with the other regression results confirms that m_N is systematically larger than 1-Hz $m_b(Lg)$. However, the offset is smaller than what EBEL (1982) and KIM (1998) found, and it may depend on magnitude. Comparing the two regression lines, offsets are 0.20, 0.14, and 0.09 mu for M_w values of 5, 4, and 3.

KIM (1998) reports an offset of 0.37 mu for M_L of 4.0, which corresponds to M_w of ~3.5; this is three times larger than my estimate. While I did not make a point of collecting data for this purpose, there are 15 "high-frequency" $m_b(Lg)$ values and 30 1-Hz values for events with M_w 3.4–4.0, and I computed the average differences, $E[m_b(Lg) - M_w]$, for these events. The results are 0.488 ± .026 mu and 0.384 ± 0.039 mu for high frequency and 1 Hz, respectively, and the difference is 0.104 ± 0.047 mu. I infer that CGS's m_N is systematically offset from Nuttli's 1-Hz $m_b(Lg)$, but the offset may not be as large as previously reported. Indeed, considering data scatter, high frequency magnitudes do not stand out in Figure 2.2, with the exception of one event with M_w of ~4.2, which plots low compared to events of similar magnitude.

Regional magnitudes based on *Pn* waves were not available for ENA, so I show $m_b(P)$ scaling for 63 events in Figure 2.3 instead. Teleseismic magnitudes were taken

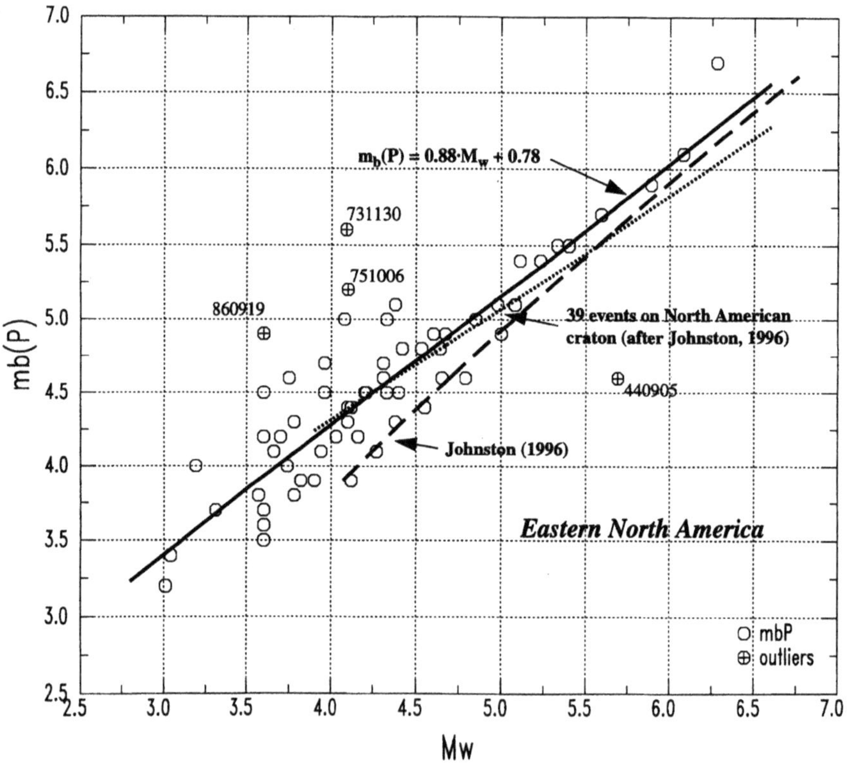

Figure 2.3

Scaling of $m_b(P)$ with M_w for events in ENA. Only 63 of the 103 events had reported teleseismic magnitudes. Four events are considered outliers on the scaling plot, and they were excluded from regression analysis giving solid line. Long-dash line is quadratic regression result for stable cratons from JOHNSTON (1996). Dotted line is linear regression result for 39 earthquakes with $m_b(P)$ greater than 4.5 located in the stable craton of ENA (after JOHNSTON, 1996).

from the Preliminary Determination of Epicenters (PDE), compiled by the National Earthquake Information Center (NEIC) or a source referenced in Table 2.1. Shown in Figure 2.3 with the regression result for these data are results based on a data set compiled by JOHNSTON (1996) for Precambrian cratons worldwide and m_b greater than 4.8. Many observations in this study plot above the quadratic relationship suggesting ENA earthquakes have larger m_b for the M_o than Johnston's average. This is also reflected to a degree in Johnston's data set (see Fig. 5 in JOHNSTON, 1996). Below 4.8, almost all of Johnston's observations are from ENA, but they were down-weighted because of concerns over network magnitude bias related to data censoring practices (JOHNSTON, 1996). RINGDAL (1986) demonstrated that this can lead to over estimation of $m_b(P)$ at small magnitudes. It may be overly cautious to down-weight events as large as 5.0, especially those located in well-instrumented areas of the Northern Hemisphere. I ran a linear regression analysis taking just those events in Johnston's data set from the North American craton with m_b greater than 4.5. The result is plotted in Figure 2.3, and shows good agreement with the observations in this study, especially for smaller magnitudes.

2.2. Western United States (WUS) Data Set

Data. Events selected are located west of 110°W longitude on land in the United States or in areas just offshore. The main areas sampled are Nevada, California, and western Utah, and most events date from late 1970s to mid-1990s (see Table 2.2). Other than geographical location, my other main selection criteria were availability of M_o and high-quality waveform data upon which I made measurements of $m_b(Lg)$ and $m_b(Pn)$. Waveform data come from the digital network of four broadband stations run by the Lawrence Livermore National Laboratory for over 30 years (JARPE, 1989). These stations are plotted on the map in Figure 2.4 along with earthquake epicenters. All events in Table 2.2 had waveform data from this network except events in 1966. Measurements of $m_b(Lg)$ for these four events were provided by Dr. Robert Herrmann (pers. comm., 1992).

The Livermore network went through a number of changes in sensor and recording instrumentation over this time period. Digitization rates and system response characteristics were accounted for in processing of the data. As discussed below, magnitudes were measured near 1 Hz for about 4.5 orders of M_o on high-quality broadband data for distances usually less than 1000 km.

Lg magnitudes. Measurements of $m_b(Lg)$ were made following Nuttli's generalized method (NUTTLI, 1986a). Measurement windows were determined for group velocities of 3.6 and 3.2 km/s. Amplitudes were measured off simulated WWSSN short-period records using Nuttli's practice of taking the 3rd peak for both upswings and downswings and averaging the two amplitudes. Wave periods were measured and if they did not fall between 0.55 and 1.45 s, the measurement was discarded (KIM, 1998; I used a somewhat more liberal low frequency limit than Nuttli's). In

Table 2.2

Magnitudes for Western United States earthquakes

yrmodahrmnsc	m_b P	m_b Pn	M_L	m_b Lg	M_w Avg	#1	#2	#3	#4	#5	Region
660513172555	4.6	–	4.5[a]	**4.25**[b]	**3.98**	3.98[c]	–	–	–	–	central CA
660702120833	3.7	–	3.6[d]	**3.74**[b]	**3.91**	3.91[e]	–	–	–	–	central CA
660803123904	4.1	–	3.9[d]	**3.76**[b]	**4.01**	4.01[e]	–	–	–	–	central CA
661027120602	4.2	–	4.2[d]	**4.35**[b]	**4.33**	4.33[e]	–	–	–	–	central CA
710805175817	4.3	**3.98**	4.2[a]	**4.37**	**4.30**	4.30[f]	–	–	–	–	Mass Mtn NV
790812113119	–	**3.18**	3.6[g]	**3.72**	–	–	–	–	–	–	southern NV
790816033744	–	**2.96**	3.7[g]	**3.43**	–	–	–	–	–	–	southern NV
791225141710	–	**3.67**	4.2[a]	**3.99**	**3.91**	3.91[h]	–	–	–	–	near NTS NV
800115202822	–	**3.63**	3.9[a]	**3.82**	**3.88**	3.88[i]	–	–	–	–	southern Sierras
800225234332	–	**3.86**	3.9[d]	**3.97**	**3.96**	3.96[i]	–	–	–	–	southern Sierras
800408001341	–	**4.62**	4.7[a]	**4.87**	**4.70**	4.70[i]	–	–	–	–	W-central NV
800428135534	–	**3.94**	4.1[a]	**3.87**	**4.32**	4.32[i]	–	–	–	–	N-western NV
800524100336	5.0	**4.44**	4.4[j]	**4.66**	**4.46**	4.46[i]	–	–	–	–	central UT
800527145057	5.7	**5.79**	6.2[a]	**6.16**	**6.01**	5.92[k]	6.03[l]	6.02[m]	6.08[n]	–	Mammoth Lakes CA
801025003059	–	**2.99**	3.8[d]	**3.51**	**3.55**	3.55[h]	–	–	–	–	near NTS NV
801128182113	4.9	**4.94**	5.2[a]	**5.32**	**4.75**	4.75[i]	–	–	–	–	northern Sierras
810405054040	4.2	**4.27**	4.5[j]	**4.36**	**4.21**	4.21[i]	–	–	–	–	southern UT
810429115552	–	**3.83**	4.2[a]	**3.78**	**3.78**	3.78[i]	–	–	–	–	W-central NV
810514051104	4.5	**3.97**	3.5[j]	**3.69**	–	–	–	–	–	–	central UT (mine)
811119180155	–	**3.76**	3.5[g]	**3.71**	**3.87**	3.87[i]	–	–	–	–	central NV
811201161850	–	**4.02**	4.5[a]	**4.63**	**4.27**	4.27[i]	–	–	–	–	W-central NV
811219205652	–	**4.12**	4.4[a]	**4.20**	**4.03**	4.03[i]	–	–	–	–	W-central NV
820124154407	–	**4.09**	4.3[a]	**4.08**	**4.06**	4.06[i]	–	–	–	–	S-western NV
820316084700	–	**3.48**	3.7[d]	**3.77**	–	–	–	–	–	–	near NTS NV
820512192924	–	**3.49**	4.0[a]	**3.90**	**3.95**	3.95[i]	–	–	–	–	Pahranagat SZ NV
820524121327	4.7	**4.13**	3.8[j]	**4.26**	**4.08**	4.08[i]	–	–	–	–	central UT
820529130225	4.3	**4.56**	3.9[a]	**4.15**	**4.57**	4.57[i]	–	–	–	–	Geysers CA
820621064338	4.1	**4.10**	4.2[a]	**3.97**	**4.27**	4.27[i]	–	–	–	–	Cascades CA
820706021043	4.1	**4.30**	4.7[a]	**4.62**	**4.29**	4.29[i]	–	–	–	–	Pahranagat SZ NV
820903185823	4.3	**4.15**	4.0[a]	**4.11**	**4.34**	4.34[i]	–	–	–	–	northern CA
820924074024	5.0	**4.99**	5.4[a]	**5.35**	**5.04**	5.04[i]	–	–	–	–	S-western NV
821001013335	–	**4.16**	3.9[a]	**4.12**	**4.10**	4.10[i]	–	–	–	–	S-western NV
830223111020	–	**3.90**	3.9[g]	**3.78**	**3.79**	3.79[i]	–	–	–	–	Lake Mead NV
830604113740	–	**3.44**	3.6[g]	**3.52**	**3.61**	3.61[i]	–	–	–	–	Pharanagat SZ NV
831209085841	4.3	**4.20**	3.6[j]	**4.33**	**4.28**	4.28[i]	–	–	–	–	central UT
840802110134	–	**3.49**	3.7[g]	**3.76**	**3.61**	3.61[i]	–	–	–	–	Pharanagat SZ NV
840908061640	5.0	**5.32**	–	**5.23**	**5.08**	5.04[i]	5.13[k]	–	–	–	central ID
840917191107	–	**3.38**	3.7[g]	**3.71**	**3.73**	3.73[i]	–	–	–	–	E-central NV
841123180825	5.6	**5.54**	6.1[a]	**5.90**	**5.87**	5.94[o]	5.85[k]	5.82[p]	–	–	Round Valley CA
850124112721	4.3	**4.31**	5.3[a]	**4.94**	**4.70**	4.46[i]	4.95[p]	–	–	–	E-central CA
851210061025	–	**3.70**	3.7[g]	**3.66**	–	–	–	–	–	–	Pahranagat SZ NV

Table 2.2

Continued

yrmodahrmnsc	m_b P	m_b Pn	M_L	m_b Lg	M_w						Region
					Avg	#1	#2	#3	#4	#5	
880814200303	5.5	**4.54**	5.3[j]	**5.48**	**5.21**	5.21[q]	–	–	–	–	central UT
880831164516	5.1	**5.50**	5.4[d]	**4.94**	**4.75**	4.75[q]	–	–	–	–	Mexicali MX
890119065328	5.2	**5.23**	5.2[a]	**5.37**	**5.02**	5.05[q]	4.98[r]	–	–	–	Malibu CA
890130040622	5.0	**4.80**	5.4[j]	–	**5.24**	5.12[q]	5.37[k]	–	–	–	central UT
891019101435	4.6	**4.66**	4.6[a]	**4.66**	**4.72**	4.75[q]	4.69[s]	–	–	–	Loma Prieta CA
901024061520	5.4	**5.28**	5.7[a]	**5.64**	**5.36**	5.29[q]	5.29[t]	5.31[u]	5.54[p]	5.37[k]	Lee Vining CA
910628144354	5.8	**5.44**	5.4[d]	**5.71**	**5.64**	5.57[q]	5.56[v]	5.69[w]	5.66[x]	5.70[k]	Sierra Madre CA
920423045023	5.7	**6.29**	6.1[d]	–	**6.21**	6.22[q]	6.25[y]	6.22[k]	6.19[w]	6.16[u]	Joshua Tree CA
920504161949	4.4	**5.04**	–	**5.08**	**4.80**	4.80[q]	4.79[w]	–	–	–	Joshua Tree CA
920506023843	4.4	**4.72**	4.9[a]	**4.51**	**4.66**	4.72[q]	4.59[w]	–	–	–	Joshua Tree CA
920518154417	4.6	**4.91**	–	**4.76**	**4.78**	4.81[q]	4.75[w]	–	–	–	Joshua Tree CA
920628115734	6.2	**6.54**	–	**7.24**	**7.36**	7.32[q]	7.42[y]	7.36[k]	7.36[t]	7.36[u]	Landers CA
920629101422	5.6	**5.53**	6.2[a]	**6.27**	**5.70**	5.68[q]	5.70[z]	5.62[z]	5.73[y]	5.75[aa]	Skull Mtn NV
920629103102	4.3	**4.66**	–	**5.04**	**4.76**	4.70[q]	4.81[ab]	–	–	–	Skull Mtn NV
920629155239	–	**3.89**	–	**4.29**	**4.15**	4.15[h]	–	–	–	–	Skull Mtn NV
920629170116	–	**3.81**	–	**4.17**	**3.95**	3.95[h]	–	–	–	–	Skull Mtn NV
920630143811	5.1	**5.20**	4.8[d]	–	**5.11**	5.04[q]	5.18[w]	–	–	–	Landers CA
920630160624	–	**3.50**	3.5[g]	**3.73**	**3.50**	3.50[h]	–	–	–	–	Skull Mtn NV
920705065412	–	**4.38**	4.4[g]	**4.82**	**4.41**	4.39[q]	4.43[aa]	–	–	–	Skull Mtn NV
920705084838	–	**3.05**	–	**3.29**	**3.31**	3.31[h]	–	–	–	–	Skull Mtn NV
920711181416	5.3	**5.09**	5.7[d]	**5.61**	**5.24**	5.27[q]	5.22[w]	5.24[y]	–	–	Garlock CA
920809080706	–	**3.09**	3.2[g]	**3.42**	**3.41**	3.41[h]	–	–	–	–	Skull Mtn NV
920902102620	5.7	**5.82**	5.8[j]	**5.60**	**5.55**	5.44[q]	5.68[z]	5.33[z]	5.59[y]	5.69[k]	St George UT
920913114620	4.0	**4.06**	4.4[a]	**4.44**	**4.26**	4.32[q]	4.21[ab]	–	–	–	Skull Mtn NV
920914053550	–	**2.83**	3.0[g]	**2.94**	**3.11**	3.11[h]	–	–	–	–	Skull Mtn NV
920919230447	4.5	**4.69**	4.9[a]	**4.69**	**4.70**	4.74[q]	4.83[z]	4.53[z]	–	–	Geysers CA
920925211223	–	**3.00**	3.2[g]	**3.29**	**3.34**	3.34[h]	–	–	–	–	Skull Mtn NV
920925211638	–	**3.17**	3.1[g]	**3.44**	**3.51**	3.51[h]	–	–	–	–	Skull Mtn NV
921008122356	–	**2.86**	–	**3.22**	**3.34**	3.34[h]	–	–	–	–	Skull Mtn NV
930116062934	4.8	**4.92**	5.3[a]	**5.18**	**5.05**	5.08[q]	5.11[z]	4.93[z]	5.07[z]	5.06[ac]	Gilroy CA
930210214835	4.8	**4.41**	5.1[a]	**5.02**	**4.64**	4.69[q]	4.53[z]	4.64[z]	4.70[ac]	–	Pyramid Lake NV
930325133435	5.5	**5.83**	5.7[ad]	**5.10**	**5.62**	5.58[q]	5.55[t]	5.69[ac]	5.68[k]	–	Scotts Mill OR
930425092950	5.0	**5.16**	4.6[z]	**5.29**	**4.81**	4.81[q]	4.94[z]	4.67[z]	4.81[ac]	–	Cataract Ck AZ
930429082100	5.5	**5.68**	6.2[j]	**5.95**	**5.33**	5.27[q]	5.48[z]	5.20[z]	5.34[ac]	5.34[k]	Cataract Ck AZ
930517232049	6.0	**5.84**	6.2[d]	**6.13**	**6.11**	6.08[q]	6.10[ac]	6.08[t]	6.12[u]	6.18[k]	Eureka Valley CA
930518010306	4.6	**4.90**	5.3[a]	**5.10**	**5.10**	5.08[q]	5.12[t]	–	–	–	Eureka Valley CA

Table 2.2

Continued

yrmodahrmnsc	m_b P	m_b Pn	M_L	m_b Lg	M_w						Region
					Avg	#1	#2	#3	#4	#5	
930518234853	5.0	**4.93**	5.2[a]	**5.19**	**5.00**	4.95[q]	4.93[t]	5.04[u]	5.08[w]	–	Eureka Valley CA
930518235740	4.5	**4.58**	4.7[a]	**4.67**	–	–	–	–	–	–	Eureka Valley CA
930519141322	4.9	**5.21**	5.2[a]	**5.13**	**4.91**	4.94[q]	4.90[t]	4.95[u]	4.84[w]	–	Eureka Valley CA
930520201414	4.3	**4.32**	4.6[d]	**4.62**	**4.71**	4.73[q]	4.64[t]	4.77[w]	–	–	Coso Hot Springs CA
930528044740	4.6	**4.76**	5.2[d]	**5.34**	**4.81**	4.80[q]	4.83[t]	4.83[u]	4.84[w]	4.75[ac]	Bakersfield CA
930811223304	4.6	**5.09**	4.7[a]	**5.15**	**4.87**	4.90[q]	5.01[t]	4.74[u]	4.86[w]	4.84[ac]	Alum Rock CA
930923062110	4.3	**3.92**	4.8[a]	**4.15**	**4.66**	4.66[q]	4.65[t]	–	–	–	Klamath Falls OR
931114122534	4.1	**4.28**	4.9[a]	**4.44**	**4.86**	4.86[q]	4.86[t]	4.88[u]	4.87[w]	4.83[ac]	Parkfield CA
931204221519	5.2	**5.47**	5.4[a]	**5.69**	**5.44**	5.41[q]	5.46[t]	5.39[x]	5.48[k]	–	Klamath Falls OR
940117123055	6.4	**6.14**	6.7[a]	**6.77**	**6.73**	6.64[q]	6.77[t]	6.73[k]	6.70[ac]	6.81[ae]	Northridge CA
940118004308	5.4	**4.86**	5.5[a]	**5.75**	**5.20**	5.32[q]	5.21[t]	5.08[ae]	5.21[p]	–	Northridge CA
940320212012	5.2	**4.91**	5.3[d]	**5.48**	**5.33**	5.32[q]	5.22[t]	5.37[u]	5.39[ae]	5.37[k]	Northridge CA
940406190104	4.4	**4.70**	5.0[a]	**4.70**	**4.58**	4.67[q]	4.59[t]	4.54[u]	4.50[ae]	–	San Bernadino CA
940607133003	4.9	**5.15**	5.0[g]	**5.42**	**5.10**	4.99[q]	5.16[t]	5.16[p]	–	–	Borah Peak ID
940616162427	4.5	**5.00**	5.3[a]	**5.12**	**4.80**	4.88[q]	4.73[t]	4.78[u]	4.79[af]	–	Landers CA
940912235709	4.9	**5.01**	5.4[a]	**5.14**	**5.20**	5.19[q]	5.30[t]	5.15[u]	5.16[p]	–	Dbl Spring Flat NV
941114012823	4.2	**4.53**	4.3[a]	**4.32**	**4.64**	4.69[q]	4.62[t]	4.63[u]	4.63[p]	–	Cape Mendocino CA
941220102747	4.5	**4.69**	5.0[a]	**4.80**	**5.05**	5.10[q]	5.02[t]	5.04[u]	–	–	Parkfield CA
941226141029	5.1	**5.70**	5.3[a]	**5.16**	**5.44**	5.31[q]	5.44[u]	5.46[p]	5.56[k]	–	Cape Mendocino CA

[a] Berkeley; [b] HERRMANN, R. (pers. comm., 1992); [c] BAKUN (1984); [d] Pasadena; [e] WYSS and BRUNE (1968); [f] PATTON (1982); [g] NEIC; [h] BOLT and HERRAIZ (1983) method (unpublished results); [i] PATTON and ZANDT (1991); [j] PECHMANN, J. pers. comm., 1998; based on Wood-Anderson recordings at DUG and SLC; [k] Harvard CMT, best double-couple solution; [l] WALLACE *et al.* (1982); [m] BARKER and LANGSTON (1983); [n] JULIAN and SIPKIN (1985); [o] PRIESTLEY *et al.* (1988); [p] Berkeley M_o using BOLT and HERRAIZ (1983) method; see PDE catalog; [q] MAYEDA and WALTER (1996) coda-based M_o; [r] KANAMORI (1990); [s] BECK and PATTON (1991); [t] PASYANOS *et al.* (1996); [u] DREGER, D. (pers. comm., 1995); [v] KANAMORI *et al.* (1993); [w] THIO and KANAMORI (1995); [x] Laboratoire de Geophysique, Papeete, French Polynesia; TALANDIER *et al.* (1987); [y] RITSEMA and LAY (1993); [z] ROMANOWICZ *et al.* (1993); [aa] WALTER (1993); [ab] WALTER, W. (pers. comm., 1995); [ac] RITSEMA and LAY (1995); [ad] Seattle; [ae] THIO and KANAMORI (1996); [af] THIO, H. K. (pers. comm., 1998).

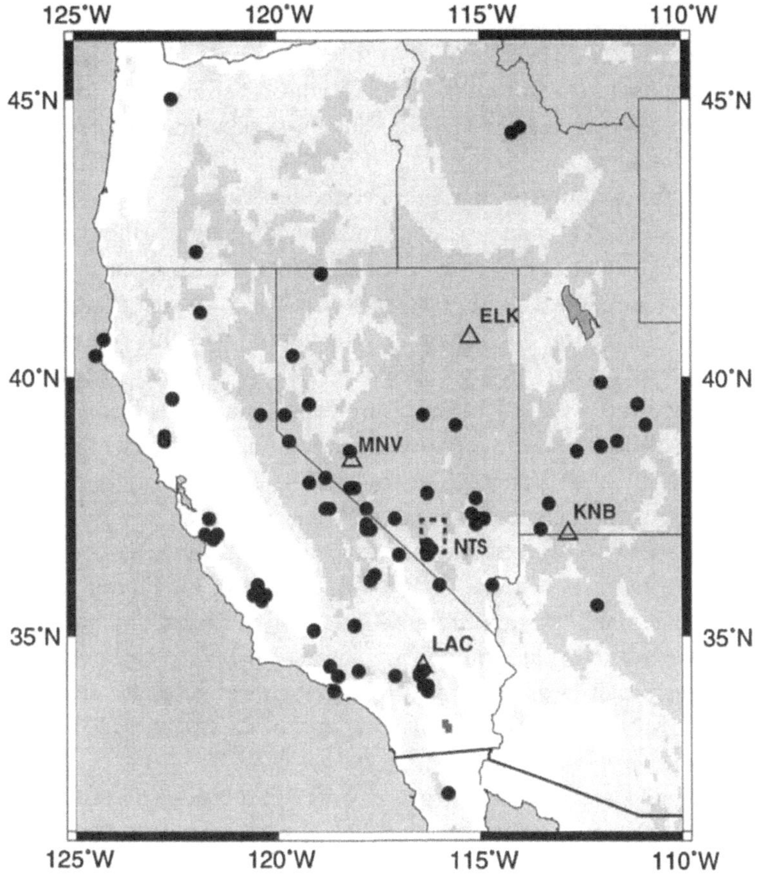

Figure 2.4
Map of WUS showing locations of events and the sites of the Lawrence Livermore National Laboratory broadband seismic stations. Shading is topography. Waveform data from these stations were used to compute $m_b(Lg)$ and $m_b(Pn)$ analyzed in this study. Also shown for reference is the Nevada Test Site.

addition, I measured RMS amplitude and RMS frequency over the full window length and calculated magnitudes based on these measures (PATTON, 1988). RMS frequencies were checked for satisfying the same frequency limits as the 3rd peak frequencies. I adopted a group velocity, U, of 3.45 km/s for the RMS measurements, while group delay times of 3rd-peak measurements were used to compute U of that measurement (U is only used in the computation of attenuation coefficient, as discussed below).

Nuttli's general $m_b(Lg)$ formula calls for making amplitude corrections based on attenuation rates appropriate for the region of propagation. This allowance for region-specific corrections was novel at the time Nuttli developed $m_b(Lg)$. The procedure uses a power-law Q model of the form, $Q = Q_0 f^\eta$, where f is frequency, η

is a frequency-independent exponent, and Q_0 is the Q at 1 Hz. Attenuation coefficient is computed using the familiar formula, $\gamma = \pi / Q \cdot U \cdot T$, where T is period. For NTS explosions, NUTTLI (1986a) and PATTON (1988) used HERRMANN (1980)s coda Q method to estimate Q_0 and η for WWSSN and Livermore stations, respectively. This worked well for explosions at NTS, but PRIESTLEY and PATTON (1997) found that adopting the same Q_0 values obtained by Nuttli and Patton led to overestimation of $m_b(Lg)$ for earthquakes located near the test site.

In this study, I used Q_0 and η based on published results for the Basin and Range from XIE and MITCHELL (1990) as a starting model for all stations in the Livermore network. I then developed station corrections, which accommodate station site effects and, to first order, differences in attenuation rates for individual stations. Starting values of Q_0 and η were 270 and 0.4, respectively. Magnitudes were computed using 3rd peak and RMS measurements, and events that had network m_b determined for all four stations were retained for further analysis discussed below. There were 23 such events for RMS $m_b(Lg)$ and 14 for 3rd-peak $m_b(Lg)$.

I determined station corrections, dm_{ij}, of the form, $dm_{ij} = sc_j + b_j \cdot \Delta_{ij}$, where $dm_{ij} = m_{ij} - \overline{m}_i$, and $\overline{m}_i$ is network magnitude for the i-th event, m_{ij} and Δ_{ij} are magnitude and epicentral distance for the i-th event and j-th station, and sc_j and b_j are site correction and perturbation on attenuation factor for the j-th station, respectively. The sc_j and b_j are unknowns. The inversion for 8 unknowns (4 stations with 2 parameters each) was performed by least-squares subjected to the constraint that $\Sigma\Sigma dm_{ij} = 0$. The Lagrange-multiplier method of CLAERBOUT (1976) was used to solve the linear-system-with-constraints problem.

Inversion results are based on measurements of m_b residuals available for events with all four stations reporting. Station-correction relationships determined by this method along with magnitude residual data are displayed in Figure 2.5. Each station has 37 observed residuals. Mina station (MNV) shows the largest corrections for site and distance effects, while Landers (LAC) shows the smallest corrections.

These station-correction relationships were used to adjust station measurements of $m_b(Lg)$ based on the initial Q model of XIE and MITCHELL (1990). Table A.1 in the Appendix provides adjusted station values and final network values for all Livermore-network events in the data set. Network values are determined from a simple average of adjusted station magnitudes. Note that I did not measure $m_b(Lg)$ if epicentral distances were less than 100 km.

Because of the robust nature of RMS measurements, most of the reported values of $m_b(Lg)$ in Tables 2.2 and A.1 are based on RMS amplitude. I found in many instances that 3rd-peak amplitudes had to be discarded because their associated frequency failed the frequency-range test, whereas the RMS frequency was still within limits. This resulted in more RMS observations, which was important for some events. However, RMS amplitudes are systematically smaller than 3rd-peak amplitude, and a relationship must be found that converts RMS-based m_b into an equivalent 3rd-peak m_b so that comparisons can be made with Nuttli's magnitudes

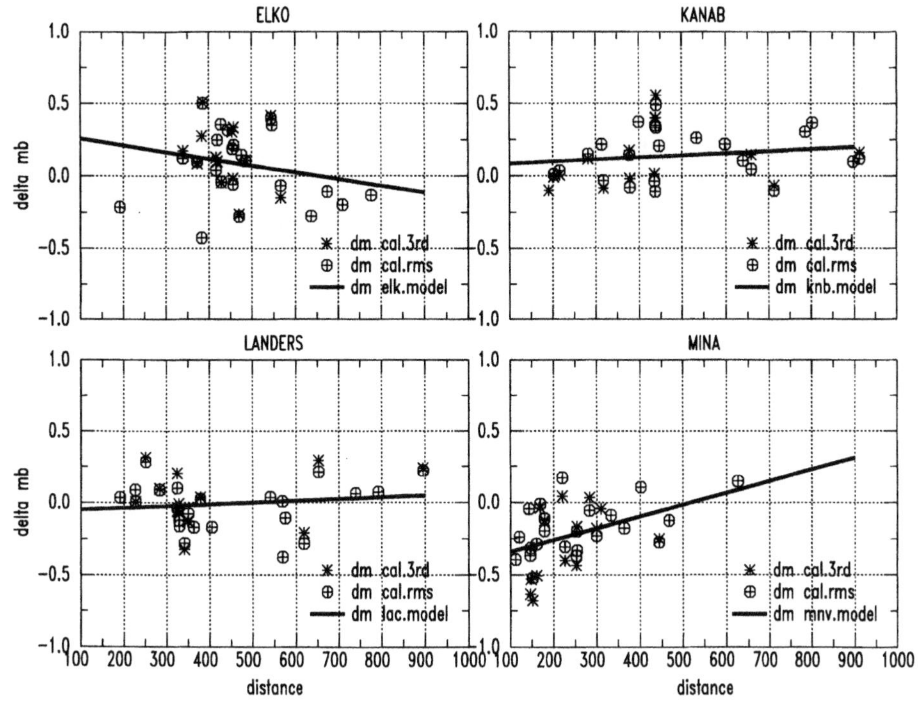

Figure 2.5
Station corrections applied to $m_b(Lg)$ for Livermore network stations. Solid lines were obtained from a linear inversion with a constraint that $\Sigma\Sigma dm_{ij} = 0$. Different symbols are used for magnitudes based on 3rd peak amplitudes and RMS amplitudes.

for ENA. Figure 2.6 shows a graph that compares network $m_b(Lg)$'s based on the two amplitude measures. They scale one-to-one, and observed offset provides an estimate of systematic difference between the 3rd-peak and RMS m_b. The offset for this data set is 0.229 mu, which is in good agreement with 0.22 mu determined for NTS explosions (PATTON, 1988) and 0.28 mu for central Asian events (PRIESTLEY and PATTON, 1997). In addition to these systematic effects, there are also differences in the apparent decay rates of RMS and 3rd-peak amplitudes with distance (PRIESTLEY and PATTON, 1997). This arises due to increasing duration of the measurement window with distance, which affects measurement of RMS amplitude as *Lg* disperses. A correction for this was obtained by comparing observed differences in decay rates. RMS amplitudes were adjusted accordingly in the $m_b(Lg)$ formula, and all RMS $m_b(Lg)$ values listed in Tables 2.2 and A.1 have been converted to equivalent 3rd-peak values using this correction and adding 0.229 mu.

Pn magnitudes. Measurements of $m_b(Pn)$ were made following the procedures of DENNY *et al.* (1987), specifically using equation (7) in their paper. Their coefficient on distance corrects for attenuation of *Pn* amplitudes in WUS, and the scale was calibrated so that the expected value of $m_b(Pn)$ is consistent with the expected value

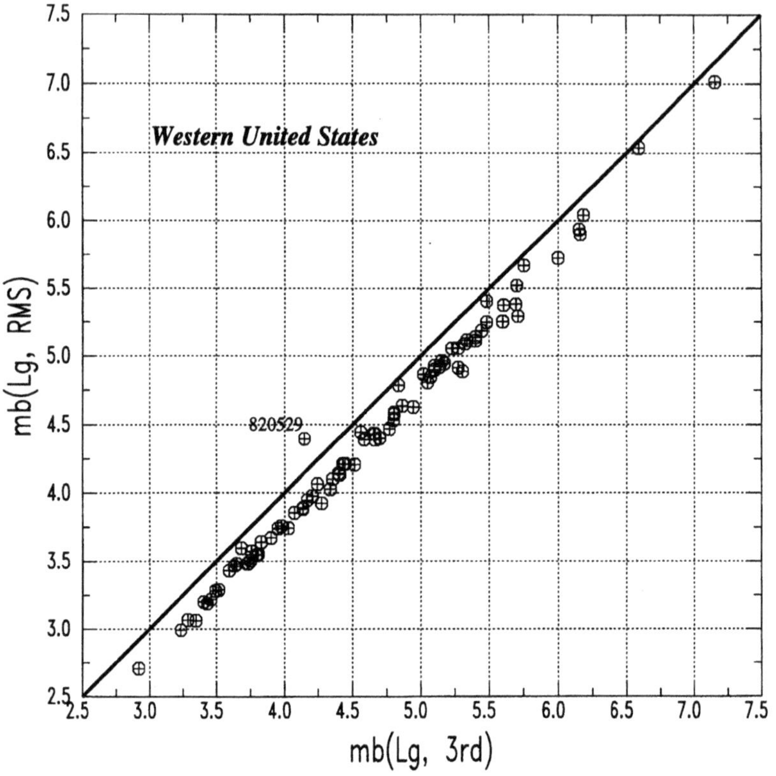

Figure 2.6

Comparison of final $m_b(Lg)$ values based on 3rd peak and RMS amplitudes; line represents unity. One notable flyer is 820529. In this case, there were three stations which gave passable 3rd peak $m_b(Lg)$ values, and the fourth station (MNV) had too long of a wave period. On the other hand, only one station (MNV) gave a passable RMS $m_b(Lg)$ value. Thus, there was no overlap of stations between the two measurements, and their $m_b(Lg)$ values ended up much more discrepant than they usually are.

of the teleseismic magnitude (i.e., $E[m_b(Pn) - m_b(P)] = 0$). Measurements of amplitude are made of simulated WWSSN short-period records, which limited the frequency range within one octave of 1 Hz for most events.

Station and network measurements of $m_b(Pn)$ are provided in Table A.1. Stations with epicentral distances less than the *Pn-Pg* crossover distance (~200 km for WUS) were not used. Magnitudes for individual stations have been adjusted for station corrections (DENNY *et al.*, 1987), and the network value is obtained by taking a simple average of all station m_b available.

Magnitude relationships. The $m_b(Pn)$ formula developed by Denny *et al.* should provide magnitudes consistent with teleseismic m_b determinations. In addition, HERRMANN and NUTTLI (1982) found evidence for equivalence between $m_b(Lg)$ and M_L for a limited data set in WUS. Recently, KIM (1998) provides evidence in support of equivalence of $m_b(Lg)$ and M_L scales in ENA. Here I make comparisons between

$m_b(Pn)$ and $m_b(P)$ and between $m_b(Lg)$ and M_L to test magnitude equivalence for events in this data set. These comparisons also serve a quality control function, as large discrepancy between equivalent magnitudes may identify unreliable estimates.

Comparisons between $m_b(Lg)$ and M_L are shown for 82 events in Figure 2.7. M_L magnitudes were taken from a variety of sources identified by event in Table 2.2. The results scale one-to-one over the entire magnitude range, and the offset is insignificant, −0.011 ± 0.020 mu. Events with $|m_b(Lg) - M_L|$ greater than 0.3 mu are identified on the plot in Figure 2.7. One or the other magnitude may be poorly determined, however it should be pointed out that this test may not detect events where both magnitudes are anomalously high or low for some reason.

Figure 2.8 shows a graph comparing $m_b(P)$ and $m_b(Pn)$ for 61 events in the WUS. All measurements of $m_b(P)$ are taken from the PDE catalog. The comparison shows that these magnitudes scale one-to-one, and $E[m_b(Pn) - m_b(P)]$ equals +0.050 ± 0.041 mu. Events with $|m_b(Pn) - m_b(P)|$ greater than 0.4 mu are identified on the graph. By comparing these events with those showing large discrepancies on the

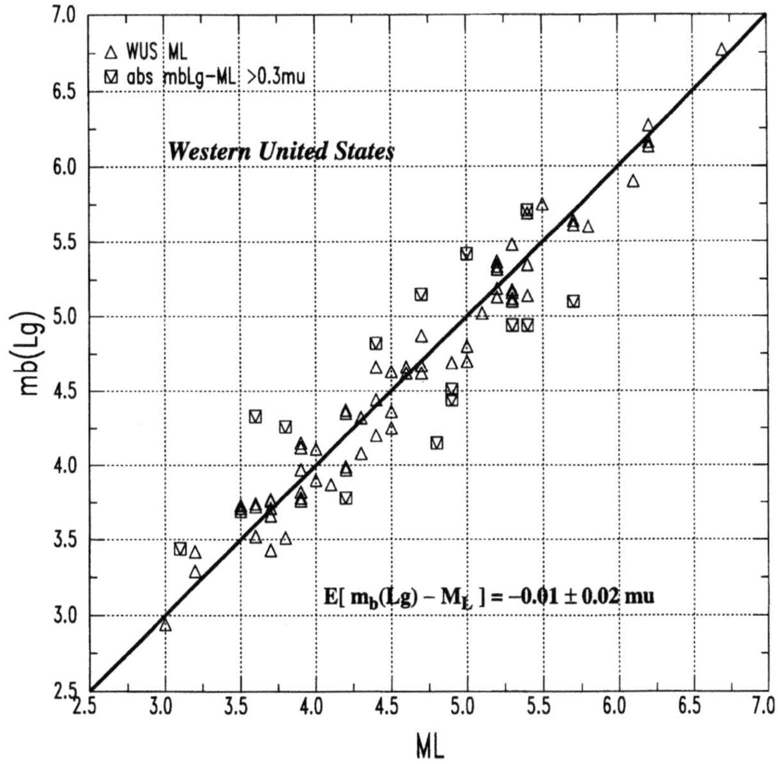

Figure 2.7
A plot of M_L versus $m_b(Lg)$ for 82 events in WUS; line represents unity. Inverted triangles enclosed in squares are events where difference $|m_b(Lg) - M_L|$ exceeds 0.3 mu.

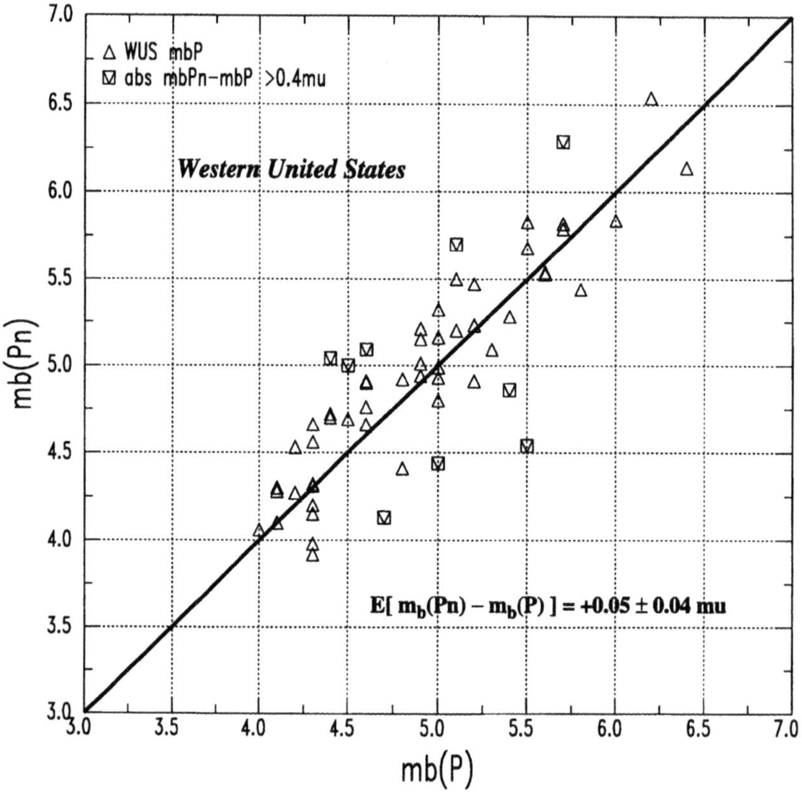

Figure 2.8
A plot of $m_b(P)$ versus $m_b(Pn)$ for 61 events in WUS; line represents unity. Inverted triangles enclosed in squares are events where difference $|m_b(Pn) - m_b(P)|$ exceeds 0.4 mu.

$m_b(Lg) - M_L$ plot, I could identify events that may have had unnoticed gain problems on digital recordings since both $m_b(Pn)$ and $m_b(Lg)$ would have been affected by gain errors in the same way. Out of 18 events, the results showed that 14 were negatively correlated, which is inconsistent with discrepancies caused by errors in the gain correction. I dismissed the role of gain errors for the remaining four events.

WUS scaling relationships for M_w:$m_b(Lg)$ are shown in Figure 2.9 for 87 events. M_w for these events were obtained from regional and teleseismic modeling studies of long-period radiation, and literature references are provided in Table 2.2 footnotes. When more than one estimate of M_w was available, I took a simple average for use in subsequent analysis. A least-squares regression line for all 87 data points is shown with the data points. For comparison, I show two other relationships based on M_L. Having established the equivalence between M_L and $m_b(Lg)$, these comparisons should be of some interest. CHUNG and BERNREUTER (1981) based their results on observations for events in California and a few events from the California-Nevada border region. The number of observations in their study is small for M_L 2–4, but

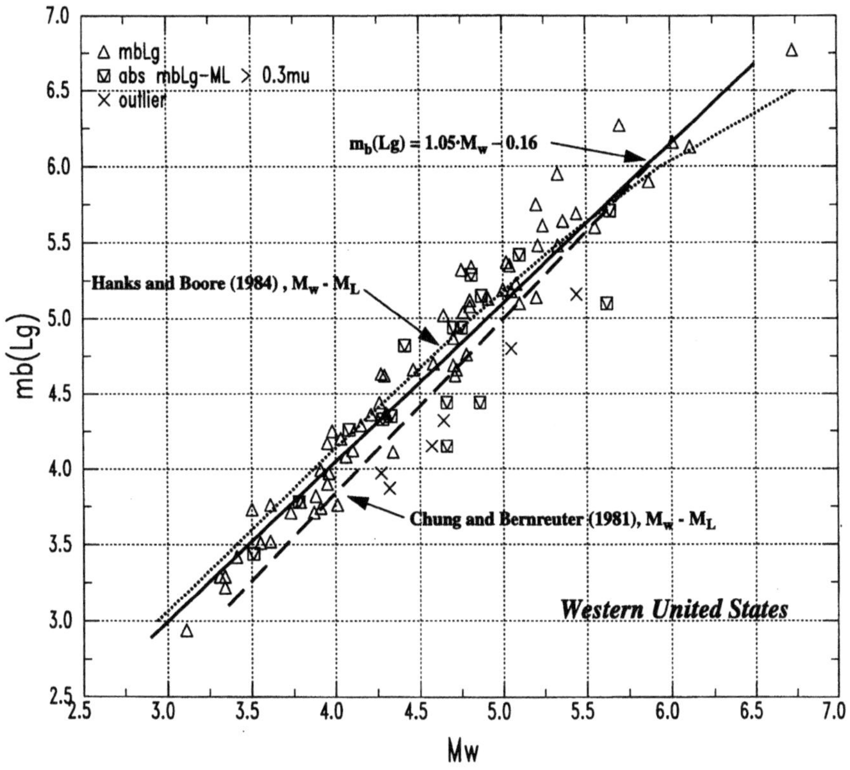

Figure 2.9

Scaling of $m_b(Lg)$ with M_w for events in WUS. Shown are data points for 86 events (Landers mainshock plots off-scale), where inverted triangles enclosed in squares are events which gave large differences in Figure 2.7. Events plotted with x's refer to outlier events located in anomalous source regions or have $|m_b(Lg) - M_L|$ large enough to suspect $m_b(Lg)$ estimates (see text). Solid line is regression result for all 86 events; curved, dotted line represents the relationship between M_w and M_L for earthquakes in central California (HANKS and BOORE, 1984); long-dash line is result of regressing log M_o versus M_L for earthquakes located throughout California (CHUNG and BERNREUTER, 1981).

increases significantly below M_L 2 (see their Fig. 6). A considerably larger M_o–M_L data set for central and northern California was analyzed by HANKS and BOORE (1984). Their sampling is ample for M_L 0.5–5 and limited above M_L 5. In any case, both of these published relationships show reasonably good agreement with $M_w - m_b(Lg)$ data from this study.

In Figure 2.9, I selected a different symbol to indicate events with large discrepancies between M_L and $m_b(Lg)$. Some of these events may have poorly determined M_L or $m_b(Lg)$ values because of propagation effects (e.g., structural blockages and/or inadequate corrections for propagation on long paths) or possibly source effects, although source effects would be expected to affect both magnitudes similarly. Events with large $[m_b(Lg) - M_L]$ residuals and which plot far from the regression curve in Figure 2.9 are particularly suspicious for their $m_b(Lg)$. For example, events 920506 (Joshua Tree, CA), 930325 (Scott's Mill, OR), 930923 (Klamath Falls, OR), and 931114 (Parkfield, CA) have small $m_b(Lg)$'s for their M_w and all four have large $[m_b(Lg) - M_L]$ discrepancies (−0.39, −0.60, −0.65, and −0.46 mu, respectively). M_L estimates for these events give far better agreement with the regression line.

In addition to these events, six other events are also outliers, but their $m_b(Lg)$ values were not classified as highly inconsistent with M_L. In fact, three of the six (Cape Mendocino events of 941114 and 941226 and Geyser event of 820529) are located in areas where events produce anomalously low M_L estimates for their M_o (Dreger, pers. comm. 1999; PASYANOS *et al.*, 1997). The other three events (800428, 820621, and 941220) have M_L values larger than their $m_b(Lg)$ values, but $[m_b(Lg) - M_L]$ differences (−0.23, −0.23, and −0.20 mu, respectively) fell below the 0.3 mu threshold.

Scaling of $m_b(Pn)$ with M_w is shown in Figure 2.10, and outlier events (e.g., $|m_b(Pn) - m_b(P)| > 0.4$ mu) are identified for reference. Very few of those events appear to be outliers on this plot, which suggests that $m_b(P)$ estimates for those events are suspect (e.g., noise and/or path effects). Regression line was obtained by a fit to all of the data, and is compared with the results of PATTON and WALTER (1994). The present data set is substantially larger than the one compiled by Patton and Walter, which had limited sampling at small and large magnitudes.

2.3. Central Asian (CA) Data Set

Data. Events in this data set are located in the Tien Shan fold belt and neighboring geologic structures of western China, Mongolia and eastern Kazakhstan. The map in Figure 2.11 shows epicenters of these events, as well as the location of the Chinese Digital Seismic Network (CDSN) station Urumqi (WMQ), and Chinese and former Soviet test sites at Lop Nor (LPN) and Eastern Kazakhstan (KTS), respectively. While subcrustal earthquakes are known to occur in this region, the vast majority of events in this data set are at crustal depths, which can reach thicknesses greater than 60 km (FIELDING *et al.*, 1994). The events are listed in Table 2.3, and I split them into six geographical regions which are shown on the map and identified for each event on the table. This grouping is somewhat arbitrary, but it will serve the purpose for analysis later in the paper.

Magnitudes for 74 events in Table 2.3 range from about 3.8 to more than 6.5. This range is smaller than the range available for other tectonic regions, as minimum magnitudes reach close to 3.0 for both WUS and ENA. However, there is a better representation of large events from CA. Measurements of regional magnitudes and M_o were available for 59 events from the studies of PRIESTLEY and PATTON (1997) and PATTON (1998). I computed $m_b(Pn)$, $m_b(Lg)$ and M_o values for an additional 15 events from the study of CONG *et al.* (1996). Final M_o values for these events were obtained by averaging estimates of CONG *et al.* and estimates obtained using the

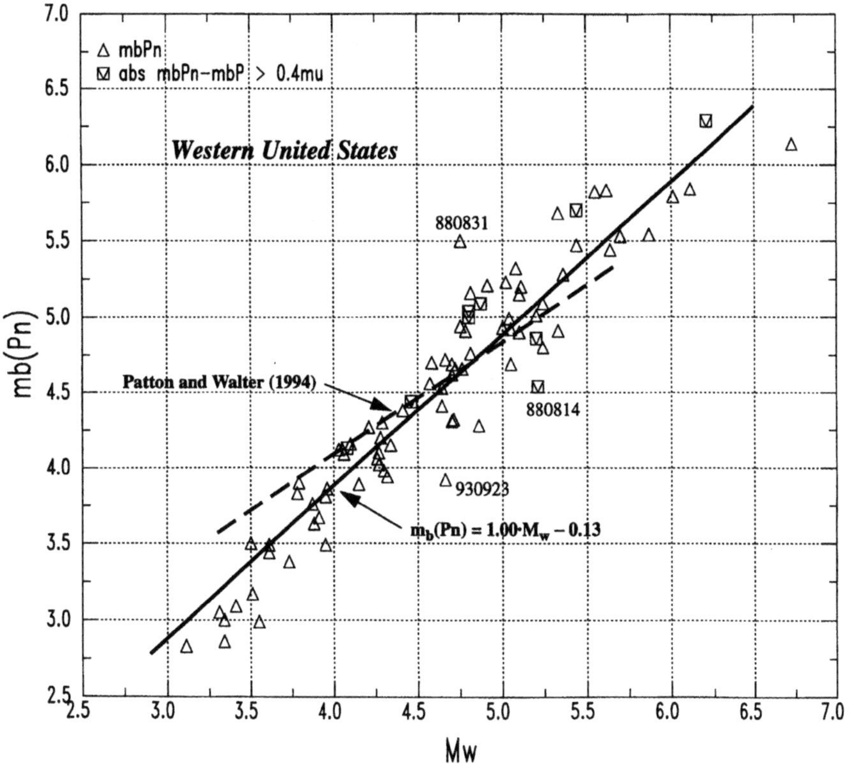

Figure 2.10
Scaling of $m_b(Pn)$ with M_w for events in WUS. Shown are data points for 86 events (Landers mainshock plots off-scale); inverted triangles enclosed in squares are events which gave differences $|m_b(Pn) - m_b(P)|$ greater than 0.4 mu. In addition, I identify a few events that appear to be outliers on the scaling plot. Solid line is regression result for all 86 events; long-dash line is M_w:$m_b(Pn)$ scaling relationship from PATTON and WALTER (1994).

BOLT and HERRAIZ (1983) method developed for this region of CA by Priestley and Patton and improved by PATTON (1998). For three events (900121, 93100223, 931007), M_0 estimates differed considerably, and I chose to use the estimates obtained by the Bolt and Herraiz method.

The $m_b(Pn)$ scale was calibrated against $m_b(P)$ reported by the International Seismological Centre (ISC) using a procedure similar to DENNY *et al.* (1987), and the m_b values in Table 2.3 are based on peak-to-peak amplitude measurements (see Table 2a in PRIESTLEY and PATTON, 1997). Measurements of $m_b(Lg)$ are based on RMS amplitudes, converted to equivalent 3rd-peak values by adding 0.281 mu.

Magnitude relationships. Figure 2.12 shows $[m_b(Pn) - m_b(P)]$ data plotted against M_w. The average difference equals 0.060 ± 0.039 mu, and omitting seven outlying events with differences greater than 0.5 mu, reduces the average to only 0.03 mu. As expected, the average difference between $m_b(Pn)$ and $m_b(P)$ is small.

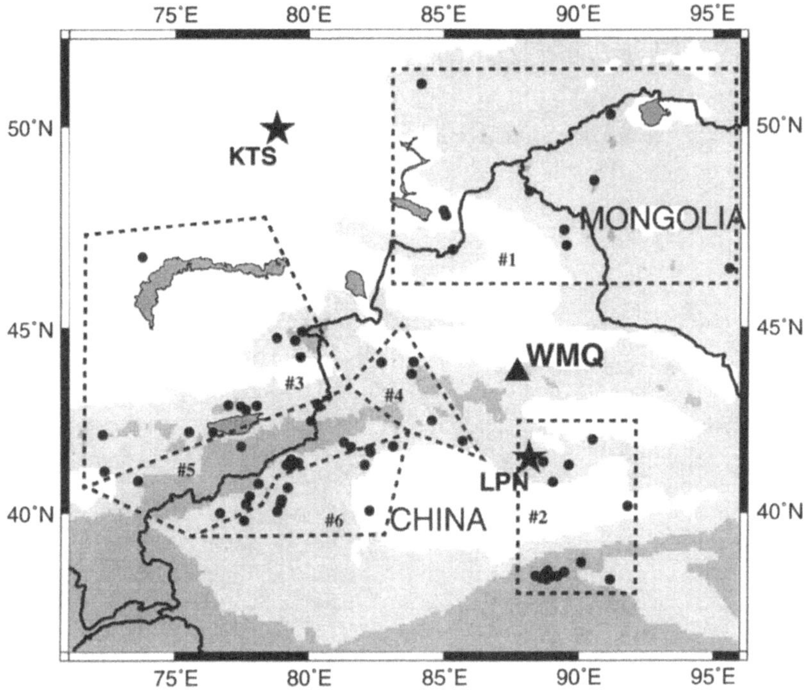

Figure 2.11
Map showing epicentral locations of events in this study, the location of CDSN station (WMQ) and the location of the former Soviet test site at Semipalatinsk, Kazakhstan (KTS) and the Chinese test site at Lop Nor (LPN). Shading is topography. Polygons are source areas referred to in the text and in Table 2.3: 1) Siberian Fold Belt, 2) Eastern Tarim Basin, 3) Northern Tien Shan, 4) Central Tien Shan, 5) Tien Shan Fold Belt, and 6) Northwestern Tarim Basin.

Scaling of $m_b(Pn)$ with M_w is shown in Figure 2.13, where outlying events in Figure 2.12 are identified. Only a few of these events are considered outliers on the $m_b(Pn)$ scaling plot. These events (881215, 900121, 930526) arguably have suspect $m_b(Pn)$ estimates caused by source and/or path effects. Scaling of $m_b(Lg)$ is shown in Figure 2.14. The data are well behaved with no outlying events and no apparent increase in scatter with decreasing M_w, which is apparent in the *Pn* data. Signal-to-noise ratios were good for both regional phases for most events.

2.4. Data Set Summary

Comparisons of scaling relationships are summarized on three graphs in Figures 2.15 and 2.16. Figure 2.15 shows scaling of $m_b(Lg)$, where data points and regression lines are plotted for all three tectonic regions. Figure 2.16a exhibits scaling of $m_b(P)$, where data points and associated regressions lines are shown for WUS and CA, and just the regression line without data points for ENA. Figure 2.16b shows

Table 2.3

Magnitudes for Central Asian earthquakes

yrmodahrmnsc	m_b P	m_b Pn	m_b Lg	M_w	Region
870105225246	5.8	5.77	6.07	5.52[a]	Tien Shan Fold Belt
870124080921	5.8	6.08	6.61	6.18[a]	Tien Shan Fold Belt
870124085935	4.3	4.18	4.64	4.27[b]	Tien Shan Fold Belt
870125061057	4.5	4.26	4.24	4.01[b]	Tien Shan Fold Belt
870125224914	4.5	4.15	4.06	3.90[b]	Tien Shan Fold Belt
870126052841	4.4	4.11	4.23	4.01[b]	Tien Shan Fold Belt
870128000141	4.6	4.33	4.01	3.94[b]	Tien Shan Fold Belt
870214230315	4.4	4.36	4.33	4.17[b]	Tien Shan Fold Belt
870223223107	4.6	4.46	4.49	4.19[b]	Tien Shan Fold Belt
870225195635	5.7	5.61	6.16	5.74[a]	Eastern Tarim Basin
870303094131	5.1	4.80	5.28	5.03[a]	Tien Shan Fold Belt
870404174514	4.0	4.42	4.09	4.16[b]	Tien Shan Fold Belt
870510201933	4.5	4.11	4.30	4.47[b]	Northern Tien Shan
870805102418	4.8	4.99	5.30	4.94[b]	N. Western Tarim Basin
870918215840	5.4	5.37	5.27	5.35[a]	S. Siberian Fold Belt
870920035407	4.6	4.72	4.68	4.51[b]	Northern Tien Shan
871016183049	4.8	4.38	4.44	4.56[b]	Central Tien Shan
871217121723	5.1	4.28	4.83	4.53[b]	N. Western Tarim Basin
871222001639	5.8	5.71	6.11	5.33[a]	Eastern Tarim Basin
880208174920	4.4	4.66	4.33	4.35[b]	Central Tien Shan
880301154539	4.3	4.19	4.00	3.89[b]	Tien Shan Fold Belt
880315155523	4.4	4.30	4.40	4.28[b]	Northern Tien Shan
880325020756	4.5	4.64	4.71	4.53[b]	Northern Tien Shan
880401012716	4.5	4.43	4.64	4.75[b]	S. Siberian Fold Belt
880502021326	4.9	5.18	5.18	4.93[b]	N. Western Tarim Basin
880525000523	4.7	5.10	4.64	4.37[b]	Tien Shan Fold Belt
880525182155	5.2	5.29	5.52	5.22[b]	Central Tien Shan
880617133043	5.2	5.61	5.82	5.40[a]	Northern Tien Shan
880630152513	5.0	4.70	5.31	4.82[b]	S. Siberian Fold Belt
880723073810	5.4	6.12	6.43	5.80[a]	S. Siberian Fold Belt
880927161136	4.5	4.57	4.48	4.19[b]	Northern Tien Shan
881215064049	5.2	4.66	5.52	5.17[a]	S. Siberian Fold Belt
881221082105	5.4	4.92	5.43	4.72[b]	Northern Tien Shan
890105174222	4.2	4.41	4.35	4.54[b]	Eastern Tarim Basin
890212234914	4.6	4.71	4.96	4.65[b]	S. Siberian Fold Belt
890215012330	4.3	4.69	4.50	4.56[b]	Central Tien Shan
890314084202	4.5	4.48	4.19	4.13[b]	N. Western Tarim Basin
890323123850	4.4	4.66	4.31	4.16[b]	Tien Shan Fold Belt
890508000314	4.5	4.75	4.56	4.41[b]	Northern Tien Shan
891024133328	4.6	4.32	4.50	4.21[b]	N. Western Tarim Basin
891024133827	4.7	4.38	4.44	4.24[b]	N. Western Tarim Basin
891201173243	4.3	4.33	4.08	4.07[b]	N. Western Tarim Basin
900121075331	4.6	5.56	5.33	4.83[b]	Eastern Tarim Basin
900202140427	4.3	4.31	4.41	4.17[c]	Northern Tien Shan
900503100219	4.7	5.18	4.65	4.46[c]	Northern Tien Shan
900614124728	6.1	6.40	6.39	6.47[a]	S. Siberian Fold Belt
900803091504	6.0	6.30	6.24	6.17[a]	S. Siberian Fold Belt
900919080558	4.4	4.71	4.69	4.21[c]	Eastern Tarim Basin

Table 2.3

Continued

yrmodahrmnsc	m_b P	m_b Pn	m_b Lg	M_w	Region
901024233815	5.2	4.85	5.56	5.00[a]	Central Tien Shan
901103172514	5.1	4.74	5.41	4.88[c]	Eastern Tarim Basin
901112122850	5.8	6.14	6.32	5.89[a]	Northern Tien Shan
901201180929	5.0	4.88	5.55	4.99[a]	Tien Shan Fold Belt
910225143028	5.5	6.17	6.27	5.92[a]	N. Western Tarim Basin
910819060551	5.5	5.49	5.31	5.24[a]	S. Siberian Fold Belt
920610023701	4.1	4.54	4.67	4.21[c]	Eastern Tarim Basin
921224050947	5.2	5.28	5.65	5.39[a]	Northern Tien Shan
930217020026	5.0	5.27	5.53	4.87[c]	Eastern Tarim Basin
930526141119	4.2	4.82	4.32	4.08[c]	Eastern Tarim Basin
931002084232	6.1	6.26	6.61	6.06[a]	Eastern Tarim Basin
931002092013	4.8	4.40	4.90	4.41[c]	Eastern Tarim Basin
931002094319	5.6	5.69	6.00	5.27[c]	Eastern Tarim Basin
931002172333	5.5	5.59	5.93	5.13[c]	Eastern Tarim Basin
931002191643	3.7	4.16	4.43	3.99[c]	Eastern Tarim Basin
931002235002	4.7	5.03	4.90	4.58[b]	Eastern Tarim Basin
931007032701	4.9	5.28	5.41	4.86[b]	Eastern Tarim Basin
931012204922	4.7	4.66	4.88	4.40[c]	Eastern Tarim Basin
931230142402	5.7	5.51	5.60	5.46[a]	Northern Tien Shan
940509091412	4.8	5.01	4.78	4.89[a]	N. Western Tarim Basin
940823141831	5.0	5.07	4.82	4.92[a]	N. Western Tarim Basin
950926043903	5.2	5.18	5.52	4.95[a]	Tien Shan Fold Belt
951101122926	5.3	4.91	5.23	4.94[a]	Tien Shan Fold Belt
960118093351[d]	5.2[d]	5.37	5.42	4.81[a]	Tien Shan Fold Belt
960312184342[d]	5.6[d]	5.60	5.67	5.54[a]	S. Siberian Fold Belt
960319150026[d]	5.7[d]	6.60	6.78	6.21[a]	Tien Shan Fold Belt

[a] intermediate-period surface waves, PATTON (1998).
[b] method of BOLT and HERRAIZ (1983); see PRIESTLEY and PATTON (1997).
[c] log average M_0 determined by CONG *et al.* (1996) and BOLT and HERRAIZ (1983) method.
[d] PDE catalog, magnitude.

$m_b(Pn)$ data and regression results for WUS and CA. From visual inspection of Figures 2.15 and 2.16, it is apparent that $m_b(Lg)$ observations are the most homogeneous among the three magnitudes. All 103 ENA events (including high frequency measurements) and all 74 CA events are plotted. For WUS, 76 (out of a total of 87) events are plotted, where 10 events were excluded because of $m_b(Lg) - M_L$ comparisons or for other reasons mentioned above. In addition, the Landers mainshock with 7.4 M_w and 7.2 $m_b(Lg)$ was omitted. The Northridge mainshock is the largest event on the plot with 6.7 M_w and 6.8 $m_b(Lg)$. Consistency between the scaling relations is greatest at larger magnitudes ($M_w > 4.0$). Regression results for ENA and WUS reflect an increasing separation between m_b at small magnitudes ($M_w < 4.0$).

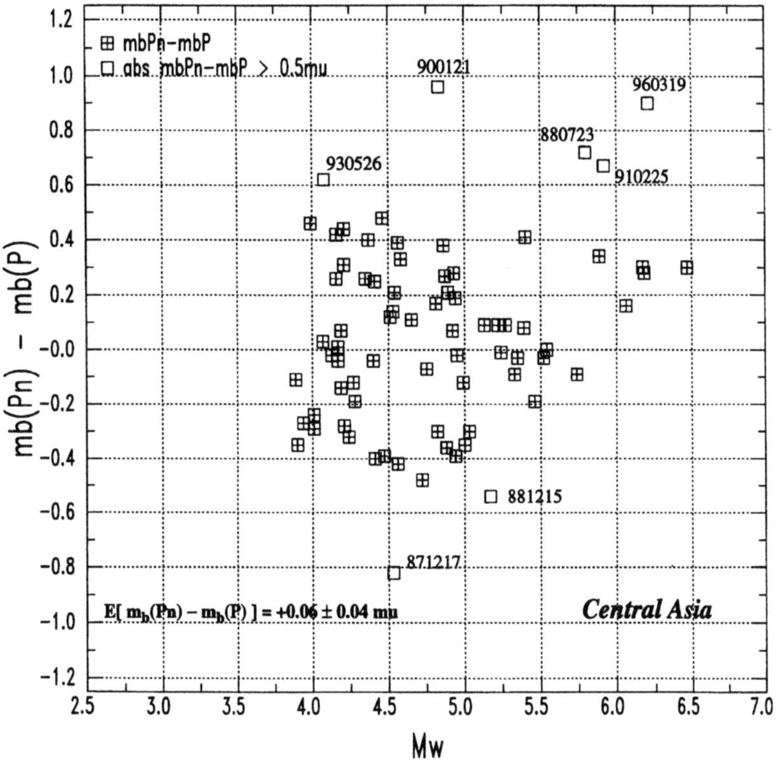

Figure 2.12
Differences, $[m_b(Pn) - m_b(P)]$, plotted against M_w for all 74 events in the CA data set. Outliers (i.e., differences > 0.5 mu) are plotted with open squares with events identified.

Observations of $m_b(P)$ scaling in Figure 2.16a are perhaps the most heterogeneous in appearance of the three magnitudes. Regression line for ENA is computed for 59 out of 63 events with reported teleseismic m_b values. Scaling results for CA are based on 70 events, 4 events being removed because of magnitude saturation. Other events suggest effects of magnitude saturation too. Measurements of $m_b(P)$ for WUS introduce the most heterogeneity because they are offset from observations for ENA and CA. WUS regression line is based on 49 events located in the Basin and Range province and California (13 events located in Utah, Arizona, Idaho, and northern Oregon were removed). For a given M_w, $m_b(P)$ for California and Basin and Range earthquakes are smaller by as much as 0.4 mu compared to events in ENA and CA. These results illustrate the well-known regional bias in teleseismic m_b for WUS (see REGIONAL $m_b(P)$ BIAS section).

In Figure 2.16b, differences in $m_b(Pn)$ scaling for WUS and CA are greater than they are for *Lg*, yet $m_b(Pn)$ scaling coefficients are not significantly different at the 95% confidence level (see next section and DISCUSSION). Regression results for CA are based on 71 of 74 events, while WUS results are based on 82 of 86 events with

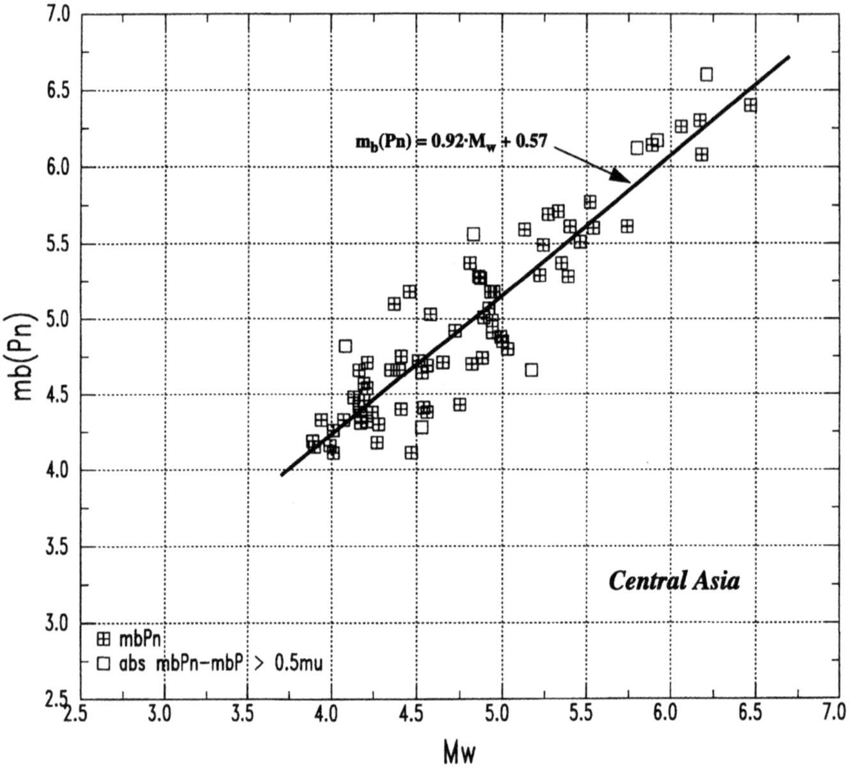

Figure 2.13
Scaling of $m_b(Pn)$ with M_w for all events in CA data set. Outliers identified in Figure 2.12 are plotted as open squares. Solid line is regression result obtained by fitting all 74 events.

$m_b(Pn)$ available. One of the WUS events removed is the Landers mainshock with 7.4 M_w and 6.5 $m_b(Pn)$. This event and the Northridge mainshock are probably showing effects of magnitude saturation. Measurements of $m_b(Pn)$ for CA events show more scatter than WUS, which is most evident at small magnitudes where sample sizes are the largest. *Pn* amplitudes can be affected by source radiation pattern (ZHAO and EBEL, 1991), which may explain why single-station measurements of $m_b(Pn)$ at WMQ show more scatter than $m_b(Lg)$. There seems to be little evidence for radiation pattern effects on 1-Hz *Lg* waves.

3. Statistical Analysis of Regional m_b Scaling

The purpose of this section is to place inter-region comparisons of m_b scaling on solid statistical ground. We would like to ascertain whether the relationships for one region are coincidence with those in another. If coincidence can be established from statistical tests of the data, this will serve to answer questions related to

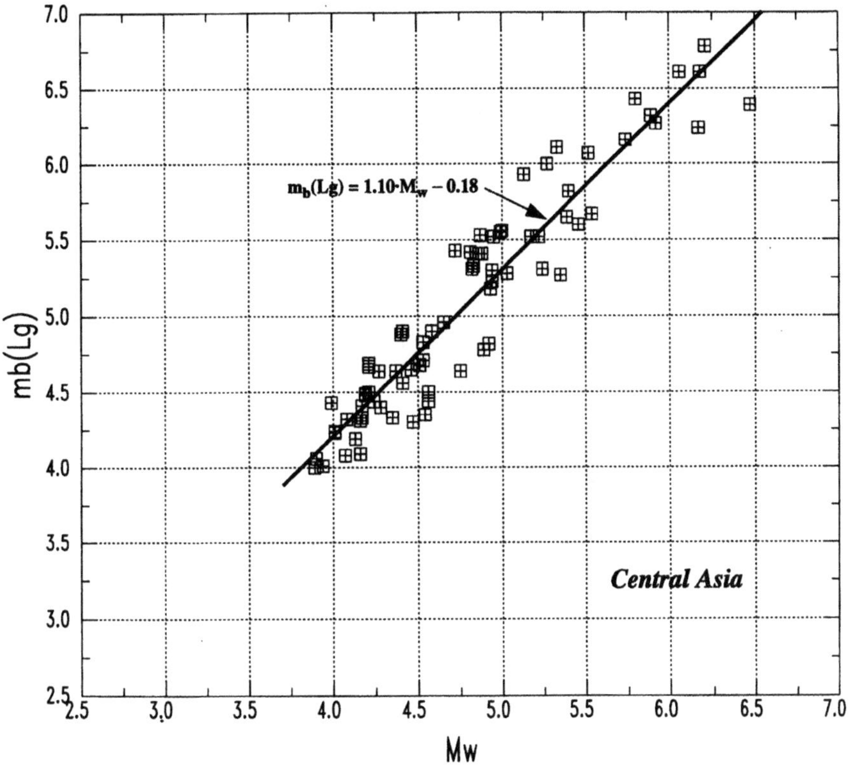

Figure 2.14
Scaling of $m_b(Lg)$ with M_w for all events in CA data set. Solid line is regression result obtained by fitting all 74 events.

transportability of m_b scales, such as the validity of Nuttli's calibration equation for continental regions in general.

Method. The statistical method is summarized in Figure 3.1. In the figure I show two end-members models: one is the most general, where scaling is unique for each region considered; that is, slopes and intercepts take on significantly different values. The other end-member is the most stringent model, where scaling is coincident in all regions; in that case estimates of slopes and intercepts are the same, statistically speaking. These end-members are referred to as Model A and Model B, respectively. One can view Model B as a "unifying" model in the sense that, for the present application, it would unite continental regions with a transportable m_b scale. A statistical test comparing the models can be devised in the form of a null hypothesis, H_0: Model B explains observations of M_w versus m_b for different regions as good as an alternative model, in this case Model A. This test takes the form of a standard F-test (WEISBERG, 1980, section 7.4) involving the ratio of variances, where the ratio is small if the null hypothesis is true. Residual sum of squares (RSS) of the alternative model (RSS_A) is the total RSS from regressions for each region, and like-wise the

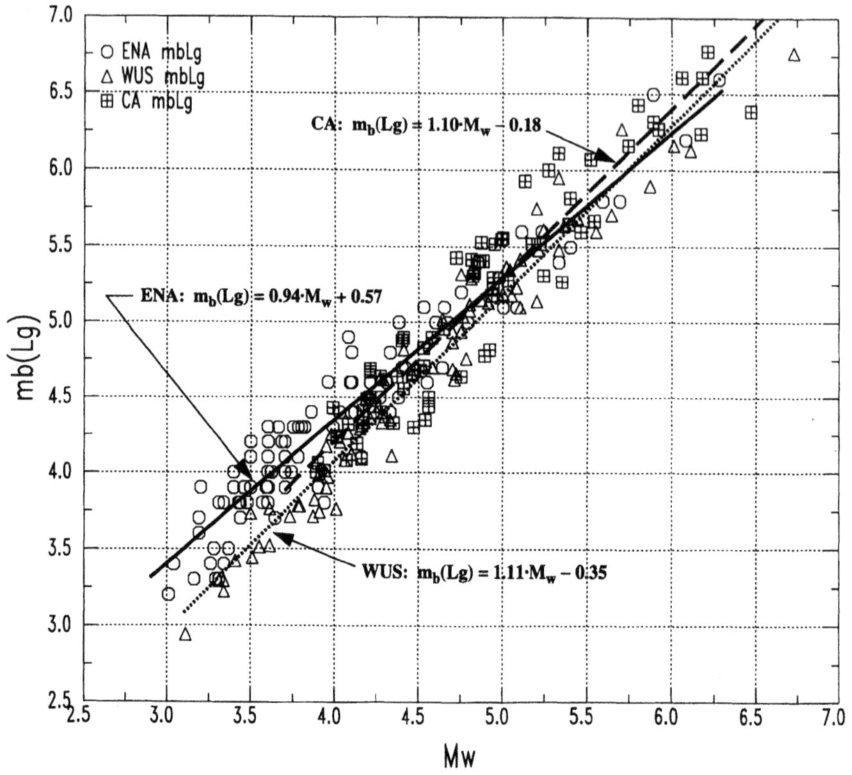

Figure 2.15
Scaling of $m_b(Lg)$ with M_w for three tectonic regions. Regression lines for each region were obtained by fitting the plotted data; 103 events for ENA, 74 events for CA, and 76 events for WUS.

degrees of freedom (*DF*) of the alternative model (DF_A) is the total *DF* from regressions for each region. The computed *F* value ($F_{B:A}$) is compared with critical values from tabulated $F(DF_B - DF_A,\ DF_A)$ distributions for selected confidence levels. Quoted critical values are for 95% level of confidence.

Each magnitude scale, $m_b(P)$, $m_b(Pn)$, and $m_b(Lg)$, will be tested below using data presented in Figures 2.15 and 2.16 of the previous section. I will first consider two-region comparisons, for which there are three possibilities for ENA, WUS, and CA data. After reviewing the two region results, I consider three-region comparisons for $m_b(P)$ and $m_b(Lg)$. Comparison tests are made as a function of M_w range, where I held the high limit fixed at 6.5 and allowed the lower limit to vary. This will identify a size range where scaling is coincident.

Application. Linear regression results for all cases are summarized in Table 3.1. This table identifies magnitude, region, and M_w range fit by least-squares. The number of data points in each regression may be obtained by adding 2 to the *DF*. One standard deviation error estimates are provided on the table in parentheses next to slope and intercept values.

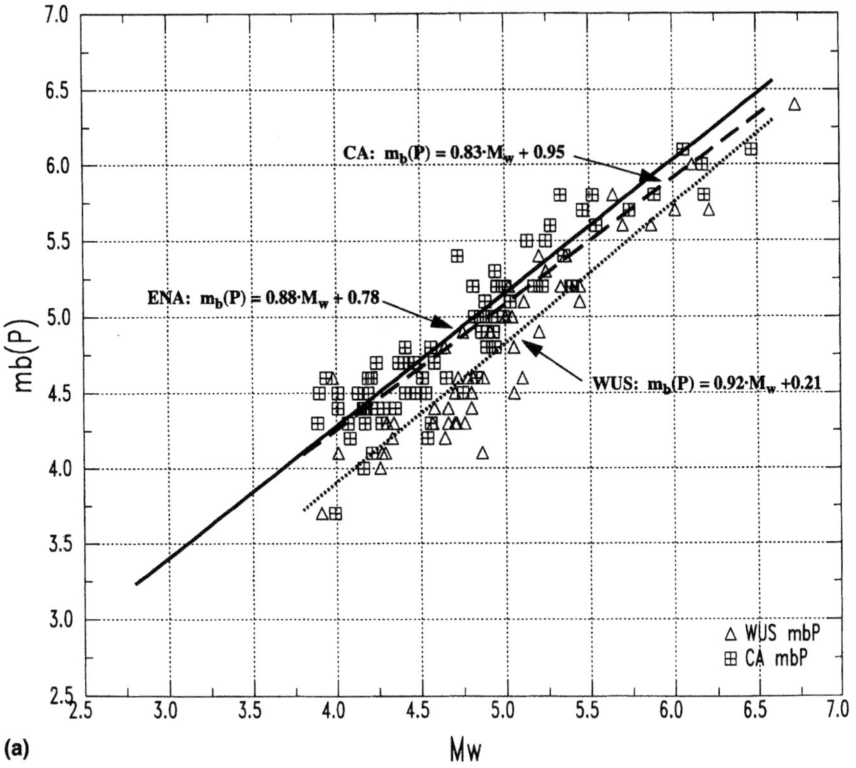

Figure 2.16
Scaling of $m_b(P)$ and $m_b(Pn)$ with M_w. (a) $m_b(P)$; regression lines were obtained using measurements plotted for WUS and CA. Regression line for ENA spans M_w range of the data in Figure 2.3. (b) $m_b(Pn)$; regression lines were obtained using a total of 82 events for WUS and 71 events for CA.

Hypothesis test results are summarized in Table 3.2 and graphically in Figure 3.2. There are a total of 17 test cases split between $m_b(P)$, $m_b(Pn)$, and $m_b(Lg)$. I quote two critical values in Figure 3.2, one for the two-region comparisons and the other for the three-region comparisons, based on 2 and 120 *DF* and 4 and 120 *DF*, respectively (values do not vary much for *DF* in the denominator between 60 and ∞). The null hypothesis can be rejected at the 95% confidence level for *F* values above these critical values. That is, the data are inconsistent with coincident scaling relationships because of scaling differences and/or regional offsets (biases).

Results for $m_b(P)$ demonstrate that the null hypothesis can be rejected for all two-region comparisons involving WUS, regardless of the M_w range (cases 1.a–c). Furthermore, the null hypothesis is rejected for both three-region cases tried. This confirms what has been known from previous research concerning anomalously small *P* waves from sources located in WUS and recorded elsewhere, or vice versa. A regional bias is present in WUS such that teleseismic *P*-wave amplitudes are reduced

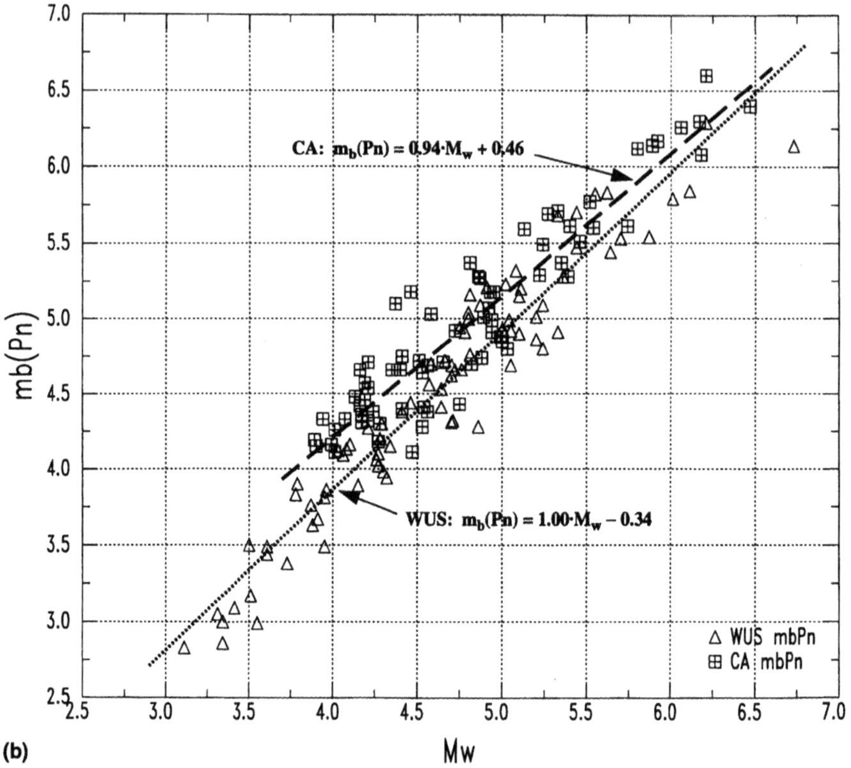

Figure 2.16b

by 0.3–0.4 mu (see REGIONAL $m_b(P)$ BIAS section), which is clearly seen for the data plotted in Figure 2.16a. Hypothesis tests are positive for ENA and CA in the 3.9–6.5 M_w range (case 1.d), and a single scaling relationship is acceptable within the scatter of the observations.

Only two cases were considered for $m_b(Pn)$ data, and both show that the null hypothesis can be rejected for WUS and CA. This may not be surprising since $m_b(Pn)$ scales were calibrated using $m_b(P)$ data which failed most hypothesis tests too. In the 4.5–6.5 M_w range, the F value for $m_b(Pn)$ is considerably smaller than it is for $m_b(P)$. This reflects differences in scaling for $m_b(P)$ and $m_b(Pn)$, as seen in Figure 2.16. It should be noted that the $m_b(P)$ data set for WUS includes events located in California and active areas of the Basin and Range only. On the other hand, the $m_b(Pn)$ data set was not winnowed, except for a few outliers discussed in the previous section.

Lg results provide the strongest evidence for magnitude scaling differences between regions, yet the data are consistent with unified scaling relationships for limited ranges of M_w. F values for ENA-WUS comparisons in the 3.0–6.5 M_w range (case 3.a) are the largest of all tests. Clearly, these data support different magnitude

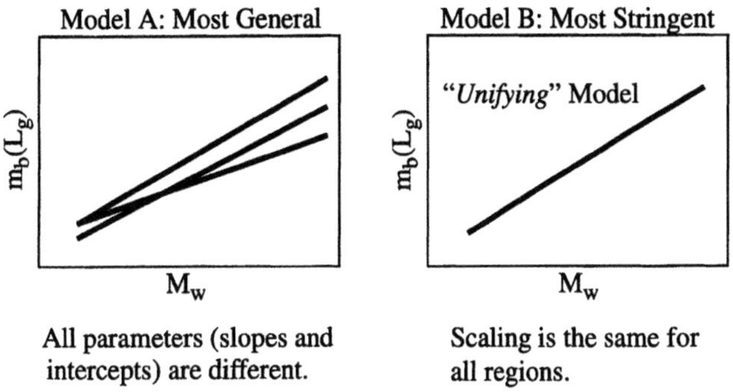

Null Hypothesis, H_0: Model B provides as good an explanation of the data as Model A does.

F - test:

$$F_{B:A} = \frac{(RSS_B - RSS_A) \,/\, (df_B - df_A)}{RSS_A \,/\, df_A}$$

where -
RSS ≡ residual sum of squares
df ≡ degrees of freedom

Figure 3.1
Statistical method used to test differences in magnitude scaling for different tectonic regions.

scaling relationships for ENA and WUS. Removing high-frequency $m_b(Lg)$ from the ENA data set (case 3.b) has no effect on this conclusion. However, test results change rapidly as the M_w range is decreased. While the null hypothesis can be rejected for ENA and WUS in the 4.0–6.5 M_w range (case 3.c), test results are positive in the 4.5–6.5 M_w range (case 3.d) and positive for tests between WUS and ENA and CA (cases 3.e and 3.f). Three-region test results reject the null hypothesis for the 4.0–6.5 M_w range (case 3.g), but the data are consistent with a unified model for M_w greater than 4.2 or 4.3. Thus, three-region test results suggest that $m_b(Lg)$ scaling is coincident in all regions for M_w greater than about 4.3.

Unified scaling relationships for *Lg* waves were obtained by regressing data for all three regions, and results are given in Table 4.1. In addition to M_w ranges of 4.25–6.5 and 4.5–6.5, I give results for M_w range 4.0–6.5 too, even though it is not strictly valid to unify over this range. However, it is clear that scaling results (slope and intercept) do not differ much whether M_w 4.0 or 4.25 is chosen as the cutoff, and thus the scaling results are not sensitive to the exact M_w where unifying is valid. Standard deviation of the residuals ($\sqrt{[RSS/DF]}$) is found to be 0.20, 0.19, 0.27, and 0.23 mu for ENA, WUS, CA, and unified data sets, respectively, for M_w range 4.5–6.5.

Table 3.1

Linear regression results for earthquakes

Magnitude	Region	M_w range	*DF*	*RSS*	Slope (err)	Intercept (err)
$m_b(P)$	ENA	3.9–6.5	40	2.506	0.904 (0.066)	0.641 (0.305)
		4.5–6.5	17	0.454	1.055 (0.073)	−0.176 (0.375)
	WUS	3.9–6.5	46	2.666	0.924 (0.066)	0.201 (0.324)
		4.5–6.5	37	1.941	1.034 (0.084)	−0.378 (0.428)
	CA	3.9–6.5	68	3.449	0.828 (0.045)	0.947 (0.214)
		4.5–6.5	40	2.165	0.876 (0.074)	0.689 (0.382)
$m_b(Pn)$	WUS	3.0–6.5	79	3.656	1.081 (0.034)	−0.472 (0.156)
		3.9–6.5	66	3.124	1.036 (0.046)	−0.244 (0.223)
		4.5–6.5	44	2.509	0.939 (0.084)	0.266 (0.431)
	CA	3.9–6.5	69	3.631	0.938 (0.042)	0.463 (0.206)
		4.5–6.5	42	2.066	1.044 (0.063)	−0.116 (0.330)
$m_b(Lg)$	ENA	3.0–6.5	101	4.007	0.945 (0.028)	0.574 (0.116)
		3.0–6.5	82	3.441	0.964 (0.030)	0.480 (0.126)
		4.0–6.5	42	1.568	0.931 (0.051)	0.632 (0.238)
		4.5–6.5	19	0.726	0.954 (0.083)	0.520 (0.430)
	WUS	3.0–6.5	73	2.314	1.129 (0.030)	−0.442 (0.134)
		4.0–6.5	52	1.935	1.071 (0.051)	−0.151 (0.246)
		4.5–6.5	36	1.356	0.965 (0.080)	0.405 (0.411)
	CA	4.0–6.5	68	3.984	1.097 (0.047)	−0.177 (0.229)
		4.5–6.5	44	3.264	1.066 (0.077)	−0.008 (0.403)
	Unified	4.0–6.5	166	8.206	1.041 (0.029)	0.071 (0.142)
		4.25–6.5	132	6.700	1.043 (0.037)	0.060 (0.183)
		4.5–6.5	103	5.635	1.023 (0.048)	0.167 (0.247)

4. *Regional Extensions of M_s:m_b Discriminants*

Methods for routinely detecting surface waves at regional distances are currently a focus of research and development effort in the verification community. The hope is to extend M_s:m_b discrimination to small magnitudes using regional measurements. Presuming technical challenges confronting routine measurement of regional M_s can be solved, there are still questions about how well M_s:m_b will serve as a discriminant at small magnitudes. Theoretical studies have addressed this issue for teleseismic monitoring (STEVENS and DAY, 1985) and regional monitoring (PATTON and WALTER, 1993), but little is known about its performance from practical experience using regional data. In this section, I determine M_s:$m_b(Lg)$ relationships in order to shed light on discrimination potential at small magnitudes. Actual performance evaluation must await the development of technologies for M_s determination.

Table 3.2

Hypothesis tests results

Magnitude	Regions	Case #	M_w range	RSS_A (DF)	RSS_B (DF)	$F_{B:A}$ (DF_N, DF_D)
$m_b(P)$	ENA ⇔ WUS	1.a	3.9–6.5	5.172 (86)	7.649 (88)	20.6 (2, 86)
		1.b	4.5–6.5	2.395 (54)	3.621 (56)	13.8 (2, 54)
	WUS ⇔ CA	1.c	4.5–6.5	4.106 (77)	5.594 (79)	14.0 (2, 77)
	ENA ⇔ CA	1.d	3.9–6.5	5.955 (108)	6.071 (110)	1.1 (2, 108)
	ENA, WUS, CA	1.e	3.9–6.5	8.621 (154)	11.71 (158)	13.8 (4, 154)
		1.f	4.5–6.5	4.560 (94)	6.567 (98)	10.3 (4, 94)
$m_b(Pn)$	WUS ⇔ CA	2.a	3.9–6.5	6.755 (135)	8.854 (137)	21.0 (2, 135)
		2.b	4.5–6.5	4.575 (86)	5.180 (88)	5.7 (2, 86)
$m_b(Lg)$	ENA ⇔ WUS	3.a	3.0–6.5	6.321 (174)	9.002 (176)	36.9 (2, 174)
		3.b	3.0–6.5	5.755 (155)	7.934 (157)	29.3 (2, 155)
		3.c	4.0–6.5	3.503 (94)	3.976 (96)	6.3 (2, 94)
		3.d	4.5–6.5	2.082 (55)	2.124 (57)	0.6 (2, 55)
	WUS ⇔ CA	3.e	4.0–6.5	5.919 (120)	6.207 (122)	2.9 (2, 120)
	ENA ⇔ CA	3.f	4.0–6.5	5.552 (110)	5.829 (112)	2.7 (2, 110)
	ENA, WUS, CA	3.g	4.0–6.5	7.487 (162)	8.002 (166)	3.9 (4, 162)
		3.h	4.5–6.5	5.346 (99)	5.635 (103)	1.3 (4, 99)
		3.i	4.3–6.5	6.256 (121)	6.605 (125)	1.7 (4, 121)

First, I present $m_b(Pn)$ and $m_b(Lg)$ scaling observations for nuclear explosions and determine their scaling relationships. I draw upon published m_b measurements for NTS explosions and explosions detonated at the Soviet test sites in Kazakhstan (KTS) and Novaya Zemlya (NZ), and two other explosions, one that occurred at the Chinese test site, Lop Nor (LPN), and another, Salmon, in ENA. Salmon was detonated in a salt dome in Hattiesburg, MS.

Measurements of 1-Hz $m_b(Lg)$ for KTS and NZ are available from NUTTLI (1986b; 1987; 1988) and from PRIESTLEY and PATTON (1997); for NTS from NUTTLI (1986a) and PATTON (1988); and for Salmon and Lop Nor explosions from NUTTLI (1986a) and PRIESTLEY and PATTON (1997), respectively. I used M_o estimates from published papers and my own unpublished results for NTS explosions. Explosion M_o for KTS are available from STEVENS (1986), EKSTRÖM and RICHARDS (1994), and PRIESTLEY and PATTON (1997; also a Lop Nor explosion). Explosion M_o for NZ are obtained from measurements of the static level of the reduced displacement potential ψ_∞ using the results of BURGER *et al.* (1986) and COHEE and LAY (1988). For this purpose, I assumed a compressional velocity α of 6.0 km/s and density ρ of 2.5×10^3 kg/m^3 in the equation, $M_o = 4 \cdot \pi \cdot \rho \cdot \alpha^2 \cdot \psi_\infty$ (AKI *et al.*, 1974). The Salmon M_o was computed using ψ_∞ estimates from DENNY and GOODMAN (1990). It should be pointed out that explosion M_o is for the isotropic

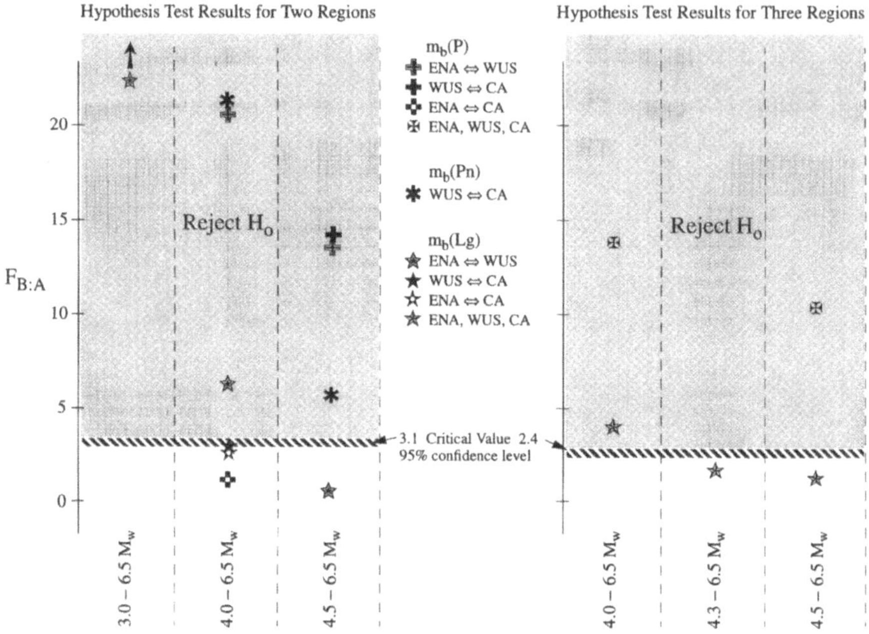

Figure 3.2
The results of statistical test discussed in the body of the paper. Results for two-region tests are shown as a function of M_w range on the left panel, and results for three-region tests are shown on the right.

Table 4.1

Linear regression results for explosions

Magnitude	Test site	M_w range	*DF*	*RSS*	Slope (err)	Intercept (err)
$m_b(Pn)$	NTS	3.0–5.5	78	1.206	1.371 (0.026)	−0.682 (0.116)
		3.4–5.5	76	1.192	1.362 (0.029)	−0.642 (0.128)
	KTS	3.5–5.0	23	0.606	1.279 (0.079)	0.044 (0.349)
$m_b(Lg)$	NTS	3.0–5.5	94	1.491	1.223 (0.024)	0.205 (0.110)
		3.4–5.5	92	1.416	1.201 (0.026)	0.309 (0.118)
	KTS + NZ	3.5–5.8	74	2.430	1.149 (0.047)	0.499 (0.221)
	Unified	3.4–5.8	168	3.957	1.174 (0.025)	0.412 (0.114)

source, and in most of the work above, corrections were made to eliminate effects on amplitudes due to seismic radiation by deviatoric sources, such as tectonic release. While these corrections may not be completely successful, it does point out a difference between M_0 and M_s measurements, the latter having no corrections for source mechanism whatsoever. The lack of such corrections may introduce scatter and/or bias into M_s determinations. A complete list of explosions used in this study is available upon request.

$m_b(Lg)$ results. Plots of earthquake and explosion M_w:$m_b(Lg)$ relationships are shown in Figures 4.1 and 4.2 for tectonic and stable regions, respectively. NTS explosions were detonated in a variety of environments, both above and below the water table and in different rock types. PATTON and WALTER (1993) found that M_w:m_b data demonstrated little dependence on testing environment, and observations in Figure 4.1 are consistent with that finding. Note that a single-burst, 1-kt chemical explosion, the Non-Proliferation Experiment (NPE) on Rainier Mesa, falls squarely within the nuclear explosions. Regression results for NTS explosions are provided in Table 4.1. I plot two earthquake regression results in Figure 4.1, one for

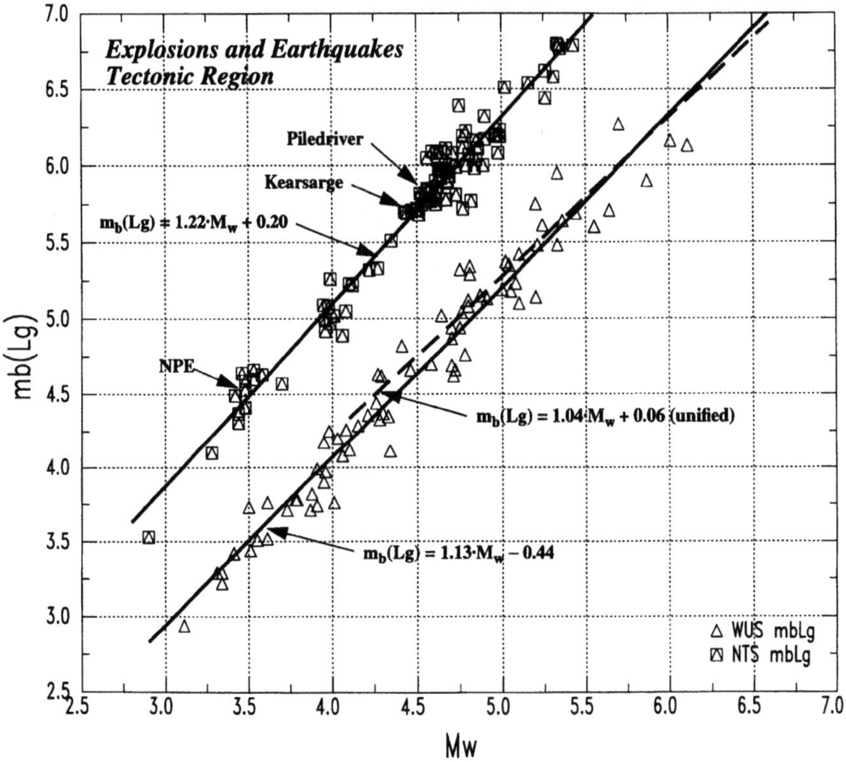

Figure 4.1

M_w:$m_b(Lg)$ scaling for earthquakes and explosions in tectonic regions. Earthquake data are plotted as triangles, and explosions as triangles enclosed in squares. 96 NTS explosions are plotted, along with the regression line for that data; 75 WUS earthquakes are plotted, along with the regression line for that data and the unified M_w:$m_b(Lg)$ relationship for M_w 4.25–6.5, given in Table 3.1. NPE (Non-Proliferation Experiment) is the chemical, 1-kt explosion detonated on Rainier Mesa, Kearsarge is the US JVE (Joint Verification Experiment) nuclear explosion for seismic yield calibration, and Piledriver is a famous nuclear explosion conducted in granite at NTS.

M_w 3.0–6.5 and the other for the unified relationship. The scaling of $m_b(Lg)$ appears to be almost the same for WUS earthquakes and NTS explosions, where the slopes are 1.13 ± 0.03 and 1.22 ± 0.02, respectively.

Results for stable regions are shown in Figure 4.2. Eleven NZ explosions plot with M_w greater than 5.0, and I identify one KTS explosion occurring on 27 Dec. 1980, which has a small M_w compared to Nuttli's $m_b(Lg)$ value. This explosion has a large deviatoric source component (EKSTRÖM and RICHARDS, 1994) and M_0 could be underestimated in spite of the corrections for tectonic release. The style of deviatoric source mechanisms at KTS tends to reduced amplitudes of Rayleigh waves and hence lower apparent source strength (EKSTRÖM and RICHARDS, 1994). Other KTS explosions with large deviatoric source mechanisms may also have reduced M_0. In any case, results in Figure 4.2 indicate that scaling coefficients for earthquakes and explosions are less similar for stable regions than they are for tectonic regions. This

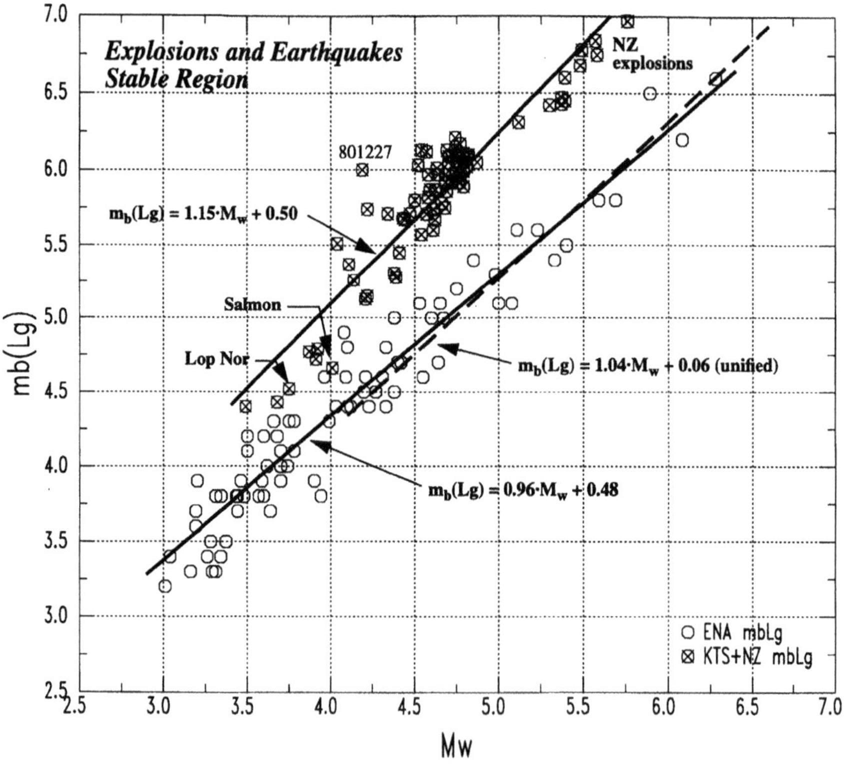

Figure 4.2
M_w:$m_b(Lg)$scaling for earthquakes and explosions in stable continental regions. Earthquake data are plotted as circles, and explosions as x's enclosed in circles. 76 nuclear explosions, 63 located at KTS and 11 at NZ are plotted, along with regression line for all explosion data; 84 ENA earthquakes are plotted, along with regression line for that data and the unified M_w:$m_b(Lg)$ relationship for M_w 4.25–6.5, given in Table 3.1.

apparent difference between stable and tectonic regions is not caused by $m_b(Lg)$ scaling for explosions, as the $m_b(Lg)$ scaling rate for Soviet tests is 1.15 ± 0.05, very close to the scaling rate for NTS explosions. Rather, the cause is the difference between scaling rates for earthquakes in stable and tectonic regions, 0.96 ± 0.03 versus 1.13 ± 0.03, which becomes significant at magnitudes less than M_w 4.0 (see Figure 2.15). This result has implications for the ability to discriminate between earthquakes and explosions at small magnitudes.

Scaling results for explosions in tectonic and stable regions are very similar, which begs statistical testing to determine if one scaling curve can represent all test sites. For NTS explosions in the M_w range 3.4–5.5 and explosions detonated at test sites in stable areas for M_w range 3.5–5.8, $F_{B:A}$ is 2.4 for 2 and 166 *DF*. The null hypothesis cannot be rejected at the 95% confidence level (critical value of 3.1), and thus a unified scaling relationship for $m_b(Lg)$ is acceptable within the scatter of the data for all continental test sites. This is an extraordinary result and is consistent with uniform 1-Hz *Lg* excitation in different crustal structure, just as was found for earthquakes earlier in the paper. Furthermore, a unified relationship supports the results of PATTON and WALTER (1993) who showed for NTS explosions that effects of source medium impedance and source function in different testing environments tend to cancel or at least act in opposite directions mitigating the net impact on $M_w{:}m_b$ relationships.

$m_b(Pn)$ results. Plots of $M_w{:}m_b(Pn)$ scaling relationships for tectonic regions are given in Figure 4.3. Different rock types and emplacement conditions are represented by explosions in this data set. Just as in the case of *Lg* waves, $M_w{:}m_b(Pn)$ data show little effects of explosions detonated in different materials and/or above and below the water table. However, explosion scaling rates for $m_b(Pn)$, 1.37 ± 0.03, are significantly higher than they are for $m_b(Lg)$, 1.22 ± 0.02, based on data for M_w 3.0–5.5. For earthquakes, the scaling rate for $m_b(Pn)$, 1.08 ± 0.03, is not substantial different than the scaling rate for *Lg* waves, 1.13 ± 0.03. Convergence of natural and man-made sources at small magnitudes is a consequence of higher scaling coefficients for nuclear explosions.

Results for stable regions are provided in Figure 4.4. The data set is small for explosions, just 24 KTS explosions and 1 LPN explosion from the study of PRIESTLEY and PATTON (1997). Explosion scaling rates are similar to what they are at tectonic test sites, 1.28 ± 0.08 for KTS versus 1.36 ± 0.03 for NTS. Since $m_b(Pn)$ measurements were not available for ENA, I used measurements for CA earthquakes. The validity of this substitution is certainly questionable as many areas of CA express vigorous tectonic activity. However, tectonic stresses are generally characterized by horizontal, maximum compressive stresses in both ENA and CA. Furthermore, based on statistical tests in the previous section (case 1.d in Table 3.2), $m_b(P)$ scaling is similar for ENA and CA, and the comparison between $m_b(P)$ scaling in ENA and $m_b(Pn)$ scaling in CA is remarkably close (Fig. 4.4). The $m_b(Pn)$ scaling rate for CA earthquakes is only 0.94 ± 0.04, which is lower than it is

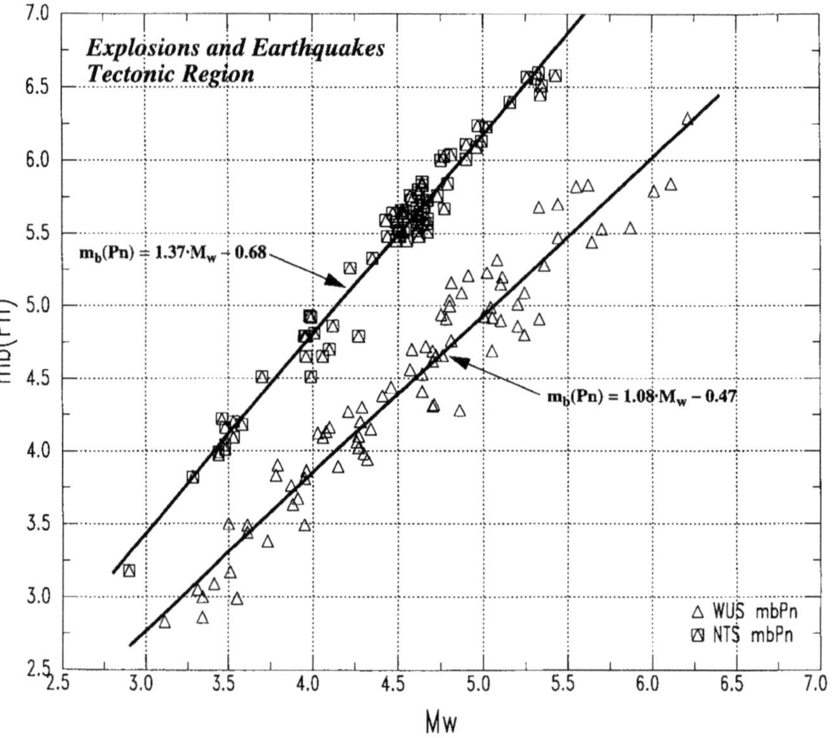

Figure 4.3
M_w:$m_b(Pn)$ scaling for earthquakes and explosions in tectonic regions. Earthquake data are plotted as triangles, and explosions as triangles enclosed in squares. 80 NTS explosions are plotted along with regression line for that data; 81 WUS earthquakes are plotted along with regression line for that data.

in WUS. Convergence of earthquake and explosion populations is readily apparent in the $m_b(Pn)$ results for stable regions.

While $m_b(Pn)$ scaling coefficients for NTS and KTS explosions are similar, the scaling relationships are offset by ~0.3 mu. NTS data have uniformly smaller $m_b(Pn)$ values for a given M_w. Testing the hypothesis of coincident scaling, I found that the null hypothesis can be rejected unequivocally, as $F_{B:A}$ equals 53.9 with 2 and 99 *DF*. This result is not surprising in light of the $m_b(P)$ test site bias between NTS and KTS (NUTTLI, 1986b; MURPHY *et al.*, 1991). Once again, regional offsets are present in $m_b(Pn)$ data because of the region by region calibration against $m_b(P)$ used in developing $m_b(Pn)$ scales.

M_s:$m_b(Lg)$ relationships for continental sources. In this subsection, M_s:$m_b(Lg)$ relationships will be derived from scaling results in Figures 4.1 and 4.2 by substituting the functional form of M_w in terms of M_o and exploiting known *log* M_o:M_s relationships for earthquakes and explosions. I compare the derived M_s:m_b relationships with observations later in the paper.

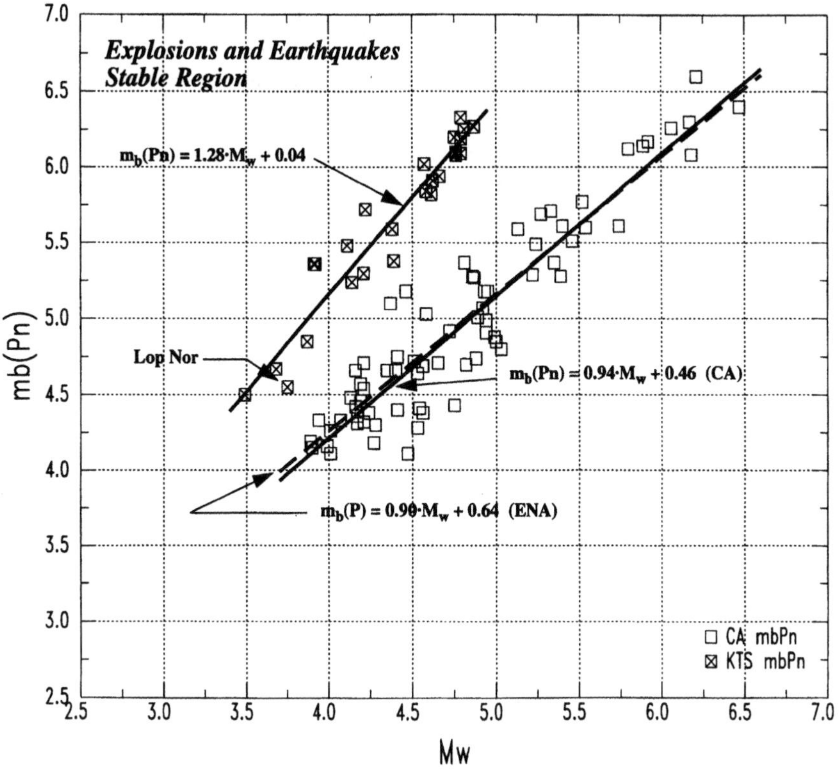

Figure 4.4

M_w:$m_b(Pn)$ scaling for earthquakes and explosions in stable continental regions. Earthquake data are plotted as squares, and explosions as x's enclosed in squares. 24 KTS explosions and 1 Lop Nor explosion are plotted along with regression line for that data; 71 CA earthquakes are plotted along with regression line for that data. Also shown as a dash line is the M_w:$m_b(P)$ relationship for ENA from Table 3.1.

Relationships between M_w and M_s are based on empirical correlations. EKSTRÖM and DZIEWONSKI (1988) developed global relationships for earthquakes using observations of CMT M_o and PDE M_s. For earthquakes with M_s less than 5.3, the relationship takes the form, $\log_{10} M_o = M_s + C$, where C is a constant. The worldwide average value for C is 19.24, but systematic differences were detected by Ekström and Dziewonski depending on event location. For a given M_o, M_s tends to be ~0.4 mu larger for continental earthquakes compared to mid-ocean ridge and transform fault earthquakes. Ekström and Dziewonski showed this for moderate-size earthquakes, M_s 5.5–6.5, and differences may be even greater for larger events. Smaller events apparently showed too much scatter to say for sure if the effect was present for M_s less than 5.5. More recently, data set compilations by JOHNSTON (1996) provide evidence that M_s for small events in stable continents are larger than the global average. The difference is roughly 0.2 mu, and this may arise because 20-s surface waves propagate more efficiently over continental paths. Current M_s

measurements only correct for a global-average attenuation rate, and do not account for regional variations in that rate. In this study, I decided to use the value 19.04 for C to convert M_o (or M_w) to M_s, emphasizing the applicability of my results to CTB monitoring of continental regions. Relationships between log M_o and M_s for nuclear explosions are not as well characterized as they are for earthquakes. This is due to the fact that there are significantly fewer M_o measurements for explosions, and the measurements span a narrower range of M_o. M_s for small explosions is difficult to measure, and reliable estimates of M_o must properly account for non-isotropic source contributions to long-period radiation. STEVENS and MCLAUGHLIN (1997) find that a single relationship, $\log_{10} M_o = M_s + (18.74 \pm 0.21)$, based on maximum likelihood estimates of M_o, can satisfy observations at NTS and KTS. However, their observations are for a narrow range of M_o.

Figure 4.5 summarizes log $M_o - M_s$ observations I have complied for nuclear explosions at NTS, KTS, NZ and Lop Nor. There are 48 NTS explosions spanning about three orders of magnitude in M_o, although the majority of data are between 10^{22} and 10^{24} dyn-cm. The M_s measurements originate from the study of WOODS and HARKRIDER (1995) and have been converted to an equivalent NEIC M_s by subtracting a constant, 0.14 mu. It is important to note the need for consistency in the M_s "baseline," since it is well known that magnitudes are arbitrary constructs and exhibit different baselines depending on the formulas used. Since the global relationships between log M_o and M_s for earthquakes are based on NEIC magnitudes, I have adhered to a NEIC baseline throughout this paper. The results should not depend on this choice of baseline, as long as it is consistently adhered to. KTS and Lop Nor observations are generally consistent with those for NTS; on the other hand, NZ observations seem to show low M_s values for the estimated M_o (see discussion above about how M_o were obtained for NZ explosions). Assuming that M_s is proportional to log M_o, the observations for NTS, KTS, and Lop Nor yield a constant of 18.64 ± 0.03, in good agreement with STEVENS and MCLAUGHLIN (1997).

On the other hand, NTS observations in Figure 4.5 indicate that M_s scaling is not one-to-one with log M_o, but is a significantly higher rate. KTS observations also suggest a higher scaling rate, although those observations do not span the M_o range that NTS observations do. Several possible causes for a higher scaling rate are effects on M_s due to systematic variations in coupling and tectonic release (MURPHY, 1977; PATTON, 1991). A least-squares fit to NTS data yields the scaling relationship, $M_s = (1.12 \pm 0.04) \cdot \log M_o - (21.49 \pm 0.83)$. This relationship seems to satisfy most of the observations for smaller explosions at NTS and KTS.

I used unified $M_w{:}m_b(Lg)$ relations to derive $M_s{:}m_b(Lg)$ relationships for nuclear explosions and earthquakes shown in Figure 4.6. Also shown are $M_s{:}m_b(Lg)$ relationships for earthquakes located in stable and tectonic regions based on ENA and WUS results. There are two relations for explosions, one based on M_s scaling one-to-one with log M_o and another based on the regression line for NTS data in Figure 4.5, where M_s scales as $1.12 \cdot \log M_o$. Earthquake relationships reflect

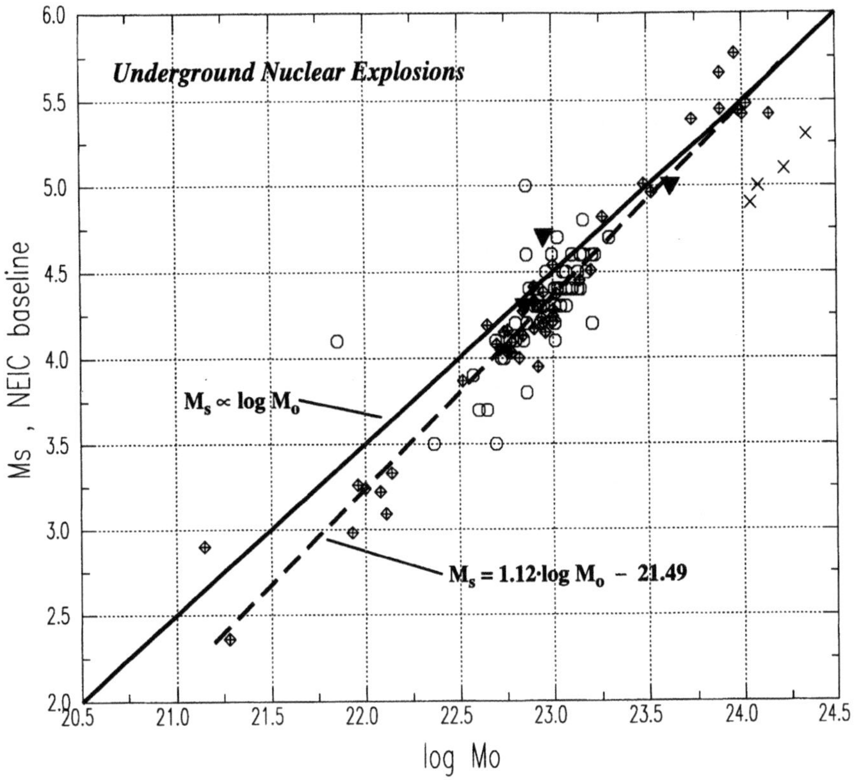

Figure 4.5

Log $M_0 - M_s$ observations for nuclear explosions at NTS (diamonds), KTS (circles), NZ (x's) and Lop Nor (inverted triangles). The observations are compiled from various sources: for NTS, M_s from WOODS and HARKRIDER (1995), M_0 from STEVENS (1986) and PATTON (1991; unpublished results); for KTS, M_s from NEIC catalog, M_0 from EKSTRÖM and RICHARDS (1994); for NZ, M_s from NEIC catalog, M_0 from Ψ_∞ measurements of BURGER *et al.* (1986) and COHEE and LAY (1988); for Lop Nor, M_s from NEIC catalog, M_0 from XIE and PATTON (1999). Note that M_s measurements of Woods and Harkrider were reduced by 0.14 mu to be compatible with NEIC M_s. Solid line indicates M_s scaling one-to-one with log M_0, while the dash line is a least-squares fit to all of NTS observations.

regional differences in $m_b(Lg)$ scaling in tectonic and stable continents. At large magnitudes, however, the scaling merges and there are no significant differences for earthquakes with M_w greater than 4.2 ($M_s \sim 3.0$). For M_s between 3.0 and 5.5, explosion relationships do not depend strongly on the choice of M_s scaling with log M_0, and the means of earthquake and explosion populations are separated by nearly a constant 0.6–0.7 mu. Below M_s 3.0, the relationships suggest that earthquake and explosion populations might converge. Separation between the populations is region-dependent as well as magnitude-dependent. Estimates of the separation are quite uncertain because the slope of explosion M_s:$m_b(Lg)$ relationships is not well constrained by present knowledge of M_s scaling with explosion M_0, especially at small yields.

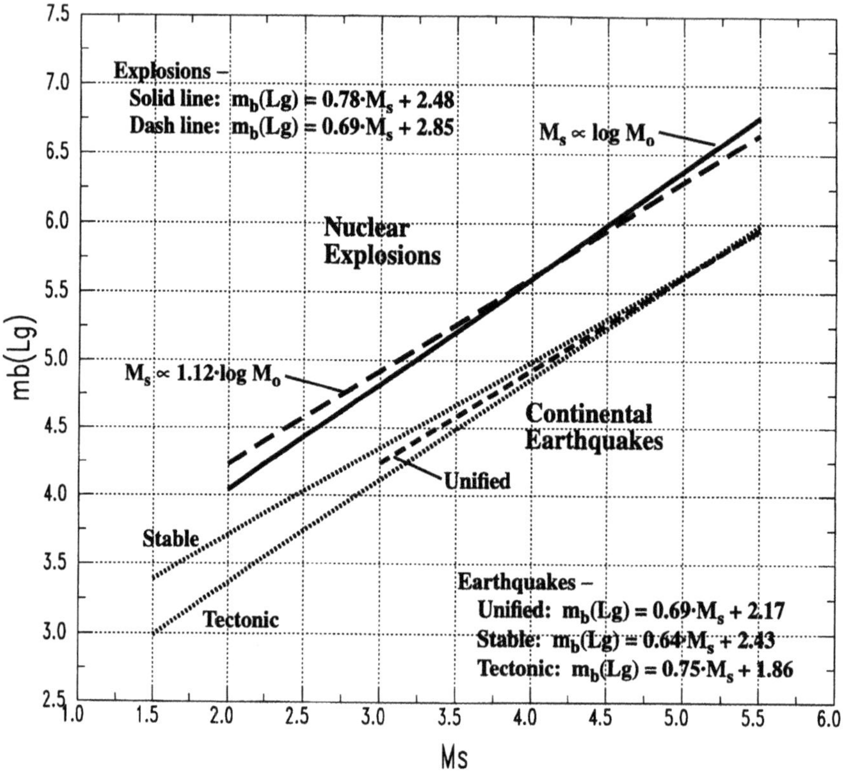

Figure 4.6
M_s:$m_b(Lg)$ scaling relationships for earthquakes and explosions. Solid line is relationship for explosions if M_s is proportional to log M_o, and long-dash line is the relationship if M_s scales as 1.12 log M_o, as seen for NTS explosions (Fig. 4.5). Dotted lines are relationships for continental earthquakes located in stable and tectonic regions, and the unified relationship is plotted over the M_s range of applicability as short-dash line.

5. *Regional $m_b(P)$ Bias*

Regional variations in $m_b(P)$ have been investigated for many years (see CHUNG and BERNREUTER, 1981 and DOUGLAS and MARSHALL, 1996 for reviews). As discussed in the *Introduction*, this issue came to the forefront of nuclear monitoring research when controversies arose over estimates of seismic yield for nuclear explosions detonated on foreign test sites. It remains important for accurate yield estimation to this day.

WUS has been known for some time to be particularly anomalous with respect to $m_b(P)$ variations. Early studies by GUYTON (1964), BOOTH *et al.* (1974), Nuttli and co-workers (NUTTLI and ZOLLEWEG, 1974; NUTTLI *et al.*, 1979) among others showed that high attenuation in the crust and upper mantle under WUS affected ground motions both near and at great distances from the source. CHUNG and

BERNREUTER (1981) summarized early measurements and concluded that $m_b(P)$ bias between the eastern and western United States is about 1/3 mu, i.e., $m_b(P)^{\text{west}} = m_b\ (P)^{\text{east}} - 0.33$ mu.

The work of NUTTLI (1986ab, 1987, 1988) established the best estimates of $m_b(P)$ bias for test sites around the world. This was accomplished by taking differences between $m_b(P)$ and $m_b(Lg)$ for a given explosion, and computing the mean for many explosions. Nuttli argued that differences in $E[m_b(P) - m_b(Lg)]$ arose from regional variations in upper-mantle structure under test sites affecting the attenuation of P-wave amplitudes to distant stations. Underlying this method is the assumption that $m_b(Lg)$ is transportable from one region to the next.

NUTTLI (1986a) obtained a bias estimate of -0.31 ± 0.02 mu for NTS based on an average of 86 $[m_b(P) - m_b(Lg)]$ observations for explosions with $m_b(Lg)$ ranging from 4.75–6.5. This implies that $m_b(P)$ for a 150-kt NTS explosion is smaller by 0.31 mu on average compared to $m_b(P)$ for the same yield explosion detonated in the central United States under the same emplacement conditions. Nuttli's test-site bias is in good agreement with estimates of $m_b(P)$ differences between ENA and WUS based on earthquake studies (CHUNG and BERNREUTER, 1981). Bias for the Shagan River test area of KTS (again with respect to ENA) was estimated to be $+0.04 \pm 0.02$ mu; the net bias between NTS and Shagan River is 0.35 ± 0.03 mu (NUTTLI, 1986b). This bias estimate was later confirmed by the JVE (Joint Verification Experiment) conducted by the United States and the former Soviet Union (DOUGLAS and MARSHALL, 1996). It is believed to be one of the major factors that contributed to overestimation of seismic yields for nuclear tests at Shagan River.

It is noteworthy that Nuttli averaged $m_b(Lg)$ estimates from several stations in his studies of test-site bias. This reduced the chance of single-station measurements affecting the result due to receiver site effects, for example. Also, $m_b(P)$ magnitudes were taken from the PDE or ISC, and the results did not depend upon which catalog teleseismic magnitudes were drawn from (NUTTLI, 1986b, 1988). However, the test-site bias obtained for the Degelen Mountain test area of KTS, $+0.27 \pm 0.03$ mu (NUTTLI, 1987), is very large compared to the bias estimated for the near-by Shagan River test area. This estimate is apparently affected by overestimates of catalog $m_b(P)$ for smaller Degelen Mountain explosions (PRIESTLEY and PATTON, 1997). For 22 KTS explosions (i.e., both Shagan River and Degelen Mountain test areas), $E[m_b(P) - m_b(Lg)]$ is $+0.18 \pm 0.02$ mu if ISC $m_b(P)$ are used versus $+0.06 \pm .02$ mu if maximum-likelihood estimates (MLE) of $m_b(P)$ are used. For smaller explosions, MLE estimates make a significant difference. Priestley and Patton found no difference in m_b bias between Shagan River and Degelen test areas.

In the previous sections of this paper, I presented $m_b(Lg)$ scaling relationships supporting transportability of $m_b(Lg)$ to different tectonic regions. These relationships validate Nuttli's assertion that 1-Hz Lg Green's functions are essentially invariant for different continental structures. Thus, $m_b(Lg)$ is transportable provided accurate corrections can be made for effects of propagation (attenuation, scattering, geometrical spreading). Furthermore, the fact that explosion *and* earthquake scaling relationships are transportable implies that invariance of *Lg* Green's functions is valid over a substantial range of source depths within the crustal waveguide. These results suggest a similar approach that Nuttli took with explosions may be taken with earthquake observations to estimate variations of $m_b(P)$ bias for seismogenic zones, thus covering considerably larger regions than nuclear test sites.

One precaution extending Nuttli's approach to earthquake data is the effect of source radiation pattern, which is not an issue for nuclear explosions. There is little evidence for radiation pattern affecting amplitudes of *Lg* waves. However, the effect is well known for *P* waves, and seismologists have exploited variations in *P*-wave amplitudes for information about earthquake source mechanism for decades. Such studies have relied upon long-period data, and quantitative knowledge about radiation pattern effects on short-period, teleseismic observations has been quite limited. Recently, SCHWEITZER and KVÆRNA (1999) investigated the correlation of magnitudes reported by the Prototype International Data Center (PIDC; Arlington, VA) with radiation patterns predicted by Harvard Centroid Moment Tensor (CMT) solutions. Based on over 10,000 magnitude determinations, the radiation pattern signature in the data was found to be only 24% of the predicted signature from source theory. Differences between magnitudes corrected for radiation pattern and standard m_b were always within 0.15 mu. The reason for such a muted effect is believed to be related to focusing, defocusing and scattering of seismic energy along the ray path. This will smear out radiation pattern effects for short-period data more so than for long-period data used in source studies (SCHWEITZER and KVÆRNA, 1999).

Thus, a small radiation pattern effect is present in $m_b(P)$ observations. Nuttli's calibration equation, which is empirically based, will *de facto* have included this effect to some degree. For sizeable data sets then, I expect radiation pattern should have little or no impact on bias estimates determined from earthquakes, and the following results support this view.

Table 5.1 summarizes the results applying earthquake observations of $[m_b(P) - m_b(Lg)]$ to estimate regional $m_b(P)$ bias. There were several considerations in the selection of earthquakes used for these estimates. At large enough magnitudes, 1-Hz amplitudes may saturate due to source scaling. At small magnitudes, there is concern about overestimation of catalog $m_b(P)$ due to data censoring practices. On the other hand, I could not afford to be overly selective with limited numbers of events. Thus, I screened data samples for systematic effects at large and small magnitudes and took the liberty of removing outliers when it appeared warranted. The results are compared with estimates of test-site bias in Figure 5.1.

The results for WUS are based on 29 earthquakes located in California and the Basin and Range, and are in excellent agreement with the findings of CHUNG and BERNREUTER (1981) and with those of Nuttli for NTS. This is a successful test of the method applied to earthquake measurements for an area that is well-characterized. ENA shows a small negative bias. Subdividing events into those

Table 5.1

Regional $m_b(P)$ Bias Estimates from Earthquake Data

Region	Number of events	$E[m_b(P) - m_b(Lg)]$
California and Basin-Range of WUS	29	−0.33 ± 0.03
Eastern North America (ENA)	52	−0.07 ± 0.03
New Madrid Seismic Zone	22	−0.06 ± 0.03
Outside of New Madrid Seismic Zone	30	−0.08 ± 0.05
Central Asia (CA)	62	−0.12 ± 0.04
Tien Shan Fold Belt (+ Central region)	21	0.00 ± 0.07
Northern Tien Shan Region	12	−0.10 ± 0.07
Northwestern Tarim Basin	8	0.04 ± 0.10
Southern Siberian Fold Belt	6	−0.14 ± 0.10
Eastern Tarim Basin	15	−0.37 ± 0.05
Eastern Tarim Basin (North of 40°N)	5	−0.32 ± 0.11

located inside or outside the New Madrid seismic zone (see Figure 2.1) made only a small difference, and the entire region appears to be uniform with respect to $m_b(P)$ bias. For the New Madrid seismic zone in central U.S., the result is −0.055 ± 0.034 mu, which is not significantly different from zero at the 95% confidence level. Using 62 events with $m_b(Lg)$ less than 5.5 in CA, $m_b(P)$ bias is significantly less than zero, but not significantly different from ENA. Because heterogeneity of earth structure is so great in CA, I subdivided the region into zones for the purpose of estimating bias on a finer distance scale, and those zones are shown in Figure 2.11. The results for each zone have larger error bars because fewer events were involved in the estimates. In 4 of the 5 zones, bias estimates are not significantly different from zero. Only the eastern Tarim Basin shows a bias significantly different from zero, and excluding earthquakes located on the northern fringe of the Tibetan Plateau did not make a difference. It is interesting to note that the bias estimated from earthquakes in eastern Tarim Basin is consistent with one Lop Nor explosion (see Fig. 5.1), and the size of the bias is comparable to Nuttli's estimate for NTS.

Several considerations should be kept in mind interpreting the CA results. First, all measurements of $m_b(Lg)$ are for a single station. Second, a Q model with power law dependence, $Q = Q_o \cdot f^\eta$, and $Q_o = 400$, $\eta = 0.6$, was used to compute $m_b(Lg)$ for all events. The parameter values are reasonable for this region as a whole (MITCHELL *et al.*, 1997), however, variations of Lg Q_o are likely to exist on smaller length scales. Recently, Q variations have been reported in the Tien Shan region (MARTYNOV *et al.*, 1999). Thus, it is possible that variations in $[m_b(P) - m_b(Lg)]$ are not due to $m_b(P)$ alone, but could reflect changes in propagation efficiencies of both Lg and teleseismic P waves. For example, if Lg Q_o is in fact high ($\sim$1000) for paths southeast of WMQ, the estimated bias for the eastern Tarim Basin becomes vanishingly small for source regions located only 500 km away.

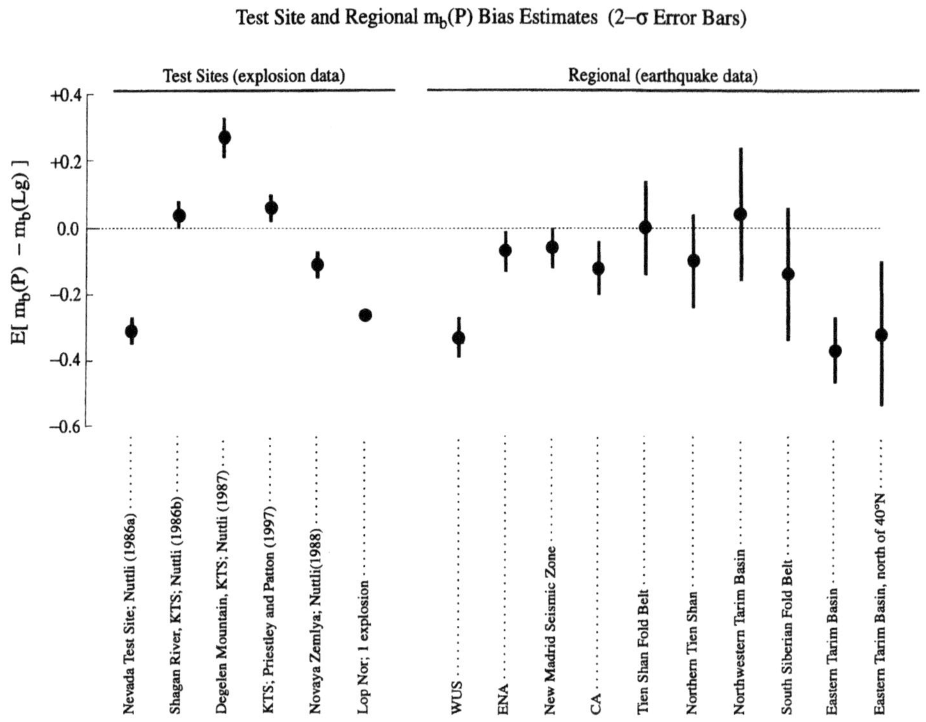

Figure 5.1
Estimates of $m_b(P)$ bias for selected test sites and tectonic regions. Test site estimates were obtained in previous studies reference on the figure. The Lop Nor estimate is based on measurements for a single event reported by PRIESTLEY and PATTON (1997). All of the regional estimates were obtained in this study, as discussed in the text. Values of the bias are given in Table 5.1. The error bars are 2-σ estimates.

6. *Discussion*

Transportability. As discussed above, transportability of $m_b(Lg)$ depends on (1) accurate corrections for Lg attenuation on a regional basis and (2) invariance of the 1-Hz Lg Green's functions with respect continental earth structures. In this study, I investigated scaling relationships for three regions of the earth with very different tectonic character and crust/upper mantle structures. Corrections for attenuation of Lg waves were done with care and are believed to be accurate. Comparisons with M_L for earthquakes in WUS and ENA demonstrates consistency with $m_b(Lg)$, and supports my claim that propagation corrections were done properly. Lg scaling relationships showed no significant differences between continental regions above M_w 4.2. There are significant differences between stable (ENA) and tectonic (WUS) regions for smaller magnitudes, and they may well arise from intrinsic scaling properties of the source function related to source dimensions and faulting mechanics, which are unique to particular tectonic regions (see BOATWRIGHT, 1994; BOORE and ATKINSON, 1987 for discussions on source dimension and corner frequency scaling in ENA). Differences at small magnitudes do not detract from my conclusions that 1-Hz *Lg* Green's functions are essentially invariant for very different continental structures and that $m_b(Lg)$ is indeed transportable.

Magnitude saturation. As mentioned earlier in the paper, JOHNSTON (1996) tested for effects of magnitude saturation on a data set of $m_b(Lg)$ observations for the North American craton. Using the results of linear, quadratic and cubic regressions, Johnston ran statistical tests which proved negative at the 99% confidence level (i.e., no evidence for saturation in $m_b(Lg)$). The data set for CA provides another opportunity to look for effects of saturation because of the number of relatively large events.

Based on $m_b(Lg)$ and $m_b(P)$ observations in Figure 6.1 for the same CA events, the scaling behavior appears to be different for *Lg* and *P* waves at large magnitudes. I ran analysis of variance tests on the improvement of a quadratic model over a linear, and the results were not significant even at 90% confidence level. Thus, a linear model satisfies the apparent scaling behavior of $m_b(Lg)$ well. On the other hand, $m_b(P)$ observations start to show a roll-over effect for M_w as small as 5.0–5.5. The effect is pronounced at larger M_w where the difference between $m_b(Lg)$ and $m_b(P)$ is generally more than 0.5 mu. I ran analysis of variance tests on these data like I did with $m_b(Lg)$, and surprisingly the tests were negative at even lower levels of significance. A quadratic model does not seem to fit the rapid roll-over seen in $m_b(P)$ data very well, and perhaps higher-order models would do better. Such models were not considered in this study.

In my opinion, $m_b(P)$ observations in Figure 6.1 show effects of magnitude saturation above M_w ~5.0, as expected based on well-accepted theories on earthquake source scaling (e.g., BRUNE, 1970). The puzzle is why there is little tendency in observations of $m_b(Lg)$ to show effects of saturation for moderate size events (M_w 5.5–6.5), not just for this data set but for other regions as well. The largest WUS event is the Landers mainshock with $m_b(Lg)$ 7.24 and M_w 7.36. The unified $m_b(Lg)$ scaling relationship developed in this paper predicts 7.74 for M_w of 7.36, which is 0.5 mu larger than observed. This could be saying that $m_b(Lg)$ saturates at a larger M_w than $m_b(P)$ for some reason. This is an important issue for future research, although it does not impact my results for small magnitude M_s:m_b discrimination.

Regional M_s:m_b relationships and teleseismic observations. Several M_s:$m_b(P)$ data sets were compared with M_s:$m_b(Lg)$ relationships to test the notion that differences (offsets) between $m_b(P)$ observations and unified $m_b(Lg)$ relationships are directly related to regional bias in $m_b(P)$. The first data set consists of ENA earthquakes. Recall that a bias estimate for ENA was obtained earlier in the paper from direct comparison of $m_b(P)$ and $m_b(Lg)$ measurements for 52 earthquakes with the result, $E[m_b(P) - m_b(Lg)] = -0.07 \pm 0.03$ mu (see Table 5.1). I searched ISC and PDE catalogs for events located in the stable areas of the eastern United States and Canada between 30° to 55°N latitude and 60° to 110°W longitude. The M_s range was restricted between 3.5 and 4.7 to minimize effects of data censoring and m_b

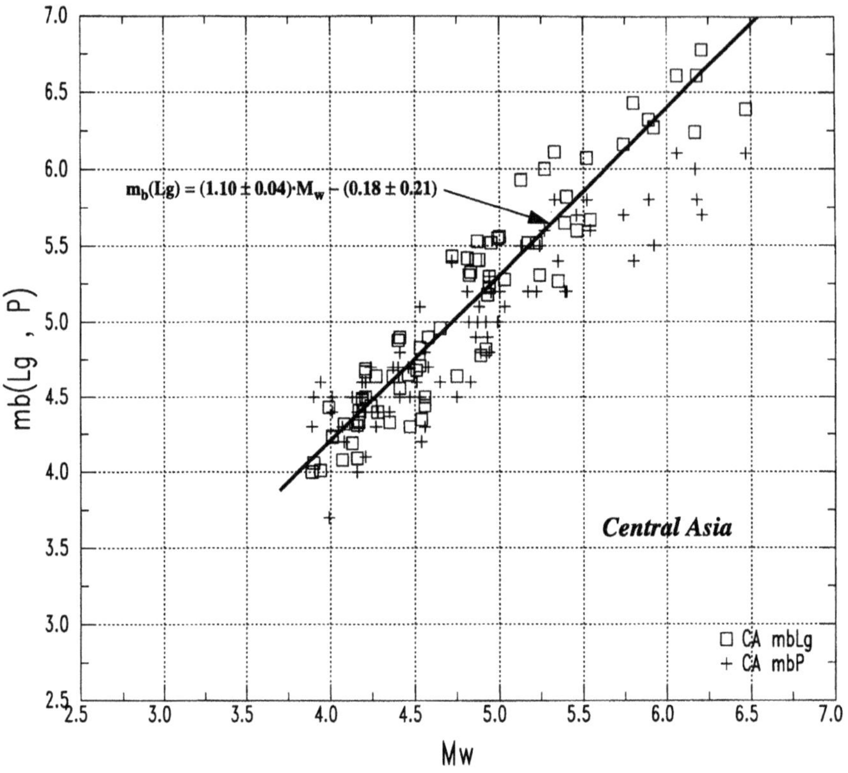

Figure 6.1
Scaling of $m_b(Lg)$ and $m_b(P)$ measurements for all 74 CA events in this study. Regression line is based on all of the $m_b(Lg)$ measurements.

saturation. The search yielded 15 events going back to 1979, and I used ISC and PDE reportings for several events in common, giving a total of 26 M_s:m_b data points. The medium grey-shaded area in Figure 6.2 represents a data cloud where these observations plot on the graph. Values of $m_b(Lg)$ were calculated using the unified relationship, $m_b(Lg) = 0.69 \cdot M_s + 2.17$, and catalog M_s values. The $[m_b(P)^{\text{observed}} - m_b(Lg)^{\text{calculated}}]$ values were averaged for the 26 data points. The average is -0.11 ± 0.06 mu, which is not significantly different from zero at the 95% confidence level and compares well with the ENA bias estimate obtained in this paper from direct comparison of $m_b(P)$ and $m_b(Lg)$ values.

A second M_s:$m_b(P)$ data set consists of nuclear explosions detonated at the Shagan River test site. NUTTLI (1986b) estimated the test-site bias for Shagan River (Balapan) to be $+0.04 \pm 0.02$ mu by direct comparison of $m_b(P)$ and $m_b(Lg)$ measurements for 62 explosions. A search of PDE catalogs between 1973 and 1989 yielded 49 Balapan explosions with M_s in the range 4.0 to 5.0. The data cloud of M_s:$m_b(P)$ observations for these explosions is shown by the dark-grey area in Figure 6.2. Values of $m_b(Lg)$ were calculated using the unified relationship for explosions, $m_b(Lg) = 0.69 \cdot M_s + 2.85$, and catalog M_s values. Then, $[m_b(P)^{observed} - m_b(Lg)^{calculated}]$ values were averaged for all 49 explosions. The average is $+0.09 \pm 0.03$ mu, which compares well with the bias estimate obtained by Nuttli from direct comparison of $m_b(P)$ and $m_b(Lg)$ values.

The light-shaded area in Figure 6.2 is a data cloud of NEIC M_s:$m_b(P)$ observations for ~650 earthquakes located in the tectonic areas of southern Asia between 20° to 55°N latitude and 65° to 110°E longitude. I computed centroid values by binning M_s observations into 0.5 mu bins, and computing mean $m_b(P)$ and M_s values for each bin. The results indicate that the centroids are offset from the unified $m_b(Lg)$ relationship, suggesting that $m_b(P)$ values are generally smaller than $m_b(Lg)$ by an average of about 0.3–0.4 mu for events in southern Asia. The offset between magnitudes is smallest for the bin M_s 3.5 to 4.0 and largest for the bin M_s 5.5 to 6.0.

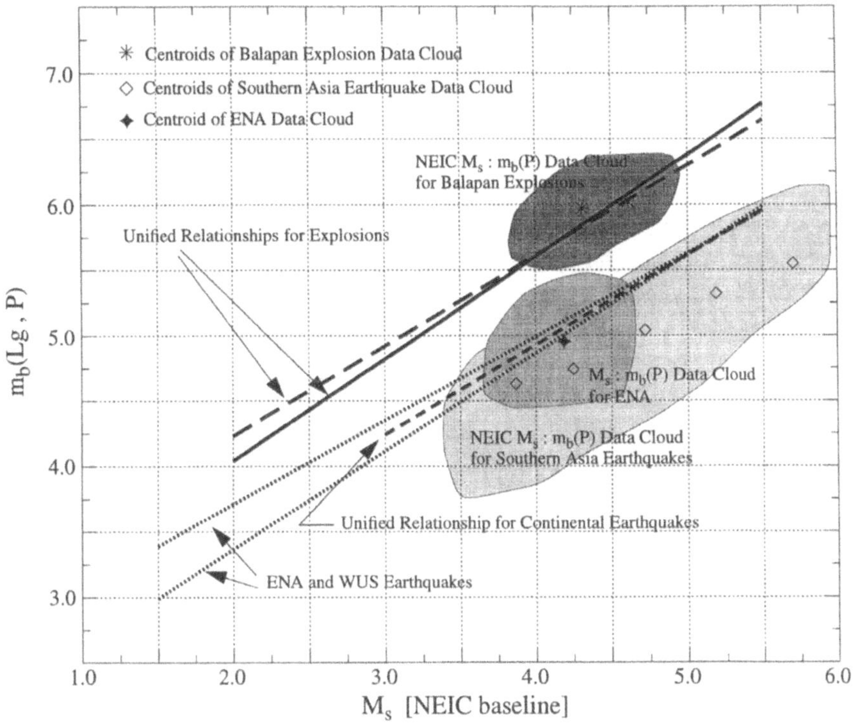

Figure 6.2

Comparison of M_s:$m_b(Lg)$ relationships with M_s:$m_b(P)$ observations shown by data clouds for clarity. Dark-shaded area is data cloud for nuclear explosions detonated at Balapan; medium-shaded cloud is for ENA earthquakes; and light-shaded cloud is for southern Eurasian earthquakes (see text for details). Open diamonds are centroid estimates of the Eurasia data cloud obtained by binning the data into 0.5 mu bins (e.g., M_s 3.5 to 4.0, M_s 4.0 to 4.5, etc.) and computing mean m_b and M_s for each bin. Black diamond is the centroid of the ENA data cloud, and asterisks are centroids for the data cloud of Balapan explosions in two bins, 4.0 to 4.5 and 4.5 to 5.0. Equations for M_s:$m_b(Lg)$ relationships are given in Figure 4.6.

For small events, $m_b(P)$ estimates (and M_s for that matter) tend to be overestimated due to data censoring practices (RINGDAL, 1986), and this would cause the trend of M_s:$m_b(P)$ observations to "flatten" for smaller events, just as the centroid for bin M_s 3.5 to 4.0 shows in Figure 6.2. As discussed in the previous subsection, $m_b(Lg)$ does not show the effects of saturation that $m_b(P)$ does, and increasing offsets for bins with M_s between 5 and 6 are probably related to magnitude saturation in $m_b(P)$.

Summarizing, these results indicate that comparisons of M_s:$m_b(P)$ observations with the unified M_s:$m_b(Lg)$ relationships provide information regarding the regional bias in teleseismic m_b. However, care must be taken with m_b measurements for large and small events. The range M_s 4.0 to 5.0 is probably the least affected by problems due to data censoring and m_b saturation. The results for ENA earthquakes and Balapan explosions produce average offsets between centroids of the observations and the $m_b(Lg)$ relationships in good agreement with previous determinations of m_b bias from direct comparisons of $m_b(P)$ and $m_b(Lg)$ measurements. For southern Asia, M_s:$m_b(P)$ observations have offsets that suggest $m_b(P)$ values are smaller than $m_b(Lg)$ by 0.3 to 0.4 mu. One interpretation of this result is the existence of pervasive regional m_b bias in tectonic areas of southern Asia due to effects of upper-mantle attenuation. Furthermore, lateral variations of earth structure implies that some areas in southern Asia may have m_b bias significantly larger than the average. Such large m_b bias, if it exists, could have impact on both detection estimates and on the ability to discriminate at small magnitudes.

Scaling coefficients for regional magnitudes. Figure 6.3 summarizes scaling coefficients for M_w:m_b relationships in this study (see Tables 3.1 and 4.1). Plotted are slopes for earthquake and explosion data sets with M_w ~3.9–6.5 and ~3.4– ~5.8, respectively (the exact range depends on the phase and region). In general, slopes are greater for explosions than earthquakes for both phases, *Pn* and *Lg*. Slope estimates for *Pn* and *Lg* waves are not significantly different for earthquakes, but slope estimates for *Pn* waves are significantly greater than estimates for *Lg* waves in the case of nuclear explosions. The *Pn* scaling coefficient for explosions is 1.352 ± 0.027, highest of all, while its earthquake counterpart is the lowest at 0.983 ± 0.031. Unified *Lg* data sets for earthquakes and explosions have scaling coefficients, 1.041 ± 0.029 versus 1.174 ± 0.025, respectively, which are not as dissimilar as they are for *Pn* waves.

Of particular interest is the observed difference in scaling rates for *Pn* and *Lg* waves from nuclear explosions. This unexpected result could provide clues about the nature of explosion source functions. I use the term "effective" source functions to refer to the combined effects of source function and near-source propagation. For example, near-source scattering of *Rg* waves into *S* waves, which become trapped in the crust, is a viable mechanism for generating low-frequency (<2 Hz) *Lg* waves from explosions. This mechanism was first proposed by GUPTA *et al.* (1992) and is supported by observational studies (PATTON and TAYLOR, 1995; GUPTA *et al.*, 1997; MYERS *et al.*, 1999). In general, these studies find that the *Rg* scattering mechanism is

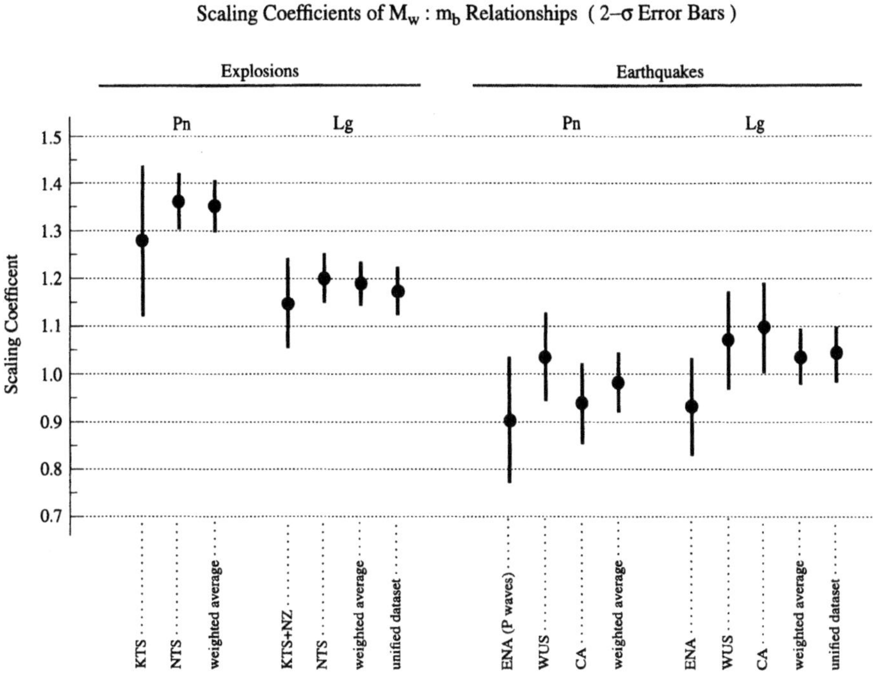

Figure 6.3
Estimate of scaling coefficients for *Pn* and *Lg*. Explosion results are plotted by test site on the left, and earthquakes results by tectonic region on the right . Values of these coefficients may be found on Tables 3.1 and 4.1. The error bars are 2-σ estimates. Note that $m_b(P)$ results for ENA earthquakes were not used in computing the weighted mean for *Pn* slope.

most efficient for frequencies in the neighborhood of 1 Hz, and that near-field *Rg*-to-*P* scattering is not as important in the generation of regional *P* waves. So there are reasons to believe that the *effective Lg* source function may differ from its *Pn* counterpart in ways that could explain the difference in observed scaling rates.

Figure 6.4 shows hypothetical explosion source functions for *Pn* and *Lg* waves. For the purpose of discussion, I use source spectra from the Mueller-Murphy model (MUELLER and MURPHY, 1971) computed for yields of 100 and 1 kt at standard depths of burial (i.e., $120 \cdot W^{0.33333}$ m, where W is yield). At low frequencies, these spectra are proportional to the static level of the reduced displacement potential (RDP) Ψ_∞, which scales as $W^{0.76}$. The bump in the *Pn* source function is related to overshoot in the RDP, which describes motions on the elastic radius of a spherical explosion source (DENNY and JOHNSON, 1991). Note that the bump shifts to higher frequencies as yield decreases and eventually moves out of the 1-Hz band.

On the other hand, I assume that the source spectrum of *Lg* waves is a combination of effects of the Mueller-Murphy model and near-source *Rg*-to-*S* scattering. Based on observational results, the *Rg*-to-*S* scattering function is narrowband, peaking near 1 Hz as shown in Figure 6.4. Thus, the *Lg* effective source function for a 100-kt explosion is peaked near 1 Hz for two reasons, (1) dynamics of motion on the elastic radius (overshoot) and (2) near-source scattering of *Rg* waves. With this in mind, the key difference in scaling behavior between *Pn* and *Lg* waves is revealed by the following analysis.

Rg-to-*S* scattering is a property of the source medium and does not scale with yield as the source function does. As shown in Figure 6.4, the scattering function is unchanged for 100-kt and 1-kt explosions, except for the amplitude. The bump on the *Lg* source spectrum for the 1-kt explosion, while diminishing because of

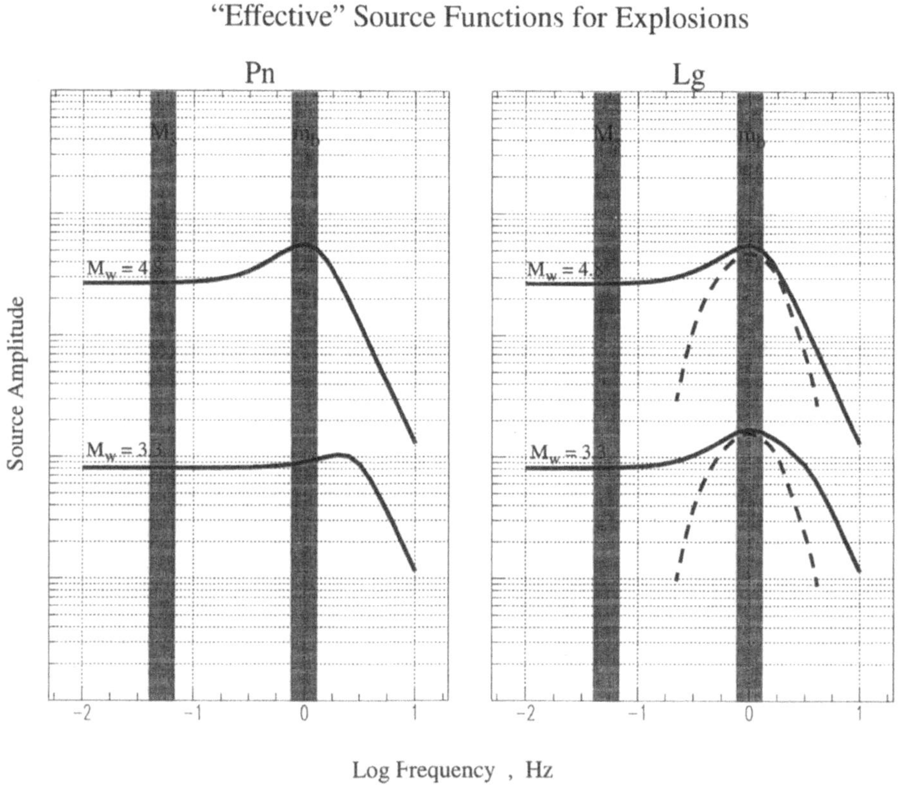

Figure 6.4
Effective source functions for explosions with yields of 100 and 1-kt (solid) and *Rg*-to-*S* scattering functions (dash; see text). The realizations of explosion spectra are based on the Mueller–Murphy model (MUELLER and MURPHY, 1971) for explosions in granite with α, 5 km/s, β, 3 km/s, and ρ, 2.5 g/cm^3. The 1-kt Lg source function is a hybrid, where the source spectrum for 100 kt is shifted straight down and sutured with the 1-kt source function for *Pn* waves at frequencies near 3 Hz. Vertical shading represents approximate frequency bands where time-domain amplitudes of surface waves, *Pn* waves, and *Lg* waves are measured for calculating magnitudes.

overshoot scaling to higher frequencies, remains significant. To see the impact this might have on observed scaling coefficients, let A be the low-frequency amplitude of the Lg source function for the 1-kt explosion and $A + B$ be the amplitude at 1 Hz, where B is the amplitude contributed by Rg-to-S scattering. In the case of Pn waves, if A' is the amplitude of its source function at low frequencies, then the amplitude at 1 Hz is also A' since the spectrum is flat at low frequencies (see Fig. 6.4). Now, for a large yield explosion, like 100 kt, the low frequency level will scale as $W^{0.76}$ for both Pn and Lg source functions. Let K' be an overshoot factor; then the amplitude of the Pn source function at 1 Hz is $K'A'W^{0.76}$, and $[K'AW^{0.76} + BW^c]$ is the 1-Hz amplitude of the Lg function. Note that the overshoot factor K' is greater than 1 and assumed to be the same for both phases, and c is a scaling coefficient on the amplitude of scattered Rg waves. For Rg waves produced by the explosion and/or by spall-related mechanisms (PATTON, 1990), the scaling coefficient c is approximately the same as the low-frequency scaling rate, and the Lg amplitude reduces to $[K' \cdot A + B]W^{0.76}$ at 1 Hz.

To relate the above model to scaling coefficient, $\Delta m_b/\Delta M_w$, note that ΔM_w is proportional to the change in log spectral amplitude at low frequencies, which is the same for Pn and Lg source functions, $0.76 \cdot \log W$. Thus, differences in scaling coefficients are driven by Δm_b which is proportional to the change in log amplitude of 1-Hz Pn waves, $\log K' + 0.76 \cdot \log W$, on one hand and to the change in log amplitude of 1-Hz Lg waves, $\log K + 0.76 \cdot \log W$, where $K = (K' \cdot A + B)/(A + B)$, on the other. Clearly, the scaling coefficients for Pn and Lg waves are not the same for this model, and $|\Delta m_b(Pn)/\Delta M_w|$ is greater than $|\Delta m_b(Lg)/\Delta M_w|$ since $K' > K > 1$.

Under this model, scaling coefficients of Pn and Lg waves are expected to be different for nuclear explosions, because their respective source functions are not the same and do not scale similarly. Indeed, their effective source functions are controlled by different physical processes in the 1-Hz frequency band. The scaling coefficient for Pn waves is greater because the bump on the source spectrum shifts to higher frequencies as yield decreases and eventually moves out of the 1-Hz band, while the bump on the effective source spectrum for Lg waves remains significant at small yields because of the amplitude contribution from Rg-to-S scattering.

7. Conclusion

To my knowledge, this is the first study to rigorously test Nuttli's claim of transportability. The central U. S. calibration equation which equates 1-Hz Lg waves of 110 microns amplitude with a magnitude 5.0 events is at the heart of this claim. For this to be true in any continental region, 1-Hz Lg Green's functions must be invariant with respect to changes in continental earth structures. In this study, statistical tests of M_w:$m_b(Lg)$ scaling relationships for earthquakes in three diverse regions validates Nuttli's calibration at the 95% confidence level and furthermore shows $m_b(Lg)$ scaling is the same within the scatter of the data for M_w greater than ~4.2. Some of the greatest variations of tectonics and crustal structures worldwide are represented by landforms of central Asia, eastern North America and western United States. It does not require a great "leap of faith" to argue that transportability of $m_b(Lg)$ holds for all continental regions.

Another condition on transportability is the ability to make accurate corrections for Lg attenuation on a regional basis. In a broad regional sense, this condition was met in this study, but it is important to point out that Lg Q_0 can vary significantly on regional and subregional scales. Among the three continental regions, data scatter was largest on M_w:$m_b(Lg)$ scaling plots for central Asia, where lateral heterogeneity of earth structures is most significant. Assuming one Q model for the entire region of central Asia is too simplistic. In addition to possible variations of intrinsic Q, the presence of deep sedimentary basins and variations in crustal thickness and topography can cause scattering Q to vary from path to path. At the extreme, propagation of 1-Hz energy in the Lg phase can be blocked completely. Because of its importance for discrimination reliability and for reducing the scatter in $m_b(Lg)$ observations, more work is needed to map propagation blockages and variations of Q_0 on finer length scales than are resolvable with current tomographic data sets.

The importance of $m_b(Lg)$ transportability for earthquake studies as well as treaty monitoring cannot be overstated. In earthquake hazard studies, $m_b(Lg)$ can provide a standard measure of source size for comparing ground motions between different sites when M_o measurements are not available. This could have significant impact on enlarging data sets of ground motion comparisons for small earthquakes. Improvements in CTBT monitoring are also advanced by transportability of $m_b(Lg)$ and include the following: (1) M_w:$m_b(Lg)$ relationships unify observations of natural and man-made sources located in diverse tectonic and geological settings; this unification should help simplify discrimination; (2) M_s:$m_b(Lg)$ relationships shed light on discrimination behavior at small magnitudes and as a function of source location (e.g., small earthquakes in stable and tectonic regions), and (3) M_s:$m_b(Lg)$ relationships lend themselves to investigations of regional $m_b(P)$ bias. Knowledge of regional bias variations is important for yield estimation and M_s:m_b discrimination. If carefully applied, earthquake observations of $[m_b(P) - m_b(Lg)]$ can provide information about these variations. Furthermore, comparisons of M_s:$m_b(P)$ observations with unified M_s:$m_b(Lg)$ relationships may offer a quick way to survey large continental areas. Such surveys will be useful for targeting source regions for intensive calibration studies that might require measurements of $m_b(Lg)$ for specific groups of events and investigations of upper-mantle P-wave attenuation.

Acknowledgments

It is a pleasure to acknowledge my colleagues' assistance with the project and manuscript preparation. Measurements of $m_b(Lg)$ and $m_b(Pn)$ for NTS explosions and WUS earthquakes began while I was at Lawrence Livermore National Laboratory (LLNL). I have drawn upon waveform and measurement databases for $m_b(Pn)$, which Marv Denny, Eileen Vergino, Bill Walter and others were involved in developing. I am grateful to LLNL colleagues, Stan Ruppert and Kevin Mayeda, for providing LLNL waveform data for some of the more recent WUS earthquakes used in this study. Mike Pasyanos and Doug Dreger (through Kevin Mayeda) made M_0 measurements available of many WUS earthquakes. In addition to his M_0 data, Doug Dreger of UC Berkeley was helpful answering my questions about Berkeley M_L. Hong Kie Thio of Woodward-Clyde, Pasadena, provided M_0 data for many southern California earthquakes. Jim Pechmann and colleagues at the University of Utah provided information on M_L observations for Utah and Arizona earthquakes. Hans Hartse and Aaron Velasco of Los Alamos National Laboratory (LANL) searched digital catalogs for M_s:m_b data used in section 6 of this study. Hans and Mike Pasyanos provided technical reviews which improved the manuscript. Jeff Stevens of Maxwell Technologies was helpful with discussions about the PIDC M_s:m_b data set and assisted with a PIDC data set for ENA. I benefited from many helpful discussions with Steve Taylor and Scott Phillips of LANL, which helped to focus (and defocus when necessary) my thoughts on various aspects of this work. This work was performed under the auspices of the Department of Energy, DOE, by the Los Alamos National Laboratory under contract W-7405-ENG-36. This is contribution #0005 of the Los Alamos Seismic Research Center.

Table A.1

Station and network $m_b(Pn)$ and $m_b(Lg)$ values for Western United States earthquakes

yrmodahrmnsc	m_b Pn	$m_b(Pn)$				m_b Lg	$m_b(Lg)$				Region
		ELK	KNB	LAC	MNV		ELK	KNB	LAC	MNV	
710805175817	**3.98**	3.842	4.155	3.992	3.947	**4.37**	4.220	4.405	4.480	4.383	Mass Mtn NV
790812113119	**3.18**	3.158	2.655	3.725	×[a]	**3.72**	3.708	3.721	3.715	×	southern NV
790816033744	**2.96**	2.963	2.427	3.484	×	**3.43**	3.561	3.382	3.336	×	southern NV
791225141710	**3.67**	3.565	3.516	3.937	Δ[b]	**3.99**	3.897	3.987	4.089	3.972	near NTS NV
800115202822	**3.63**	3.490	4.175	3.236	3.636	**3.82**	4.111	3.567	3.929	3.692	southern Sierras
800225234332	**3.86**	S/N[c]	4.262	3.889	3.415	**3.97**	4.301	3.782	3.992	3.802	southern Sierras
800408001341	**4.62**	5.060	4.314	4.489	Δ	**4.87**	BD[d]	4.699	4.726	5.172	W-central NV
800428135534	**3.94**	4.572	3.748	S/N	3.489	**3.87**	3.684	3.859	BD	4.063	N-western NV
800524100336	**4.44**	×	4.236	4.730	4.346	**4.66**	×	4.476	4.714	4.794	central UT
800527145057	**5.79**	5.939	5.627	5.798	Δ	**6.16**	6.199	6.252	5.979	6.220	Mammoth Lakes CA
801025003059	**2.99**	2.965	3.019	S/N	Δ	**3.51**	3.480	3.353	3.542	3.665	near NTS NV
801128182113	**4.94**	5.272	4.413	5.144	Δ	**5.32**	4.965	5.054	5.521	5.741	northern Sierras
810405054040	**4.27**	3.891	Δ	4.789	4.129	**4.36**	4.506	Δ	4.236	4.344	southern UT
810429115552	**3.83**	3.959	3.537	3.984	Δ	**3.78**	3.917	3.657	3.480	4.051	W-central NV
810514051104	**3.97**	3.747	3.976	4.190	S/N	**3.69**	3.230	3.887	3.817	3.843	central UT (mine)
811119180155	**3.76**	3.622	3.732	3.929	Δ	**3.71**	3.298	3.973	3.759	3.808	central NV
811201161850	**4.02**	×	3.839	4.191	Δ	**4.63**	×	4.499	4.759	Δ	W-central NV
811219205652	**4.12**	4.158	3.970	4.247	Δ	**4.20**	4.000	4.291	4.321	Δ	W-central NV
820124154407	**4.09**	4.165	3.966	4.147	Δ	**4.08**	4.306	4.128	3.902	3.991	S-western NV
820316084700	**3.48**	3.464	3.418	3.561	3.464	**3.77**	3.782	3.551	4.068	3.689	near NTS NV
820512192924	**3.49**	3.525	3.093	3.692	3.635	**3.90**	4.283	3.816	3.646	3.854	Pahranagat SZ NV
820524121327	**4.13**	3.853	4.016	4.582	4.054	**4.26**	3.992	T_s[e]	4.444	4.332	central UT
820529130225	**4.56**	4.528	4.182	4.622	4.915	**4.15**[f]	4.108	4.078	4.256	T_1[g]	Geysers CA
820621064338	**4.10**	3.795	4.116	4.281	4.211	**3.97**	3.921	3.943	4.200	3.817	Cascades CA
820706021043	**4.30**	3.810	×	4.619	4.463	**4.62**	4.575	×	BD	4.656	Pahranagat SZ NV
820903185823	**4.15**	3.865	3.937	4.266	4.530	**4.11**	3.862	4.044	4.182	4.347	northern CA
820924074024	**4.99**	4.963	4.945	5.063	Δ	**5.35**	5.225	5.262	5.550	Δ	S-western NV
821001013335	**4.16**	4.435	3.891	4.153	Δ	**4.12**	4.200	4.010	4.146	Δ	S-western NV
830223111020	**3.90**	3.793	3.880	3.800	4.137	**3.78**	3.888	T_s	3.669	3.782	Lake Mead NV

830604113740	**3.44**	3.361	2.679	3.718	4.004	**3.52**	3.491	3.456	3.463	3.657	Pahranagat SZ NV
831209085841	**4.20**	×	×	4.685	3.713	**4.33**	×	×	4.188	4.471	central UT
840802110134	**3.49**	3.183	Δ	3.669	3.609	**3.76**[f]	3.916	3.566	3.654	3.890	Pahranagat SZ NV
840908061640	**5.32**	5.616	4.759	5.390	5.514	**5.23**[f]	5.331	5.158	T_1	5.193	central ID
840917191107	**3.38**	×	3.172	3.578	3.395	**3.71**	×	3.641	3.520	3.973	E-central NV
841123180825	**5.54**	×	5.723	5.355	Δ	**5.90**	×	5.936	5.940	5.824	Round Valley CA
850124112721	**4.31**	4.370	BD	4.257	Δ	**4.94**[f]	4.877	BD	5.012	Δ	E-central CA
851210061025	**3.70**	×	3.145	3.960	3.985	**3.66**	×	3.722	3.573	3.681	Pahranagat SZ NV
880814200303	**4.54**	×	4.539	×	×	**5.48**	×	5.482	×	×	central UT
880831164516	**5.50**	×	5.500	×	×	**4.94**	×	4.942	×	×	Mexicali MX
890119065328	**5.23**	5.017	5.645	×	5.035	**5.37**	5.309	5.557	×	5.239	Malibu CA
890130040622	**4.80**	×	4.802	×	×	–	×	BD	×	×	central UT
891019101435	**4.66**	4.380	4.687	5.289	4.285	**4.66**	4.502	4.855	4.673	4.624	Loma Prieta CA
901024061520	**5.28**	5.584	5.166	5.104	BD	**5.64**	5.755	5.778	5.371	BD	Lee Vining CA
910628144354	**5.44**	×	5.695	Δ	5.183	**5.71**[f]	×	5.706	BD	5.713	Sierra Madre CA
920423045023	**6.29**	×	6.591	Δ	5.994	–	×	BD	BD	BD	Joshua Tree CA
920504161949	**5.04**	×	BD	×	5.037	**5.08**	×	BD	×	5.076	Joshua Tree CA
920506023843	**4.72**	×	BD	Δ	4.719	**4.51**	×	BD	BD	4.513	Joshua Tree CA
920518154417	**4.91**	×	BD	Δ	4.909	**4.76**	×	BD	BD	4.762	Joshua Tree CA
920628115735	**6.54**	×	6.801	×	6.284	**7.24**	×	7.244	×	BD	Landers CA
920629101422	**5.53**	×	5.964	×	5.100	**6.27**	×	6.436	×	6.106	Skull Mtn NV
920629103102	**4.66**	×	4.909	×	4.420	**5.04**	×	5.038	×	5.035	Skull Mtn NV
920629155239	**3.89**	×	4.173	×	3.602	**4.29**	×	4.497	×	4.084	Skull Mtn NV
920629170116	**3.81**	×	4.080	×	3.542	**4.17**	×	4.227	×	4.120	Skull Mtn NV
920630143811	**5.20**	×	5.236	Δ	5.157	–	×	T_1	BD	T_1	Landers CA
920630160624	**3.50**	×	3.834	×	3.159	**3.73**	×	3.699	×	3.768	Skull Mtn NV
920705065412	**4.38**	×	4.623	×	4.137	**4.82**	×	5.113	×	4.521	Skull Mtn NV
920705084838	**3.05**	×	3.314	×	2.790	**3.29**	×	3.242	×	3.339	Skull Mtn NV
920711181416	**5.09**	4.983	5.204	×	×	**5.61**	5.636	5.589	×	×	Garlock CA
920809080706	**3.09**	×	3.495	×	2.680	**3.42**	×	3.497	×	3.332	Skull Mtn NV
920902102620	**5.82**	×	Δ	6.135	5.514	**5.60**	×	Δ	5.532	5.678	St George UT
920913114620	**4.06**	3.795	4.652	×	3.732	**4.44**	4.084	4.633	×	4.608	Skull Mtn NV
920914053550	**2.83**	S/N	3.081	×	2.586	**2.94**	2.685	3.041	×	3.085	Skull Mtn NV
920919230447	**4.69**	×	×	×	4.694	**4.69**	×	×	×	4.692	Geysers CA
920925211223	**3.00**	2.987	3.328	×	2.699	**3.29**	3.143	3.192	×	3.546	Skull Mtn NV

Table A.1

Continued

yrmodahrmnsc	m_b Pn	$m_b(Pn)$				$m_b(Lg)$	m_bLg				Region
		ELK	KNB	LAC	MNV		ELK	KNB	LAC	MNV	
920925211638	**3.17**	2.960	3.561	×	2.995	**3.44**	3.189	3.626	×	3.519	Skull Mtn NV
921008122356	**2.86**	2.905	3.401	×	2.277	**3.22**	3.154	3.353	×	3.149	Skull Mtn NV
930116062934	**4.92**	×	4.827	5.541	4.401	**5.18**	×	5.155	4.973	5.399	Gilroy CA
930210214835	**4.41**	4.129	4.878	×	4.235	**5.02**	4.726	4.951	×	5.369	Pyramid Lake NV
930325133435	**5.83**	5.448	5.511	6.357	5.991	**5.10**	T_1	T_1	T_1	5.102	Scotts Mill Or
930425092950	**5.16**	×	Δ	5.245	5.072	**5.29**	×	5.525	5.135	5.200	Cataract Ck AZ
930429082100	**5.68**	×	Δ	5.801	5.560	**5.95**	×	6.165	5.931	5.765	Cataract Ck AZ
930517232049	**5.84**	×	5.593	6.082	Δ	**6.13**	×	6.019	6.527	5.840	Eureka Valley CA
930518010306	**4.90**	4.800	4.929	4.974	Δ	**5.10**	5.168	5.274	4.933	5.010	Eureka Valley CA
930518234853	**4.93**	4.853	4.935	4.989	Δ	**5.19**	5.663	T_1	4.600	5.314	Eureka Valley CA
930518235740	**4.58**	4.446	4.612	4.676	Δ	**4.67**	4.773	4.864	4.631	4.427	Eureka Valley CA
930519141322	**5.21**	5.006	5.410	5.213	Δ	**5.13**	4.957	5.463	5.008	5.094	Eureka Valley CA
930520201414	**4.32**	×	×	4.415	4.227	**4.62**	×	×	4.811	4.429	Coso Hot Springs CA
930528044740	**4.76**	×	×	4.648	4.871	**5.34**	×	×	5.274	5.410	Bakersfield CA
930811223304	**5.09**	4.835	5.250	5.848	4.437	**5.15**	5.060	5.281	5.043	5.221	Alum Rock CA
930923062110	**3.92**	3.445	×	×	4.401	**4.15**	4.035	×	×	4.270	Kalmath Falls OR
931114122534	**4.28**	×	×	4.790	3.770	**4.44**	×	×	4.359	4.523	Parkfield CA
931204221519	**5.47**	×	5.469	×	×	**5.69**	×	5.687	×	×	Klamath Falls OR
940117123055	**6.14**	×	6.710	5.749	5.973	**6.77**	×	6.876	6.660	T_1	Northridge CA
940118004308	**4.86**	4.901	5.120	4.860	4.559	**5.75**	5.827	5.713	5.712	T_1	Northridge CA
940320212012	**4.91**	5.128	5.474	4.400	4.620	**5.48**	5.407	5.545	5.555	5.400	Northridge CA
940406190104	**4.70**	×	4.965	Δ	4.427	**4.70**	×	4.909	Δ	4.489	San Bernadino CA
940607133003	**5.15**	×	5.013	×	5.281	**5.42**	×	5.393	×	5.437	Borah Peak ID
940616162427	**5.00**	×	4.996	Δ	×	**5.12**	×	5.117	Δ	×	Landers CA
940912235709	**5.01**	5.550	4.843	4.635	Δ	**5.14**	5.355	5.075	4.751	5.395	Dbl Spring Flat NV
941114012823	**4.53**	×	×	×	4.533	**4.32**	×	×	×	4.319	Cape Mendocino CA
941220102747	**4.69**	×	4.889	×	4.493	**4.80**	×	4.760	×	4.850	Parkfield CA
941226141029	**5.70**	×	×	×	5.697	**5.16**	×	×	×	5.157	Cape Mendocino CA

[a] Data not available; [b] For *Pn*, distance ≤*Pn-Pg* cross-over distance; for *Lg*, distance too short; [c] *Pn* signal failed *S/N* test; [d] Bad data for *Pn* or *Lg* (i.e., clipped, spikes, dropouts, etc); [e] Wave period <0.55 sec; [f] Based on 3rd peak amplitude measurement; [g] Wave period >1.45 sec.

References

AKI, K., BOUCHON, M., and REASENBERG, P. (1974), *Seismic Source Function for an Underground Nuclear Explosion*, Bull. Seismol. Soc. Am. *64*, 131–148.

ATKINSON, G. M. (1993), *Earthquake Source Spectra in Eastern North America*, Bull. Seismol. Soc. Am. *83*, 1778–1798.

BAKUN, W. H. (1984), *Seismic Moments, Local Magnitudes, and Coda Duration Magnitudes for Earthquakes in Central California*, Bull. Seismol. Soc. Am. *74*, 439–458.

BARKER, J. S., and LANGSTON, C. A. (1983), *A Teleseismic Body-wave Analysis of the May 1980 Mammoth Lakes, California, Earthquakes*, Bull. Seismol. Soc. Am. *73*, 419–434.

BASHAM, P. W., WEICHERT, D. H., and BERRY, M. J. (1979), *Regional Assessment of Seismic Risk in Eastern Canada*, Bull. Seismol. Soc. Am. *69*, 1567–1602.

BECK, S. L., and PATTON, H. J. (1991), *Inversion of Regional Surface-wave Spectra for Source Parameters of Aftershocks from the Loma Prieta Earthquake*, Bull. Seismol. Soc. Am. *81*, 1726–1736.

BENT, A. L. (1992), *A Re-examination of the 1925 Charlevoix, Quebec, Earthquake*, Bull. Seismol. Soc. Am. *82*, 2097–2114.

BOATWRIGHT, J. (1994), *Regional Propagation Characteristics and Source Parameters of Earthquakes in Northeastern North America*, Bull. Seismol. Soc. Am. *84*, 1–15.

BOATWRIGHT, J., and CHOY, G. L. (1992), *Acceleration Source Spectra Anticipated for Large Earthquakes in Northeastern North America*, Bull. Seismol. Soc. Am. *82*, 660–682.

BOLT, B. A., and HERRAIZ, M. (1983), *Simplified Estimation of Seismic Moment from Seismograms*, Bull. Seismol. Soc. Am. *73*, 735–748.

BOORE, D. M., and ATKINSON, G. M. (1987), *Source Spectra for the 1988 Saguenay, Quebec, Earthquakes*, Bull. Seismol. Soc. Am. *77*, 440–467.

BOOTH, D. C., MARSHALL, P. D., and YOUNG, J. B. (1974), *Long- and Short-period P-wave Amplitudes from Earthquakes in the Range 0°–114°*, Geophys. J. R. astr. Soc. *39*, 523–537.

BRUNE, J. N. (1970), *Tectonic Stress and the Spectra of Seismic Shear Waves from Earthquakes*, J. Geophys. Res. *75*, 4997–5009.

BURGER, R. W., LAY, T., WALLACE, T. C., and BURDICK, L. J. (1986), *Evidence of Tectonic Release in Long-period S Waves from Underground Nuclear Explosions at the Novaya Zemlya Test Sites*, Bull. Seismol. Soc. Am. *76*, 733–755.

CHAEL, E. P. (1987), *Spectral Scaling of Earthquakes in the Miramichi Region of New Brunswick*, Bull. Seismol. Soc. Am. *77*, 347–365.

CHOY, G. L., BOATWRIGHT, J., DEWEY, J. W., and SIPKIN, S. A. (1983), *A Teleseismic Analysis of the New Brunswick Earthquake of January 9, 1982*, J. Geophys. Res. *88*, 2199–2212.

CHUNG, D. H., and BERNREUTER, D. L. (1981), *Regional Relationships among Earthquake Magnitude Scales*, Rev. Geophys. Space Phys. *19*, 649–663.

CLAERBOUT, J. F., *Fundamentals of Geophysical Data Processing* (McGraw-Hill, San Francisco, California 1976).

COHEE, B. P., and LAY, T. (1988), *Modeling Teleseismic SV Waves from Underground Explosions with Tectonic Release: Results for Southern Novaya Zemlya*, Bull. Seismol. Soc. Am. *78*, 1158–1178.

CONG, L., XIE, J., and MITCHELL, B. J. (1996), *Excitation and Propagation of Lg from Earthquakes in Central Asia with Implications for Explosion/Earthquake Discrimination*, J. Geophys. Res. *101*, 27,779–27,789.

DENNY, M. D., and GOODMAN, D. M. (1990), *A Case Study of the Seismic Source Function: Salmon and Sterling Reevaluated*, J. Geophys. Res. *95*, 19,705–19,723.

DENNY, M. D., and JOHNSON, L. R. (1991), *The Explosion Seismic Source Function: Models and Scaling Laws Reviewed.* In *Explosion Source Phenomenology*, Am. Geophys. Monograph *65*, 1–24.

DENNY, M. D., TAYLOR, S. R., and VERGINO, E. S. (1987), *Investigation of m_b and M_s Formulas for the Western United States and Their Impact on the M_s/m_b Discriminant*, Bull. Seismol. Soc. Am. *77*, 987–995.

DOUGLAS, A., and MARSHALL, P. D. (1996) *Seismic Source Size and Yield for Nuclear Explosions.* In *NATO Proc. on Monitoring a Comprehensive Test-Ban Treaty* (eds. Husebye, E. S. and Dainty, A. M.) (Kluwer Academic Publishers, Hingham Massachusetts 1996) pp. 309–353.

EBEL, J. E. (1982), *M_L Measurements for Northwestern United States Earthquakes*, Bull. Seismol. Soc. Am. *72*, 1367–1378.

EBEL, J. E., SOMERVILLE, P. G., and MCIVER, J. D. (1986), *A Study of the Source Parameters of Some Large Earthquakes of Northeastern North America*, J. Geophys. Res. *91*, 8231–8247.

EKSTRÖM, G., and DZIEWONSKI, A. M. (1988), *Evidence of Bias in Estimations of Earthquake Size*, Nature *332*, 319–323.

EKSTRÖM, G., and RICHARDS, P. G. (1994), *Empirical Measurements of Tectonic Moment Release in Nuclear Explosions from Teleseismic Surface Waves and Body Waves*, Geophys. J. Int. *117*, 120–140.

FIELDING, E., BARAZANGI, M., ISACKS, B., and SEBER, D. (1994), *A network-accessible geological and geophysical database for Eurasia, north Africa, and the Middle East: Digital database development for the Middle East and north Africa.* In *Proc. of the 16th Annual Seismic Research Symposium* (eds. Cipar, J. J., Lewkowicz, J. P., and McPhetres, J. M.), Report PL-TR-94-2217, Phillips Laboratory, Hanscom AFB, Massachusetts.

GLASSMOYER, G., and BORCHERDT, R. D. (1990), *Source Parameters and Effects of Bandwidth and Local Geology on High-frequency Ground Motions Observed for Aftershocks of the Northeastern Ohio Earthquake of 31 January 1986*, Bull. Seismol. Soc. Am. *80*, 889–912.

GUPTA, I. N., CHAN, W. W., and WAGNER, R. A. (1992), *A Comparison of Regional Phases from Underground Nuclear Explosions at East Kazakh and Nevada Test Sites*, Bull. Seismol. Soc. Am. *82*, 352–382.

GUPTA, I. N., ZHANG, T. R., and WAGNER, R. A. (1997), *Low-frequency Lg from NTS and Kazakh Nuclear Explosions – Observations and Interpretations*, Bull. Seismol. Soc. Am. *87*, 1115-1125.

GUYTON, J. W. (1964), *Systematic deviations of magnitude from body waves at seismograph stations in the United States.* In *Proc. VESIAC Conference on Seismic Event Magnitude Determination*, AD441592, University of Michigan, Ann Arbor, Michigan.

HANKS, T. C., and BOORE, D. M. (1984), *Moment-magnitude Relations in Theory and Practice*, J. Geophys. Res. *89*, 6229–6235.

HANKS, T. C., and KANAMORI, H. (1979), *A Moment-magnitude Scale*, J. Geophys. Res. *84*, 2348–2350.

HASEGAWA, H. S. (1983), *Lg Spectra of Local Earthquakes Recorded by the Eastern Canada Telemetered Network and Spectral Scaling*, Bull. Seismol. Soc. Am. *73*, 1041–1061.

HASEGAWA, H. S., and WETMILLER, R. J. (1980), *The Charlevoix Earthquake of 19 August 1979 and its Seismo-tectonic Environment*, Earthquake Notes *51*, 23–37.

HERRMANN, R. B. (1979), *Surface Wave Focal Mechanisms for Eastern North American Earthquakes with Tectonic Implications*, J. Geophys. Res. *84*, 3543–3552.

HERRMANN, R. B. (1980), *Q Estimates Using the Coda of Local Earthquakes*, Bull. Seismol. Soc. Am. *70*, 447–468.

HERRMANN, R. B., and KIJKO, A. (1983), *Short-period Lg Magnitudes: Instrument, Attenuation, and Source Effects*, Bull. Seismol. Soc. Am. *73*, 1835–1850.

HERRMANN, R. B., and NUTTLI, O. W. (1982), *Magnitude: The Relation of M_L to m_{bLg}*, Bull. Seismol. Soc. Am. *72*, 389–397.

HERRMANN, R. B., LANGSTON, C. A., and ZOLLEWEG, J. E. (1982), *The Sharpsburg, Kentucky, Earthquake of 27 July 1980*, Bull. Seismol. Soc. Am. *72*, 1219–1239.

HORNER, R. B., STEVENS, A. E., HASEGAWA, H. S., and LEBLANC, G. (1978), *Focal Parameters of the July 12, 1975, Maniwaki, Quebec, Earthquake – An Example of Intraplate Seismicity in Eastern Canada*, Bull. Seismol. Soc. Am. *68*, 619–640.

HORNER, R. B., WETMILLER, R. J., and HASEGAWA, H. S. (1979), *The St-Donat, Quebec, Earthquake Sequence of February 18–23, 1978*, Can. J. Earth Sc. *16*, 1892–1898.

JARPE, S. P. (1989), *Lawrence Livermore National Laboratory Nevada Test Site Regional Seismic Network*, Report No. LLL-TB-100, Lawrence Livermore National Lab., University of California.

JOHNSTON, A. C. (1996), *Seismic Moment Assessment of Earthquakes in Stable Continental Regions – I. Instrumental Seismicity*, Geophys. J. Int. *124*, 381–414.

JULIAN, B. R., and SIPKIN, S. A. (1985), *Earthquake Processes in the Long Valley Caldera Area, California*, J. Geophys. Res. *90*, 11,155–11,169.

KANAMORI, H. (1990), *Pasadena Very-broad-band System and Its Use for Real-time Seismology*, U. S. Geol. Surv. Open File Report *90–98*.

KANAMORI, H., MORI, J., HAUKKSSON, E., HEATON, T. H., HUTTON, L. K., and JONES, L. M. (1993), *Determination of Earthquake Energy Release and M_L Using Terrascope*, Bull. Seismol. Soc. Am. *83*, 330–346.

KIM, W.-Y. (1998), *The M_L Scale in Eastern North America*, Bull. Seismol. Soc. Am. *88*, 935–951.

LI, Y., DOLL, C., and TOKSÖZ, N. M. (1995), *Source Characterization and Fault Plane Determinations for $m_{bLg} = 1.2$ to 4.4 Earthquakes in the Charlevoix Seismic Zone, Quebec, Canada*, Bull. Seismol. Soc. Am. *85*, 1604–1621.

MARTYNOV, V. G., VERNON, F. L., MELLORS, R. J., and PAVLIS, G. L. (1999), *High-frequency Attenuation in the Crust and Upper Mantle of the Northern Tien Shan*, Bull. Seismol. Soc. Am. *89*, 215–238.

MAYEDA, K., and WALTER, W. R. (1996), *Moment, Energy, Stress Drop, and Source Spectra of Western United States Earthquakes from Regional Coda Envelopes*, J. Geophys. Res. *101*, 11,195–11,208.

MITCHELL, B. J., PAN, Y., XIE, J., and CONG, L. (1997), *Lg Coda Q Variation across Eurasia and Its Relation to Crustal Evolution*, J. Geophys. Res. *102*, 22,767–22,779.

MUELLER, R. A., and MURPHY, J. R. (1971), *Seismic Characteristics of Underground Nuclear Detonations, Part 1: Seismic Spectrum Scaling*, Bull. Seismol. Soc. Am. *61*, 1675–1692.

MURPHY, J. R. (1977), *Seismic Source Functions and Magnitude Determinations for Underground Nuclear Explosions*, Bull. Seismol. Soc. Am. *67*, 135–158.

MURPHY, J. R., STEVENS, J. L., O'NEILL, D. C., BARKER, B. W., MCLAUGHLIN, K. L., and MARSHALL, M. E. (1991), *Development of a Comprehensive Seismic Yield Estimation System for Underground Nuclear Explosions*, Phillips Laboratory Report No. *PL-TR-91-2161*.

MYERS, S. C., WALTER, W. R., MAYEDA, K., and GLENN, L. (1999), *Observations in Support of Rg Scattering as a Source for Explosion S Waves: Regional and Local Recordings of the 1997 Kazakhstan Depth of Burial Experiment*, Bull. Seismol. Soc. Am. *89*, 544–549.

NABELEK, J. L. (1984), *Determination of Earthquake Source Parameters from Inversion of Body Waves*, Ph.D. Thesis, Mass. Inst. of Technol., Cambridge, Massachusetts.

NABELEK, J., and SUAREZ, G. (1989), *The 1983 Goodnow Earthquake in the Central Adirondacks, New York: Rupture of a Simple, Circular Crack*, Bull. Seismol. Soc. Am. *79*, 1762–1777.

NUTTLI, O. W. (1973), *Seismic Wave Attenuation and Magnitude Relations for Eastern North America*, J. Geophys. Res. *78*, 876–885.

NUTTLI, O. W. (1986a), *Yield Estimates of Nevada Test Site Explosions Obtained from Seismic Lg Waves*, J. Geophys. Res. *91*, 2137–2151.

NUTTLI, O. W. (1986b), *Lg Magnitudes of Selected East Kazakhstan Underground Explosions*, Bull. Seismol. Soc. Am. *76*, 1241–1251.

NUTTLI, O. W. (1987), *Lg Magnitudes of Degelen, East Kazakhstan, Underground Explosions*, Bull. Seismol. Soc. Am. *77*, 679–681.

NUTTLI, O. W. (1988), *Lg Magnitudes and Yield Estimates for Underground Novaya Zemlya Nuclear Explosions*, Bull. Seismol. Soc. Am. *78*, 873–884.

NUTTLI, O. W., and ZOLLEWEG, J. E. (1974), *The Relation between Felt Area and Magnitude for Central United States Earthquakes*, Bull. Seismol. Soc. Am. *64*, 73–85.

NUTTLI, O. W., BOLLINGER, G. A., and GRIFFITHS, D. W. (1979), *On the Relation between Modified Mercalli Intensity and Body-wave Magnitude*, Bull. Seismol. Soc. Am. *69*, 893–909.

PASYANOS, M. E., DREGER, D., and ROMANOWICZ, B. (1996), *Toward Real-time Estimation of Regional Moment Tensors*, Bull. Seismol. Soc. Am. *85*, 1255–1269.

PASYANOS, M. E., DREGER, D., and ROMANOWICZ, B. (1997), *Differences in High and Low Frequency Observations of Earthquakes at the Geysers, California Geothermal Region*, Seismol. Res. Lett. Abstr. *68* (2) 291–334.

PATTON, H. J. (1982), *Measurements of Rayleigh-wave Phase Velocities in Nevada: Implications for Explosion Sources and the Massachusetts Mountain Earthquake*, Bull. Seismol. Soc. Am. *72*, 1329–1349.

PATTON, H. J. (1988), *Application of Nuttli's Method to Estimate Yield of Nevada Test Site Explosions Recorded on Lawrence Livermore National Laboratory's Digital Seismic System*, Bull. Seismol. Soc. Am. *78*, 1759–1772.

PATTON, H. J. (1990), *Characterization of Spall from Observed Strong Ground Motions on Pahute Mesa*, Bull. Seismol. Soc. Am. *80*, 1326–1345.

PATTON, H. J. (1991), *Seismic moment estimation and the scaling of the long-period explosion source function.* In *Explosion Source Phenomenology*, Am. Geophys. Monograph *65*, 171–183.

PATTON, H. J. (1998), *Bias in the Centroid Moment Tensor for Central Asian Earthquakes: Evidence from Regional Surface Wave Data*, J. Geophys. Res. *103*, 26,963–26,974.

PATTON, H. J., and TAYLOR, S. R. (1995), *Analysis of Lg Spectral Ratios from NTS Explosions: Implications for the Source Mechanisms of Spall and the Generation of Lg Waves*, Bull. Seismol. Soc. Am. *85*, 220–236.

PATTON, H. J., and WALTER, W. R. (1993), *Regional Moment: Magnitude Relations for Earthquakes and Explosions*, Geophys. Res. Lett. *20*, 277–280.

PATTON, H. J., and WALTER, W. R. (1994), *Correction to "Regional Moment: Magnitude Relations for Earthquakes and Explosions" by Howard J. Patton and William R. Walter.* Geophys. Res. Lett. *21*, p. 743.

PATTON, H. J., and ZANDT, G. (1991), *Seismic Moment Tensors of Western US Earthquakes and Implications for the Tectonic Stress Field*, J. Geophys. Res. *96*, 18,245–18,259.

PRIESTLEY, K. F., and PATTON, H. J. (1997), *Calibration of $m_b(Pn)$, $m_b(Lg)$ Scales and Transportability of the M_0:m_b Discriminant to New Tectonic Regions*, Bull. Seismol. Soc. Am. *87*, 1083–1099.

PRIESTLEY, K. F., SMITH, K. D., and COCKERHAM, R. S. (1988), *The 1984 Round Valley, California Earthquake Sequence*, Geophys. J. Int. *95*, 215–235.

RINGDAL, F. (1986), *Study of Magnitudes, Seismicity, and Earthquake Detectability Using a Global Network*, Bull. Seismol. Soc. Am. *76*, 1641–1659.

RITSEMA, J., and LAY, T. (1993), *Rapid Source Mechanism Determination of Large ($M_w \geq 5$)* Earthquakes in the Western United States, Geophys. Res. Lett. *20*, 1611–1614.

RITSEMA, J., and LAY, T. (1995), *Long-period Regional Wave Moment Tensor Inversion for Earthquakes in the Western United States*, J. Geophys. Res. *100*, 9853–9864.

ROMANOWICZ, B., DREGER, D., PASYANOS, M., and UHRHAMMER, R. (1993), *Monitoring of Strain Release in Central and Northern California*, Geophys. Res. Lett. *20*, 1643–1646.

SCHWEITZER, J., and KVÆRNA, T. (1999), *Influence of Source Radiation Patterns on Globally Observed Short-period Magnitude Estimates (m_b)*, Bull. Seismol. Soc. Am. *89*, 342–347.

SHI, J., KIM, W.-Y., and RICHARDS, P. G. (1998), *The Corner Frequencies and Stress Drops of Intraplate Earthquakes in the Northeastern United States*, Bull. Seismol. Soc. Am. *88*, 531–542.

SHIN, T.-C., and HERRMANN, R. B. (1987), *Lg Attenuation and Source Studies Using 1982 Miramichi Data*, Bull. Seismol. Soc. Am. *77*, 384–397.

SOMERVILLE, P. G., MCLAREN, J. P., SAIKIA, C. K., and HELMBERGER, D. V. (1990), *The 25 November 1988 Saguenay, Quebec, Earthquake: Source Parameters and the Attenuation of Strong Ground Motion*, Bull. Seismol. Soc. Am. *80*, 1118–1143.

STEVENS, J. L. (1986), *Estimation of Scalar Moments from Explosion-generated Surface Waves*, Bull. Seismol. Soc. Am. *76*, 123–151.

STEVENS, J. L., and DAY, S. M. (1985), *The Physical Basis of m_b : M_s and Variable Frequency Magnitude Methods for Earthquake/Explosion Discrimination*, J. Geophys. Res. *90*, 3009–3020.

STEVENS, J. L., and MCLAUGHLIN, K. L. (1997), *Improved Methods for Regionalized Surface Wave Analysis.* Maxwell Technologies Final Report No. *MFD-TR-97-15887*, Phillips Laboratory Report No. *PL-TR-97-2135.*

STREET, R. L. (1976), *Scaling Northeastern United States/Southeastern Canadian Earthquakes by Their Lg Waves*, Bull. Seismol. Soc. Am. *66*, 1525–1537.

STREET, R. L. (1984), *Some Recent Lg Phase Displacement Spectral Densities and Their Implications with Respect to the Prediction of Ground Motions in Eastern North America*, Bull. Seismol. Soc. Am. *74*, 757–762.

STREET, R. L., and TURCOTTE, F. T. (1977), *A Study of Northeastern North American Spectral Moments, Magnitudes, and Intensities*, Bull. Seismol. Soc. Am. *67*, 599–614.

STREET, R. L., HERRMANN, R. B., and NUTTLI, O. W. (1975), *Spectral Characteristics of the Lg Wave Generated by Central United States Earthquakes*, Geophys. J. R. astr. Soc. *41*, 51–63.

STREET, R., TAYLOR, K., JONES, D., HARRIS, J., STEINER, G., ZEKULIN, A., and ZHANG, D. (1993), *The 4.6 m_{bLg} Northeastern Kentucky Earthquake of September 7, 1988*, Seismol. Res. Lett. *64*, 187–199.

TALANDIER, J., REYMOND, D., and OKAL, E. A. (1987), *M_m: Use of a Variable-period Mantle Magnitude for the Rapid One-station Estimation of Teleseismic Moments*, Geophys. Res. Lett. *14*, 840–843.

TAYLOR, K. B., HERRMANN, R. B., HAMBURGER, M. W., PAVLIS, G. L., JOHNSTON, A., LANGER, C., and LAM, C. (1989), *The Southeastern Illinois Earthquake of 10 June 1987*. Seismol. Res. Lett. *60*, 101–110.

THIO, H. K., and KANAMORI, H. (1995), *Moment-tensor Inversions for Local Earthquakes Using Surface Waves Recorded at TERRAscope*, Bull. Seismol. Soc. Am. *85*, 1021–1038.

THIO, H. K., and KANAMORI, H. (1996), *Source Complexity of the 1994 Northridge Earthquake and its Relation to Aftershock Mechanisms*, Bull. Seismol. Soc. Am. *86*, S84–S92.

VOSS, J. A., and HERRMANN, R. B. (1980), *A Surface Wave Study of the June 16, 1978 Texas Earthquake*, Earthquake Notes *51*(1), 3–14.

WALLACE, T., GIVEN, J., and KANAMORI, H. (1982), *A Discrepancy between Long- and Short-period Mechanisms of Earthquakes near the Long Valley Caldera*, Geophys. Res. Lett. *9*, 1131–1134.

WALTER, W. R. (1993), *Source Parameters of the June 29, 1992 Little Skull Mountain Earthquake from Complete Regional Waveforms at a Single Station*, Geophys. Res. Lett. *20*, 403–406.

WEISBERG, S., *Applied Linear Regression* (John Wiley and Sons, Inc., New York, NY 1980).

WETMILLER, R. J., ADAMS, J., ANGLIN, F. M., HASEGAWA, H. S., and STEVENS, A. E. (1984), *Aftershock Sequences of the 1982 Miramichi, New Brunswick, Earthquakes*, Bull. Seismol. Soc. Am. *74*, 621–653.

WOODS, B. B., and HARKRIDER, D. G. (1995), *Determining Surface-wave Magnitudes from Regional Nevada Test Site Data*, Geophys. J. Int. *120*, 474–498.

WU, F. (1984), Studies of July 1981, *Cornwall, Canada and October 7, 1983, Goodnow Earthquakes Using Digital Accelerograms*, Abstr. 54th Annual Meeting, Eastern Section Seismol. Soc. Am., St Louis, MO, p. 4.

WYSS, M., and BRUNE, J. N. (1968), *Seismic Moment, Stress, and Source Dimensions for Earthquakes in the California-Nevada Region*, J. Geophys. Res. *73*, 4681–4694.

XIE, J. (1998), *Spectral Inversion of Lg from Earthquakes: A Modified Method with Applications to the 1995, Western Texas Earthquake Sequence*, Bull. Seismol. Soc. Am. *88*, 1525–1537.

XIE, J., and MITCHELL, B. J. (1990), *Attenuation of Multiphase Surface Waves in the Basin and Range Province, Part I: Lg and Lg Coda*, Geophys. J. Int. *102*, 121–137.

XIE, J., and PATTON, H. J. (1999), *Regional Phase Excitation and Propagation in the Lop Nor Region of Central Asia and Implications for P/Lg Discriminants*, J. Geophys. Res. *104*, 941–954.

ZHAO, X., and EBEL, J. E. (1991), *Radiation Pattern of Crustal Phases of New England Earthquakes*, Geophys. J. Int. *106*, 647–655.

(Received June 29, 1999, revised December 10, 1999, accepted January 5, 2000)

Pure appl. geophys. 158 (2001) 2017–2039
0033–4553/01/112017–23 $ 1.50 + 0.20/0

Pure and Applied Geophysics

Shallow Velocity Structure at the Shagan River Test Site in Kazakhstan

JESSIE L. BONNER,[1] D. CRAIG PEARSON,[2] W. SCOTT PHILLIPS[2] and STEVEN R. TAYLOR[2]

Abstract—During 1997 and 1998, twelve chemical explosions were detonated in boreholes at the former Soviet nuclear test site near the Shagan River (STS) in Kazakhstan. The depths of these explosions ranged from 2.5 to 550 m, while the explosive yield varied from 2 to 25 tons. The purpose of these explosions was for closure of the unused boreholes at STS, and each explosion was recorded at local distances by a network of seismometers operated by Los Alamos National Laboratory and the Institute of Geophysics for the National Nuclear Center (NNC). Short-period, fundamental-mode Rayleigh waves (*Rg*) were generated by these explosions and recorded at the local stations, resultingly the waves exhibited normal dispersion between 0.2 and 3 seconds. Dispersion curves were generated for each propagation path using the Multiple Filter Analysis and Phase Match Filtering techniques. Tomographic maps of *Rg* group velocity were constructed and show a zone of relatively high velocities for the southwestern (SW) region of the test site and slow propagation for the northeastern (NE) region. For 0.5 sec *Rg*, the regions are separated by the 2.1 km/sec contour, as propagation in the SW is greater than 2.1 km/sec and less in the NE region. At 1.0 sec period, the 2.3 km/sec contour separates the two regions. Finally, for 1.5 and 2.0 sec, the separation between the two regions is less distinct as velocities in the NE section begin to approach the SW except for a low velocity region (<2.1 km/sec) near the center of the test site. Local geologic structure may explain the different regions as the SW region is composed predominantly of crystalline intrusive rocks, while the NE region consists of alluvium, tuff deposits, and Paleozoic sedimentary rocks. Low velocities are also observed along the Shagan River as it passes through the SW region of the test site for shorter period *Rg* (0.5–1.0 sec). Iterative, least-squares inversions of the *Rg* group velocity dispersion curves show shear-wave velocities for the southwestern section that are on average 0.4 km/sec higher than the NE region. At depths greater than 1.5 km the statistical difference between the models is no longer significant. The observed group velocities and different velocity structures correlate with *P*-wave complexity and with spatial patterns of magnitude residuals observed from nuclear explosions at STS, and may help to evaluate the mechanisms behind those observations.

Key words: *Rg*, shear wave velocity structure, Shagan River Test Site.

Introduction

The properties of geological materials change quickly near the earth's surface. If these changes in geologic structure affect seismic generation from shallow (near-

[1] Southern Methodist University, Dept. of Geological Sciences, P.O. Box 0395, Dallas, TX 75275, U.S.A. E-mail: bonner@passion.isem.smu.edu
[2] Los Alamos National Laboratory, EES-3, MS C335, Los Alamos, NM 87545, U.S.A. E-mails: cpearson@lanl.gov, wsp@lanl.gov, taylor@lanl.gov

surface) and deep (up to 1-km) explosions differently, we may be able to develop techniques to discriminate between large industrial blasts and an underground nuclear test. For the present study, we attempt to determine the lateral variations in the velocity structure at the former Soviet test site near the Shagan River in Kazakhstan (STS) by using the properties of explosion generated *Rg*. *Rg* was observed on the local seismograms near the test site from a series of chemical explosions (Fig. 1) detonated at varying depths within the upper crust during 1997 and 1998. The purpose of the explosions was for the closure of the unused boreholes at the former test site as part of a joint project between the U.S. Department of Energy (DOE), the U.S. Defense Special Weapons Agency (DSWA), and the National Nuclear Center of the Republic of Kazakhstan (NNC). The explosions were recorded by a dense local network operated by Los Alamos National

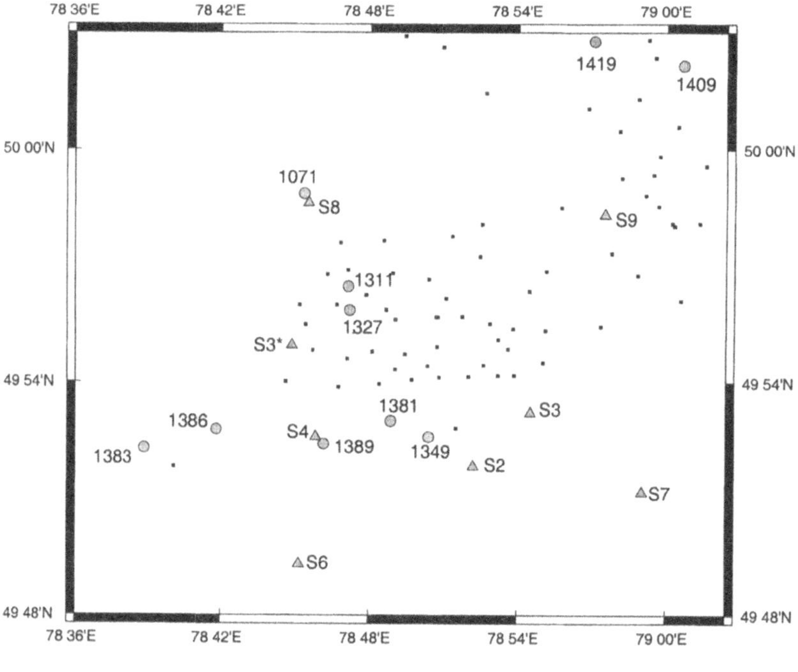

Figure 1
Event and station locations for the borehole closures at the Shagan River test site (STS) in Balapan. Between 1965 and 1989 at least 101 presumed nuclear explosions were recorded from this test site. Most of these events (located by remote sensing methods in THURBER *et al.*, 1994) are also shown as small squares. The DOB experiment was designed and implemented using boreholes 1311, 1349, 1381, and 1071 (circles). In 1998 ten additional boreholes were sealed, of which eight (8) were sealed with explosives. A collaborative agreement between LANL and the Institute of Geophysical Research (NNC) was reached for the deployment of local instrumentation (triangles) to record all hole closures. Site S3 was moved after the DOB experiment to the location S3*. The result of this collaboration is a database of more than 11 explosions of various size and depth emplacement recorded at between 5 and 8 seismic stations.

Laboratory and by a regional network operated by Lamont-Doherty Earth Observatory.

For explosions detonated on or near the earth's surface, *Rg* is often the dominant arrival at local to near-regional distances. *Rg* group velocities are controlled by the velocity variations in the upper few kilometers of crust along the propagation path. The phase is highly attenuated in regions of complex terrain, and is most prominent on paths comprised of low velocity sediments or weathered rock (KOCAOGLU and LONG, 1993). *Rg* is seldom recorded at distances greater than 300 km, thus limiting its usefulness as a depth discriminant. However, the influence of *Rg* scattering into *Lg* and *Lg* coda from explosions has been noted and can make an important contribution to explosion seismograms (e.g., GUPTA *et al.*, 1992; PATTON and TAYLOR, 1995; MAYEDA and WALTER, 1996; MEYERS *et al.*, 1999). The dispersive properties of *Rg* waves have been used to estimate shallow crustal velocity structure in a variety of regions (Table 1).

Because *Rg* group velocities are controlled by the velocity variations in the upper few kilometers of crust, the surface geology of the path can offer insight into the characteristics of the propagation. The former test site has a complex structural and geologic setting created by the interaction of the Siberian and Kazakhstan continental plates and intervening island arcs (ZONENSHAIN *et al.*, 1990). The region has undergone two periods of deformation as a result of the Caledonian and Hercynian orogenies. The Caledonian orogeny (early Paleozoic) is represented by varying degrees of faulting, folding, and metamorphism (NALIVKIN, 1960). One of these faults, the Chinrau, dissects the former test site along a NW-SE trend and

Table 1

Short-period Rayleigh Wave Group Velocities and Periods

Group velocity (km/sec)	Period range (sec)	Region
1.5–3.0	0.5–5.0	Illinois Basin-Ozark Uplift (MCEVILLY and STAUDER, 1965)
1.7–2.6	0.4–2.0	Cincinnati Arch (HERRMANN, 1969)
1.2–3.1	0.2–1.5	NY-NJ Border region (ANDERSON and DORMAN, 1973)
2.5–3.0	0.5–1.8	Sweden (BATH, 1975)
2.0–3.3	0.5–1.8	Southern New England (KAFKA and DOLIN, 1985)
2.6–3.7	0.4–1.6	Southeastern Maine (KAFKA, 1990)
1.9–2.7	0.2–1.5	Eastern Tennessee (YAO and DORMAN, 1992)
1.6–2.4	0.5–1.5	Northern Alabama (KOCAOGLU and LONG, 1993)
0.9–1.3	0.7–3.0	Central Texas (BONNER, 1993)
1.7–3.0	0.2–3.0	Shagan River Test Site, Balapan (This Study)

appears to show evidence of recent offset on SPOT imagery (LEITH, 1987). It is the later Hercynian orogeny (Carboniferous) that folded the Carboniferous-aged sedimentary rocks and implaced the granites and granodiorites at the test site (Fig. 2). The result of the compressive forces associated with these two orogenies is a NW-SE trending tectonic fabric. Quaternary sediments, including sands, clays, and gravel now cover a large portion of the test site. As can be inferred from Figure 2, the test site is not in a homogeneous body of rock. In fact, the site can be divided into two regions—a northeastern (NE) region composed of alluvial deposits overlying folded and faulted Paleozoic sedimentary rocks and a southwestern (SW) quadrant comprised of a large granodiorite body that intruded into the surrounding sedimentary rocks. This intrusive complex is encountered in boreholes 1327 and 1311, and can be inferred to connect beneath the alluvium. The lithology of this pluton has been referred to as granite and granodiorite (DAVIS and BERLIN, 1992) and granodiorite (BOROVIKOV, 1972). For the purpose of this paper, we refer to the pluton as crystalline rock. The results of this paper show how these variations in the shallow geology cause a large range of *Rg* group velocities.

Data Collection

This study involves seismograms from ten explosions at STS (Table 2). The explosions were conducted to seal the unused boreholes at the test site. The shots at boreholes 1071, 1311, 1349, and 1381 registered an explosive weight of 25 tons and were emplaced at depths of 50 m, 50 m, 300 m, and 550 m, respectively. At the intervals of the explosives emplacement, neogenic clays (1311) or crystalline rock (1071, 1349 and 1381) were observed. The primary scientific objective of this series of explosions was to determine the effect of depth of burial on the discrimination of nuclear explosions. A secondary objective was to calibrate the Kazakhstan seismic network, with special emphasis on the IRIS GSN station at Makanchi (MAK) and the IMS auxiliary stations Kurchatov (KUR) and Aktyubinsk (AKT). Results of the regional studies can be found in GLENN and MEYERS (1997) and MEYERS *et al.* (1999).

In addition to the DOB study, eight additional explosive borehole closures were recorded (Table 2). The depth of the explosives emplacement for these closures ranged from 2.5 to 46 meters with 2 ton yields. Boreholes 1071 and 1383 were detonated in granoseyenites, while the six other explosive closures in 1998 were conducted in neogenic clays.

A seismic network was deployed at local distances by Los Alamos National Laboratory to record the explosions, and the network was maintained for the duration of the program by the Institute of Geophysical Research (NNC). Seismograms were recorded on L4-C-3D, three-component, velocity sensors connected to Refraction Technology (REFTEK) data loggers. Origin times for each event were obtained by a GPS-based timing system employed at each hole for

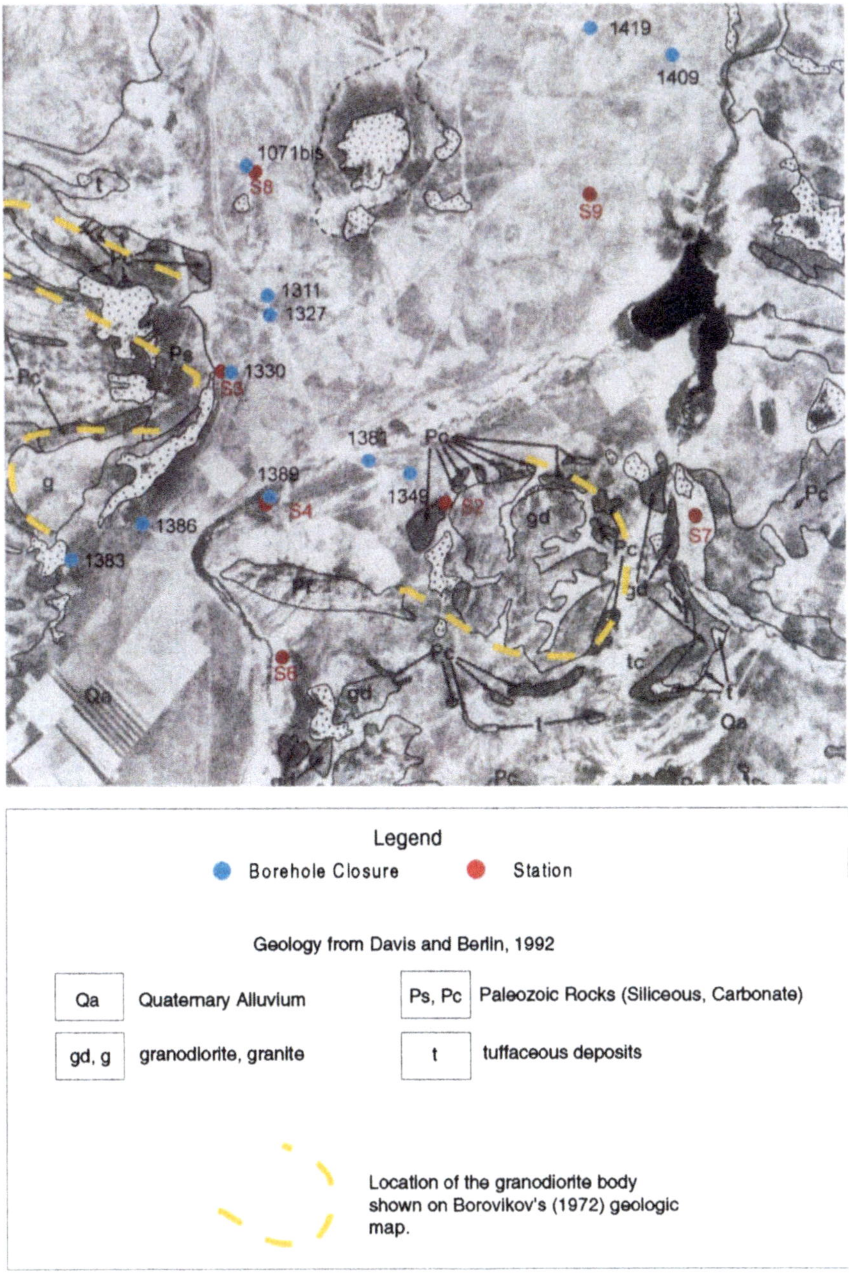

Figure 2
Land Sat image of the STS with superposed geologic units (modified from DAVIS and BERLIN, 1992). The locations of the boreholes and seismic stations for this study are also shown. The approximate outline of the granodiorite body shown on BOROVIKOV's (1972) regional scale geologic map is also shown. The granodiorite body (or similar lithology) is mapped in both 1311 and 1327 boreholes.

Table 2

Explosion Characteristics at Borehole Closures

Borehole	Latitude	Longitude	Explosive weight (kg)	Depth of charge emplacement from surface (m)	Date	Origin Time (UTC)
1071	49.9810	78.7559	25000	28	9/17/98	07:19:40.551
1311	49.9412	78.7860	25000	50	8/3/97	08:07:20.039
1327	49.9129	78.7871	2028	20	7/14/98	05:11:35.570
1349	49.8794	78.8493	25000	550	8/5/97	07:30:15.126
1381	49.8837	78.8147	25000	300	8/31/97	07:08:39.179
1383	49.8724	78.6477	2028	14	8/15/98	02:40:59.116
1386	49.8801	78.6921	2028	20	7/13/98	10:44:56.363
1389	49.8786	78.7601	2028	9	8/15/98	05:05:11.156
1409	50.0357	79.0114	2028	13	8/14/98	04:28:52.815
1419	50.0576	78.9387	2028	2.5	8/14/98	05:39:24.970

shot-break time. The origin times were checked for consistency using an accelerometer and data acquisition system placed near the test borehole. Seven sites were selected at various azimuths to the proposed locations for the DOB experiment (Table 3). For the 1998 series of hole closures, the seismic stations reoccupied the same DOB locales, except for S3 which was relocated because of technical problems. For the REFTEK/L4-C-3D sites, the data was sampled at 500 Hz, and the recorder was set up to use a STA/LTA trigger. The data were then transferred to a workstation at LANL for processing and analysis.

Data Analysis

The Multiple Filter Analysis technique (MFA), first developed by DZIEWONSKI *et al.* (1969), and implemented using a set of computer programs written by HERRMANN (1988), provides a fast, efficient method of analyzing signals that may

Table 3

Coordinates for the Balapan Network

Station	Latitude	Longitude
S2	49.8636	78.8701
S3[1]	49.8874	78.9087
S3	49.9494	78.7483
S4	49.8766	78.7636
S6	49.8217	78.7531
S7	49.8526	78.9841
S8	49.9777	78.7590
S9	49.9719	78.9587

[1] Location valid for Hole Closures 1311, 1349, and 1381.

have one or more dispersed arrivals. In this method, a set of narrow band Gaussian filters is applied to the input spectra at certain period intervals. The data are transformed back to the time domain, and the envelope of each filtered time signal is searched for the occurrence of the four largest envelope amplitudes. Each of these amplitude maxima occurs at an associated time from the origin of the event, and when combined with the distance from the source, a group velocity can be estimated for each one. These amplitudes are then contoured, normalized, and displayed on a plot of group velocity versus period.

The MFA technique was applied to the seismic data recorded from the hole closures at the test site. The diagrams produced by the MFA are often problematic at long periods due to the broadening of contour lines. This makes it difficult to correctly infer the line that marks the maximum amplitudes of the envelope function. However, these problems are not incurred with the Phase Match Filtering technique (PMF). The dispersion curves determined by the MFA process then served as the input to the PMF technique (HERRIN and GOFORTH, 1977). The PMF procedure starts with an initial estimate of dispersion, generally from the MFA, and by an iterative technique finds and applies a filter that is phase-matched to the particular mode or arrival of surface wave, *Rg* in this study. The advantage of using PMF is that the final results are less contaminated by higher modes and multi-pathing arrivals, thus the complex spectrum of the primary signal can be removed from multi-path arrivals. More importantly for this study, a more accurate estimate of the group velocity dispersion is obtained.

Several different types of uncertainty estimates have been assigned to group velocity obtained by the MFA process. DER *et al.* (1970) used the uncertainty principle to obtain an error of ± 0.01 km/sec, while BRAILE and KELLER (1975) and YU and MITCHELL (1979) used ± 0.03 km/sec for uncertainties obtained through the multiple filter analysis technique. JIN and HERRIN (1980) also adopted this value, but mention that the PMF is considered more accurate than the MFA. For this study we adopt ± 0.03 km/sec error in the dispersion estimates. We obtained this value by applying a ± 0.01 km/sec error for the MFA and PMF process and by considering a 0.1 sec error in the event origin time that would produce on average a ± 0.02 km/sec error in the dispersion estimates. Corrections for the sphericity and the gravity of the earth are smaller than the uncertainty of the data for the path lengths in this study, and are not considered.

Results

Rg Group Velocity Dispersion Tomography

The dispersion analysis of the path from the explosion at hole 1409 to station S3 is shown in Figure 3, and is representative of the dispersion in the NE extents of

1409 to S3

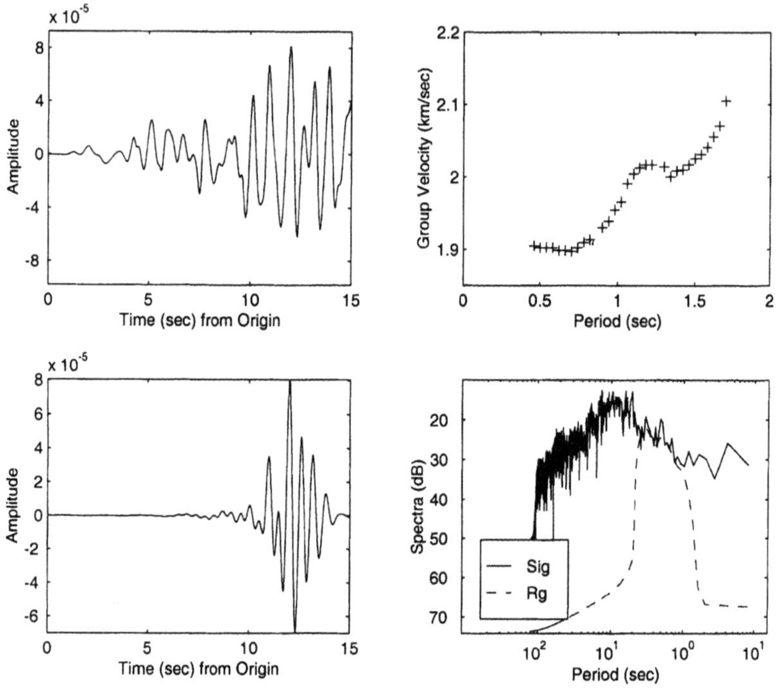

Figure 3
Results of processing *Rg* waves at station S3 for the borehole explosion 1409. (Upper Left) Low-pass (3 Hz) filtered data. (Upper Right) Dispersion results from MFA and PMF processing. (Lower Left) PMF extracted *Rg*. (Lower Right) Spectra calculated for the entire signal (solid line) as compared to the PMF extracted *Rg* (dashed).

the test site (Fig. 2). This region is characterized on the surface by alluvium that may be as thick as 1 km in some regions (DAVIS and BERLIN, 1992) overlying Paleozoic sedimentary rocks. In the upper left corner of Figure 3, the first 15 seconds of the low-passed (3 Hz) raw data are presented. Through use of the MFA technique, the dispersion curve (upper right) was estimated and refined with the PMF technique. The PMF extracted *Rg* is shown in the lower left, and the signal and extracted *Rg* spectra are shown in the lower-right plot. Figure 4 shows similar data for the path from explosion 1383 to S7—a path representative of the SW section of the test site. This region is comprised predominantly of the intrusive crystalline rocks as shown in Figure 2. The dispersion curve for path 1383 to S7 has group velocities that are on average 0.6 km/sec faster than 1409 to S3, suggesting that significant lateral variations in the geology are influencing the *Rg* velocities. Comparison of seismograms in Figures 3 and 4 illustrates that signals confined mainly to the SW portion of the STS have simpler waveforms relative to those to the NE.

1383 to S7

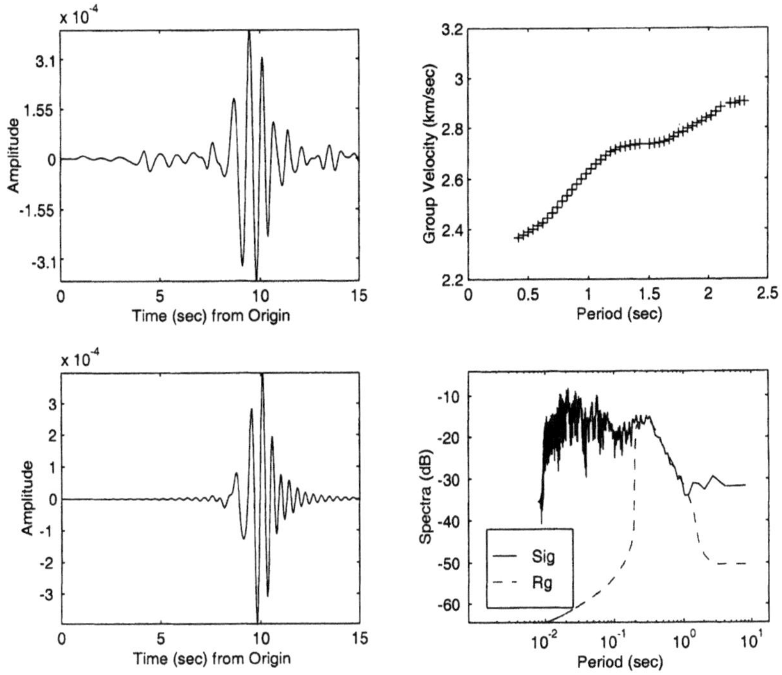

Figure 4
Results of processing *Rg* waves at station S7 for the borehole explosion 1383. (Upper Left) Low-pass (3 Hz) filtered data. (Upper Right) Dispersion results from MFA and PMF processing. (Lower Left) PMF extracted *Rg*. (Lower Right) Spectra calculated for the entire signal (solid line) as compared to the PMF extracted *Rg* (dashed).

The emplacement depth for the explosives had significant effects on the generation of *Rg*. *Rg* generated by the 25 ton, 50 m (1071 and 1311) explosions exhibited the greatest amplitudes and bandwidth observed during the two-year program of detonations. At 3 seconds period (*T*), these waves are sensitive to the velocity variations in the upper 2 to 3 km of the crust. *Rg* amplitudes were decreased significantly for the 300 m explosion (1381), and the maximum period observed in the dispersed wavetrains shifted from 3 to 2 sec, thus limiting the resolution depth of the velocity inversions to 2 km as will be discussed later. *Rg* amplitudes were considerably decreased by the 550 m explosion (1349) and had no significant signal-to-noise ratio for this study. These observations were noted in MEYERS *et al.* (1999) in an attempt to explain explosion generated *S* waves at regional and local distances. Two additional explosions in 1998 had decreased *Rg* generation. These sites were 1389 and 1419, which had the shallowest emplacement depths (9 and 2.5 m, respectively) of any explosions during the entire program. It is worth noting that these shallow depths may have directed energy upward resulting in poor coupling and decreased *Rg* generation.

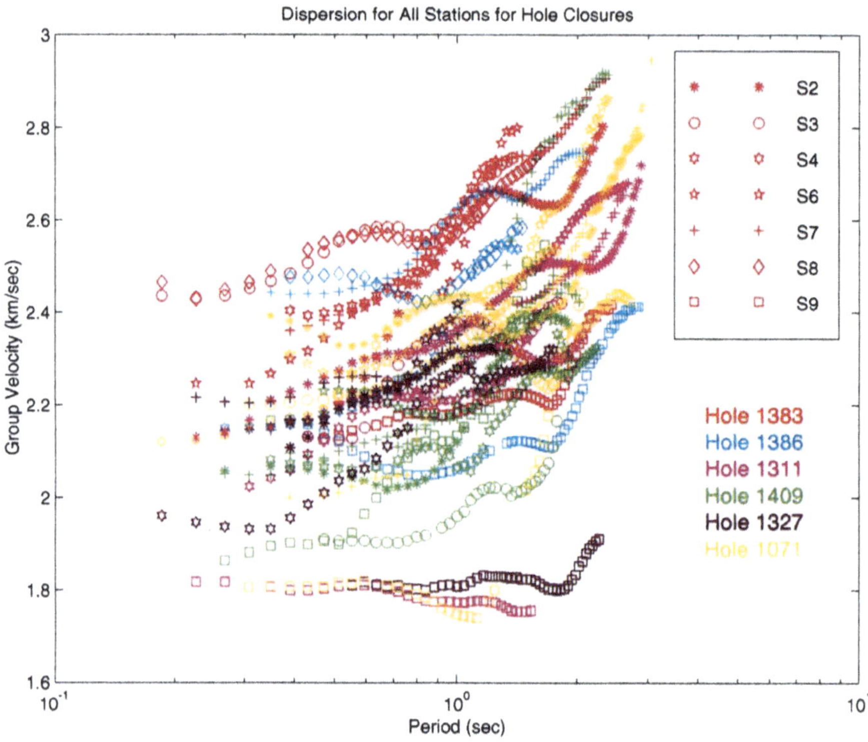

Figure 5
Rg group velocity dispersion curves for all station-event pairs processed in this study.

The dispersion curves for all paths considered in this study are shown in Figure 5. Group velocities range from 1.7 to 2.9 km/sec, and have periods between 0.2 and 3 seconds. A better way to show the same data is through the use of group velocity tomography maps for individual periods. Tomography was performed assuming straight ray propagation, parameterizing the model with an array of constant velocity circles. We used first-difference regularization to control the effects of noise on the final image. Regularization parameters were chosen to optimize resolution and covariance; further details may be found in Phillips *et al.* (1999). Tomographic inversion produced variance reduction of 80%, 94%, 78% and 66%, relative to a flat model, for periods of 0.5, 1.0, 1.5 and 2.0 seconds, respectively.

The group velocity tomography maps presented in Figures 6–9 show a relatively high-velocity region to the southwest (SW), and a low-velocity region in the northeast (NE). We suspect that the faster velocities in the SW region are related to

▶

Figure 7
Tomographic map of *Rg* velocities at $T = 1$ sec.

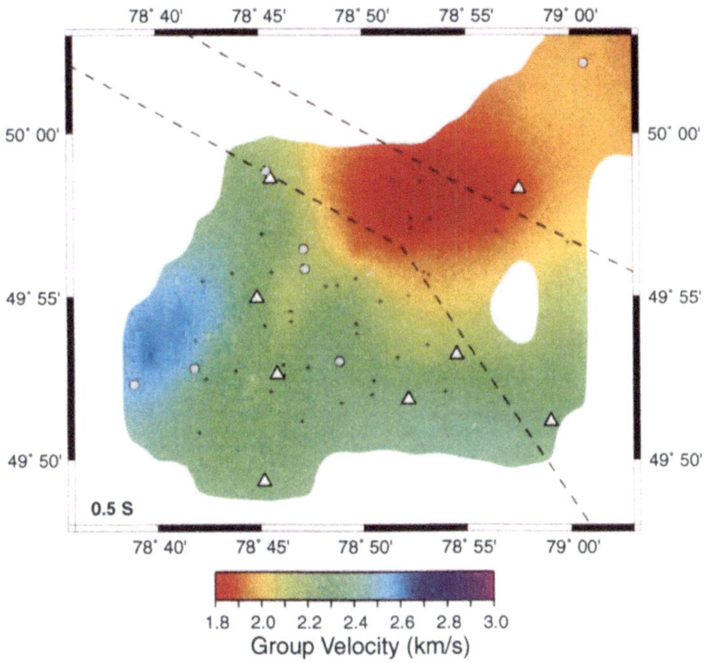

Figure 6
Tomographic map of *Rg* velocities at $T = 0.5$ sec. The stations and explosions are shown by triangles and circles respectively, while the midpoints for each path are shown as plusses. The dashed lines show the subregions of the STS as determined by magnitude residuals ($m_b - m_{Lg}$) from RINGDAL *et al.* (1992).

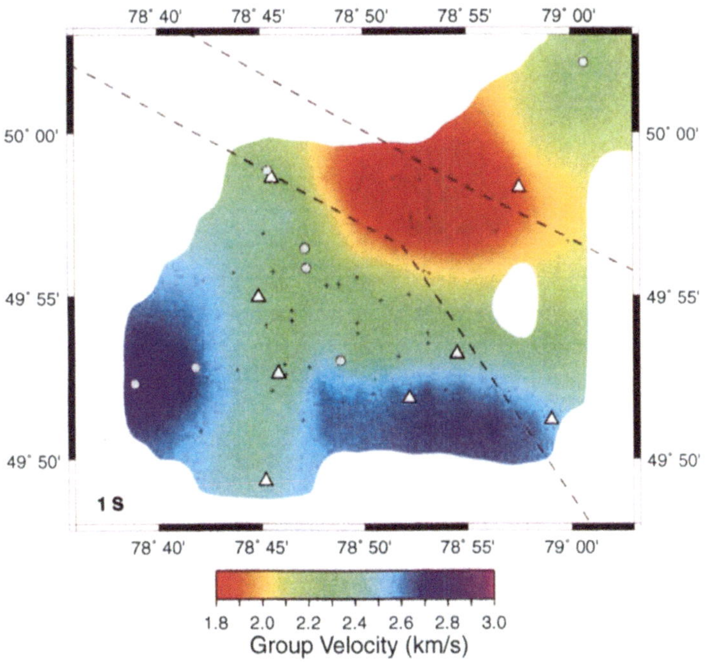

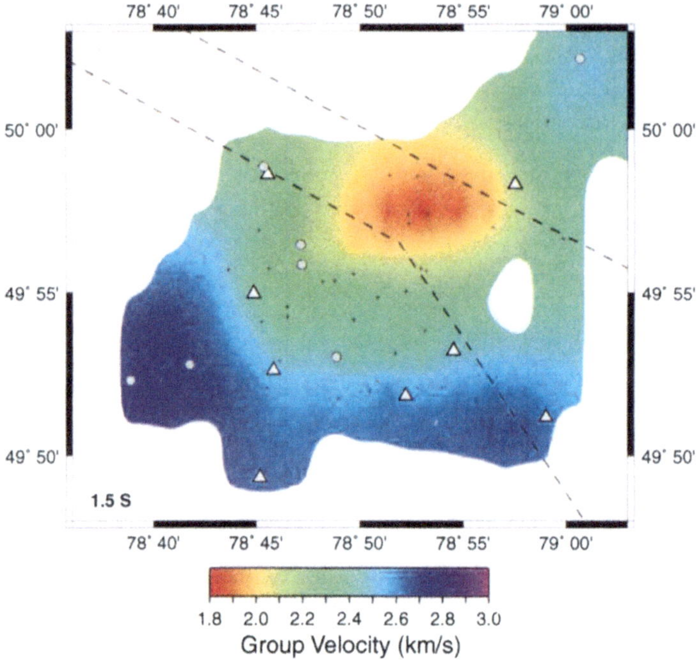

Figure 8
Tomographic map of *Rg* velocities at $T = 1.5$ sec.

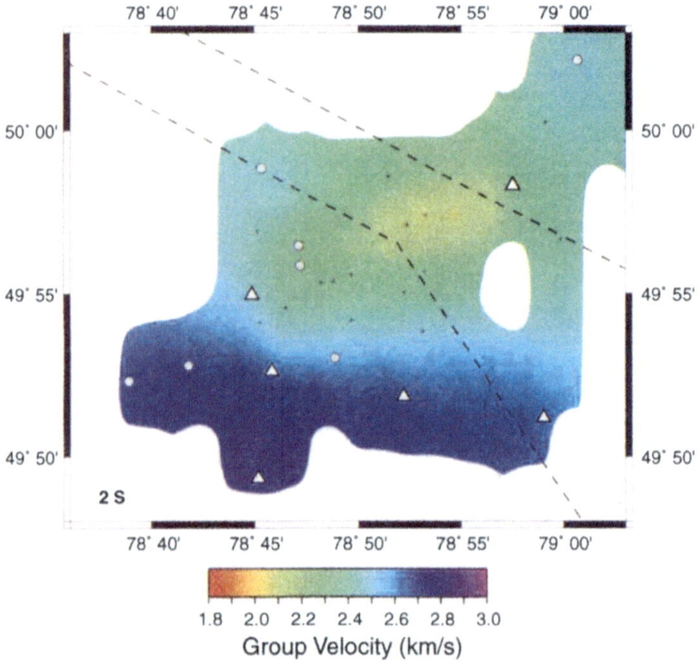

crystalline rocks at or near the surface. Logs for test hole 1383 (NNC FIELD REPORT, 1996) indicate crystalline rocks to within 5 m of the surface. At shorter periods of 0.5 and 1 sec (Figs. 6 and 7), the low velocities associated with the transition zone follow the Shagan River towards the southwest, and may be related to alluvium overlying the intrusive crystalline rocks near this section of the test site. Lithologic logs from boreholes in this region confirm that alluvium and weathered rock with thickness as great as 350 m overlie the crystalline rocks. For all period ranges, a region of low velocities is observed in the northeast. Lithologies for these boreholes include thick sections of alluvium and unconsolidated tuffs that overlie the Paleozoic sedimentary section.

Group Velocity Inversion

We used least-squares inversion technique (HERRMANN, 1988) to invert group velocity dispersion curves for the shallow shear wave velocity structure within the different regions of the STS. Paths within each region of the test site were inverted using both differential and stochastic inversions for different starting velocity models. We considered both inversion types to smooth through the intrinsic effects associated with the inversion schemes and because a constrained inversion was not possible due to lack of knowledge of the structure at depth. Stochastic inversions have a tendency to introduce low velocity zones that are invalid, while a differential inverse may sometimes smooth over the major discontinuities in the model. In all inversions, a damping factor (λ) was chosen such that the inversion results were stable. If λ is chosen too small, a low velocity layer may be overcompensated by an adjoining fast layer in the model. On the other hand, a value of λ too large will not allow the model to find a solution that fits the observed dispersion. Thus, the trade-off between resolution and λ is critical, and each inversion was completed with several different damping constraints until the estimated shear velocity structure converged. Each inversion consisted of 40 iterations.

In an iterative, least-squares inversion, the resulting model may be highly dependent upon the initial estimate of the crustal velocity structure. To insure this was not the case, a variety of starting models was used to characterize the STS dispersion data. The first set of starting models consisted of thin layers with a constant velocity of 2.61 km/sec down to a depth of 5 km. The velocity was obtained from laboratory measurements typical of the grano-seyenites from the region (NNC FIELD REPORT, 1996). By using this velocity, we were able to determine whether the resulting models attained velocities faster or slower than the laboratory measured grano-seyenitic shear-wave velocity. A half space with a shear-wave velocity of 3.5 km/sec was placed at a depth of 5 km.

◄

Figure 9
Tomographic map of *Rg* velocities at $T = 2.0$ sec.

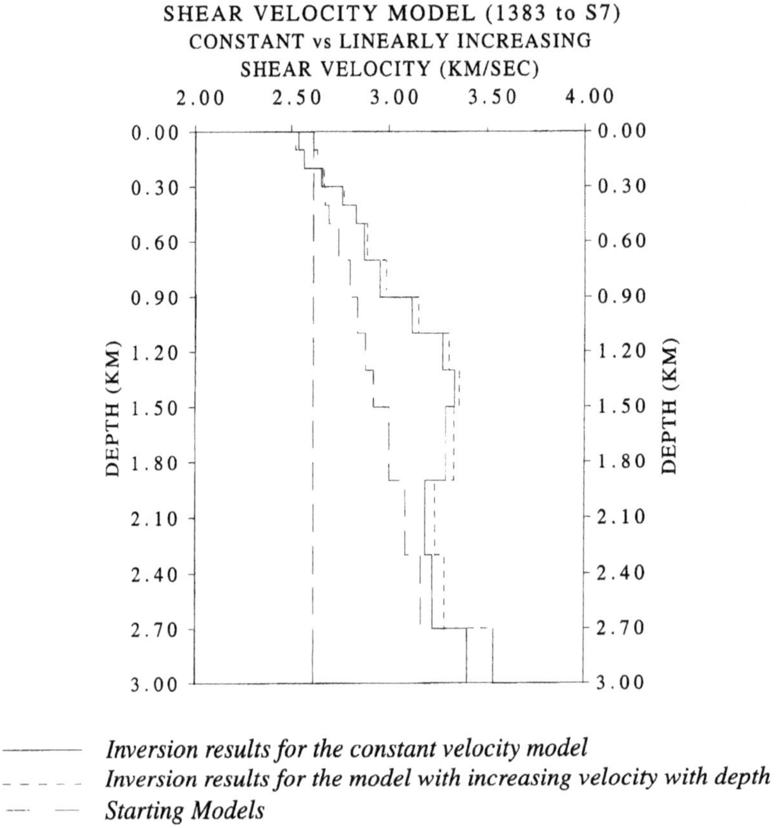

Figure 10
Inversion for shear-wave velocity structure for the event-station pair of 1383 and S7 using two different starting models. After 40 iterations, the inversions from both starting models converge to a similar solution.

The second set of starting models consisted of a linearly increasing velocity with depth from an initial value of 2.61 km/sec. For each kilometer depth increase, the shear velocity was increased 0.18 km/sec. Layer thickness varied from 100 m between 0 and 0.5 km depth, 200 m between 0.5 and 1.5 km, and 400 m between 1.5 and 5 km. The half space was again placed at 5 km with a velocity of 3.5 km/sec. The results of the two sets of inversions performed on the path from explosion 1383 to S7 are shown in Figure 10. The inverted velocity structures are similar, even though the starting models are different. In both cases, the velocity of the upper 0.2 km is approximately 0.1 km/sec less than the measured value of the grano-seyenites in the region, suggesting weathering and/or overestimation of the *in situ* seismic velocities by the laboratory measurements. The shear-wave velocities from the inversion increase with depth to approximately 1.2 km where the shear-wave velocity is approximately 3.30 km/sec. A low velocity zone (LVZ) occurs at a depth between

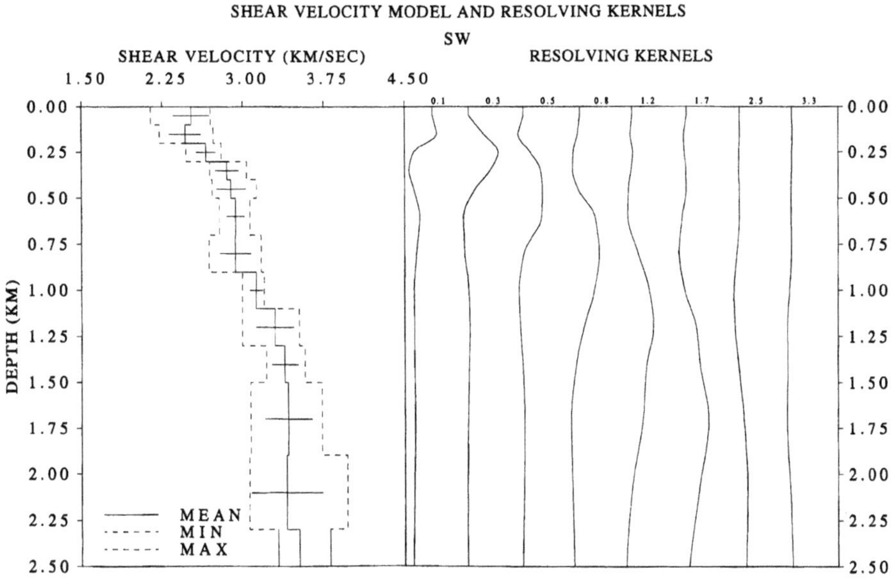

Figure 11
Inversion for shallow shear-wave velocity structure in the SW region of the STS. The range of all velocity models that fit the observed dispersion are shown by the dashed lines, and the solid line shows the average velocity structure. The final velocity models show little deviation from the starting model at depths greater than 2.5 km because of loss of resolution as noted in the normalized resolution kernels.

1.50 km and 2.70 km, however, the extent of the LVZ cannot be determined because of lack of resolution at these depths.

The minimum, maximum, and average shear-wave velocity structures from all inversions for paths in the SW and transition regions of the test site are shown in Figure 11. Also included in the range of velocity models are inversions that either considered (causal) or neglected (non-causal) the effects of Q on the group velocity dispersion. The minimum velocity was obtained by inverting the paths that predominantly sample the rocks for the transition zone between the fast and slow regions (1327 and 1311 to S6, S7, and S4). The maximum velocity curve is the result of the path from shot 1383 to S3 and S8. These paths are confined in the granodiorite and have the greatest group velocities for the entire *Rg* bandwidth. The average velocity structure exhibits three distinct layers—a 250 m thick section with velocities averaging 2.45 km/sec, a layer between 250 m and 900 m with velocities averaging 2.95 km/sec, and a third layer to a depth of 2.5 km with shear-wave velocities of approximately 3.3 km/sec. For the path of 1383 to S7, the dispersion obtained from the inversion compared with the observed is presented in Figure 12.

The final velocity model displays minor deviation from the starting model at depths greater than 2.5 km because of loss of resolution. This interpretation is supported by resolution kernels that peak near the depths of the corresponding layers for depths less than 2.5 km. At greater depths within the model, the maximum in the

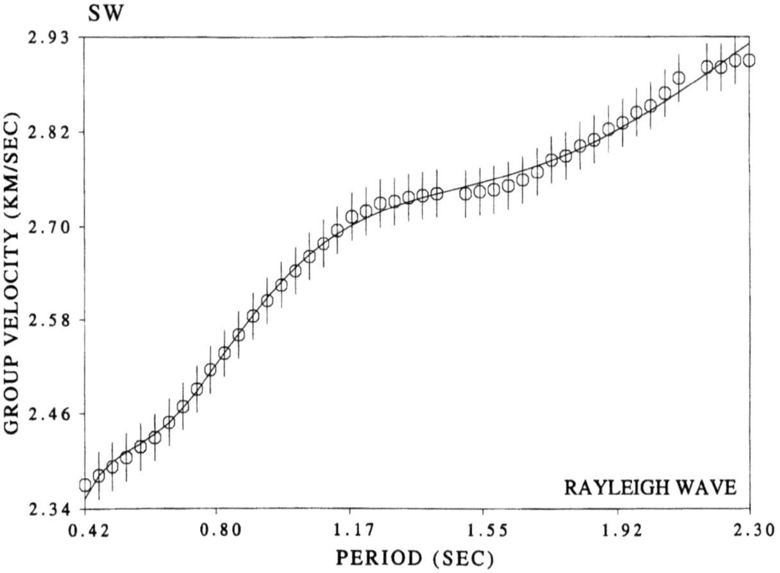

Figure 12
Dispersion results for the inversion of the 1383 to S7 dispersion curve. The observed data are shown as circles with error bars, while the theoretical dispersion from the inversion is shown as the line.

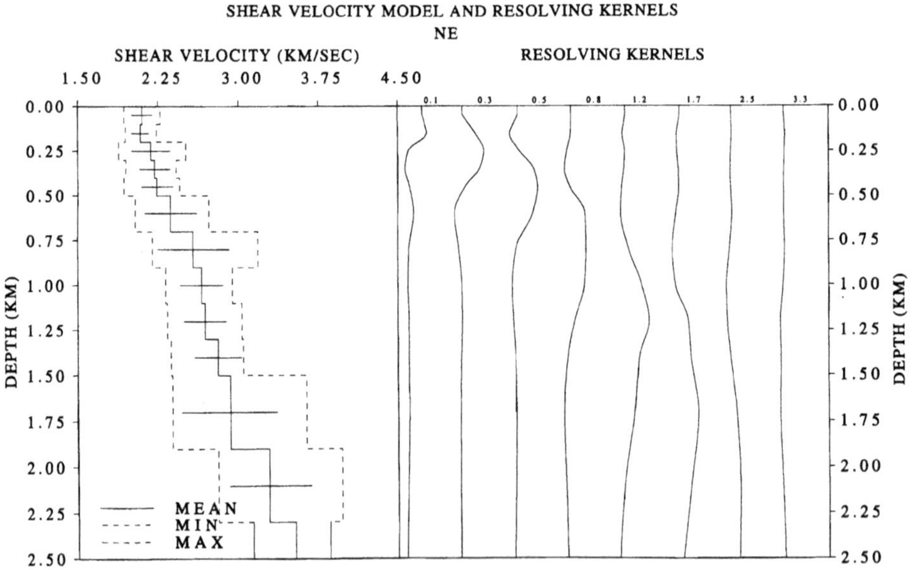

Figure 13
Inversion for shallow shear-wave velocity structure in the NE region of the test site. The range of all possible velocity models is shown, and the model loses resolution at depths greater than 2 km.

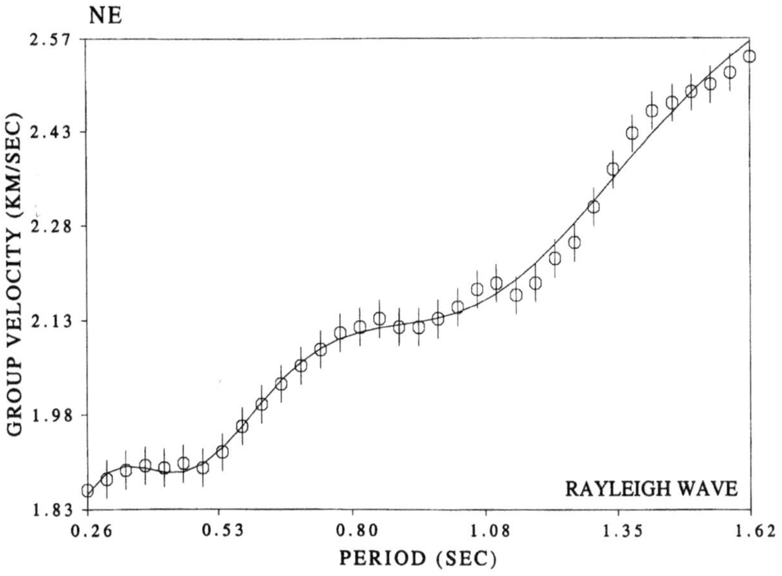

Figure 14
Dispersion results for the inversion of the 1409 to S3 dispersion. The observed data are shown as circles with error bars, while the theoretical dispersion from the inversion is shown as the line.

resolving kernels is offset to shallower depths and the magnitude of the kernel is small. However, this is expected since the longest period (2.5–3 sec) corresponds to wavelengths of approximately 7 km. Depth of penetration for *Rg* is generally assumed as 1/3 of the wavelength, thus the longest wavelengths observed in this study will be sampling approximately 2.5 km deep.

Figure 13 presents the results of the inversions in the northeast region. The minimum velocity structure was obtained from the inversion of paths such as 1327 to S9 and 1311 to S9, both having flat, low-velocity dispersion curves. The observed and theoretical dispersion for the path 1409 to S3 is shown in Figure 14. *Rg* along this path is slowed by propagation predominantly within the NE region. However, about 25% of the path occurs in the transition zone, thus creating a velocity structure with greater velocities than the average for the NE. Three distinct layers were noted in the velocity structure for the SW and transitional regions, while the suite of inversions in the NE region show the average velocity structure to be increasing with depth without prominent discontinuities to a depth of approximately 2 km. Below this depth the resolution is compromised, as *Rg* waves in this region have smaller wavelengths.

Discussion

As discussed above, the trends from the tomographic inversion appear to delineate two distinct zones separated by a NW-SE trending boundary. Relatively high velocities are observed to the SW and lower velocities to the NE. The boundary separating the two zones roughly coincides with the Chinrau fault. Sequences of tuffs and alluvium exist to the northeast of this boundary overlying Paleozoic sedimentary rocks, and to the southwest crystalline rock is predominant. Differences between the NE and SW regions have been known for some time on the basis of teleseismic *P*-wave spectral and waveform differences from nuclear explosions (MARSHALL *et al.*, 1985; STEWART, 1988; BACHE *et al.*, 1986). Additionally, more recent studies have indicated NE and SW contrasts in *Lg* amplitudes (RINGDAL *et al.*, 1992; JIH and WAGNER, 1992). Specifically, MARSHALL *et al.* (1985) observed that *P* waves from the Shagan Test Site (STS) nuclear explosions fell into two distinct classes. NE STS *P* waveforms are more complex and have higher corner frequencies than those from SW explosions of similar magnitude. The *P* waveforms from the NE explosions exhibit a few cycles of ringing not observed on the simple waveforms from the SW. This is consistent with the observed complexity of the NE *Rg* waves relative to those in the SW discussed earlier (Figs. 3 and 4). The observation of longer duration waveforms in the NE relative to SW STS conflicts with the spectral observations of MARSHALL *et al.* (1985). However, in comparing spectra for the two different regions at the UK array stations in BACHE *et al.* (1986), the difference in the corner frequencies is not strongly apparent. It was suggested that changes in source material properties were the cause of the observed differences, although very little geologic information was available at the time. STEWART (1988) also noted that the NE explosion *P* waveforms registered a smaller, longer period, first negative pulse (presumably *pP*) than those from the SW. It was suggested that the greater *P-pP* times and longer period *pP* pulses were due to greater scaled depth of burial and/or slower uphole velocities in the northeast.

BARKER and MURPHY (1992) and MCLAUGHLIN *et al.* (1992) noted that the teleseismic *P*-wave magnitude anomalies across the STS depended somewhat on both azimuth and takeoff angle. Based on different back-projection modeling techniques, they suggested that the magnitude variations are due to energy focusing and defocusing by an anomalous structure at a depth of about 100 km situated 50 km northwest of STS. While this may be a factor contributing to the variability of the measurements, we feel that the close-in observations presented here, suggest that near-source material properties play a significant role in affecting waveform variations across the STS.

More recent studies have also delineated NE/SW differences in *Lg* amplitudes across the STS. RINGDAL *et al.* (1992) made maps of $m_b(P) - m_b(Lg)$ amplitude variations across the STS suggesting as much as 0.15 magnitude units between the NE and SW regions (with the SW gaining larger values). At this time, more geologic information was available due to the activities associated with the TTBT and the JVE experiments. It was recognized that the NE portion of the STS was characterized by larger thicknesses of sedimentary sequences than the SW. This indicates that the NE explosions are characterized by some combination of lower amplitude *P* waves and larger amplitude *Lg* waves relative to the SW. Analysis of magnitudes from four nuclear explosions with announced yields suggested that the *P* waves were affected by the NE and SW differences, while the *Lg* waves were not.

The spatial variations in velocities from this study (Figs. 6–9 and Figs. 11 and 13) are consistent with the observations of teleseismic *P* waves and *Lg* waves from STS nuclear explosions. For example, it is expected that explosions detonated in a region with low-velocity sedimentary layers overlying crystalline basement will indicate *P* waveforms that are more complex and lower frequency than those detonated in crystalline rock. The complex waveforms and larger *pP* delays are presumably due to near-surface reverberations and lower uphole velocities. The lower corner frequencies are predicted from established explosion-source models (e.g., MUELLER and MURPHY, 1971). Using a Mueller-Murphy explosion-source model for granite, we compute Reduced Velocity Potential (RVP) spectra for two identical explosions. Each explosion produced a yield of (150 kT) and near-source compressional velocities of 5 and 3.9 km/s taken from velocity models shown in Figures 11 and 13 (representing the SW and NE portions of the STS, respectively). The results are shown in Figure 15 and it can be seen that just based on simple elastic properties, we expect explosions of identical yields from the NE to have lower corner frequencies

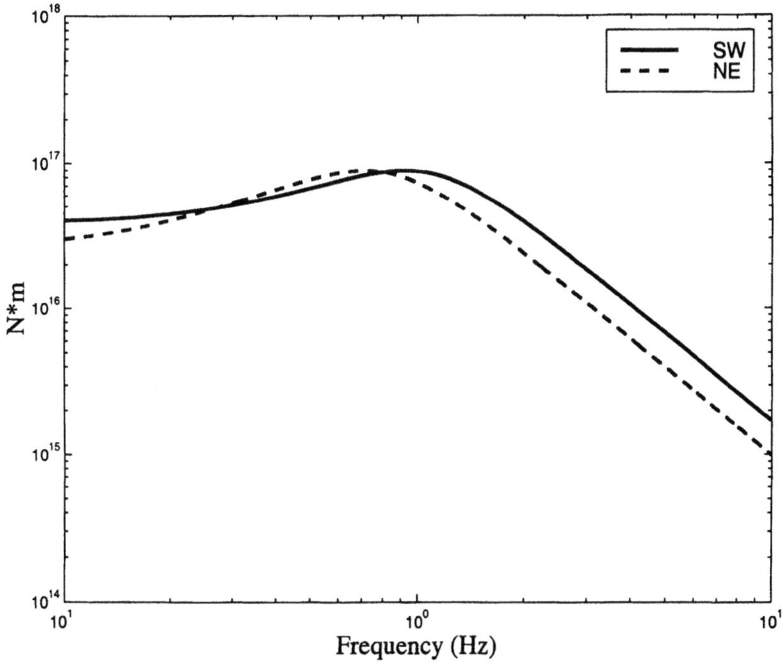

Figure 15
Reduced Velocity Potential (RVP) spectra for two identical explosions in geologic structures representative of the NE and SW regions of STS.

than those from the SW. The explosion-source corner frequency is proportional to the decay rate of the pressure-time history acting at the elastic radius, which in turn is proportional to the compressional velocity at the source. Thus the lower corner frequencies as evidenced by the longer duration *P* waves for the NE explosions may simply be due to the reduced velocities (e.g., material strength) in the near-source region.

As discussed above, it is possible that the NE STS explosions have relatively larger amplitude *Lg* waves than those from the SW. *Lg* signals from explosions are thought to be enriched by scattered *Rg* in the frequency band from about 0.5 to 3 Hz (e.g., GUPTA *et al.*, 1992; PATTON and TAYLOR, 1995). It is expected that explosions detonated in a region with a low-velocity surficial layer overlying a crystalline basement will excite relatively large-amplitude *Rg* that will scatter into *Lg* (as suggested by MEYERS *et al.*, 1999 as part of this same experiment). Thus the spatial and depth variations in velocity observed as part of this study are consistent with an enhancement of *Lg* amplitudes (partially through *Rg* to *Lg* scattering). All of the above discussion neglects the effects of secondary sources such as spall and cavity rebound occurring in the near-source region (which may be manifest as an equivalent compensated linear vector dipole (CLVD) source; PATTON and TAYLOR, 1995). These secondary source effects are very complex and will certainly depend on near-source material properties. Quantification of these effects can be undertaken through elastic and nonlinear modeling, but will be very nonunique and are beyond the scope of this paper. However, now that more quantitative information on near-source velocity structure is available from this study, it may be possible in future work to improve our understanding of observed waveform differences across the STS.

Conclusions

The results of this study show that variations in geologic structure have a significant effect on *Rg* group velocities at STS. Tomography maps of *Rg* group velocity show that the SW region of the test site is characterized by faster velocities than the NE section. As an example, *Rg* with a period of 1 sec propagates with velocities greater than 2.3 km/sec for the SW section of the test site, while propagating at less than 2.3 km/sec for the NE section. Surface geologic maps and borehole lithology logs for the region show that the slow region in the NE section of the test site correlates well with sedimentary rocks and tuff deposits overlain by alluvium. A large granodiorite body that has intruded the sedimentary rocks appears to correlate well with the fast velocities in the SW region. The inversions of the *Rg* dispersion curves give shear-wave velocities for the SW region that are on average 0.4 km/sec greater than the NE region. At depths greater than 1.5 km the standard deviations for the models begin to overlap, and the statistical difference between the models is no longer significant.

The systematic variation in the relative patterns of *P*-wave complexity *P* and *Lg* source size estimators across the test site (MARSHALL *et al.*, 1985; STEWART, 1988; RINGDAL *et al.*, 1992) correlate with the two geophysically distinct regions obtained from this study. Future studies should focus on the quantification of how the velocity structure at the test site influences the regional and teleseismic nature of *P*- and *Lg*-wave seismograms.

Acknowledgements

The authors wish to thank Brian Stump, Bob Herrmann, and Howard Patton for insightful suggestions concerning the research. We also wish to thank Diane Baker and Los Alamos National Laboratory for providing the raw and instrument corrected data along with details concerning the explosions at the test site. We wish to thank Luke Kluchko of DSWA for his support and expertise. Much of the information about the 1998 borehole closures, including borehole lithology and rock properties, was provided by the Institute of Geophysical Research of the National Nuclear Center of the Republic of Kazakhstan (NNC), and we wish to thank them for their assistance with this experiment. The research was sponsored by the Office of Nonproliferation and National Security and the Office of Research and Development for the Department of Energy under contract no. W-7405-ENG-26.

REFERENCES

ANDERSON, J., and DORMAN, J. (1973), *Local Geologic Effects on Short-period Rayleigh Waves around New York City*, Bull. Seismol. Soc. Am. *63*, 1487–1497.

BACHE, T. C., BRATT, S. R., and BUNGUM, H. (1986), *High-frequency P-wave Attenuation along Five Teleseismic Paths from Central Asia*, Geophys. J. R. astr. Soc. *85*, 505–522.

BARKER, B. W., and MURPHY, J. R. (1992), *A Lithospheric Velocity Anomaly beneath the Shagan River Test Site. Part 1. Detection and Location with Network Magnitude Residuals*, Bull. Seismol. Soc. Am. *82*, 980–998.

BATH, M. (1975), *Short-period Rayleigh Waves from Near-surface Events*, Phys. Earth Planet. Inter. *10*, 369–376.

BONNER, J. (1993), *Crustal Velocity Structure of Central Texas.*, M.S. Thesis, Baylor University, Waco, Texas, 125 pp.

BOROVIKOV, L. E. (1972), *Tectonic Map of Central and Eastern Kazakhstan, Scale 1:1000000*, Sheet 2.

BRAILE, L. W., and KELLER, G. R. (1975), *Fine Structure of the Crust Inferred from Linear Inversion of Rayleigh Wave Dispersion*, Bull. Seismol. Soc. Am. *65*, 71–83.

DAVIS, P. A., and BERLIN, G. L. (1992), *Geologic Mapping of the Semipalatinsk Region, Eastern Kazakhstan, Using Land Sat Thematic Mapper and Spot Panchromatic Data*, Los Alamos National Laboratory, LAUR-92-310, 15 pp.

DER, Z., MASSE, R., and LANDISMAN, M. (1970), *Effects of Observational Errors on the Resolution of Surface Waves at Intermediate Distances*, J. Geophys. Res. *75*, 3399–3409.

DZIEWONSKI, A., BLOCH, S., and LANDISMAN, M. (1969), *A Technique for the Analysis of Transient Seismic Signals*, Bull. Seismol. Soc. Am. *59*, 427–444.

GLENN, L., and MEYERS, S. (1997), *Depth of burial experiment at Balapan*, Lawrence Livermore National Laboratory, UCRL-JC-128313.

GUPTA, I. N., CHAN, W. W., and WAGNER, R. A. (1992), A Comparison of Regional Phases from Underground Nuclear Explosions at East Kazakh and Nevada Test Sites, Bull. Seismol. Soc. Am. 352–382.

HERRIN, E., and GOFORTH, T. (1977), *Phase-matched Filters: Application to the Study of Rayleigh Waves*, Bull. Seismol. Soc. Am. *67*, 1259–1275.

HERRMANN, R. B. (1969), *The Structure of the Cincinnati Arch as Determined by Short-period Rayleigh Waves*, Bull. Seismol. Soc. Am. *59*, 399–407.

HERRMANN, R. B., *Computer Programs in Seismology*. Vols II–IV (St. Louis University, St. Louis, Missouri. 1988).

JIH, R. S., and WAGNER, R. A. (1992), *Path-Corrected Body-wave Magnitudes and Yield Estimates of Semipalatinsk Explosions*, Teledyne-Geotech, TGAL-92-05, Alexandria, Virg., 69 pp.

JIN, D. J., and HERRIN, E. T. (1980), *Surface-Wave Studies of the Bering Sea and Alaska Area*, Bull. Seismol. Soc. Am. *70*, 2117–2144.

KAFKA, A. L., and DOLLIN, M. F. (1985), *Constraints on the Lateral Variation in the Upper Crustal Structure beneath Southern New England from Dispersion of Rg Waves*, Geophys. Res. Lett *12*, 235–238.

KAFKA, A. (1990), *Rg as Depth Discriminant for Earthquakes and Explosions: A Case Study in New England*, Bull. Seismol. Soc. Am, *80*, 373–394.

KNOPOFF, L., and SCHWAB, F. A. (1968), *Initial Phase of a Source of Rayleigh Waves*, J. Geophys. Res. *73*, 755–760.

KOCAOGLU, A., and LONG, L. T. (1993), *Tomographic Inversion of Rg Wave Group Velocities for Regional Near-Surface Structure*, J. Geophys. Res. *98*, 6579–6587.

LEITH, W. (1987), *Tectonics of eastern Kazakhstan and implications for seismic source studies in the Shagan River area.* In *9th Annual DARPA/AFGL Seismic Research Symposium*, 15–18 June 1987, pp. 34–37.

MCLAUGHLIN, K. L., MURPHY, J. R., and BARKER, B. W. (1992), *A Lithospheric Velocity Anomaly beneath the Shagan River Test Site. Part 2. Imaging and Inversion with Amplitude Transmission Tomography*, Bull. Seismol. Soc. Am. *82*, 999–1017.

MARSHALL, P. D., BACHE, T. C., and LILWALL, R. C. (1985), *Body-Wave Magnitudes and Locations of Soviet Underground Explosions at the Semilpalatinsk Test Site*, AWRE Report No. 0 16/84 (re-issue), HMSO, 87 pp.

MAYEDA, K., and WALTER, W. (1996), *Moment, Energy, Stress Drop, and Source Spectra of Western United States Earthquakes from Regional Code Envelopes*, J. Geophys. Res. *101*, 11,195–11,208.

MCEVILLY, T. V., and STAUDER, W. (1965), *Effect of Sedimentary Thickness on Short-period Rayleigh Wave Dispersion*, Geophys. *30*, 198–203.

MEYERS, S. C., WALTER, W. R., MAYEDA, K., and GLENN, L. (1999), *Observation in Support of Rg Scattering as a Source of Explosion S Waves: Regional and Local Recordings of the 1997 Kazakhstan Depth of Burial Experiment*, Bull. Seismol. Soc. Am. *89*, 544–549.

MUELLER, R. A., and MURPHY, J. R. (1971), *Seismic Characteristics of Underground Nuclear Detonations*, Bull. Seismol. Soc. Am. *61*, 1675–1692.

NALIVKIN, D. V., *The Geology of the USSR. A Short Outline (Including a 1:7,500,000 Scale Geologic Map)* (Pergamon Press, New York 1990) 170 pp.

NNC FIELD REPORT (1996), *Unused Vertical Boreholes at the Balapan Test Field of the Former Semipalatinsk Test Site; Review of Existing Data on the Boreholes Geology and Design*, National Nuclear Center of the Republic of Kazakhstan, Kurchatov City, Republic of Kazakstan.

PATTON, H. J., and TAYLOR, S. R. (1995), *Analysis of Lg Spectral Ratios from NTS Explosions: Implications for the Source Mechanisms of Spall and the Generation of Lg Waves*, Bull. Seismol. Soc. Am. *85*, 220–236.

PHILLIPS, W. S., HARTSE, H. E., TAYLOR, S. R., VELASCO, A. A., and RANDALL, G. E. (1999), *Regional Phase Amplitude Tomography for Seismic Verification*, Pure appl. geophys. in press.

RINGDAL, F., MARSHALL, P. D., and ALEWINE, R. W. (1992), *Seismic Yield Determination of Soviet Underground Nuclear Explosions at the Shagan River Test Site*, Geophys. J. Int. *109*, 65–77.

STEWART, R. C. (1988), *P-Wave Seismograms from Underground Explosions at the Shagan River Test Site Recorded at Four Arrays*, AWE Report O 4/88, Aldermaston, UK, 249 pp.

THURBER, C. H., QUIN, H. R., and SALEH, R. (1994), *Catalog of Locations of Nuclear Explosions at Balapan, Kazakhstan, 1965 to 1985*, Bull. Seismol. Soc. Am. *84*, 458–461.

YAO, P. C., and DORMAN, J. (1992), *Short-period Surface-Wave Dispersion and Shallow Crustal Structure of Central and Eastern Tennessee*, Bull. Seismol. Soc. Am., *82*, 962–979.

YU, G.-K., and MITCHELL, B. J. (1979), *Regionalized Shear Velocity Models of the Pacific Upper Mantle from Observed Love and Rayleigh Wave Dispersion*, Geophys. J. *57*, 311–341.

ZONENSHAIN, L. P., KUZMIN, M. I., and NATAPOV, L. M., *Geology of the USSR: A plate-tectonic synthesis*, (B.M. Page, ed.) (American Geophysical Union Geodynamic Series, 21 1990) 242 pp.

(Received June 28, 1999)

Pure appl. geophys. 158 (2001) 2041–2058
0033–4553/01/112041–18 $ 1.50 + 0.20/0

Pure and Applied Geophysics

The Kirovskiy Explosion of September 29, 1996: Example of a CTB Event Notification for a Routine Mining Blast

DOUGLAS R. BAUMGARDT[1] and WILLIAM LEITH[2]

Abstract—On September 29, 1996, a routine mining blast of about 390 metric tons was detonated underground at the Kirovskiy mine in the central Kola Peninsula. The United States was notified two weeks in advance that the blast was to take place and was given the date, approximate time, location and total charge. The explosion was detected and located by the prototype International Data Center (pIDC) and published in the Reviewed Event Bulletin (REB). Detailed information about the blast, including the type and depth of mining operation, the underground charge configuration, and the blasting delay pattern, is reviewed and combined with a seismological analysis of the event. The seismic analysis points to a possible associated tectonic component to the blast, consisting of a small rock burst or induced tremor, spall, or some combination of these mechanisms, that may have enhanced the shear waves, produced large *Rg* waves at low frequency, and small *Pn*/*Sn* and *Pn*/*Lg* amplitude ratios at high frequency. While these discriminants might identify the event as an earthquake, the spectral/cepstral analysis of the event clearly shows the ripple-fire delays. This event provides important confidence-building measures for both location calibration, in the form of travel-time corrections for location of mine events in this region, and for improved understanding of seismic discriminants expected for large mine blasts that may have an associated induced tectonic component (e.g., spall, mine tremor or rock burst).

Key words: Mining, explosion, ripple-fire, discrimination, confidence building.

Introduction

On September 29, 1996, a routine mining blast of about 390 metric tons was detonated underground at the "Rudnik" shaft of the Kirovskiy mining combine, located in the central Kola Peninsula. This blast was unusual, in that the United States was notified[3] two weeks in advance that the blast was to take place and was given the date, time, location and approximate total charge. The blast occurred underground at nuclear-test containment (200–300 m) depths in a region of intense mining activity, where both surface and underground mining operations occur and

[1] ENSCO, Inc., 5400 Port Royal Rd., Springfield, VA 22151, U.S.A. E-mail: doug@ensco.com
[2] U.S. Geological Survey, 926-A National Center, Reston VA 20192, U.S.A. E-mail: wleith@usgs.gov

[3] The United States was notified on September 18, 1997, by facsimile from the Russian Ministry of Defense, Special Monitoring Service, to the U.S. Deputy Assistant Secretary of Defense for Nuclear Treaty Monitoring.

the potential exists for underground cavity creation. The mining techniques employed in the region are relatively common, and therefore, they represent a potential source of network-locatable underground events.

The explosion was detected and located by the prototype International Data Center (pIDC), published in the Reviewed Event Bulletin (REB), and had a magnitude ($M_L = 3.4$) in the range of a small-scale or cavity-decoupled nuclear test. Having occurred only days after the adoption of the Comprehensive Nuclear-Test-Ban Treaty (CTBT) by the United Nations, this event is the first example of the pre-notification of an explosion under the voluntary confidence building measures of the CTBT.

Analysis of this event provides a detailed picture of a type of underground mining explosion that will be detected by the CTBT International Monitoring System (IMS). We have collected detailed information on the September 29, 1996 blast (hereafter, the "Kirovskiy blast"), including the type and depth of mining operation, the underground charge configuration, and the blasting delay pattern. We have also completed a detailed seismological analysis, evaluating spectral discriminants and performing cepstral analysis to detect ripple-fire. Our main objective in this study has been to determine if the seismic data analysis can confirm the information provided in the pre-blast notification for purposes of confidence building under CTBT.

Mining Operations

The Kirovskiy mine is located in the Khibiny Mountains of the central Kola Peninsula (Fig. 1). In this heavily mined area, where rich apatite (calcium phosphate) and related ores are excavated as the principal raw material for Russia's super-phosphate industry. Khibiny operations accounted for 70% of all phosphate mined in the former Soviet Union (LEVINE, 1989). The Kirovskiy mine is one of the largest underground mining enterprises in Russia, in terms of production capacity, and the ores are excavated from the greatest depths of underground development in the Khibiny massif.

The location uncertainty ellipse reported in the REB is plotted on Figure 1. The REB location is offset about 40 km northwest of the actual mine location, and the ellipse does not contain the Kiriovskiy mine but actually encompasses a different mining region. This points up the need for location calibration, and this event, in fact, will serve as a useful reference event for location calibration in the region.

The underground ore body is elongate, 12 km in length, with a thickness ranging from 50 to 300 m. The rock massif is distinguished by its high strength (approaching that of granites), structural irregularities of various scales (from the Saam fault, with a thickness of 2–5 m, to a network of small fractures, with spacings from 0.1 to 1 m),

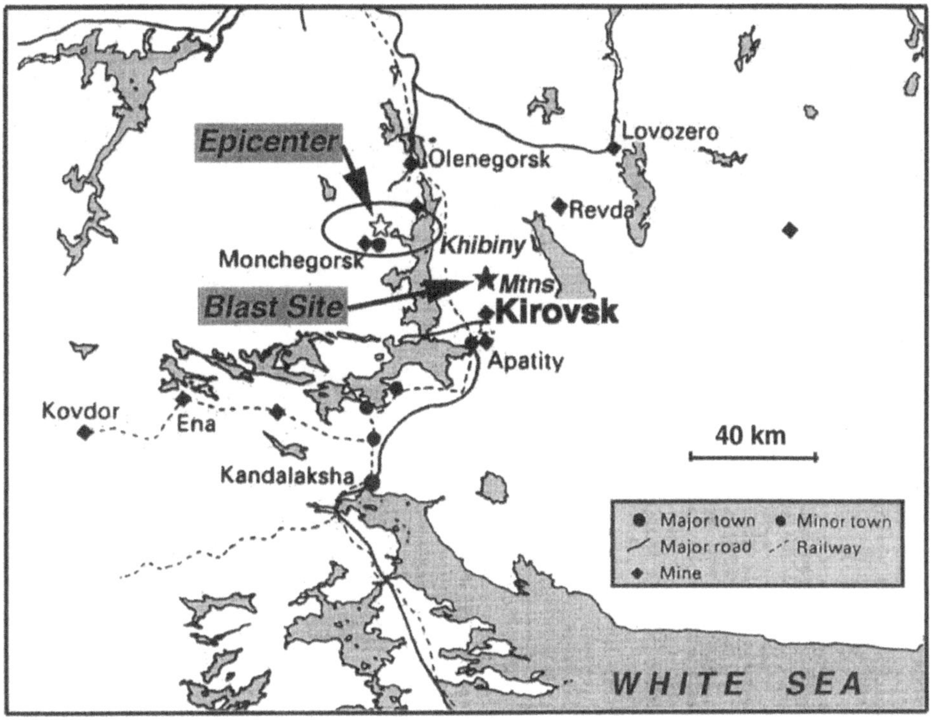

Figure 1
Map of the southwestern Kola Peninsula, showing the locations of major mines and the Kirovskiy blast of September 29, 1996. The star and ellipse, labelled "Epicenter," is the seismic location from the Reviewed Events Bulletin (REB) of the prototype International Data Center (pIDC).

and a high level of tectonic stresses, ranging from 30 to 80 Mpa. Rockbursts are well documented, and there is evidence for tectonic stress release accompanying underground mining blasts (SYRNIKOV and RYBNOV, 1997).

The ore is mined by a method known as blasthole stoping (sublevel stoping), in which multiple, large-diameter, long (up to 60 m), vertical holes are drilled for emplacing the explosives (HUSTRULID, ed., 1982, p. 95). The use of this relatively new mining method employs large charge sizes in comparison with most underground mining operations, ranging from 150 to 400 tons. The dimensions of a single excavated block are normally 50 by 50 by 70 m, which is drilled with a complex system of holes, usually numbering about 1000, defined by the particular blast design.

Shot delays are used to decrease the seismic effect of the blasts on underground openings. The charged holes are exploded in groups with up to 20 delay steps. The time interval for each delay is normally 23 msec. In this manner the blast energy is dispersed in space and in time.

Examples of Earlier Blasts

The dates of previous blasts in the mines and the regions in which they occurred are shown in Figure 2. The blast on December 6, 1989, was detonated at 9:33 AM local time (Moscow) with a total charge of 320 tons (including 80 tons at the horizon +322 m to the north; see Fig. 2). The charges were exploded simultaneously, with normal delay intervals between the blast holes. The block dimensions in plan view are 30 by 90 m, with a height of 90 m. The thickness of undisturbed rock above the block was about 400 m. An extremely high level of aftershocks occurred in the 72 hours after the detonation. Dates of blasts at the Kirovskiy mine in 1989 are given in Table 1.

The blast of December 5, 1993, at 9:50 AM local time (Moscow) was situated on the hanging side of the ore body under previously collapsed covering rocks. The

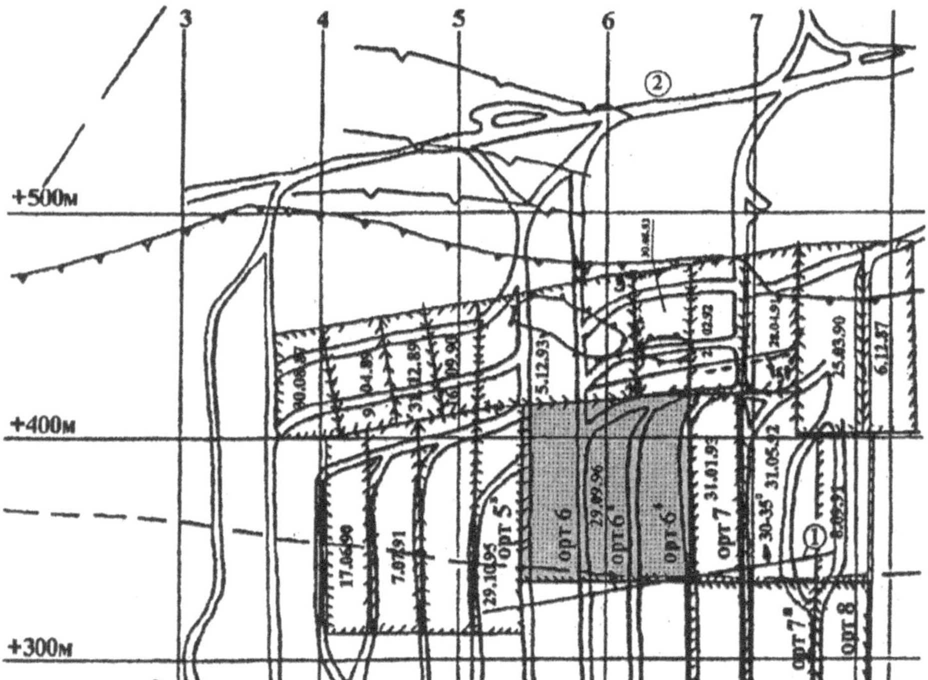

Figure 2
Plan view of the underground configuration at the Rudnik mine, showing the location of the September 29, 1996, Kirovskiy mine blast (shaded region). This is a partial view of the hauling horizon (at the "pillar-block"); note the locations of previous, similar blasts at the mine (dates). The horizontal lines indicate the hauling horizons in meters north. The September 29, 1996, event was detonated at hauling horizon +322 m north. Numbers in circles refer to: 1 – fault, 2 – east haulage drift.

Table 1

Dates, times, total charges, and magnitudes of underground mining explosions at the Kirovskiy mine in 1989

Date (d/m/y)	Time (hr/min)	Charge (tons)	Magnitude (M_L)
15.01.89	07:20	100	2.5
26.02.89	06:34	220	3.1
09.04.89	08:46	160	2.7
30.04.89	09:55	259	2.8
30.07.89	08:50	159	2.9
26.11.89	09:35	180	2.6

thickness of rock mass over the block was about 500 m. The block dimensions in the plan view (see Fig. 2) are approximately 80 by 55 m and the height was 70 m (for a total volume of ~310,000 m^3). The total charge was 255 tons. The blast was carried out in 16 stages, each with a 23 msec delay; i.e., total duration of the blast was approximately 350 msec. The displacements and geoelectric potential measurement results for this blast were given by SYRNIKOV *et al.* (1996).

The Kirovskiy Blast of September 6, 1996

On September 29, 1996 at 10:05:46 AM, local time (Moscow), the Kirovskiy mine blast, for which the United States received advance notification, was detonated with a total charge of 389 tons and 18 delays. Local instrumentation of the site, put out by the Russians, provided information about the blast pattern and blast-wave shock-front velocities in the mine. A plan view of the blasting design at hauling horizon +252 is given in Figure 2, which also shows the locations and dates of previous blasts. The zone depicted shows the massif in a domain limited in the plane by zones to the north and south in which, before 1989, the ore had already been excavated. This is the so-called "pillar-block" (KOZYREV and PANIN, 1993).

The vertical section of the block for this blast (charge-hole groups) is shown in Figure 3. The distribution of charges in delay steps is presented in Table 2. This blast successfully separated the pillar block from the ore-handling side and did not create any significant damage to the mine. Maximum velocity values in this case do not exceed 0.25 km/sec.

The rock broken by the detonation of the first group of blast holes provides an attenuating effect with respect to the subsequent groups. The velocities of the blast wave at the hauling horizon were relatively small, ranging from 1 km/sec, directly against the block, to 0.1–0.2 km/sec further away. In those areas where the instruments were located, the seismic measurements were characterized by velocities in the range of 0.1–0.3 km/sec, with frequencies in the range of 20–80 Hz.

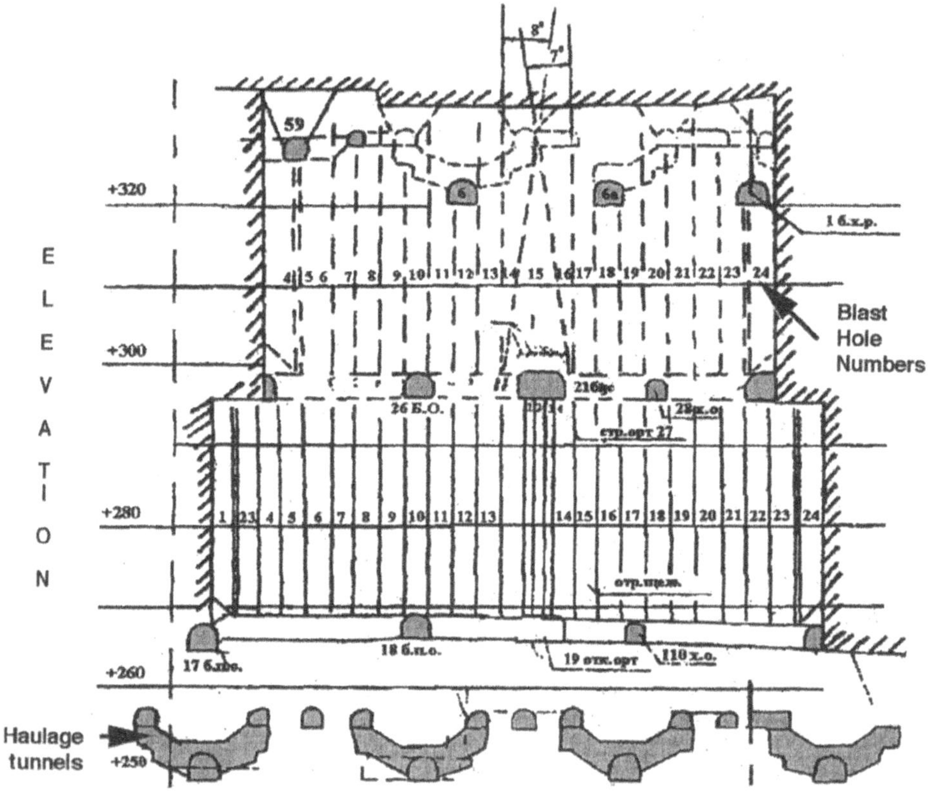

Figure 3
Cross section of the drillhole and delay pattern for the September 29, 1996 Kirovskiy mine blast. Numbered holes refer to the charge-delay sequence given in Table 2. The block is approximately 100 m across (see Fig. 2), and the surface elevation is about 800 m.

Regional Seismic Waveform Characteristics

Figure 4 is a map showing the location of the Kirovskiy blast and the nearest regional stations that recorded the best signals from the event. Six regional seismic stations recorded the event and reported to the prototype International Data Center (pIDC). The pIDC estimated the local magnitude (M_L) to be 3.4. Figure 5 shows a record section of vertical-component, short-period waveforms for the stations shown in Figure 4. The waveforms have been filtered in two bands, 2–4 Hz and 4–6 Hz, using 3rd-order Butterworth filters. The plots are in distance order, that ranges from 382 km for ARCESS, to 1280 km for NORESS. Lines mark the primary phases, *Pn*, *Sn* and *Lg*.

Obviously, the best signals were recorded at the stations closest to the event. For the more distant stations, Hagfors (HFSA1) at 1248 km and NORESS (NRA0) at 1279 km, the signal-to-noise ratios were low, although all three phases are visible in the low frequency band (2–4 Hz). In the higher band (4–6 Hz), only *Pn* and *Sn* were

Table 2

Delay sequence for the underground blast on September 29, 1996. The delay interval is 23 msec. The upper level holes were detonated first (see Fig. 3)

Delay no.	Holes fired	Charge increment (tons)
0	1,2,23,24	29100
1	3,22	22478
2	4,5,23,24	30357
3	5,21	18221
4	6,6	17447
5	20,22	17911
6	7,7	16786
7	19,21	17430
8	8,8	17447
9	18,20	17351
10	9,9	14790
11	17,19	17077
12	10,10	15366
13	16,18	16312
14	11,11	15900
15	15,17	16398
16	12,12	16561
17	13,14	18221
18	13,14,14′,16,	34055
	15,15′	16287
	Total Charge = 389,175	

recorded and *Lg* has been strongly attenuated. However, it is notable that for all distance ranges, the amplitudes of the shear waves (*Sn* and *Lg*) are larger than the *Pn* wave amplitudes.

ARCESS recordings had the highest signal-to-noise ratios and therefore were chosen for more in-depth analysis. Figure 6 displays the results of a bandpass-filter analysis of the seismogram recorded by the ARA0 array element of ARCESS. The filters were applied in successive bands covering the range from 0.5 to 2.5 Hz, which is best for *Rg* detection, and 7 other frequency bands up to 8–10 Hz. Also shown on the plot are the times of the identified main regional phases.

The analysis in Figure 6 clearly identifies 5 main phases, *Pn*, *Pg*, *Sn*, *Lg*, and *Rg*. As mentioned above, the shear waves (*Sn*, *Lg*, and *Rg*) have the highest amplitudes in certain frequency bands. *Rg* is obviously large at low frequency but decays sharply with frequency. Particularly striking are the energetic *Sn* and early *Lg* arrivals in the high-frequency (8–10 Hz) band, which we discuss in more detail in the next section.

Discrimination Waveform Feature Analysis

Waveform features were extracted from the ARCESS waveforms in the Intelligent Seismic Event Identification System (ISEIS) using techniques described

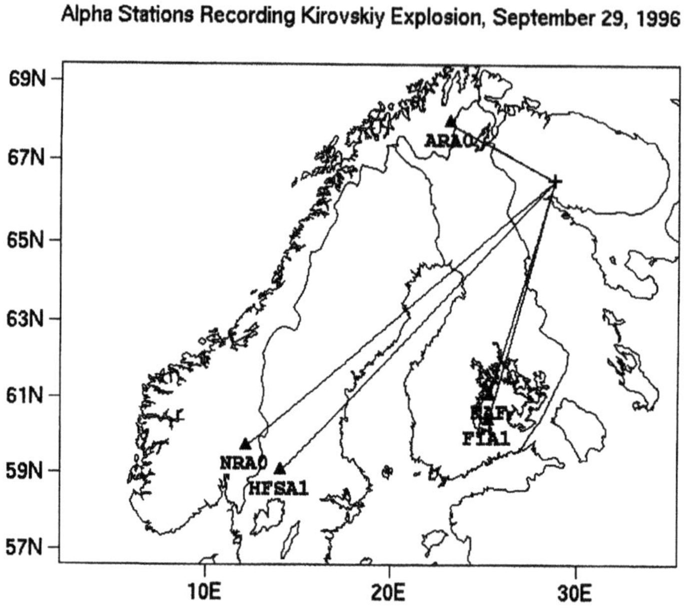

Figure 4
Coastline map showing the locations of the Kirovskiy blast and the seismic stations which recorded the best signals.

by BAUMGARDT *et al.* (1991). Two important discriminants are the *Pn*/*Sn* and *Pn*/*Lg* amplitude ratios in multiple frequency bands. Previous discrimination studies (e.g., BAUMGARDT and YOUNG, 1990; RYALL *et al.*, 1996; KIM *et al.*, 1997; HARTSE *et al.*, 1997) indicate that these ratios are higher for blasts than earthquakes, presumably because earthquakes are shear sources and produce stronger shear waves than explosions that are primarily compressional sources.

To determine amplitude ratios for regional arrays like ARCESS, phase amplitudes are first measured from incoherent beams, computed by time-averaging rms amplitudes in one second window on each array-element trace, after applying bandpass filters, and then averaging the amplitudes across all the waveforms. Examples of ARCESS incoherent beams for the Kirovskiy blast in 9 bandpass-filter bands are shown in Figure 7. The phase picks made on the waveforms are superimposed on the incoherent beams. These incoherent beams provide smoothed envelope measurements for the waveforms, and the waveform features pointed out in Figure 6 can be clearly discerned in the incoherent beams. Most notable are the large low-frequency *Rg* amplitudes and the large *Sn* and *Lg* amplitudes that persist across the entire frequency band.

Amplitude ratios for *Pn*/*Sn* and *Pn*/*Lg* were computed from rms amplitude estimates for each phase in nine narrow frequency bands. The maximum rms amplitude estimates were taken from the incoherent beams in time windows beginning with the phase picks shown in Figure 7. This same analysis was also done

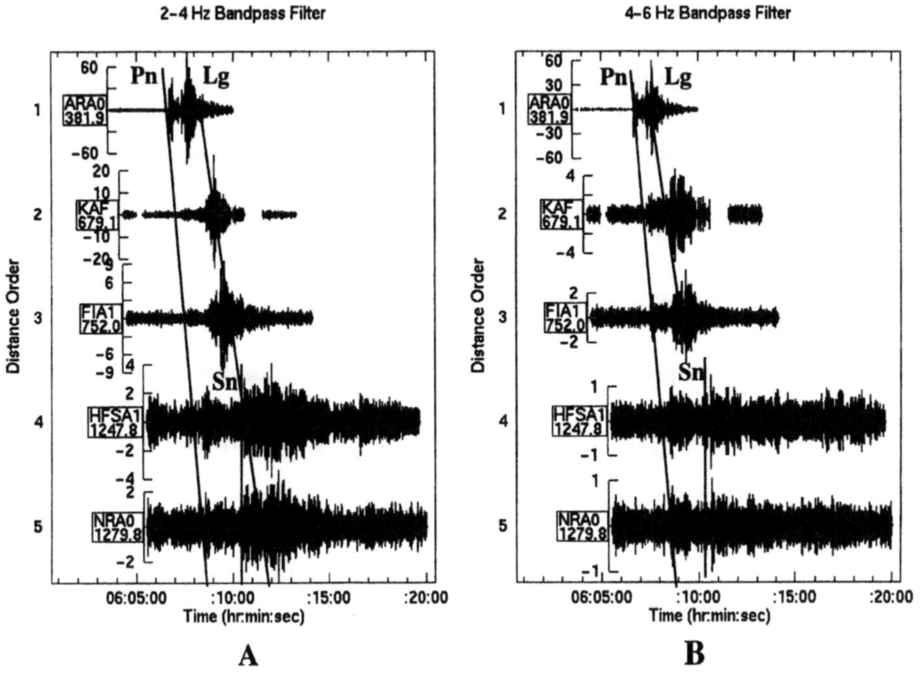

Figure 5
Waveforms from the September 29, 1996, Kirovskiy blast recorded at the six nearest stations. Waveforms are plotted in distance order. (a) 2–4 Hz filter, (b) 4–6 Hz filter. Note that *Lg* only appears at ARCESS and KAF at high frequency (4–6 Hz). At Hagfors (HFSA1) and NORESS (NRA0), *Sn* is the dominant shear wave at high frequency. Amplitudes in these waveforms have been tract-normalized.

for nearby reference events, including other mine blasts on the Kola Peninsula, earthquakes in northern Norway, and earthquakes and mine blasts in northern Sweden. Figure 8 shows scatter plots of *Pn*/*Sn* and *Pn*/*Lg* amplitude ratios versus distance in two filter bands, 2–4 and 8–10 Hz, where the earthquakes are plotted with square symbols, the blasts with circle symbols, and the Kirovskiy blast with an asterisks.

The distance plots in Figure 8 reveal significant scatter in the mine blast and earthquake points, with considerable overlap in the earthquake and blast points, particularly in the 2–4 Hz band. The separation of points is somewhat better in the 8–10 Hz band, however many mine blasts have low ratios comparable to earthquakes. The Kirovskiy blast *Pn*/*Sn* and *Pn*/*Lg* ratios are very low in the 8–10 Hz band. Distance and magnitude corrections have not been applied to these data, but the results would not be significantly changed since the earthquakes and blasts were at about the same distance and magnitude.

Figure 9 shows the measurements of *Pn*/*Sn* and *Pn*/*Lg* ratios as a function of frequency in all 9 filter bands. Each point represents the average ratio for each class (earthquake – square, blast – circle, and Kirovskiy event – asterisks) and the error

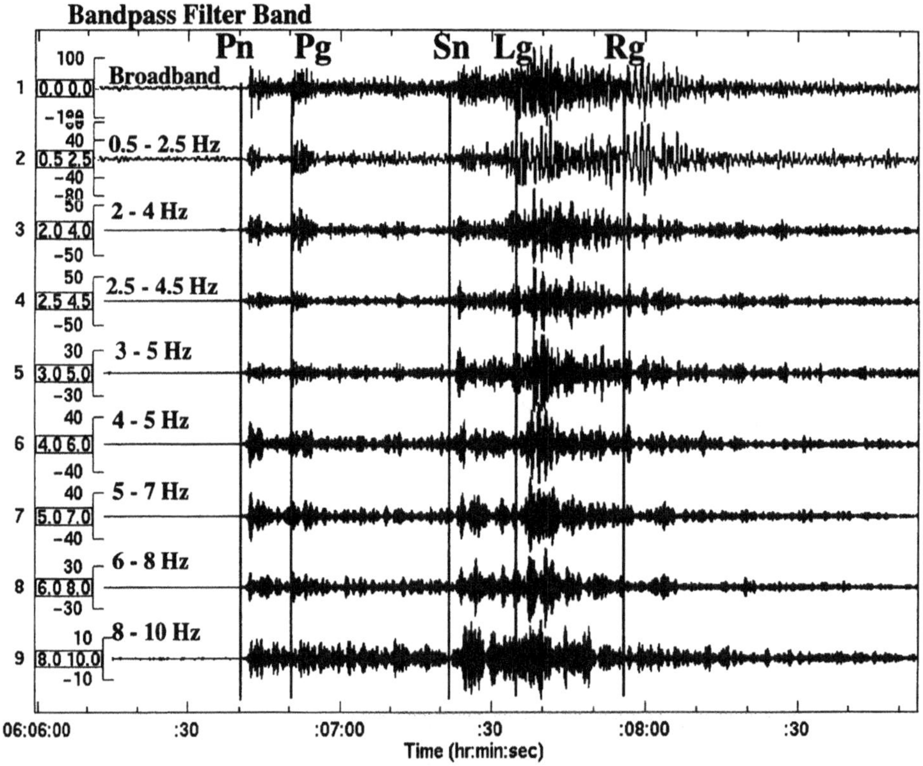

Figure 6

Bandpass-filter analysis of the waveforms recorded at the ARA0 vertical component of ARCESS from the September 29, 1996, event with major phases indicated on the top broadband trace. Note large *Sn* and *Lg* wave amplitudes relative to the *Pn* across the entire frequency band and especially in the 8–10 Hz band. Amplitudes in these waveforms have been trace-normalized.

bar through each point indicates one standard deviation. The asterisk symbols indicate the September 29, 1996 Kirovskiy blast and are connected by a line.

These plots indicate generally good discrimination between blasts and earthquakes, with earthquakes having generally lower ratios than blasts. Figure 8 has overlapping points in the 8–10 Hz bands for all points plotted as a function of distance.

In most frequency bands the *Pn*/*Lg* and *Pn*/*Sn* ratios for the Kirovskiy blast sequence are comparable with most Kola mine blasts. Figure 10 shows a striking dip in the *Pn*/*Sn* ratio for the Kirovskiy blast sequence in the frequency band above 8 Hz that falls into the earthquake category. This dip is caused by the large *Sn* amplitudes in this frequency band discussed earlier. A similar but less dramatic high-frequency dip appears in the *Pn*/*Lg* ratios. BAUMGARDT (1995) and RYALL *et al.* (1996) have demonstrated that mine blasts on the Kola Peninsula generally have ratios which strongly overlap measurements for earthquakes and may make clear discrimination of mine blasts and earthquakes problematic. The Kirovskiy blast might have been identified as an earthquake on the basis of *Pn*/*Sn* and *Pn*/*Lg* ratio discriminants in

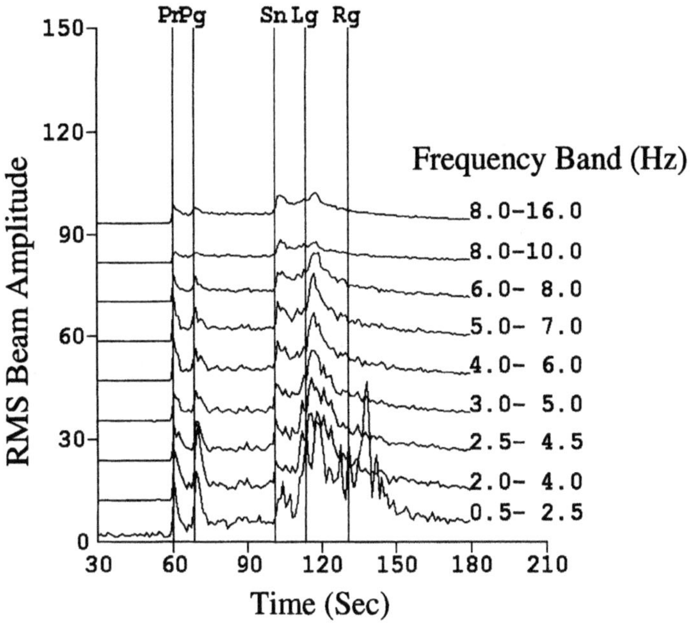

Figure 7
Incoherent beams of the September 29, 1996, event in 9 filter bands recorded at the ARCESS array. The beams result from average RMS amplitudes in 1 second time windows, computed on the bandpass-filtered vertical-component traces, averaged across the all array vertical channels. The beams have been shifted for viewing purposes and the vertical scale is in relative units only. The regional phases, *Pn*, *Pg*, *Sn*, *Lg*, and *Rg*, shown here were picked on waveforms.

the 8–10 Hz band. However, looking at the overall frequency trend in Figure 9, the Kirovskiy blast clearly falls in the blast group.

Many mines in the Khibiny Massif and Apatity region have a history of rock bursts. It is possible that many explosions in these regions may have had associated rock bursts or induced mine tremors (e.g., BAUMGARDT, 1995). Figure 2 shows that the mine has at least one fault in the same haulage horizon, labeled by the circled number 1, and other, perhaps smaller, faults and fractures may exist in the vicinity of the blast. Recent discrimination studies of mine tremors and rock bursts throughout the world (e.g., BENNETT *et al.*, 1994) have shown that discrimination features of these events more closely resemble earthquakes than nuclear explosions which lack an associated tectonic component. We suggest that the Kirovskiy blast falls in the category of combined mine blast and mine tremor, where the tremor may have been induced by the blast and the enhanced shear waves may have resulted from this induced tectonic component.

As another possible mechanism for the enhanced shear waves, recently proposed source models for cast blasting in mines have incorporated a mechanism for rock spall (e.g., ANANDAKRISHNAN *et al.*, 1997). Although not a cast shot, the Kirovskiy blast did successfully separate the pillar block rock, and the resultant spall for this

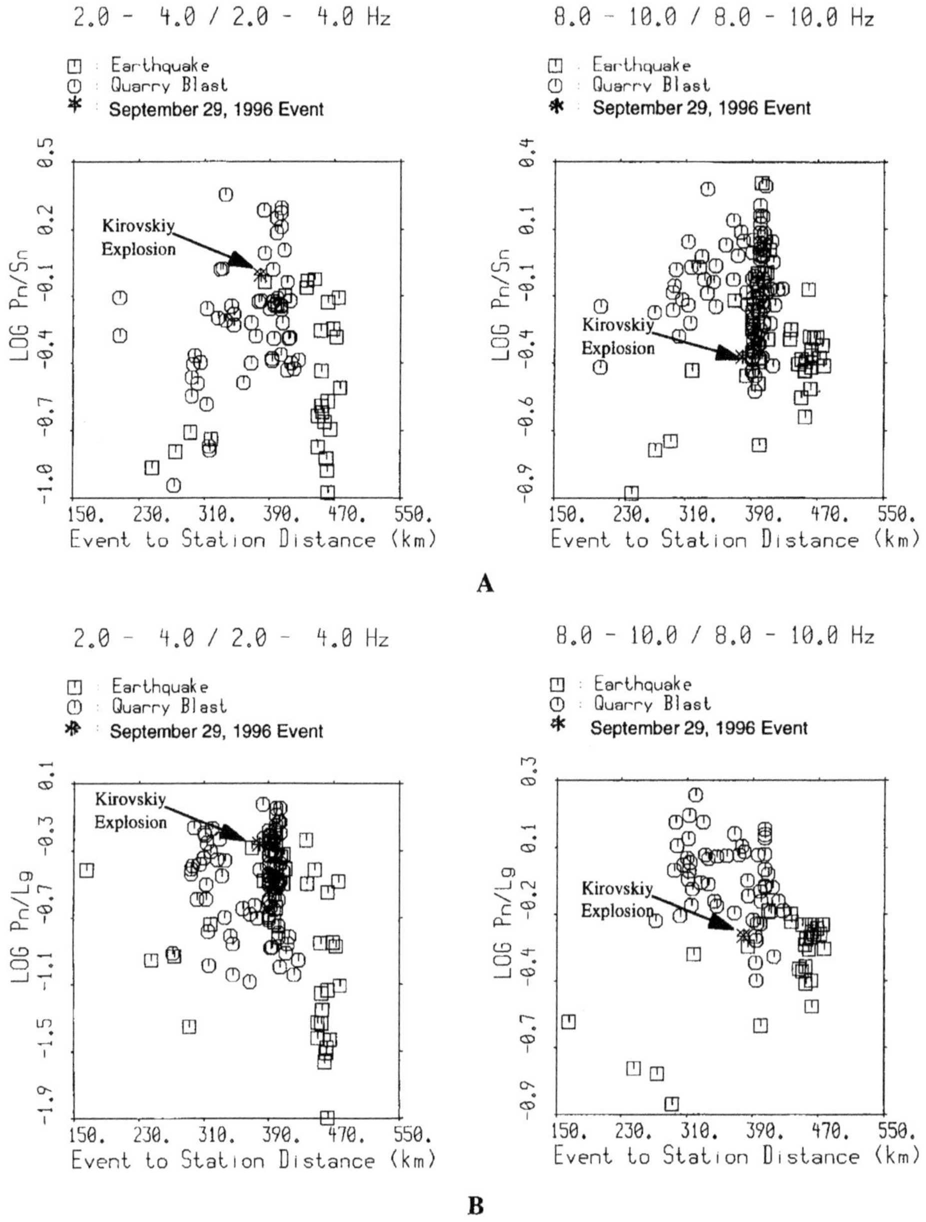

Figure 8

Scatter plots of distance versus amplitude ratios measured for mine blasts and earthquakes recorded at ARCESS in two frequency bands, 2–4 and 8–10 Hz. *Pn*/*Sn* ratios are plotted in A and *Pn*/*Lg* ratios in B. The amplitude ratios were obtained from maximum RMS incoherent-beam amplitudes in windows starting at the regional phase-pick times indicated on the incoherent beam plots shown in Figure 7. This same analysis was applied to other earthquakes and blasts northern Norway, Sweden, and on the Kola Peninsula. The square symbols represent earthquakes, circles represent blasts, and the asterisks symbol represents measurement for the Kirovskiy blast.

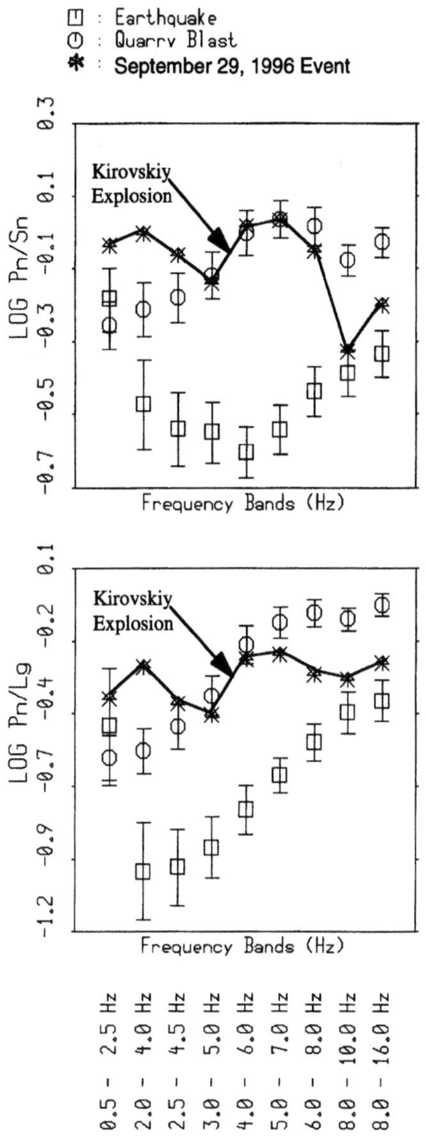

Figure 9
Plot of average values of the same amplitude ratios as in Figure 8 as a function of frequency. *Pn/Sn* amplitude ratios are plotted in the top plot and *Pn/Lg* amplitude ratios in the bottom plot. The values for the Kirovskiy blast are indicated by the asterisks and are connected by lines. The average values for explosions and earthquakes separate for most frequencies above 2 Hz, and the Kirovskiy blast ratios tend to fall more in with the other mine blasts, although below them at high frequency.

may have been significant. Therefore, another possible explanation for the small, earthquake-like *Pn/Sn* and *Pn/Lg* ratios may be that the spall mechanism excited stronger shear waves than the explosion itself.

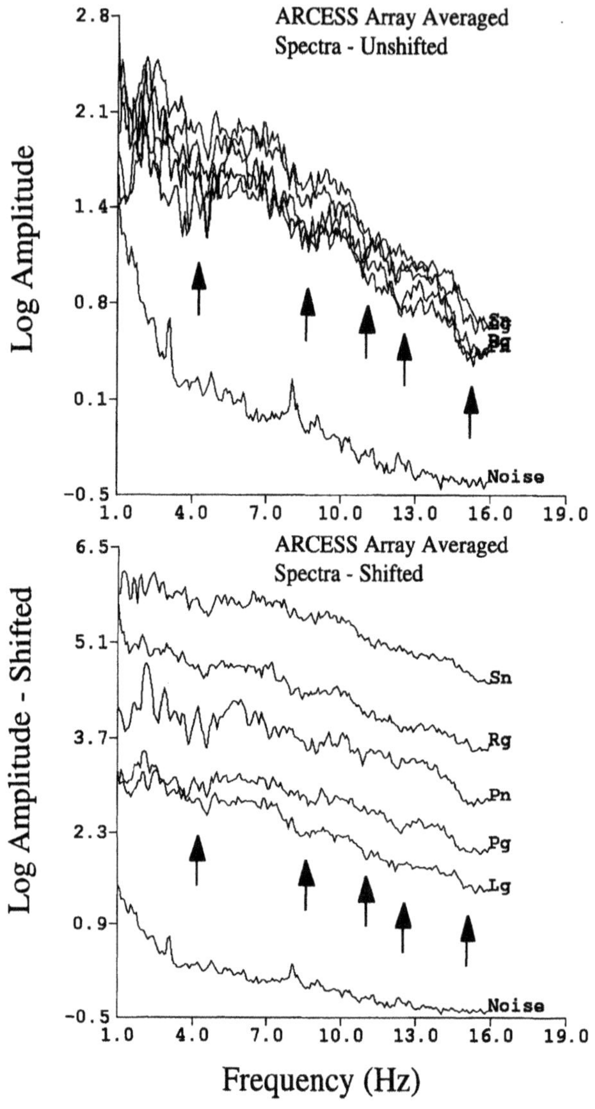

Figure 10
Array-averaged vertical-component spectra for the four regional phases and *Pn* background noises recorded at ARCESS from the September 29, 1996 Kirivoskiy blast. Each spectrum has been computed for a 10 second Parzen window for each phase, beginning with the times shown in Figure 6 on each vertical component ARCESS trace and then averaged across the array. The top plot shows the absolute spectral levels relative to the noise spectrum. The bottom plot shows the spectra shifted vertically. The arrows indicated the nulls of the spectral scalloping.

Spectral/Cepstral Analysis of Ripple-fire

Spectral and cepstral characteristics of the Kirovskiy blast data can provide information about ripple-fire sequences. BAUMGARDT and ZIEGLER (1988) have shown that ripplefire produces time-independent spectral scalloping and also consistent, time-independent cepstral peaks for all phases. The Kirovskiy shot sequence consisted of eighteen 23 msec shot delays spread over a horizontal region of about 140 m and depth range of about 60 m. The ARCESS sampling rate of 40 samples per sec, giving a Nyquist maximum bandwidth of 20 Hz, precludes resolution of 23 msec delays since these would require 1/23 msec or 43.5 Hz bandwidth. Moreover, the bandwidth of many IMS sensor systems is further reduced by the application of low-pass antialias filters that cutoff at 16 Hz. However, because the explosion consisted of a complex pattern of delays, longer apparent delays are produced that generate the spectral scalloping within the 16 to 20 Hz bandwidth.

Figure 10 shows spectra for each of the phases in the range of 1 to 16 Hz, and Figure 11 shows the cosine cepstra computed for each spectrum (BAUMGARDT and ZIEGLER, 1988). The spectra were computed by averaging across the array individual spectra on each array-element waveform, for 10 second windows, starting at the picked phase times and smoothed with a Parzen time window. The spectra are shown

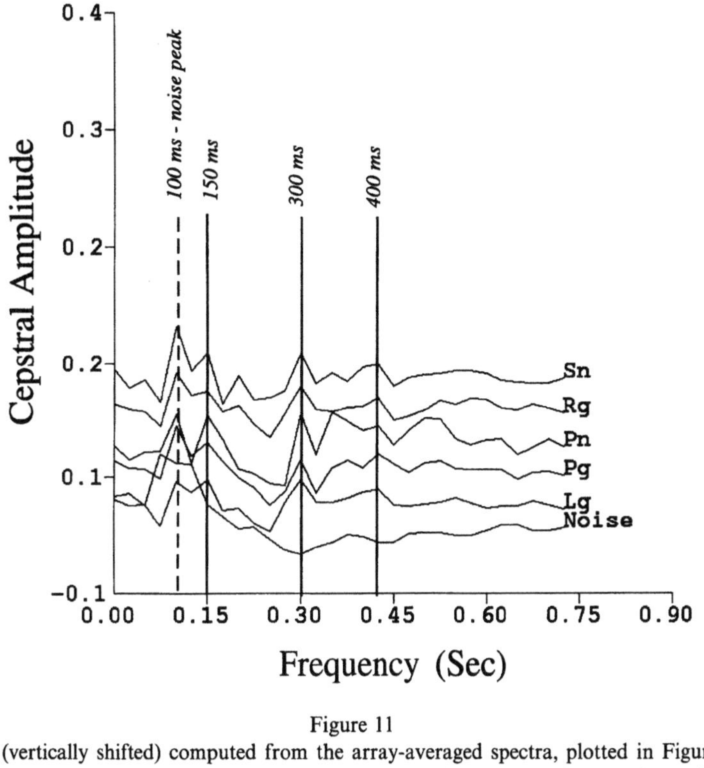

Figure 11
The cepstra (vertically shifted) computed from the array-averaged spectra, plotted in Figure 10, of the September 29, 1996 event. The cepstra were computed by taking the Fourier transform of the base-10 logarithm of each spectrum. Vertical solid lines indicate consistent time-independent cepstral peaks caused by the spectral modulations produced by the ripple-fire delays. Note that these peaks do not appear in the noise cepstrum. The dashed line indicates a peak at 100 msec in the phase cepstra but which also appears in the noise cepstra, and thus may be a processing artifact.

plotted relative to each other on the top and shifted upward, for display purposes, relative to the noise spectrum on the bottom. Cepstra were computed from the spectra by Fourier transforming the logarithm of the power spectra in the band from 1 Hz to 19 Hz, with smoothing to eliminate end effects and removal of the polynomial trend. The same analysis was also applied to the noise ahead of the *Pn* phase. The cepstra in Figure 11 have been shifted vertically for display purposes and the vertical amplitude scale only indicates relative values.

The spectra for the five detected phases in Figure 10 has scalloping not present in the noise, but the scalloping pattern is very complex. The positions of the most prominent nulls of the scalloping are indicated by arrows on both the unshifted and shifted spectra. The cepstra plotted in Figure 11 show at least three sets of consistent cepstral peaks (vertical lines solid lines). These peaks appear in all phases at the same delay time and not in the noise cepstrum. This indicates that the cepstral peaks result from spectral modulations caused by source multiplicity and not by features in the noise or processing artifacts. The delay times are 150, 300 (multiple of 150), and 400 msec. Note that there is also a large 100 msec peak, marked by the dashed line, but that peak also appears in the noise cepstrum. This peak results from a 10 Hz spectral modulation present both in the signal and the noise. Because this peak is apparent in both signal and noise, we do not consider it to be part of the ripplefire pattern of the explosion. The 400 msec time approaches 18 times 23 msec or 414 msec and gives the approximate overall shot duration.

Conclusions and Recommendations

From a CTBT verification perspective, the Kirovskiy mining blast is of special interest because:

- it is the first example of a routine mining blast where there was a government-to-government advance notification;
- the blast occurred underground at nuclear test containment depths and in a region of intense mining activity, with both surface and underground mining operations and underground cavity creation;
- the pIDC calculated a location and a relatively large magnitude ($M_L = 3.4$) for an underground blast;
- the error ellipse of the event did not contain the blast site and actually included another mining area. This has important implications for approaches to on-site inspection and points out the need for improved network calibration;
- the mining technique employed was not unique and therefore represents a potential source of network-locatable underground events.

Our analysis of regional seismic data for the event is consistent with the available information pertaining to the mining activity in the Kirovskiy mine and the event notification information provided. We find that the spectral/cepstral analysis gives delay times consistent with those expected for the various shots in the ripple-fire pattern, even though the limited bandwidth of our data precludes resolution of the shortest delay times. However, we are able to confirm the overall duration of the ripplefire sequence to be on the order of 400 msec, as specified in the event-notification report.

The seismic analysis reveals that there may be an associated tectonic component to the blast (a rock burst, induced tremor, and/or spall) indicated by small *Pn*/*Sn* and *Pn*/*Lg* amplitude ratios at high frequency. This is also consistent with independent information we have obtained about the mining region regarding the high tectonic strain and the common occurrence of mine tremors, and the fact that the explosion separated the pillar block rock. Without this information the discriminant might identify the event as an earthquake. However, the spectral/cepstral analysis of the event clearly shows the ripple-fire delays. The fact that mine blasts can resemble mine tremors or earthquakes is not a serious issue if ripple-fire can usually be detected. The issue of whether or not a single large chemical or nuclear explosion in a mine, with no ripple-firing, could also produce similar tectonic component and resultant earthquake-like *Pn*/*Sn* or *Pn*/*Lg* needs to be considered.

With respect to confidence building, this event provides an important calibration data point for location and discrimination. Travel-time corrections for this mine will allow for improved locations of future events in this region. Seismic waveform analysis confirms the earlier observation of large shear waves for previous mine blasts, presumably due to a tectonic component. The large *Rg* waves give an indication of shallow depth and an extended source, as confirmed by observations of spectral scalloping due to ripple-fire. Thus, in monitoring blasts in mines, discrimination procedures should expect earthquake-like characteristics in recorded waveforms for blasts and associated mine tremors.

Acknowledgements

We wish to thank Russian colleagues, Nikolai M. Syrnikov and Yjuri S. Rybnov, both of the Institute for Dynamics of Geospheres, Russian Academy of Sciences, Moscow, Russia, for providing most of the information concerning the Kirovskiy mine and the September 29, 1996 blast. The two reviewers of this paper, Jay Zucca and Mark Tinker, provided detailed comments which greatly improved this paper.

References

Anandakrishnan, S. S., Taylor, S. R., and Stump, B. W. (1997), *Quantification and Characterization of Regional Seismic Signals from Cast Blasting in Mines: A Linear Elastic Model*, Geophys. J. Int. *131*, 45–60.

BAUMGARDT, D. R., CARTER, S., MAXSON, M., CARNEY, J., ZIEGLER, K., and MATSON, N. (1991), *Design and Development of the Intelligent Event Identification System*, PL-TR-91-22298(I), Final Rep., vols. I, II, and III, ENSCO, Inc., Springfield, VA.

BAUMGARDT, D. R. (1995), *Case Studies of Seismic Discrimination Problems and Regional Discriminant Transportability*, PL-TR-95-2106, Final Rep., ENSCO, Inc., Springfield, VA.

BAUMGARDT, D. R., and YOUNG, G. (1990), *Regional Seismic Waveform Discriminants and Case-based Event Identification Using Regional Arrays*, Bull. Seismol. Soc. Am. *80*, 1874–1892.

BAUMGARDT, D. R., and ZIEGLER, K. (1988), *Spectral Evidence of Source Multiplicity in Explosions: Application to Regional Discrimination of Earthquakes and Explosions*, Bull. Seismol. Soc. Am. *78*, 1773–1795.

BENNETT, T. J., MARSHALL, M. E., BARKER, B. W., and MURPHY, J. R. (1994), *Characteristics of Rockbursts for Use in Seismic Discrimination*, PL-TR-94-2269, Final Report, Maxwell Laboratories, Inc., S-CUBED Div., La Jolla, CA.

HARTSE, H. E., TAYLOR, S. R., PHILIPS, W. S., and RANDALL, G. E. (1997), *Regional Event Discrimination in Central Asia with Emphasis on Western China.*

HUSTRULID, W. A., ed. (1982), *Underground Mining Methods Hand book*, Society of Mining Engineers, New York.

KIM, W. Y., AHARONIAN, V., LERNER-LAM, A. L., and RICHARDS, P. G. (1997) *Discrimination of Earthquakes and Explosions in Southern Russia Using Regional High-frequency Three-component Data from the IRIS/JSP Caucasus Network*, Bull. Seismol. Soc. Am. *87*, 569–588.

KOZYREV, A. A., and PANIN, V. I. (1993), *Effect of Large-scale Mining Operation on the Geodynamic Behavior of the Area, and Mine-induced Seismicity Manifestation.* In *Proc. ISRM Int. Symp. EUROCK '93.*

LEVINE, R. A. (1989), *Mineral Development on the Kola Peninsula, Present Conditions and Future Prospects*, Norw. Atlantic. Comm, Security Policy Lib. no. 9–1989, 22 pp.

RYALL, A. S., BAUMGARDT, D. R., FISK, M. D., and RIVIERE-BARBIER, F., *Resolving regional discrimination problems: some case histories.* In *Monitoring a Comprehensive Test-Ban Treaty* (Husebye, E. S., and Dainty, A.M. eds.) (NATO ASI Series 1996) pp. 721–741.

SYRNIKOV, N. M., KONDRATYEV, S. V., and RYBNOV, Y. S., *On some exhibitions of structure heterogeneities in rock massif as the effect of large-scale action.* In *Proc. ISRM Symp. EUROCK'96, Torino* (A.A. Balkema, Rotterdam 1996) vol. 1, pp. 483–490.

SYRNIKOV, N. M., and RYBNOV, Y. S. (1997), *On Rock Massif Instability from Blasting*, Preprint, Russian Acad. Sci., Inst. Dynam. Geospheres, 12 pp.

(Received May 28, 1999, revised/accepted December 30, 1999)

Pure appl. geophys. 158 (2001) 2059–2076
0033-4553/01/112059-18 $ 1.50 + 0.20/0

Pure and Applied Geophysics

Source Directivity, Signal Decorrelation, Spectral Modulation and Analysis of Spatio-temporal Patterns of Multiple Explosions

ZOLTAN A. DER[1] and DOUGLAS R. BAUMGARDT[1]

Abstract—Spectral modulation is used as a criterion for discriminating multiple mining explosions from earthquakes and single blasts. Features in the spectra of regional arrivals from quarry blasts may be explained by the spatio-temporal configuration of shot patterns and are explored via a model based on spatial waveform decorrelation and propagation delay (directivity) effects. The phenomenon of decreasing modulation with decreasing average phase velocities of the seismic wave arrivals can be attributed to a trend of decreasing similarities of waveforms with increasing distance between individual explosions making up the shot pattern for the respective arrivals. Rescaling of the modulation patterns with respect to frequency can be explained by the Doppler effect.

The postulated relationship between shot pattern layouts and sizes, together with their firing sequences, on one hand, and the similarities and differences in modulation characteristics among the various regional arrivals, on the other hand, can be used to derive information about the mining practices in a region that is otherwise not accessible. Moreover, such analyses could be utilized for detecting unusual activities in the CTBT monitoring system.

Key words: Mining blasts, modulation, decorrelation, propagation delays.

Introduction

Spectral modulation (also called scalloping, if regular) is a term used to describe rapid frequency-dependent fluctuations of the spectral amplitude within the seismic signal band as opposed to a smooth, gradually varying, spectrum. It was found that the spectrum generated by a single point source or earthquake is usually smooth, although there are exceptions to this rule. On the other hand, spectra of regional arrivals from multiple blasts are usually strongly modulated (e.g., BAUMGARDT and ZIEGLER, 1988; BAUMGARDT and YOUNG, 1990; KIM *et al.*, 1997). Spectra of underwater explosions are often modulated because of the presence of bubble pulses which behave as multiple sources (BAUMGARDT and DER, 1998). For seismic events on land, modulation was found to be an effective discriminant for distinguishing ripple-fired quarry blasts from large, single explosions and earthquakes. Typically, this discriminant is applied to whole seismograms in the form of sonograms (e.g., HEDLIN *et al.*, 1990, 1998; CARR and GARBIN, 1998) or sensor-averaged spectra for

[1] ENSCO Inc, 5400 Port Royal Rd, Springfield, VA 22151, USA.

the individual regional arrivals (BAUMGARDT and ZIEGLER, 1986). The spectral modulation in seismic signals generated by patterns of multiple shots tends to be similar for all regional phases from *Pn* to *Lg*. This is largely due to the identical effective time pattern of the shots.

Frequently it is assumed that the form of the spectral modulation is determined predominantly by the duration of the shot sequence. This is only true if the shot sequence has a constant amplitude envelope over the duration. Generally, changes in the details of time domain amplitude envelope within time interval that encompasses the total blast duration determine the details of modulation patterns, not just the total duration. Therefore, duration cannot simply be derived from the frequencies associated with the spectral minima. At NORESS-type regional arrays the sampling time interval is large, 25 msec, relative to the intershot time intervals of typical quarry blasts, which are commonly of the order of 10 msec or less. Consequently, the fine details of the shot patterns at this time scale are inconsequential in the spectra computed below 20 Hz. According to the digital sampling theorem (e.g., GOLDMAN 1953; BENDAT and PIERSOL, 1986) in order to determine what effective time pattern is "seen" by a distant observer, the original detailed shot amplitude pattern must be smoothed by a function of the form $\sin(2\pi Wt)/2\pi Wt$ corresponding to a bandwidth of $W = 20$ Hz and the result resampled at 40 Hz. This smoothed "effective source function" is generally much simpler than the actual shot sequence. Furthermore, by modifying the time history within the same "duration," it is possible to generate different modulation patterns whose first spectral nulls do not correspond to the inverse of the duration of the shot pattern. Only if this effective time history is reasonably close to a Gaussian shape will there be little or no modulation since the Fourier transform of a Gaussian shape is also Gaussian without any modulation. Thus multiple shot patterns can actually be designed that would produce no spectral modulation. Such examples can be computed without including propagation delays or decorrelation, to be discussed below.

The Decorrelation Model

Signal waveform decorrelation and propagation delays among individual charges can significantly influence the modulation of the spectra. Typically, waveforms from individual charges will vary with location and gradually decorrelate with mutual distance. HARRIS (1991) demonstrated that this occurs for mine blasts at a coal mine in Missouri as observed at an array in Kansas. However, he did not do the correlation analysis separately for the various arrivals. Decorrelation distance can be arbitrarily defined as the distance where the correlation coefficient has fallen to a certain fraction of unity, $1/e$, for instance. This distance is typically a few wavelengths (HARRIS, 1991). If the shot array size is considerably larger than this decorrelation distance and it is exploded progressively from one end to the other, then the modulation patterns will correspond to those of a shot sequence with duration shorter than the actual duration. This happens because pairs of the earliest and latest shots will tend to contribute little to the modulation pattern or cross-spectra that contribute to modulation as it is discussed below. Thus spectra of compound quarry explosions made up of many individual smaller explosions will inevitably manifest spectral modulation patterns except in certain special cases.

Descriptions of firing practices for blasts used for removing the overburden in coal mining at some mines indicates the use of very large shot arrays with dimensions close to a kilometer and durations of the order of seconds (KIM *et al.*, 1994). At such large shot arrays the waveforms generated by the individual charges will be strongly decorrelated. By invoking reciprocity arguments, this decorrelation will be similar to waveform decorrelation observed at small arrays. The frequency-limited correlation coefficients (coherences) of waveforms at small arrays were found to be approximately exponentially dependent on the product of distance between two observation points and frequency. Analyses of waveform decorrelation at NORESS (MYKKELTVEIT *et al.*, 1983) indicated that the waveform decorrelation is stronger for phases with low propagation velocities, at a given frequency, than for faster travelling waves. Thus *Sn* and *Lg* will decorrelate faster than fast traveling phases such as *Pn* and *Pg*. The stronger decorrelation for *Sn* and *Lg* may also be related to the more complex mode-ray structures of these phases (DER *et al.*, 1993). There are also indications that the rate of decorrelation may weakly depend on the direction of the wave propagation relative to the direction chosen in the horizontal plane (DER *et al.*, 1998). This property can be explained by a combination of spread in propagation velocity and azimuth due to dispersion and multipathing (e.g., GOSSARD and HOOKE, 1975). In this paper, for the sake of simplicity, we shall ignore this directionality and we will assume a decorrelation factor (coherence between pairs of sensors) proportional to $C(\omega) = \exp(-\alpha d\omega)$, where α is a constant determining the rate of decorrelation, ω is angular frequency and d is the intershot distance. The rate of waveform decorrelation will also depend on the degree of lateral heterogeneity in the local geologic structures and paths and thus may vary from place to place.

The second factor that must be considered for large shot arrays is the propagation delays between pairs of individual explosive charges. Such delays are not negligible compared to the absolute time lags among shots with shot patterns of the order of hundreds of meters. These delays will be added to the absolute time differences and produce a significant modification of the cross-spectra between pairs of shots. For nonsymmetrical space-time distributions of shots, this will give a directional dependence to the spectra of quarry blasts. Moreover, since the additive relative propagation delays depend on the dominant phase velocity of each arrival, the spectral modulations will not be the same for all phases. Equation (1) approximates the expected values of the cross-power spectra (the individual realizations may differ) of a given arrival, including decorrelation and propagation delays for a shot pair

$$E\left\{P_{jk}(\omega)\right\} = \exp\left(-\alpha\omega\left|\mathbf{r}_j - \mathbf{r}_k\right|\right) A_j(\omega) A_k^*(\omega) \exp i\omega\left[\tau_j - \tau_k + \left(\mathbf{r}_j - \mathbf{r}_k\right)\mathbf{s}\right] \quad (1)$$

where the real exponential factor $C(\omega) = \exp(-\alpha\omega|\mathbf{r}_j - \mathbf{r}_k|)$ is the average coherence between two waveforms corresponding to the angular frequency ω and the distance $d = |\mathbf{r}_j - \mathbf{r}_k|$, the * superscript denotes complex conjugation, and $A_i(\omega)$ are the complex spectra of the individual shots, each proportional to the shot size. The factor $C(\omega)|A_j(\omega)A_k(\omega)|$ will then be the absolute value of the cross-power spectrum between two sensors (e.g., BENDAT and PIERSOL, 1986). If the spectra of all shots are the same then the absolute values of the products of their source spectra can be substituted by the power spectrum of each sensor

$$P(\omega) = A_j(\omega)A_k^*(\omega).$$

The actual cross-spectra will include a complex exponential associated with the phase shift due to the time difference $\tau_j - \tau_k$ between the individual shots and the propagation delay for a phase between the two shots j and k, located at $\mathbf{r}_j$ and $\mathbf{r}_k$. This can be computed using the average, dominant phase slowness vector $\mathbf{s}$ of the phase in question (such as *Pn*, *Pg*, *Sn* or *Lg*). The quantities α and $|\mathbf{s}|$ will be different for different wave types emitted by the source. In the case when the shot yields are vastly different, the waveforms of the two shots may be different due to source scaling and the product $A_j(\omega)\, A_k^*(\omega)$ will also contribute a phase shift. Since, in our simulations, we assume that all shots had the same yield, this phase shift is zero.

Equation (1) gives us no information about the *individual realizations* of cross-spectra among the waveforms excited by pairs of shots since these will vary randomly. Nevertheless, for large shot arrays it can be expected that the sum of all cross-power spectra between all pairs of shots of the form described above

$$P(\omega) = \sum_{jk} E\left[P_{jk}(\omega)\right] \quad (2)$$

will approximate the actual observed spectral shapes. Since the imaginary parts of conjugate off-diagonal elements of the spectral matrix cancel in the sum, the remaining real terms with a real (*cosine*) dependence on the time lag between pairs of distinct shots give rise to the modulation in the observed spectra. In the diagonal terms of the inter-shot spectral matrix, the power spectra of the individual shots can be assumed to smoothly vary with frequency.

Because the average phase velocities and the decorrelation factors are different for different arrivals, the spectral shapes computed for the various arrivals will be identical only if the decorrelation and the propagation delays are negligible over the shot array. In such cases Equation (1) will reduce to the simpler expressions given by BAUMGARDT and ZIEGLER (1988) which describe the situation in which the modulation patterns are the same for all phases.

In Equation (1) we have introduced necessary simplifications to make the problem tractable and compensate for the lack of our knowledge of the modal composition and angular spread of the energy in regional arrivals. The spread in phase velocity, possibly due to multiple modes, introduces a smearing effect in the propagation delays that we neglected in this discussion but implicitly account for in the decorrelation coefficients. It could be easily introduced by doing multiple summations for each mode if the mode structures were known. F-K studies of phase velocities across small arrays generally show that in each arrival there exists a *dominant phase velocity* which justifies our approximations. Phase velocities measured for various regional arrivals across small arrays tend to be higher than the group velocities of each arrival, which is not surprising given their dispersive, modal nature. For instance, typical phase velocities measured for *Lg* are near 4 km/sec as opposed to typical group velocities which are in the range 3.5–3.7 km/sec.

The simple form assumed for the decorrelation (coherence) model was also necessary because in most source regions the particulars of decorrelation are not known. Seismic reciprocity as applied to regional signal decorrelation measurements at "small" arrays of a few km diameter implies that waveforms observed at a remote sensor from similar sources displaced over small distances will be similarly decorrelated. Thus, decorrelation measurements made at small arrays are directly applicable to the reciprocal problem. In order to optimize signal processing at regional arrays, signal decorrelation was measured at small arrays in Scandinavia (INGATE *et al.*, 1985; RINGDAL and HUSEBYE, 1982). The familiar optimized beam patterns at NORSAR are directly derivable from these decorrelation properties (MYKKELTVEIT *et al.*, DER *et al.*, 1988).

The combined effects of decorrelation and propagation delays should inevitably manifest themselves as observable modifications to the spectral modulation patterns of various regional arrivals from quarry blasts. The decorrelation effect will decrease the contribution from cross-spectra of sensor pairs located widely apart. This will tend to reduce the degree of modulation in the order of decreasing phase velocity (as we go from *Pn* to *Lg* in the seismogram) and also with increasing frequency. The propagation delays will essentially rescale the frequency axis for modulation in cross-spectra associated with pairs of sensors at various distances from the observer. For 2-D shot pattern layouts this modification of modulation can be quite complex.

Data Analyses

We analyze data from various quarry blasts located in the Kola peninsula and Scandinavia, recorded at the regional arrays ARCESS and NORESS in Norway. The list of events and their epicentral data are given in Table 1. These data provide typical examples of commonly observed variations in spectral modulation patterns.

Table 1

List of quarry blast events

Date	Time	N. Latitude	E. Longitude	m_b or m_L
Feb. 10, 1988	14:17:46.50	58.342	6.425	2.5
July 1, 1990	6:20:04.06	67.6695	34.0270	2.61
Aug. 7, 1990	10:55:11.5	67.3717	33.2637	2.06
Apr. 28, 1990	12:02:02.12	67.9800	32.9889	2.03
May 10, 1990	11:53:46.83	68.0230	32.9889	2.16
Feb. 9, 1990	9:32:50.93	67.6357	33.7198	2.66

In order to make the spectral analyses comparable for the various regional phases, the spectra presented in this paper were produced by an identical, uniform procedure. For a given event, all traces were inspected and only the sensor outputs with uncompromised data quality, those without clipping and spikes and other types of problems, were kept. We chose events that had very high signal-to-noise ratios (all arrivals had a signal to noise ratio $S/N \gg 5$). Regardless of the total duration of each wave group, the various regional arrivals were windowed using the same chosen window length of 12.8 sec. The Parzen window function was used on each window which has a shape that heavily tapers at both ends of each window. The position of the windows was chosen such that the main wave group of each arrival to be analyzed was close to the center of the window. After Fourier transforming and squaring, the amplitude spectra of each sensor for a given arrival the resulting power spectra were smoothed by a 5 point moving average and averaged over all the six chosen vertical component sensors of the array, including the center sensor and the rest of the sensors chosen from the C and D rings (see DER *et al.*, 1993 and numerous other publications for the layouts of ARCESS). The spectra thus computed will still include the effects of the instrument, the individual explosions in the shot pattern and attenuation. These effects result in a downward trend in the spectra with increasing frequency. The common modulation factor due to source multiplicity is superposed on this trend and can be unmistakably recognized. In cepstral calculations the trends are often removed by subtracting a linear trend or a second order polynomial fitted to the logarithm of the spectra (e.g., BAUMGARDT and ZIEGLER, 1988). This is unavoidable in computing cepstra, but is also subject to objections. We chose not to do this, assuming that most readers prefer to view power spectra without prior manipulation. In this we follow the practice in most previous papers dealing with the subject of spectral modulation.

It is our experience that this degree of smoothing and sensor averaging yielded spectra with the essential, common modulation patterns of interest. Similar procedures were applied by BAUMGARDT and ZIEGLER (1988). We have chosen the Parzen taper because it has low-side lobes and a wider main lobe than most of the commonly used tapering functions and thus also provides extra spectral smoothing.

By keeping window length, number of sensors and smoothing procedures for all phases the same, and plotting the spectra on a log(amplitude) – linear(frequency) scale, the degree of modulation in them can be directly compared. The individual spectra were plotted with a vertical offset for easier comparison. Their relative positions on these log-linear plots thus are not indicative of their relative spectral amplitudes.

By analyzing only large events recorded with high S/N ratios ($S/N \gg 5$ in the 3–12 Hz band, even for *Pn*) other problems with comparisons were also avoided. Due to the dominance of high frequencies in the spectra of most regional signals analyzed in this study, the S/N ratio was low below 2 Hz but is increased steadily with frequency to high values. Thus while we must disregard everything below 2 Hz, our conclusions based on spectra above 3–12 Hz are not appreciably affected by ambient noise.

In most cases the modulation patterns of the main seismic phases, actually the whole seismogram, are quite similar for quarry blasts (example K2182 in Fig. 1) which results in a banded appearance of their sonograms. In contrast, most earthquakes and single explosions do not display such a spectral structure; the modulations in the various phases are quite different. This phenomenon is well known and has been discussed extensively in the literature (BAUMGARDT and ZIEGLER, 1988; GITTERMAN and ECK, 1993; HEDLIN, 1998; CARR and GARBIN, 1998). It is the basis for the use of modulation as a discriminant for separating earthquakes and single explosions from multiple explosions (quarry blasts). It has also been shown that when the sizes and their timing of the individual shots are well controlled, the observed spectra tend to match the Fourier spectra of the associated explosion impulse sequences (SMITH, 1993; KIM *et al.*, 1994; CHAPMAN *et al.*, 1991) except in the cases where the designed and actual shot arrays differ considerably due to mistimed or misfired shots (STUMP and PEARSON, 1996; ANDERSON and STUMP, 1994). This indicates that the modulation patterns are indeed determined by the spatio-temporal patterns of shots and not by other factors.

A closer examination of the spectra of most quarry blasts often reveals, despite the great degree of similarities, systematic differences in the details in the modulation of seismic phases, in particular, the degree of modulation (peak-trough contrast) decreases as we go from *Pn* to *Lg* (the remaining examples in Fig. 1) and from low to higher frequencies for each phase. Such changes in the modulation from phase to phase are actually more common than the identical modulation for all arrivals, and can be seen for the majority of quarry blasts we have analyzed (DER *et al.*, 1991; DER and BAUMGARDT, 1994). This becomes obvious when the spectra are plotted and computed the uniform way as described above, however is harder to see on colorized frequency-travel time sonograms (HEDLIN *et al.*, 1990, 1998; KIM *et al.*, 1994), and naturally cannot be ascertained from binary sonograms or spectra for various phases computed with unequal window lengths.

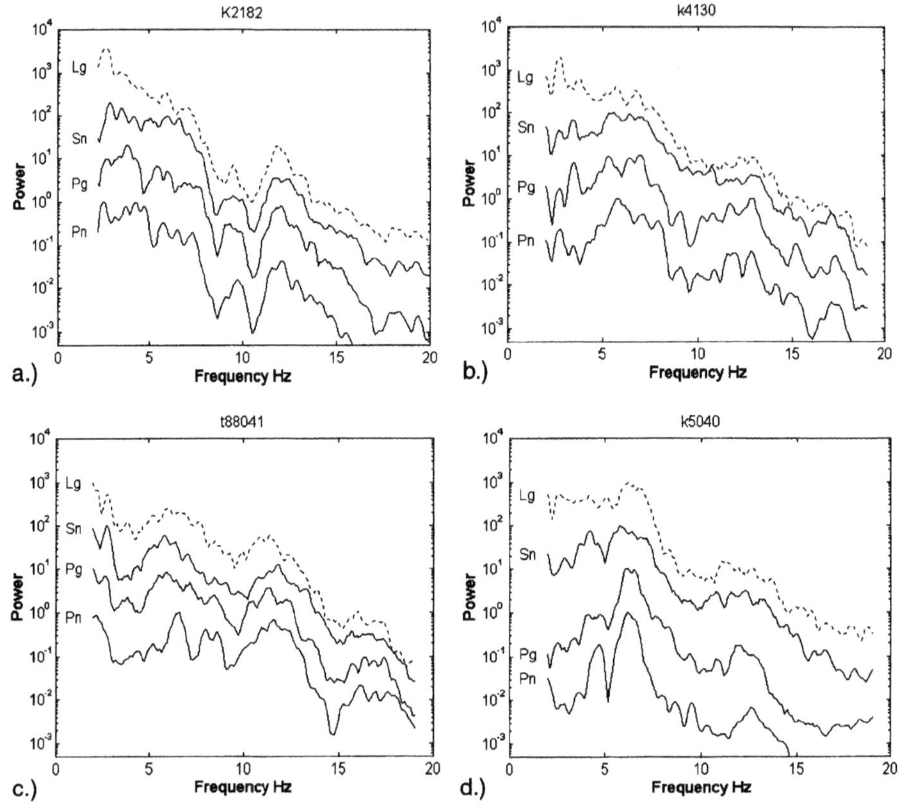

Figure 1
Spectra displaying modulation, i.e., rapid variations in the spectra relative to a smooth spectrum that would be produced by individual shot spectra, instrument and attenuation. The modulation patterns for the four wave groups, *Pn*, and *Pg*, *Sn* and *Lg*, are similar but not identical. Spectra of four quarry blast events regarded at ARCESS are arranged in the order of increasing differences in the degree of modulation from *Pn* to *Lg*. The top set shows little difference, the bottom set the largest. Most observations of quarry blast spectra conform to the latter pattern. The modulation strengths (peak-through contrast) generally decrease from *Pn* to *Lg* and with increasing frequency for each phase. Such patterns can be explained as the results of waveform decorrelation among subevents (individual charges). The dates of events in plots a–d are July 1, 1990; May 10, 1990; February 10, 1988 and February 9, 1990 (for the remaining information see Table 1).

If *Lg* has less energy at high frequencies, and thus is noisier, then the noise could obscure the spectral minima. Although this may occur, this explanation does not apply to any of the events analyzed in this paper, since events were chosen such that absolute amplitudes of *Pn* and *Lg* wave groups were considerably higher than the background noise and were at the same level in the 5–20 Hz frequency range (Fig. 2). It is certainly true, however, that all later phases are buried in the codas of the preceding arrivals, which by themselves should have the same spectral modulation and thus could not mask the modulation in the later arrivals. Nonetheless, for the simple model of a common source-function time modulation without decorrelation,

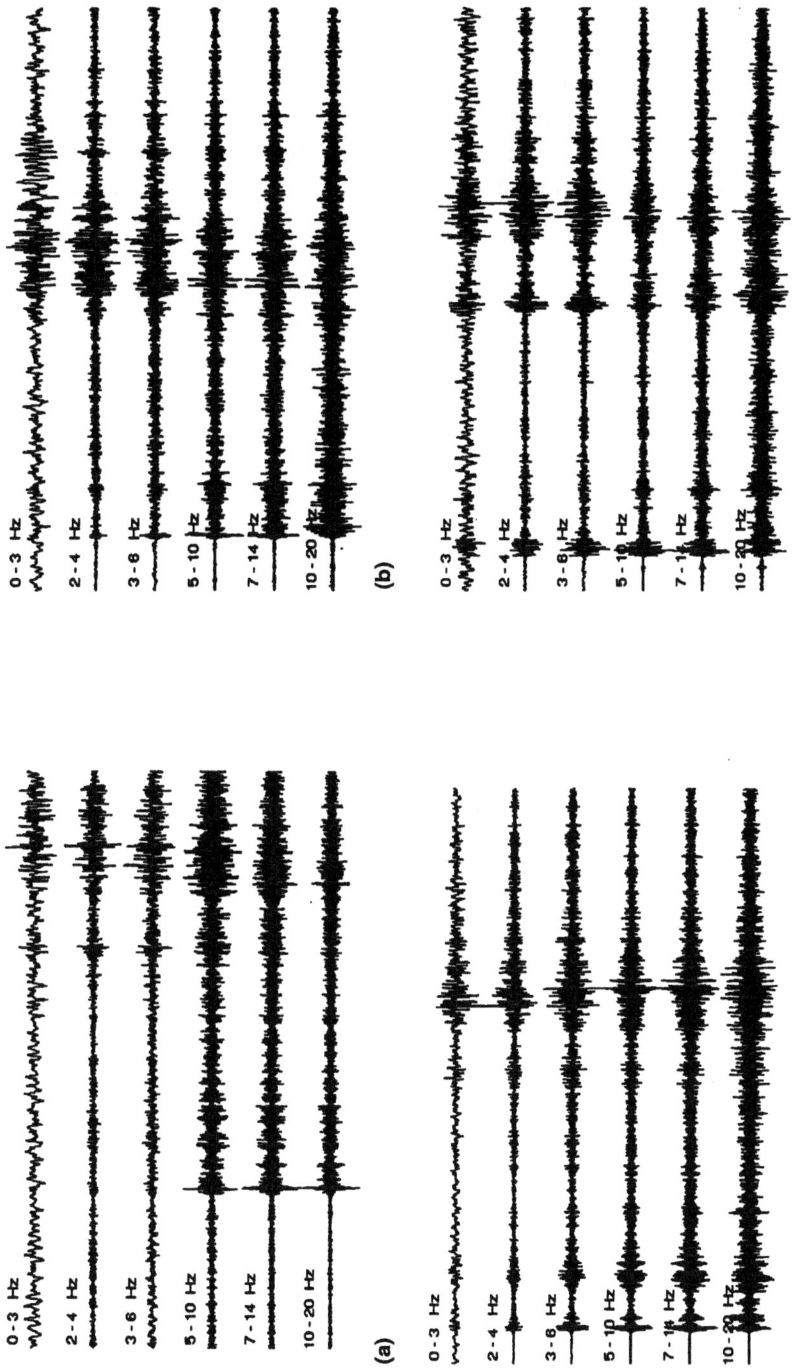

Figure 2

Waveforms of four events (marked a–d) at ARCESS (with the dates August 7, 1990; May 10, 1990, February 9, 1990; February 10, 1988; respectively, from Table 1) passed through a bank of band-pass filters. These examples show that the S/N ratios (referenced with respect to the background noise) of *Pn* are of the same order as that of *Lg* even at high frequencies (~ 6–20 Hz). Thus the variations of the modulation strengths with respect to wave slowness and frequency, found in the data, cannot be explained by differing S/N ratios.

these codas would have the same spectral structure as the arrivals themselves. Thus they could not account for the variation in the degree of modulation with time in the seismogram. One would except, based on a simple-mined modulation model, that the degree of modulation would be maintained as time progresses, assuming that the waveforms from each shot are exact replicas of each other. In addition, such spectra also show a tendency of the modulation to decrease with increasing frequency within each arrival even though the signal to background noise ratios are still high at high frequencies as implied by our theory.

Occasionally, different modulation patterns are observed for each regional phase and the spectra contain features that shift visibly with the dominant phase velocity of the energy (measurable by F-K analysis across an array) in each arrival which is around 8 km/sec for *Pn*, but are larger than 5 km/sec for *Sn* and near 4 km/sec for *Lg* (Fig. 3). It can be disputed whether this actually occurs for the top event (Fig. 3a), however the peaks and troughs definitely shift from right to left as we progress from *Pn* to *Lg*. This may be a rescaling of the frequency axis rather than a constant shift. This can be explained by propagation delay effects. Figure 3 also provides two additional good examples of the decorrelation effects, with the modulations decreasing from *Pn* to *Lg* and also with increasing frequency.

Simulations of Spectral Modulation in Recordings of Quarry Blasts

Utilizing Equations (1) and (2) it is possible to reproduce the main observed features of spectral modulations described above. Figures 4 to 6 show examples of synthetic modulated spectra for large simulated shot arrays, including both signal decorrelation and propagation delay effects. In these figures, we do not model the general decrease of amplitudes with increasing frequency, a common property of all individual explosions in the shot pattern, or any instrument response or attenuation. Of these three factors only the instrument response is known. In practical situations we do not know whether we are dealing with many small explosive charges of fewer large charges and the attenuation is poorly known along most paths. Since such parameters are not relevant to the main points we want to make, we attempt to model and discuss only the modulation shapes. In our simulations, we have assumed, somewhat arbitrarily, the phases velocities of 8.0, 6.0, 5.0 and 4.2 km/sec for *Pn*, *Pg*, *Sn* and *Lg*, respectively. The spatial decorrelation exponential factors were assumed to be 0.00234, 0.00318, 0.00955 and 0.02659 (in $(2\pi \text{ km Hz})^{-1}$), respectively, for the four arrivals. These are reasonable values for the region near NORESS, a location with moderate geological heterogeneity but some topographical relief. One can, of course, vary these to simulate various degrees of decorrelation caused by lateral heterogeneity in the source region. There is a tradeoff between distances and the decorrelation coefficients. Since with some propagating velocities a reinforcement in the energy may occur by constructive

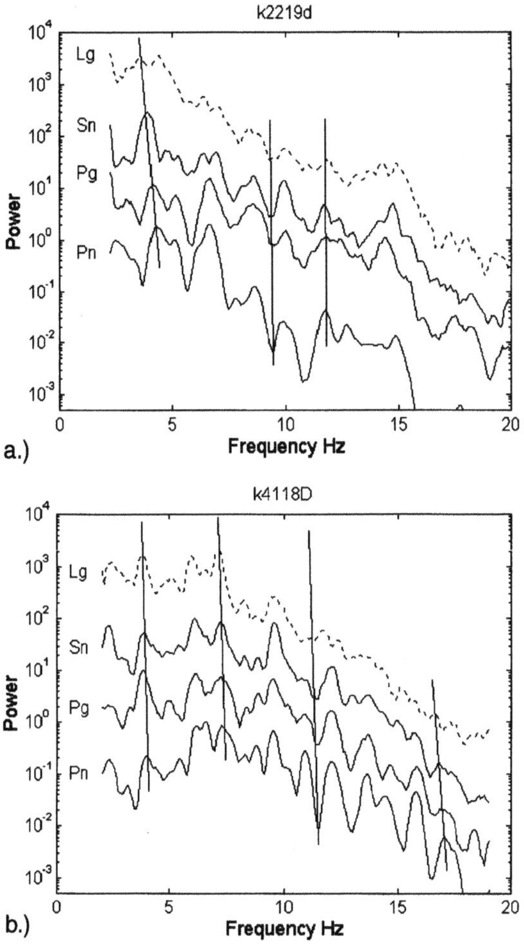

Figure 3
Spectra of two quarry blast events, August 7, 1990 (a) and April 28, 1990 (b) at ARCESS with differences in the shapes of modulation among various arrivals which may be indications of propagation delays. The apparent shifts in the position of spectral minima and maxima (actually rescaling of frequencies) indicated by lines may be due to frequency rescaling caused by variable relative propagation time delays among the various subevents. The spectra also display, prominently, the effects of decorrelation. The event dates are August 7, 1990 and April 28, 1990.

interference (similar to a shock wave) increase of simulated spectral amplitudes with frequency are seen in some cases.

Figure 4 shows spectra calculated for 12 elements a linear shot array (0.22 km long) with shot spacing of 0.02 km and an intershot delay of .045 sec. The spectra were computed both in the in-line (90° azimuth) and broadside directions. If the spatial layout of the shot pattern is elongated, the observations at the azimuths along the large dimension (inline) will be affected more by propagation delays than those in a broadside direction, resulting in rescaling of spectra. Assuming similar decorre-

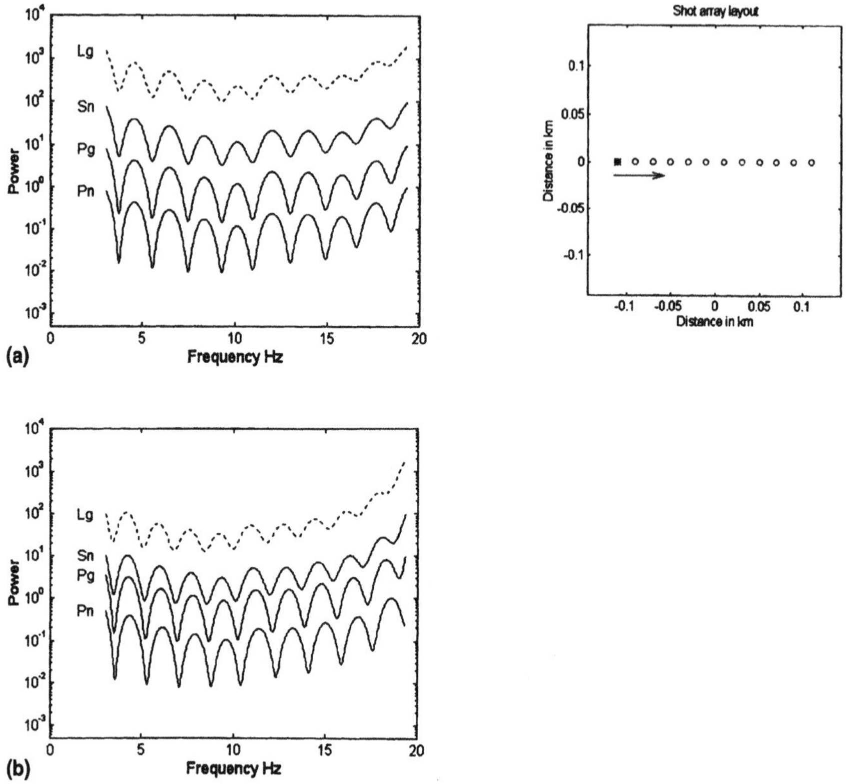

Figure 4
Simulated spectral shapes for a linear array of charges displaying the effects of waveform decorrelation and propagation delay shifts in two directions (*a.* broadside to rows as seen from the bottom, *b.* inline as seen from the left). The shot array layout is also shown, the arrow indicates the direction of the time progression of shots in each row of twelve shots.

lation in all directions, the decorrelation effects will be similar at all observing stations regardless of the azimuths. This is the reason why we show synthetic spectra in these two directions. In this example the spectra broadside show the modulation with the familiar changes due to decorrelation and the inline patterns also display propagation delays.

In Figure 5 we present similar results for two rows of shots, 12 shots in each row. In this case the delay between shots in each row was set to 0.03 sec, the delay between the two rows was 0.15 sec and the spacing between elements and rows 0.04 km, making the larger dimension of the shot array 440 m. This configuration has a time–energy distribution in which the shots in one row are initiated and in the time interval of 18–36 sec explosions overlap in time in both rows. These spectra also resemble the observed spectra in Figure 1 in which the modulation strengths decrease from *Pn* to *Lg* with increasing frequency. Doppler shifts are evident at the inline observation point.

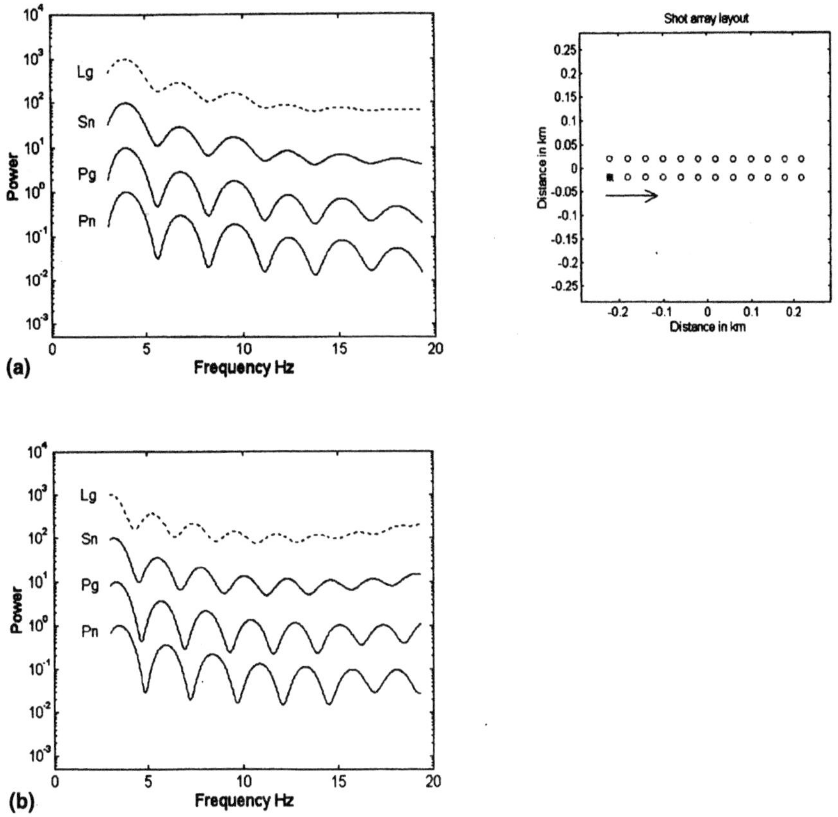

Figure 5

Simulated spectral shapes for a large elongated shot array as observed in two directions (*a.* broadside the rows as seen from the bottom, *b.* inline as seen from the left). Note the stretching of spectra along the frequency axis thus changing the positions of spectral minima and maxima. The shift is due to the differences in the average phase velocities, and hence propagation delays, of the various regional wave groups. The shot array layout is also shown, the arrow indicates the direction of the time progression of shots in each row of twelve shots.

The spectra in Figure 6 were calculated for a large shot pattern consisting of five rows with delays of 0.07 sec with 12 shots propagating along each row unilaterally, the delays between starting each row were 0.245 sec and the spacings (both between rows and the shots in each row) were 0.055 km, making up a shot array 605 m long and 220 m wide. This pattern thus approaches the sizes of shot arrays used in overburden removal in coal mining (some of these may be several times as large). This results in a long duration shot sequence with complicated modulation patterns. These spectra also display both the propagation delay (in this case a true Doppler shift) effect and the decorrelation effects.

For more complex shot patters, with bilateral propagation of shots with a fairly large spatial extent, the propagation delay effects are obscured, but the decorrelation

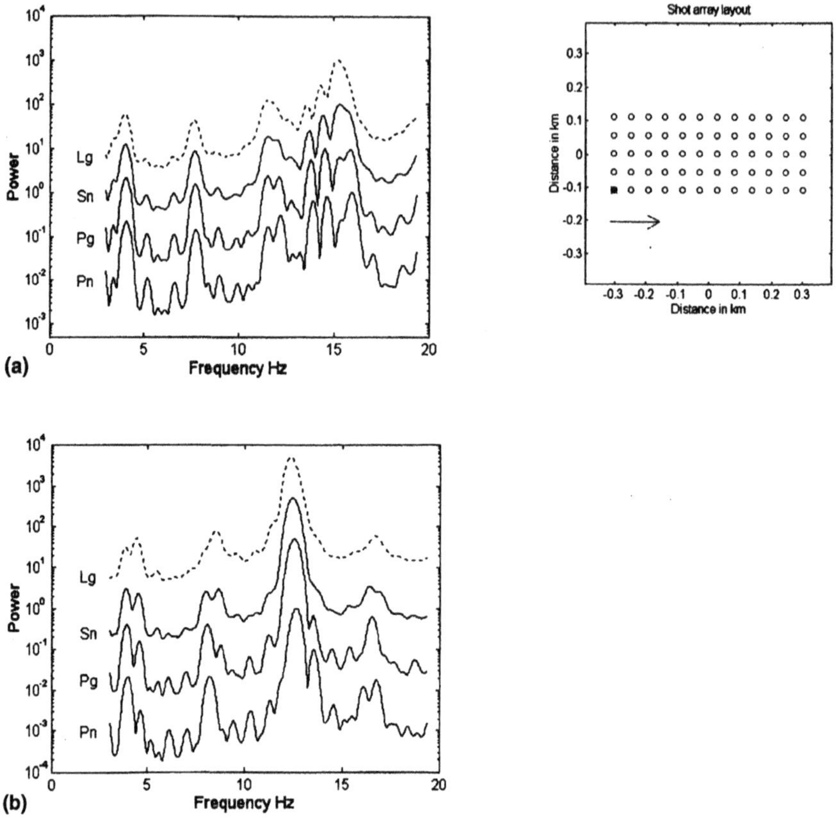

Figure 6
When the shot array is complex, the propagation delay effects may be obscured, but the decorrelation effects, the variations in modulation strengths, still remain (*a*. broadside to rows as seen from the bottom, *b*. inline as seen from the left). The shot array layout is also shown, the arrow indicates the direction of the time progression of shots in each row of twelve shots.

effects, reflected in the varying strengths of modulation remain. Such complex shot patterns combined with strong waveform dissimilarity among individual shots may be the explanation of why propagation delay effects are not observed more frequently. Many of the mining explosions in the Kola peninsula analyzed here are detonated simultaneously; a common Soviet practice at the time (LEITH *et al.*, 1997b). Such patterns, of course, would have no manifestations of any Doppler shifts.

We have found that it is possible to create large shot arrays, exploded almost simultaneously, such that the time delay patters (primarily propagation delays) will be quite different for the various arrivals. In such theoretical cases the time-independent banded structure of sonograms, commonly reported (HEDLIN *et al.*, 1990; CARR and GARBIN, 1998), would disappear, thus disabling the spectral modulation as an effective discriminant. We cannot think of any economic or political motivation to implement such shot delay patterns, however.

Discussion

While sonograms of regional seismograms of most quarry blasts reveal an apparently uniform modulation of the spectra throughout the duration of the seismograms, more detailed analyses of array-averaged spectra reveal subtle variations in the degree of modulation and shifts in the frequencies associated with the various features (peaks and troughs) of the spectra among various regional arrivals. Such subtle changes can be detected in the spectra of most quarry blasts where the signal-to-noise ratio permits such analysis. Decorrelation in the waveforms of the various individual shots with increasing mutual distance and relative phase velocity-dependent propagation delays provide a likely explanation for such phenomena. In this paper we propose a simple mathematical model for the modification of the spectra by these effects.

The extent of decorrelation effects can be assessed by invoking seismic reciprocity, thus equating the decorrelation of shot waveforms with decorrelation of waveforms recorded across small arrays. Array studies showed that phases with lower phase velocities such as *Lg* and *Sn* will decorrelate faster with increasing sensor separation than those with higher phase velocities such as *Pn* and *Pg*. Simulations thus demonstrate that decorrelation will affect the spectral modulation in the various seismic arrivals to different degrees; the strength of modulation will decrease from *Pn* to *Lg*.

The simple simulations using the mathematical formulation showed that shot patterns, even with lengths of only a few hundreds of meters, should produce clearly observable Doppler effects. The fact that such observations are rare suggests that the spatio-temporal shot patterns of many quarry blasts are more complex than in our simulations. Indeed, in many cases shot patterns are designed to reduce directivity (reinforcement) of seismic motion in order to reduce vibration-induced damage in nearby structures.

Both decorrelation effects and true Doppler shifts are likely to occur in large, propagating directional shot patterns that are used in the United States for coal mining operations. Unfortunately we could not locate array data with large sensor spacing for such shots to substantiate this. Array recordings seem to be essential for specifying the accurate patterns of modulation by sensor averaging.

Some three dimensional patterns, including several depth levels, were described in the literature (LEITH *et al.*, 1997a), although we were hesitant to use them since we do not know what effect the successive demolition of rocks at different levels will have on the spatial correlation patterns and the source spectra. Moreover, we were not able to find multiple array recordings of quarry blasts at different azimuths for quarry blasts with known shot layouts.

Nevertheless, the above examples illustrate that our generalized model for spectral modulation in quarry blasts can reproduce various observed features in the spectra of regional phases from quarry blasts. Even though unique inversions for shot patterns are not feasible, the spectral shapes of quarry blasts in conjunction with our model can be used to reveal more than the mere fact that a particular event is an earthquake or a quarry blast. In particular, directionality of the shot patterns and their areal extent may be assessed if observations along multiple azimuths are available. Such diagnostics may be useful in evaluating the possibility of hide-in-mine-blast evasion schemes in remote monitoring of possible unclear testing activities by seismic means.

Summary and Conclusions

Analyses of array-averaged spectra associated with various seismic regional arrivals from mining explosions revealed differences in modulation strength and apparent shifts in modulation patterns. These phenomena may be associated with source finiteness. The source finiteness is reflected in the effects of waveform decorrelation and propagation delays between individual pairs of shots within a shot pattern. In particular we conclude that

a) Waveform decorrelation reduces the strength of modulation with decreasing dominant phase velocity and increasing frequency.
b) Duration is not the only factor that determines the shape of modulation (especially the location of the first spectral minimum).
c) Indications of propagation delay effects can be detected in the recordings of some mining explosions.

Using simple assumption with regards to signal decorrelation and propagation delays these features can be easily simulated. Potentially these features can be used to derive information with regards to the spatial extent and layout of shot arrays. Modulation patterns of the locations of first minima in the spectra are generally not determined solely by the time duration of the shot sequences, but are complex functions of numerous factors. Lack of spectral modulation does not necessarily indicate a single explosion or the absence of ripple firing.

Based on our knowledge of array decorrelation patterns and the array phase velocities measured for the various regional arrivals, the observational findings are not surprising, actually we have been purposely looking for the manifestations of these phenomena in the spectra of regional arrivals (DER and BAUMGARDT, 1994). These effects, although subtle, provide diagnostic information about the mining practices employed and thus may be valuable in the monitoring of test ban treaties.

Acknowledgement

We thank Mark Tinker and Mike Anatolik for their careful reviews which helped greatly in improving the paper.

References

ANDERSON, D. P., and STUMP, B. W. (1994), *Utilization of Near-source Video and Ground Motion in the Assessment of Seismic Source Functions from Mining Explosions (Abstract)*. 16-th Annual Research Symposium, Phillips Laboratory-Air Force Office of Scientific Research, Thornwood Conference Center, Thornwood, N.Y., 7–9 September 1994.

BAUMGARDT, D. R., and ZIEGLER, K. A. (1988), *Spectral Evidence for Source Multiplicity in Explosions, Application to Regional Discrimination of Earthquakes and Explosions*, Bull. Seismol. Soc. Am. *78*, 1773–1795.

BAUMGARDT, D. R., and YOUNG, G. B. (1990), *Regional Seismic Waveform Discriminants and Case-based Event Identification Using Regional Arrays*, Bull. Seismol. Soc. Am. *80*, 1874–1892.

BAUMGARDT, D. R., and DER, Z. A. (1998), *Identification of Presumed Shallow Underwater Chemical Explosions Using Land-based Regional Arrays*, Bull. Seismol. Soc. Am. *88*, 581–595.

BENDAT, J. S., and PIERSOL, A. G., *Random Data: Analysis and Measurement Procedures* (Wiley and Sons, New York 1986).

CARR, D. B., and GARBIN, H. D. (1998), *Discriminating Ripple-fired Explosions with High Frequency (>16 Hz) Data*, Bull. Seismol. Soc. Am. *88*, 963–972.

CHAPMAN, M. C., BOLLINGER, G. A., and SIBOL, M. S. (1991), *Spectral Studies of the Elastic Wave Radiation from Appalachian Earthquakes and Explosions – Explosion Source Spectra Modeling Using Blaster's Logs*, Virginia Polytechnical Institute, PL-TR-91-2231. Prepared for the Phillips Laboratory, Air Force Systems Command, Hanscom AFB, MA.

DER, Z. A., SHUMWAY, R. H., and LEES, A. C. (1988), *Frequency Domain Coherent Processing of Regional Seismic Signals at Small Arrays*, Bull. Seismol. Soc. Am. *78*, 326–338.

DER, Z. A., HIRANO, M. R., ZIEGLER, K. A., and SHUMWAY, R. H. (1991), *Broadband Studies of Seismic Sources at Regional and Teleseismic Distances*, PL-TR-91-2059 (1). ENSCO Inc. Prepared for Phillips Laboratory, Hanscom AFB.

DER, Z. A., BAUMGARDT, D. R., and SHUMWAY, R. H. (1993), *The Nature of Particle Motion in Regional Seismograms and its Utilization for Phase Identification*, Geophys. J. Intl. *115*, 1012–1024.

DER, Z. A., and BAUMGARDT, D. R. (1994), *Source Diagnostics from Spectral Modulation Patterns in Regional Recording of Seismic Waves from Mine Blast* (Abstract), EOS *75*, 428.

GITTERMAN, Y., and VAN Eck, T. (1993), *Spectra of Quarry Blasts and Microearthquakes Recorded at Local Distances in Israel*, Bull. Seismol. Soc. Am. *83*, 1799–1812.

GOLDMAN, S., *Information Theory* (Prentice Hall Electrical Engineering Series, Prentice Hall, Englewood Cliffs, NJ 1953).

GOSSARD, E., and HOOKE, W. H. *Waves in the Atmosphere. Atmospheric Infrasound and Gravity Waves-their Generation and Propagation* (Elsevier, 1975).

HARRIS, D. (1991), *A Waveform Correlation Method for Identifying Quarry Explosions*, Bull. Seismol. Soc. Am. *81*, 2395–2418.

HEDLIN, M. A., MINSTER, J. B., and ORCUTT, J. A. (1990), *An Automatic Means to Discriminate between Earthquakes and Quarry Blasts*, Bull. Seismol. Soc. Am. *80B*, 2143–2160.

HEDLIN, M. A. (1998), *A Global Test of a Time-frequency Small-event Discriminant*, Bull. Seismol. Soc. Am. *88*, 973–988.

INGATE, S. F., HUSEBYE, E. S., and CHRISTOFFERSON, A. (1985), *Regional Arrays and Optimum Data Processing Schemes*, Bull. Seismol. Soc. Am. *75*, 1155–1177.

KIM, W. Y., SIMPSON, D. W., and RICHARDS, P. G. (1994), *High Frequency Spectra and Regional Phases from Earthquakes and Chemical Explosions*, Bull. Seismol. Soc. Am. *84*, 1365–1386.

KIM, W. Y., AHARONIAN, V., LERNER-LAM, A. L., and RICHARDS, P. G. (1997), *Discrimination of Earthquakes and Exposions in Southern Russiausing Regional High-frequency Three-component Dataform the IRIS/ISP Caucasian Network*, Bull. Seismol. Soc. Am. *87*, 569–588.

LEITH, W., BAUMGARDT, D., and SYRNIKOV, N. (1997a), *The Kirovdkiy explosion of 29 September 1996: Example of a CTB event notification for a routine mining blast.* In *Proc 19-th Annual Research Symposium on Monitoring a Comprehensive Test Ban Treaty.* 23–25 September 1997, Orlando FLA.

LEITH, W., ADUSHKIN, V., and SPIVAK, A. (1997b), *Large Mining Blasts from the Kursk Region, Russia*, Lawrence Livermore Laboratory, UCRL-CR-126760.

MYKKELTVEIT, S., ASTEBOL, K., DOORNBÖS, D. J., and HUSEBYE, E. S. (1983), *Seismic Array Configuration Optimization*, Bull. Seismol. Soc. Am. *73*, 173–186.

RINGDAL, F., and HUSEBYE, E. S. (1982), *Application of Arrays in the Detection, Location and Identification of Seismic Events*, Bull. Seismol. Soc. Am. *72*, S201–S224.

SMITH, A. T. (1993), *Discrimination of Explosions from Simultaneous Mining Blasts*, Bull. Seismol. Soc. Am. *83*, 160–179.

STUMP, B. W., and PEARSON, D. C. (1996), *Design and utilization of a portable seismic/acoustic calibration system.* In *Proc 18-th Annual Seismic Research Symposium on Monitoring a Comprehensive Test-Ban Treaty.* 4–6 September 1996, Annapolis, MD. PL-TR-96-2153. Phillips Laboratory, Hanscom AFB.

(Received May 7, 1999, accepted October 28, 1999)

Pure appl. geophys. 158 (2001) 2077–2101
0033–4553/01/112077–25 $ 1.50 + 0.20/0

Pure and Applied Geophysics

Seismic Source Characteristics of Soviet Peaceful Nuclear Explosions

J. R. Murphy,[1] I. O. Kitov,[2] B. W. Barker,[1] and D. D. Sultanov[2]

Abstract—During the period 1965 to 1988, the former Soviet Union (FSU) conducted over 120 peaceful nuclear explosions (PNE) at locations widely dispersed throughout the territories of the FSU. These explosions sample a much wider range of source conditions than do the historical explosions at the known nuclear test sites and, therefore, seismic data recorded from these PNE tests provide a unique resource for use in deriving improved quantitative bounds on the ranges of seismic signal characteristics which may require consideration in global monitoring of the Comprehensive Test-Ban Treaty (CTBT). In this paper we summarize the results of a detailed statistical analysis of broadband seismic data recorded at the Borovoye Geophysical Observatory from 21 of these PNE tests at regional distances extending from about 7 to 19 degrees, as well as the results of theoretical waveform simulation analyses of near-regional (Δ < 25 km) seismic data observed from a selected sample of nine of these PNE tests. The results of these analyses have been found to be consistent with those of previous teleseismic investigations in that they indicate that the seismic source coupling efficiencies are very similar for explosions in a wide variety of hardrock and water-saturated media, while explosions in water-saturated clay are observed to have significantly higher coupling efficiencies. Moreover, the scaling of the seismic source function with explosion yield and depth of burial inferred from these analyses of the Soviet PNE data are shown to be generally consistent with the predictions of the Mueller/Murphy source model. These results suggest that the Mueller/Murphy source model can provide a reasonable basis for estimating the expected variation in regional phase spectral composition over a wide range of nuclear source conditions of potential interest in CTBT monitoring.

Key words: Nuclear explosions, seismic source, Soviet PNE.

Introduction

In order to discriminate the regional seismic signals produced by underground nuclear explosions from those produced by earthquakes, rockbursts and conventional mining explosions of comparable magnitude, it is necessary to know the range of nuclear explosion signal variation that can be expected as a function of source and propagation path conditions over the entire ranges of these conditions

[1] Maxwell Technologies, Inc., Reston Geophysics Office, 11800 Sunrise Valley Drive, Suite 1212, Reston, VA 20191, USA.

[2] Institute For Dynamics of the Geospheres, Russian Academy of Sciences, 38 Leninsky Prospect, Moscow, Russia 117334.

which may be encountered in global test monitoring of the Comprehensive Test-Ban Treaty (CTBT). However, most of the empirical regional discrimination studies which have been conducted to date have focused on analyses of seismic signals recorded from underground explosions conducted at the few major nuclear test sites, and these sample a limited subset of the potential ranges of the variables of interest. This constitutes a serious limitation in that existing theoretical simulation models have not yet proven capable of fully explaining the observed characteristics of the various proposed empirical discriminants and, therefore, their extrapolation to applications in new environments is subject to considerable uncertainty.

In an attempt to overcome these limitations of previous studies, we have been analyzing regional seismic data recorded from the extensive Soviet Peaceful Nuclear Explosion (PNE) testing program. This program consisted of over 120 explosions which were conducted throughout the territories of the former Soviet Union in a wide variety of geologic emplacement media (e.g., salt, clay, sandstone, granite, limestone) and are representative of broad ranges in explosion yield and source depth of burial (SULTANOV *et al.*, 1999). Broadband regional seismic data recorded from these tests have recently become available for the first time, and these new data provide a unique opportunity to better quantify the variability of regional seismic signal characteristics over ranges of explosion source and propagation path conditions which are more representative of the diversity which may be encountered in global CTBT monitoring.

Regional Seismic Data Recorded at the Borovoye Geophysical Observatory

The Borovoye Geophysical Observatory is located in North Kazakhstan (53.08°N, 70.25°E) and is one of the oldest digitally recording seismic stations in the world, having initiated digital recording in 1966 (ADUSHKIN and AN, 1990). A variety of long-period and short-period seismic recording systems have been deployed at this station over the years (KIM and EKSTRÖM, 1996), including some relatively broadband systems which employed digitization rates in the 30–40 sample/second range. Thus, dynamic range permitting, these data provide potential resolution of seismic frequency content to 10 Hz and higher. For most of the Soviet PNE tests, these Borovoye recordings represent the highest quality regional seismic data that are available for detailed analyses.

For the purposes of the present study, data recorded at the Borovoye station from a sample of 21 Soviet PNE events located in the regional distance range extending from 7.2 to 19.1 degrees from Borovoye have been collected and analyzed. The map locations of these selected explosions with respect to the Borovoye station are shown in Figure 1, where it can be seen that they are widely dispersed throughout this region of the former Soviet Union. The source parameters of these

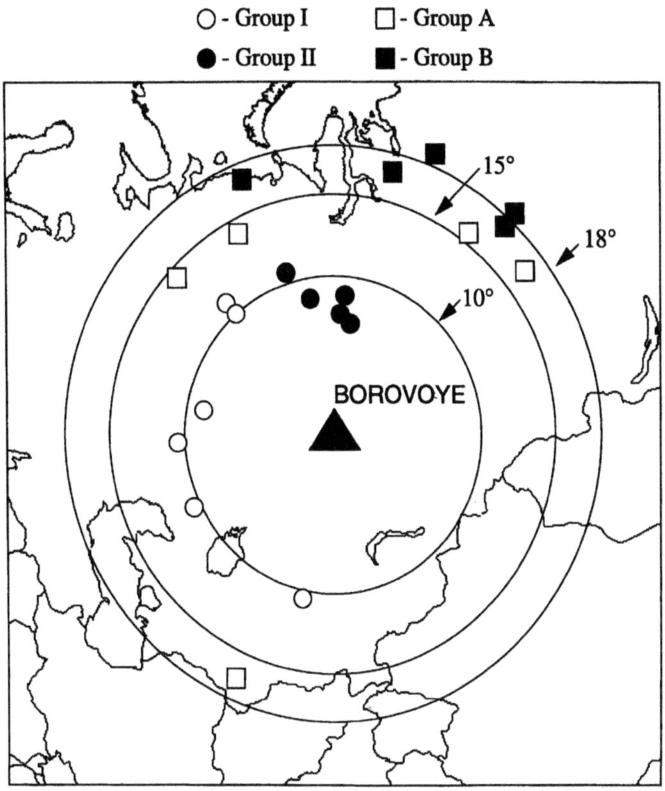

Figure 1
Map locations of selected Soviet PNE tests which were recorded at the Borovoye station in Kazakhstan.

selected PNE events are listed in Table 1, where they have been divided into four subsets (i.e., Groups I, II, A and B) on the basis of epicentral distance range and waveform similarity. The rationale for this classification will be discussed in more detail in the following sections. It can be seen from Table 1 that these selected explosions are predominantly low yield, overburied explosions in hardrock of the type which represent the greatest challenge to successful seismic monitoring of the CTBT.

All of the broadband, vertical component Borovoye data corresponding to the PNE events of Table 1 were carefully previewed by an experienced analyst to assess data quality and suitability for digital signal processing. Although the data were found to be of generally good quality (MURPHY *et al.*, 1996), close examination revealed a number of occurrences of spikes, data dropouts, and clipping which could seriously contaminate any data analysis results at high frequency. In cases where such data problems were isolated as single points, they were corrected using simple interpolation of the adjacent data points. In those few cases where the data problems were too complex to be remedied by such simple data interpolation, the time

Table 1

Source characteristics of selected Soviet PNE events (SULTANOV et al., 1999)

Date	Latitude	Longitude	W, kt	h, m	Medium	Δ,°
Group I ($\bar{\Delta} = 10.2°$)						
08/15/73	42.78	67.41	6.3	600	clay	10.5
10/26/73	53.66	55.38	10	2026	dolomite	8.3
09/02/81	60.62	55.59	3.2	2088	limestone	11.0
07/21/84	51.36	53.32	13.5	846	salt	10.6
04/19/87	60.25	57.08	3.2	2015	limestone	10.4
10/03/87	47.61	56.23	8.5	1000	salt	10.5
Group II ($\bar{\Delta} = 8.6°$)						
10/17/78	63.19	63.43	23	593	sandstone	10.7
10/04/79	60.68	71.46	21	837	clay	7.7
12/10/80	61.69	67.00	15	2485	sandstone	8.8
08/25/84	61.88	72.09	8.5	726	clay	8.9
06/18/85	60.17	72.50	2.5	2859	argillite	7.2
Group A ($\bar{\Delta} = 15.6°$)						
04/11/72	37.37	62.00	14	1720	argillite	16.7
11/01/80	60.82	97.57	8.0	720	dolomite	16.7
09/25/82	64.31	91.83	8.5	554	gabbro	15.8
08/11/84	65.03	55.19	9.5	759	clay	14.2
09/06/88	61.36	48.09	7.5	793	dolomite	14.6
Group B ($\bar{\Delta} = 18.0°$)						
07/26/77	69.58	90.38	13	879	salt	19.1
08/20/77	64.11	99.56	8.5	592	tuff	18.9
05/25/81	68.21	53.66	37.6	1511	clay	17.2
10/22/81	63.79	97.55	8.5	581	dolomite	17.7
09/04/82	69.21	81.65	16	960	sandstone	17.0

windows containing high frequency spikes were noted and avoided in subsequent spectral analyses of these data traces.

Another result of this data review was the decision to form Groups I and II. That is, it was noted that although all these explosions are distributed over the same narrow band in epicentral distance, those located north of Borovoye (i.e., Group II) consistently produce recordings at that station which have broadband P_g/P_n amplitude ratios which are significantly larger than those observed from the other PNE events at comparable distances along different azimuths (i.e., Group I). This observation is documented in Figure 2, which shows a comparison of the vertical component, broadband signals recorded at Borovoye from the 11 PNE events of Groups I and II. In this figure, the data are displayed as a function of group velocity and it can be seen that, if we associate the arrivals in the 6 to 5 km/sec window with P_g and those in the 8 to 7 km/sec window with P_n, then the broadband P_g/P_n amplitude ratios are clearly much larger for the five explosions of Group II than they are for the six explosions of Group I. Consequently, these explosions have been separated for seismic source analysis purposes. It should be noted, however, that

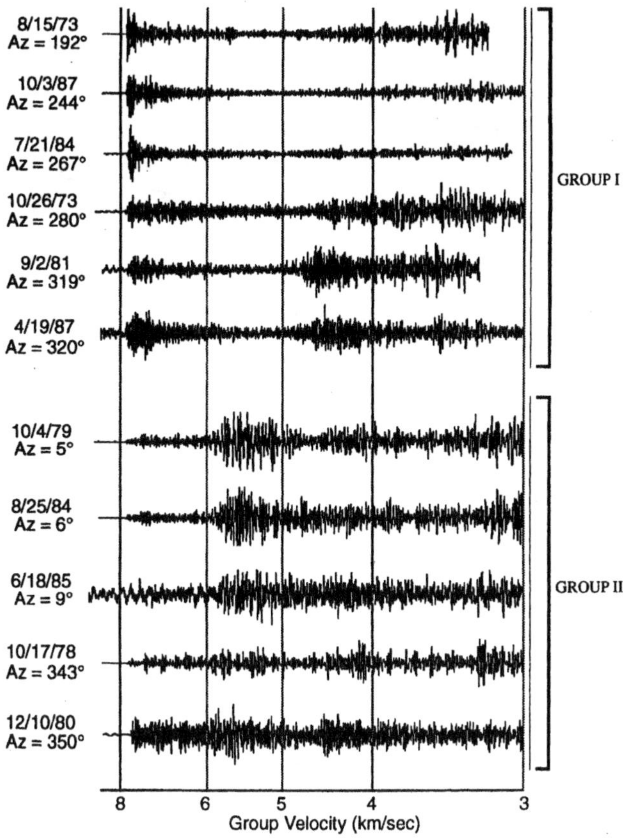

Figure 2
Vertical component regional signals recorded at Borovoye from selected Soviet PNE events in Groups I and II. Note the strong dependence of the P_g/P_n ratios on station to source azimuth (Az) for these explosions at comparable distances from the Borovoye station.

even in this relatively simple and extensively studied platform region, it has not yet been possible to correlate the seismic variability of Figure 2 with any mapped variability in crust or upper mantle structures (MURPHY *et al.*, 1997), and this suggests that regionalizations of seismic propagation characteristics based on generalized geophysical data can be subject to uncertainties which may be significant with regard to event identification.

Borovoye Regional Phase Spectral Analysis

The Borovoye digital data corresponding to the selected PNE events of Table 1 have been processed to obtain estimates of the different regional phase spectra for each event. In the first step of this processing, the data and pre-signal noise windows

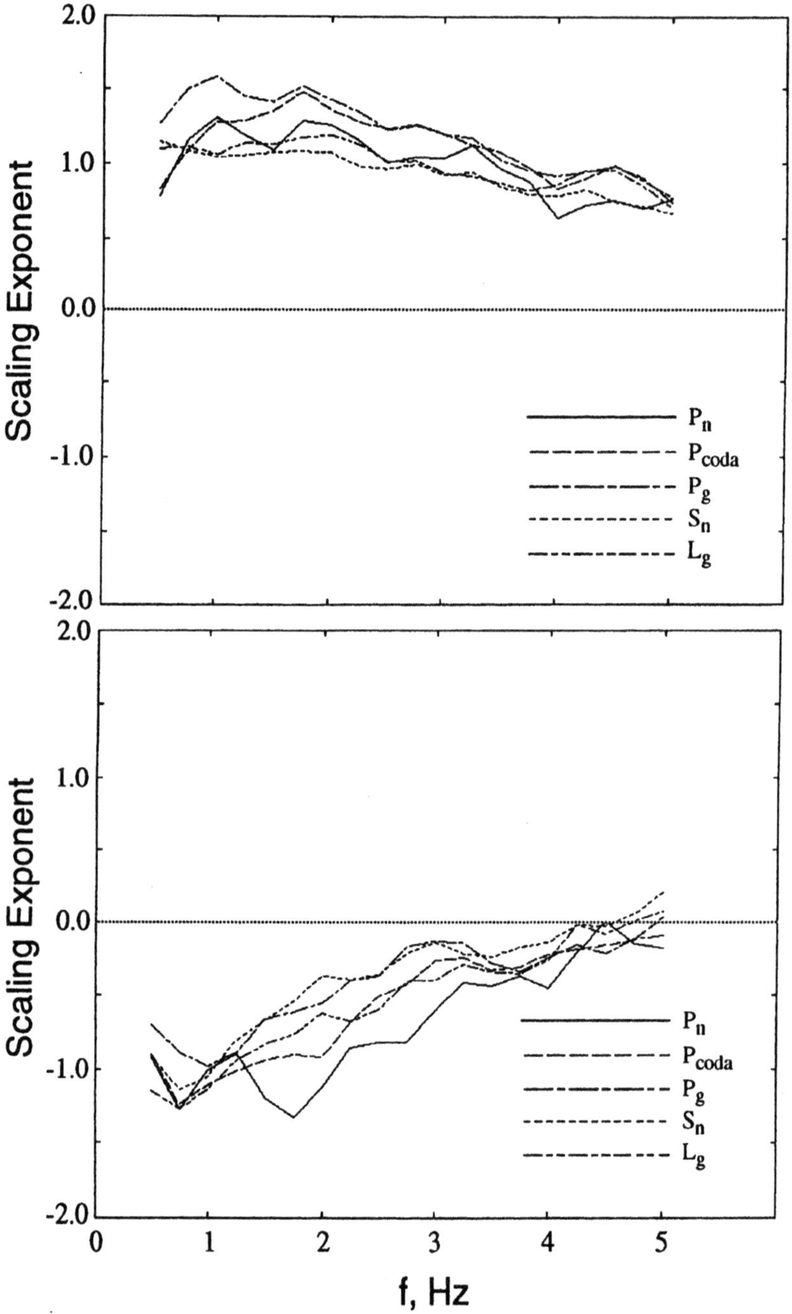

Figure 3
Comparison of yield (top) and depth (bottom) frequency-dependent scaling exponents determined for the different regional phases recorded at Borovoye from the selected PNE events.

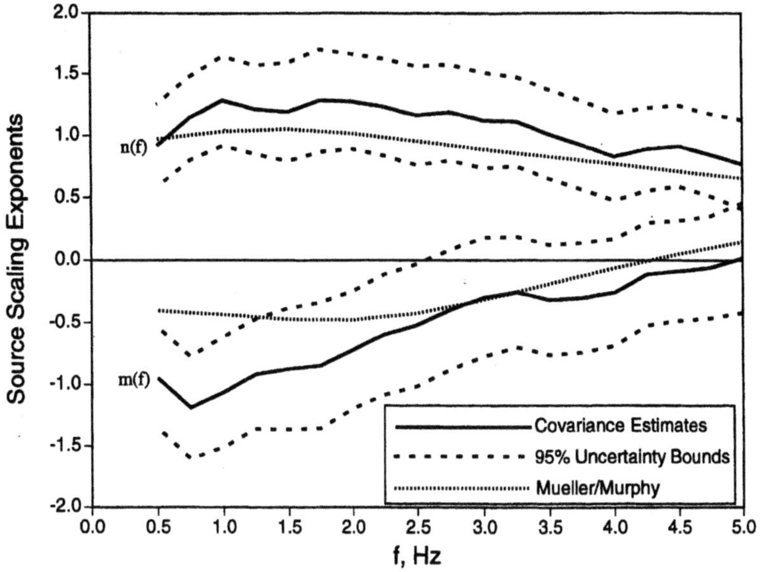

Figure 4
Comparison of the frequency-dependent yield (n) and depth (m) scaling exponents derived from the covariance analysis of Borovoye data recorded from the 21 PNE events of Groups I, II, A and B with the corresponding values predicted by the Mueller/Murphy explosion source model.

for each explosion were bandpass-filtered using a Gaussian comb of filters spaced at intervals of 0.25 Hz between 0.5 and 5.0 Hz, where each filter is characterized by a Q value of $6f_c$, with f_c the filter center frequency. Filters of this type have been used by us and a number of other investigators in previous studies (e.g., MURPHY *et al.*, 1989) and have been found to provide spectral estimates which are useful for purposes of seismic analysis. Note that, although the digitization rates for the Borovoye data are nominally high enough to support signal processing to 10 Hz or higher, we have established an upper limit of 5 Hz for the purposes of the present study. This reflects our conclusion that the combination of limitations in the dynamic range of the original data and the general decrease in the signal-to-noise ratios with increasing frequency leads to a significant increase in the uncertainty associated with the reduction of the higher frequency data.

The filter outputs for each recording were next sorted into six distinct time windows, and the spectral amplitude levels at each center frequency were estimated by computing RMS values from the instrument-corrected filter outputs in each of the designated windows. In the interest of simplicity and consistency, these regional phase time windows were defined in terms of apparent group velocity intervals as follows: (1) pre-signal noise, (2) $V_g > 7$ km/sec, (3) $6 < V_g < 7$ km/sec, (4) $5 < V_g < 6$ km/sec, (5) $4 < V_g < 5$ km/sec and (6) $3 < V_g < 4$ km/sec. For purposes of subsequent discussion these six group velocity windows will be referred to as pre-signal noise, P_n, P_{coda}, P_g, S_n and L_g, respectively. Although there are individual cases in which refined time

Table 2

Comparison of relative source coupling efficiency as inferred from average ratios of observed to predicted (Statistical Model) regional phase spectra for selected Soviet PNE events recorded at the Borovoye station

Event	Medium	W, kt	h, m	$h/W^{1/3}$ $(m/kt^{1/3})$	Average: $\frac{\text{Observed}}{\text{Predicted}}$ 0.5–5.0 Hz
09/25/82	gabbro	8.5	554	271	0.48
10/17/78	sandstone	23	593	209	0.49
09/02/81	limestone	3.2	2088	1417	0.74
07/21/84	salt	13.5	846	355	0.74
09/04/82	sandstone	16	960	381	0.79
04/19/87	limestone	3.2	2015	1367	0.81
10/03/87	salt	8.5	1000	490	0.87
08/20/77	tuff	8.5	592	290	0.89
07/26/77	salt	13	879	374	0.94
11/01/80	dolomite	8	720	360	0.95
04/11/72	argillite	14	1720	714	0.95
06/18/85	argillite	2.5	2859	2107	0.98
09/06/88	dolomite	7.5	793	405	1.09
10/22/81	dolomite	8.5	581	285	1.10
10/26/73	dolomite	10	2026	940	1.11
12/10/80	sandstone	15	2485	1008	1.12
05/25/81	clay	37.6	1511	451	1.24
10/04/74	clay	21	827	300	1.33
08/25/84	clay	8.5	726	356	1.38
08/11/84	clay	9.5	759	358	1.74
08/15/73	clay	6.3	600	325	2.09

windows could be defined which might more precisely isolate particular regional phases of interest, this simple windowing procedure was employed throughout as a test of its potential utility for routine seismic analysis purposes.

In order to evaluate the dependence of these Borovoye regional phase spectra on the characteristics of the explosion seismic source function, it has been assumed on the basis of prior experience (MUELLER and MURPHY, 1971) that this source function, denoted $S(\omega)$, can be approximated as a product of simple power-law relations of the form

$$S(\omega) \sim W^{n(\omega)} h^{m(\omega)} \quad , \tag{1}$$

where W and h are the explosion yield and depth of burial, respectively. However, since the PNE events of Table 1 represent a wide range of epicentral distance, the effects of the differences in propagation paths cannot be ignored and they must be explicitly estimated and separated from the source effects. This has been accomplished using statistical covariance techniques in which the model describing the variations in the observed spectra for each regional phase p is expressed as

$$A_{p,k}(\omega) = B_{p,k}(\omega) W^{n_p(\omega)} h^{m_p(\omega)} \quad , \tag{2}$$

where $k = 1, \ldots, 4$ denotes the designated subgroups of Table 1. The $n_p(\omega)$, $m_p(\omega)$ and $B_{p,k}(\omega)$ are simultaneously estimated from the logarithms of the observed spectral amplitude data using an analysis of variance procedure based on the least-squares fit criterion (FREUND, 1962; MURPHY and O'BRIEN, 1977). The frequency-dependent source scaling exponents derived for the five different regional phases using this procedure are compared in Figure 3 for both the yield (top) and depth (bottom) variables. It can be seen from this figure that these two sets of exponents are roughly independent of phase type, at least within the rather large estimated uncertainty bounds on the individual values. That is, these spectral data are consistent with the hypothesis that, to first order, all of these regional phases are produced from a single explosion P-wave source through linear conversion processes. As has been noted previously (MURPHY, 1977), this observation implies that the entire regional waveform, including both P-and S-wave phases, can be approximately scaled for explosion yield and depth effects using scaling laws derived from a spherically symmetric P-wave source model. This approximation greatly simplifies the estimation of the frequency-dependent effects of variations in explosion source characteristics on the regional seismic phase spectra used for event identification.

It follows that, if the yield and depth scaling exponents can be assumed to be independent of regional phase type, then the scaling model represented by equation (2) simplifies to

Table 3

Frequency-dependent regional phase scaling coefficients derived from the covariance analysis of Borovoye data recorded from the non-clay PNE events of Table 1

f(Hz)	$n(f)$	$m(f)$	$B_I(P_n)$	$B_{II}(P_n)$	$B_A(P_n)$	$B_B(P_n)$	$B_I(P_{coda})$	$B_{II}(P_{coda})$	$B_A(P_{coda})$	$B_B(P_{coda})$	$B_I(P_g)$
0.50	0.937	−0.618	1.864	1.819	1.689	1.865	1.760	1.893	1.390	1.437	1.640
0.75	1.198	−0.810	2.266	1.905	1.920	1.979	2.053	1.980	1.543	1.573	1.905
1.00	1.289	−0.792	2.112	1.845	1.761	1.723	1.874	1.810	1.392	1.226	1.751
1.25	1.214	−0.520	1.247	1.058	1.018	1.053	1.052	1.026	0.614	0.491	0.887
1.50	1.301	−0.231	0.222	−0.204	0.066	0.066	−0.015	−0.151	−0.385	−0.448	−0.178
1.75	1.360	−0.326	0.425	−0.009	0.298	0.076	0.193	0.097	−0.184	−0.349	0.020
2.00	1.298	−0.318	0.413	0.005	0.246	−0.079	0.189	0.077	−0.240	−0.516	0.042
2.25	1.228	−0.169	−0.028	−0.582	−0.321	−0.683	−0.236	−0.492	−0.701	−1.105	−0.393
2.50	1.124	−0.148	−0.060	−0.678	−0.449	−0.831	−0.202	−0.496	−0.782	−1.198	−0.397
2.75	1.206	0.039	−0.781	−1.421	−1.059	−1.460	−0.918	−1.221	−1.459	−1.803	−1.135
3.00	1.098	−0.010	−0.524	−1.135	−0.854	−1.297	−0.701	−0.933	−1.181	−1.636	−0.953
3.25	1.085	−0.076	−0.251	−0.727	−0.694	−1.192	−0.496	−0.703	−1.026	−1.617	−0.737
3.50	0.975	−0.156	0.137	−0.422	−0.443	−0.900	−0.229	−0.355	−0.809	−1.289	−0.455
3.75	0.896	−0.095	−0.049	−0.585	−0.607	−0.915	−0.428	−0.545	−0.952	−1.372	−0.625
4.00	0.809	−0.113	−0.052	−0.449	−0.520	−0.792	−0.331	−0.407	−0.845	−1.290	−0.531
4.25	0.905	0.009	−0.553	−0.999	−0.982	−1.357	−0.844	−0.931	−1.336	−1.769	−1.039
4.50	0.930	0.113	−0.894	−1.409	−1.414	−1.808	−1.234	−1.403	−1.721	−2.173	−1.410
4.75	0.885	0.173	−1.094	−1.579	−1.600	−2.059	−1.377	−1.560	−1.920	−2.394	−1.629
5.00	0.805	0.206	−1.156	−1.671	−1.681	−2.168	−1.447	−1.580	−2.035	−2.484	−1.720

Table 3

Continued

f(Hz)	$B_{II}(P_g)$	$B_A(P_g)$	$B_B(P_g)$	$B_I(S_n)$	$B_{II}(S_n)$	$B_A(S_n)$	$B_B(S_n)$	$B_I(L_g)$	$B_{II}(L_g)$	$B_A(L_g)$	$B_B(L_g)$
0.50	2.333	1.309	1.369	1.835	2.255	1.424	1.382	2.209	2.200	1.633	1.360
0.75	2.569	1.441	1.470	2.127	2.507	1.614	1.519	2.468	2.460	1.741	1.437
1.00	2.103	1.252	1.079	1.958	2.107	1.407	1.198	2.210	2.165	1.469	1.042
1.25	1.194	0.488	0.328	1.100	1.236	0.551	0.378	1.267	1.172	0.630	0.145
1.50	0.068	−0.541	−0.644	0.049	0.055	−0.471	−0.642	0.128	−0.082	−0.492	−0.932
1.75	0.281	−0.364	−0.569	0.273	0.228	−0.306	−0.594	0.256	0.141	−0.363	−0.846
2.00	0.183	−0.448	−0.737	0.282	0.170	−0.350	−0.737	0.147	0.116	−0.461	−0.974
2.25	−0.369	−0.951	−1.346	−0.227	−0.342	−0.858	−1.218	−0.396	−0.442	−1.005	−1.520
2.50	−0.469	−1.061	−1.397	−0.304	−0.395	−0.983	−1.321	−0.470	−0.523	−1.136	−1.603
2.75	−1.197	−1.733	−2.037	−1.011	−1.141	−1.702	−2.029	−1.223	−1.297	−2.012	−2.294
3.00	−0.995	−1.467	−1.882	−0.807	−0.996	−1.469	−1.823	−1.058	−1.105	−1.769	−2.092
3.25	−0.734	−1.283	−1.798	−0.647	−0.796	−1.277	−1.715	−0.903	−0.919	−1.538	−1.983
3.50	−0.413	−1.034	−1.516	−0.321	−0.501	−1.012	−1.477	−0.647	−0.602	−1.311	−1.729
3.75	−0.615	−1.223	−1.627	−0.494	−0.641	−1.166	−1.622	−0.832	−0.749	−1.572	−1.843
4.00	−0.520	−1.155	−1.524	−0.416	−0.455	−1.090	−1.511	−0.754	−0.630	−1.414	−1.698
4.25	−1.089	−1.642	−1.999	−0.891	−0.982	−1.574	−2.028	−1.276	−1.118	−1.882	−2.153
4.50	−1.519	−2.049	−2.369	−1.310	−1.330	−2.017	−2.418	−1.682	−1.495	−2.338	−2.516
4.75	−1.704	−2.247	−2.537	−1.555	−1.467	−2.232	−2.590	−1.849	−1.690	−2.511	−2.690
5.00	−1.835	−2.327	−2.611	−1.618	−1.500	−2.296	−2.629	−1.980	−1.816	−2.538	−2.755

$$A_{p,k}(\omega) = B_{p,k}(\omega) W^{n(\omega)} h^{m(\omega)} , \tag{3}$$

where the coefficients $B_{p,k}(\omega)$ represent the average spectral amplitude levels for each phase p and epicentral distance group k for some specified reference yield and depth of burial values (i.e., the frequency-dependent propagation path effects for each phase). The composite yield and depth scaling exponents obtained from the data using the model of equation (3) are plotted as a function of frequency in Figure 4, where they are compared with the corresponding values predicted for this group of explosions by the Mueller/Murphy (M/M) explosion source model for explosions in hardrock (MUELLER and MURPHY, 1971; MURPHY, 1977). It can be seen from this figure that the model predictions are generally consistent with the experimentally derived values within the 95% uncertainty bounds on those values, except at low frequency where the statistically inferred depth dependence is somewhat larger at the 95% confidence level than that predicted by the M/M model.

One simplification which was employed in the above analysis requiring assessment with regard to potential effects on the comparison of Figure 4, is the assumption that the seismic source coupling efficiency is essentially the same in all the sampled source media. Clearly, if this is not approximately true then the scaling results of Figure 4 could be biased by source medium effects. In an attempt to assess the validity of this assumption, the statistically derived scaling parameters were used together with equation (3) to predict the regional phase spectra corresponding to each of the 21 selected PNE events. Ratios of observed-to-predicted spectra were

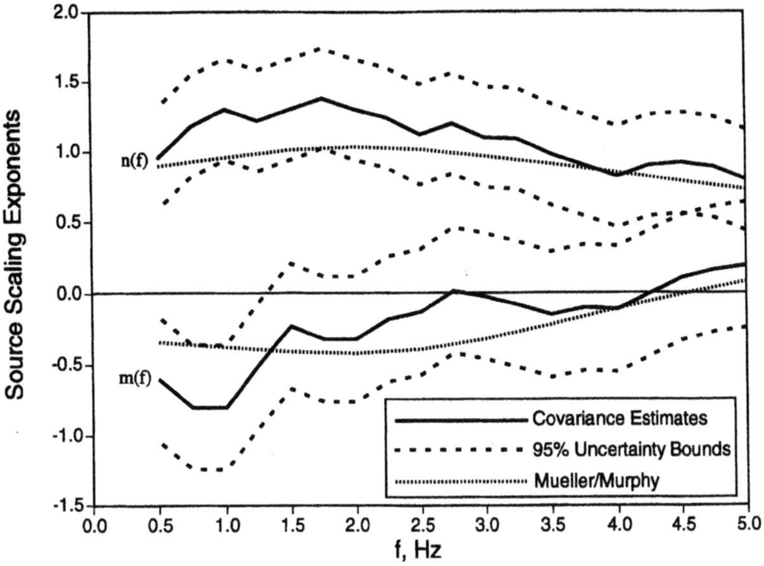

Figure 5
Comparison of the frequency-dependent yield (n) and depth (m) scaling exponents derived from the covariance analysis of Borovoye data rercorded from the non-clay PNE events of Groups I, II, A and B with the corresponding values predicted by the Mueller/Murphy explosion source model.

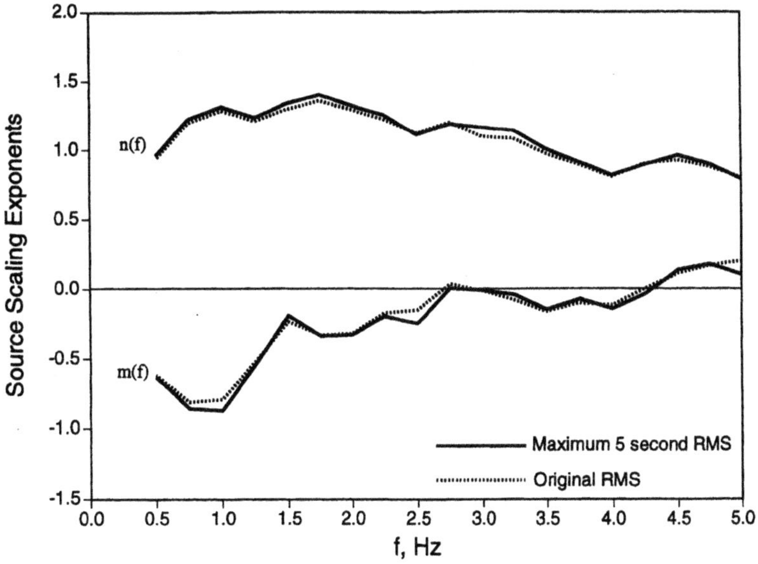

Figure 6
Comparison of the frequency-dependent yield (n) and depth (m) scaling exponents derived from the covariance analysis of Borovoye RMS spectral amplitude levels defined over short- and long-time windows.

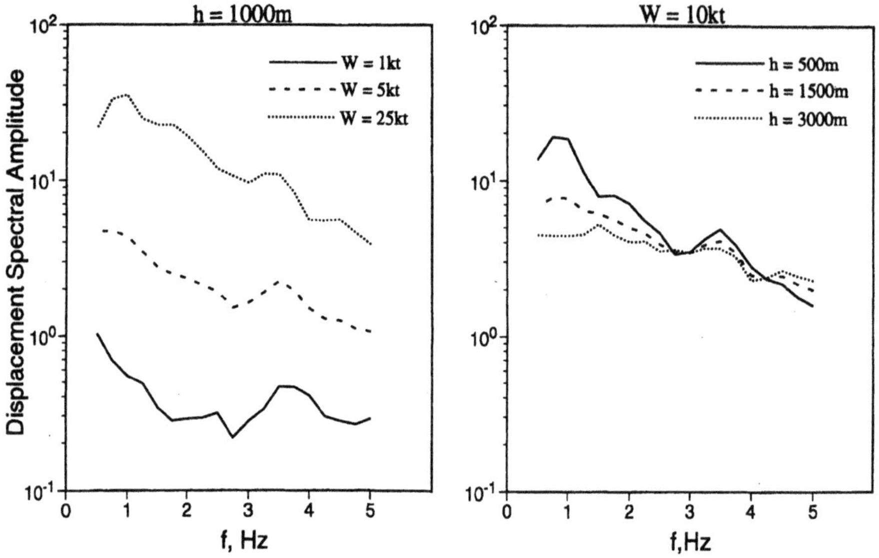

Figure 7
Illustration of the effects of explosion yield (left) and depth of burial (right) on the regional phase spectra predicted by the non-clay covariance model for the Group I P_n phase.

then computed and logarithmically averaged over the frequency band extending from 0.5 to 5.0 Hz. The resulting values for the five phases were then logarithmically averaged to obtain a measure of source coupling efficiency for each event. These final average ratio values for the 21 selected PNE events are listed in order of increasing observed/predicted ratio value in Table 2. It can be seen from this table that the values of this coupling parameter vary from about 0.5 to 2.0, with average ratios less than 1.0 corresponding to cases where the source coupling is low, and those with average ratios greater than 1.0 to cases where the source coupling is high, relative to the average values represented by the variance model predictions. Note for example

Table 4
Source parameters of Soviet PNE events used in the theoretical modeling study of near-regional seismic data

Date	Latitude	Longitude	W, kt	h, m	Medium
07/21/84	51.36	53.32	13.5	846	salt
08/10/77	50.92	110.76	8.5	499	granite
09/17/84	55.87	87.45	10	557	granite
07/18/85	65.97	40.86	8.5	772	granite
10/02/74	66.10	112.65	1.7	98	limestone
09/02/69	57.42	54.86	7.6	1212	limestone
04/11/72	37.38	62.00	14	1720	limestone
04/18/87	60.25	57.08	3.2	2055	limestone
06/18/85	60.17	72.50	2.5	2859	sandstone

Table 5

Subsurface velocity models used in the theoretical modeling study of near-regional seismic data

Medium	Thickness km	α km/sec	β km/sec	ρ gm/cm^3	Q_α	Q_β
7/21/84						
Clay	0.068	1.50	0.60	2.00	40	25
Sandstone	0.160	2.20	0.95	2.00	60	35
Anhydrite	0.107	5.00	2.78	2.70	200	150
Salt	0.365	4.00	2.20	2.20	150	100
Anhydrite	0.050	5.00	2.78	2.70	200	150
Salt	0.055	4.10	2.22	2.20	150	100
Anhydrite	0.027	5.50	2.90	2.70	200	150
Salt	3.000	4.30	2.10	2.20	150	100
Basement	–	5.20	2.95	2.30	600	500
8/10/77						
Alluvium	0.025	1.70	0.50	1.50	40	30
Sandstone	0.130	2.90	1.20	2.00	60	45
Granite	–	6.20	3.40	2.70	200	150
9/17/84						
Fractured Granite	0.050	4.20	2.20	2.50	60	40
Granite	–	5.40	3.20	2.55	200	150
7/18/85						
Clay	0.050	1.88	0.50	1.50	60	45
Argillite	0.530	4.50	2.20	2.00	60	45
Sandstone	0.270	3.60	1.80	2.00	150	100
Granite	–	6.00	3.70	2.50	200	150
10/02/74						
Marl	0.082	2.30	1.30	2.30	160	145
Limestone	–	3.80	2.30	2.50	200	150
9/02/69						
Sediments	0.040	1.00	0.30	2.20	20	10
Dolomite	0.240	3.25	1.60	2.40	100	60
Limestone	0.690	5.00	2.75	2.55	100	60
Dolomite	0.060	3.50	1.60	2.58	100	60
Limestone	0.270	5.60	3.20	2.65	100	60
Dolomite	0.060	3.50	1.60	2.47	100	60
Limestone	0.690	5.80	3.30	2.58	100	60
Dolomite	0.360	3.65	1.90	2.59	100	60
Limestone	2.500	5.50	3.25	2.60	200	100
Basement	–	5.80	3.40	2.80	200	100
4/11/72						
Clay	0.250	1.80	0.80	2.00	60	45
Sandstone	0.900	2.20	1.30	2.20	160	145
Marl	0.100	3.25	1.70	2.50	200	150
Limestone	0.550	2.85	1.50	2.30	200	150
Marl	0.450	3.20	1.70	2.50	200	150
Marl	0.450	3.50	1.75	2.50	200	150
Marl	6.000	4.50	2.30	2.60	200	150

Table 5

Continued

Medium	Thickness km	α km/sec	β km/sec	ρ gm/cm^3	Q_α	Q_β
Basement	–	6.22	3.40	2.70	400	300
4/18/87						
Sand	0.250	1.80	0.90	2.00	50	25
Clay	0.250	1.80	0.90	2.00	50	25
Clay	0.300	4.00	2.10	2.30	100	50
Dolomite	0.100	5.00	2.90	2.40	100	50
Limestone	1.600	5.70	3.20	2.60	200	100
Sandstone	–	3.00	1.70	2.60	200	100
6/18/85						
Sandstone	2.700	3.60	2.10	2.50	160	145
Clay	0.100	2.44	1.50	2.30	60	40
Sandstone	–	3.80	2.30	2.55	2.00	150

that the three explosions in sandstone are associated with low, intermediate and higher than average values, indicating that the coupling in this medium shows no systematic difference from the average. In fact, the only consistent dependence on source medium is observed for the explosions in clay, which correspond to the five highest values in this table. Moreover, it can be seen that within this clay subgroup, the offset increases with decreasing yield, consistent with the m_b/yield results reported by SULTANOV *et al.* (1993) for explosions in that medium. These results are also consistent with the alternate seismic source model for explosions in clay that was proposed by MURPHY and BARKER (1994) on the basis of their analyses of network-averaged teleseismic *P*-wave spectral data. More generally, the results of these previous teleseismic studies suggested that explosive coupling in clay represents the only significant outlier among the various source media sampled by the Soviet PNE events, a conclusion which is also consistent with the regional analysis results of Table 2.

On the basis of the results of the coupling analysis summarized in Table 2, the statistical analysis was repeated using the subset of the Borovoye spectral data which remained after the removal of the data from the five explosions in clay. The resulting revised scaling coefficients are listed in Table 3, and the associated yield and depth scaling exponents are plotted as functions of frequency in Figure 5 where they are again compared with the corresponding values predicted by the M/M source model. Note that the low-frequency depth exponents inferred from this revised analysis change significantly with respect to those shown in Figure 4 and that, in this case, the predicted M/M yield and depth scaling exponents are completely consistent with the experimentally determined values, within the statistical uncertainty of those mean values. We conclude that the Mueller/Murphy explosion source scaling model is consistent with these observed Borovoye PNE regional phase spectral data within the rather large existing uncertainty bounds. More precise data will be required in order to determine whether the observed discrepancy in the average depth scaling exponents at low frequencies is significant enough to warrant modification of that explosion source model.

One final issue to be considered concerns the definition of spectral amplitude that has been employed in the above analyses of the Borovoye data. That is, do the RMS values of the narrowband filter outputs computed over the long-time windows defined by the selected group velocity limits accurately represent the spectral amplitude levels of the individual regional phases? In an attempt to answer this question, the RMS spectral amplitude levels were estimated over 5-second windows of the filter outputs and the maximum of these 5-second values was determined for each group velocity window. These short-term maximum values were then taken to represent the spectral amplitude levels of the different regional phases and the

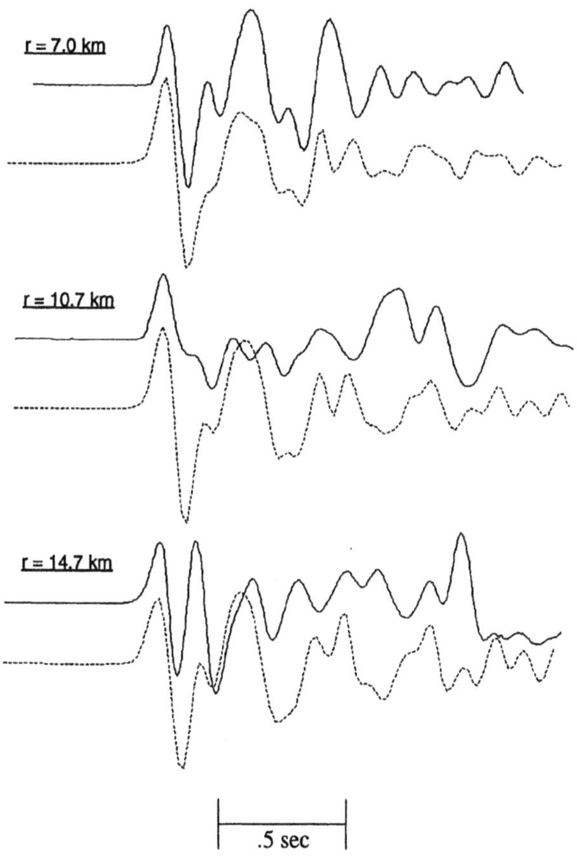

Figure 8
Comparison of observed (solid) and synthetic (dashed) near-regional, vertical component displacements for the PNE event of 7/21/84, which was a 13.5 kt explosion at a depth of 846 m in salt.

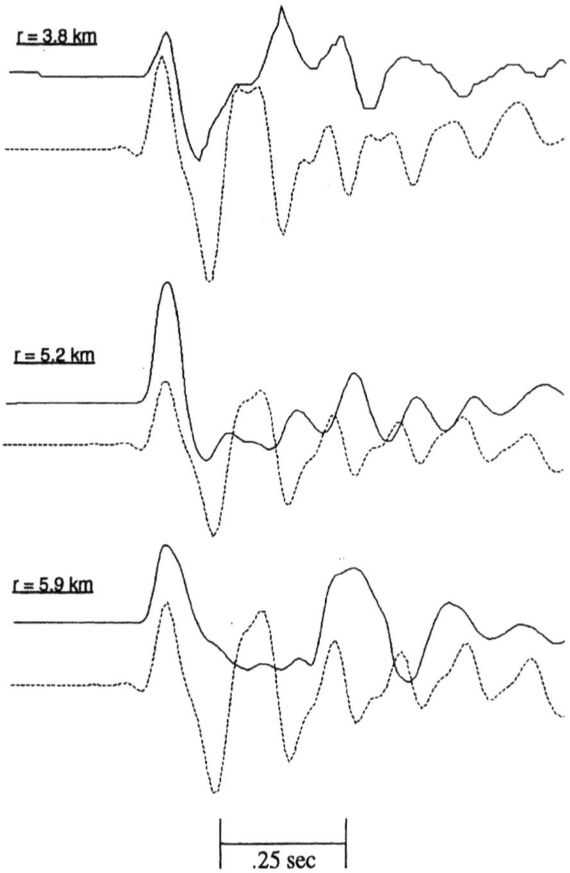

Figure 9
Comparison of observed (solid) and synthetic (dashed) near-regional, vertical component displacements for the PNE event of 8/10/77, which was an 8.5 kt explosion at a depth of 499 m in granite.

statistical analysis was repeated using the data from the non-clay explosions of Groups I, II, A and B. The results of this analysis are summarized in Figure 6 where the derived yield and depth scaling exponents as functions of frequency are compared with the corresponding values from Figure 5, which were derived using the original definition of spectral amplitude level. It can be seen from Figure 6 that these two sets of scaling exponents are remarkably similar, which indicates that the source scaling analysis results are robust and not dependent on the details of the spectral estimation procedure. More generally, these results also suggest that phase association based on rough group velocity windows may be adequate for verification purposes, which would greatly simplify semi-automated analyses of the large volumes of seismic data which must be considered in global monitoring of the CTBT.

The effects of variations in explosion yield and depth of burial on the regional phase spectra predicted by the non-clay statistical model are illustrated in Figure 7

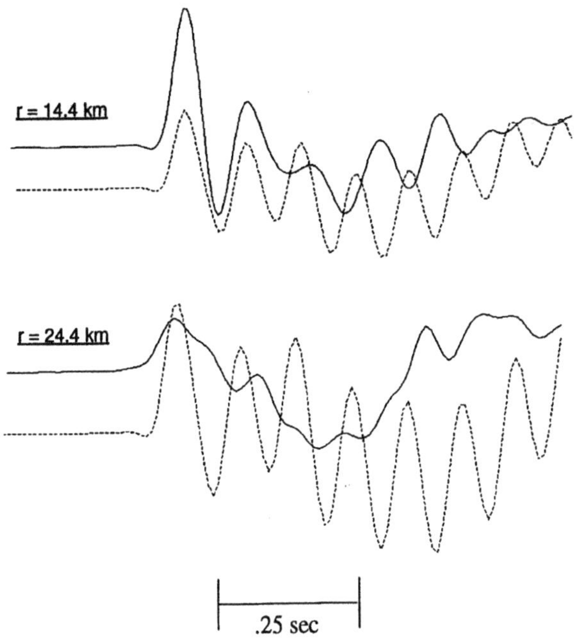

Figure 10
Comparison of observed (solid) and synthetic (dashed) near-regional, vertical component displacements, for the PNE event of 9/17/84, which was a 10 kt explosion at a depth of 557 m in granite.

for the Group I P_n phase. Note that although the overall spectral shapes depend on phase and group, the relative source effects shown here are common to all phases and groups for the model represented by equation (3). It can be seen from this figure that the low frequency ($f < 2$ Hz) regional phase spectral amplitudes are predicted to increase more rapidly with increasing explosion yield, and to decrease more rapidly with increasing source depth of burial than the higher frequency spectral amplitudes, as would be expected on the basis of the results shown in Figure 5. In particular, at a fixed yield, this source model predicts that the regional phase spectral amplitude levels at frequencies near 1 Hz can be expected to decrease by more than a factor of 4 as the source depth increases over the range from about 500 to 3000 m sampled by our Borovoye data set. This is a potentially significant reduction in seismic signal level that must be accounted for in any assessments of verification capability.

Theoretical Modeling Analyses of Near-regional Seismic Data Recorded from Selected Soviet PNE Events

In addition to the regional data recorded at the Borovoye station, near-regional ($\Delta < 25$ km) seismic data were recorded from a number of PNE events and subsequently digitized at the Institute for Dynamics of the Geospheres (IDG) of the

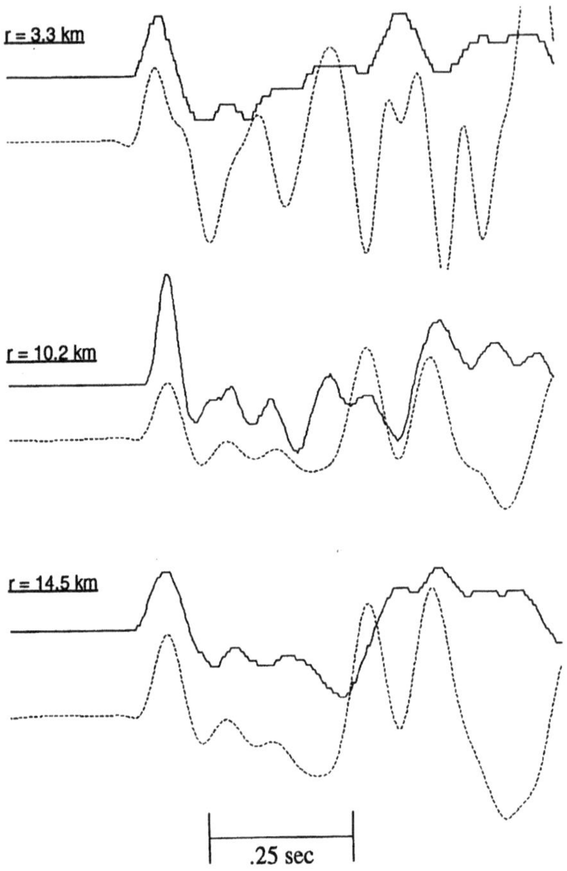

Figure 11
Comparison on observed (solid) and synthetic (dashed) near-regional, vertical component displacements for the PNE event of 7/18/85, which was an 8.5 kt explosion at a depth of 772 m in granite.

Russian Academy of Sciences. Since these data were unaffected by the variations in the long regional propagation paths to Borovoye, they provide an opportunity to test the validity of our statistical decomposition of the source and propagation path effects on the observed Borovoye data. That is, if it can be verified that these near-regional data scale with explosion yield and depth of burial in the same manner as that inferred from the analysis of the Borovoye regional data, then the confidence in the statistical decomposition will be greatly increased. Unfortunately, a direct comparison of these two source scaling estimates is not possible at this time, because there is currently little overlap between the PNE events represented in the Borovoye and near-regional data samples. However, since the source scaling inferred from the Borovoye covariance analysis was found to be generally consistent with the predictions of the Mueller/Murphy model, it is of interest to assess how well that approximate source model can account for the available near-regional PNE data.

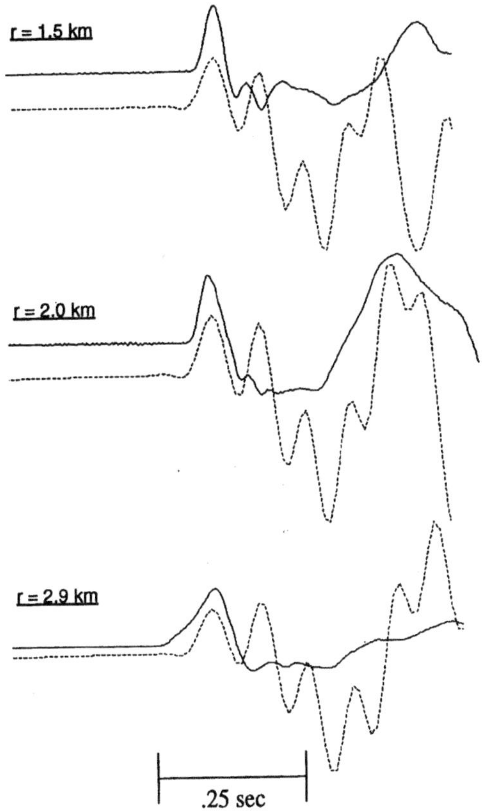

Figure 12
Comparison of observed (solid) and synthetic (dashed), near-regional, vertical component displacements for the PNE event of 10/02/74, which was a 1.7 kt explosion at a depth of 98 m in limestone.

That is, if it can be shown that the Mueller/Murphy source model can account for the observed variations in the near-regional PNE data, then that will provide additional support for the validity of the source scaling inferred from the statistical analysis of the Borovoye data. For this reason, near-regional data observed from selected PNE events have been theoretically simulated using the Mueller/Murphy explosion source model in an attempt to test the applicability of that model over the ranges of yield, depth of burial and source media represented by these explosions.

The source parameters of the nine PNE events which have been selected for theoretical simulation analysis are listed in Table 4, where it can be seen that they sample yield values ranging from 2.5 to 14 kt and depths of burial ranging from 98 to 2859 m for explosions in salt, granite, limestone and sandstone source media. As was noted previously, a comparison of Table 4 with Table 1 indicates that only two of these selected events (i.e., 7/21/84 and 6/18/85) correspond to explosions for which Borovoye regional data are currently available. However, the explosions of Table 4

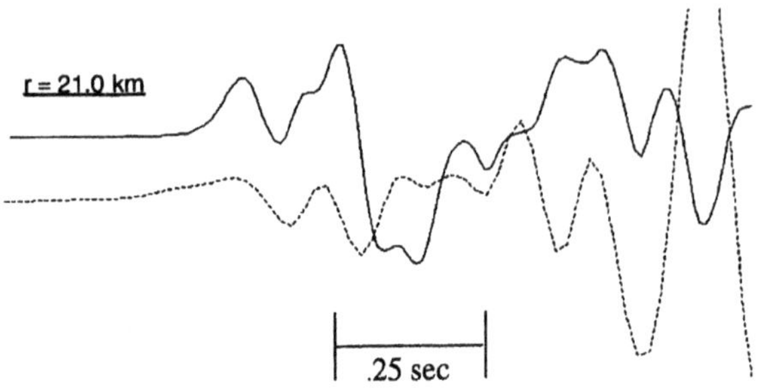

Figure 13
Comparison of observed (solid) and synthetic (dashed) near-regional, vertical component displacements for the PNE event of 9/02/69, which was a 7.6 kt explosion at a depth of 1,212 m in limestone.

do sample ranges of source parameters which are consistent with those represented in the Borovoye sample of Table 1 and, consequently, the source scaling inferred from these two data sets should be roughly comparable.

The subsurface propagation path models used in the theoretical simulation analyses are listed in Table 5 for each of the nine selected PNE events of Table 4. These plane-layered models were developed on the basis of previously unpublished *P*- and *S*-wave velocity (α, β) and density (ρ) data obtained from logging of the explosion emplacement holes, supplemented by regional geophysical data as required for the deeper structures. The listed values for the associated *P*- and *S*-wave anelastic attenuation parameters (Q_α, Q_β) were estimated for each layer using nominal values for the various rock types as a function of depth in the different source regions. Obviously, these plane-layered models provide only rough approximations to the actual propagation paths in many cases, but they are generally consistent with the type of information that might be available in a verification context. Therefore, it is relevant to attempt to assess the degree to which such simple models can be used to simulate the characteristics of the observed data.

The seismic signals corresponding to the near-regional observations from the selected PNE events of Table 4 have been theoretically simulated using the frequency-wavenumber integration algorithm developed by BOUCHON (1981). In this approach, compressional point sources equivalent to the Mueller/Murphy explosive source approximations are embedded at the appropriate depths in the associated plane-layered subsurface models for the sites listed in Table 5, and the exact theoretical seismic signals corresponding to the selected source/receiver configurations are synthesized using the referenced frequency-wavenumber integration algorithm. Since our principal interest in the present study lies in assessing the adequacy of the source approximation, the comparative analyses described below have been focused on the initial arrivals contained in the first few seconds of the observed and simulated waveforms.

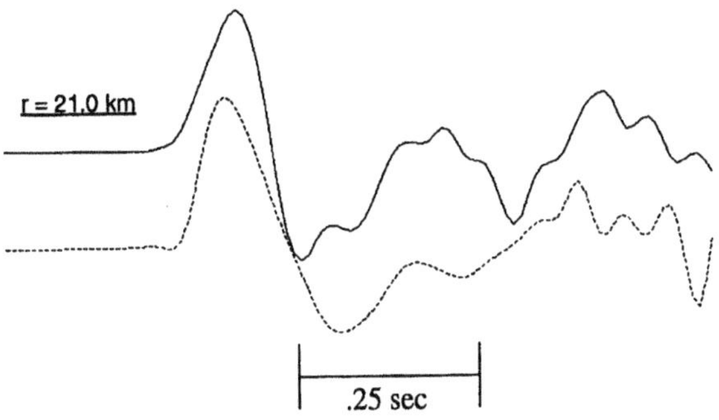

Figure 14
Comparison of observed (solid) and synthetic (dashed) near-regional, vertical component displacements for the PNE event of 4/11/72, which was a 14 kt explosion at a depth of 1720 m in limestone.

The observed and synthetic vertical component displacement waveforms for the PNE explosion of 7/21/84 are compared at the same absolute amplitude scales in Figure 8. This explosion had a yield of 13.5 kt and was detonated at a depth of 846 m in salt in the Karachaganak gas condensate field located in the Uralsk District of the former Soviet Union for the purpose of cavity creation. It can be seen from this figure that the observed and theoretically simulated waveforms agree quite well in this case with respect to overall amplitude level and frequency content for these three stations at distances of 7.0, 10.7 and 14.7 km from the source.

Similar comparisons of simulated and observed waveforms for the 8/10/77, 9/17/84 and 7/18/85 PNE explosions in granite are shown in Figures 9–11. These explosions had similar yields of 8.5, 10 and 8.5 kt, respectively, and were conducted for the purpose of providing sources of seismic waves for Deep Seismic Sounding (DSS) experiments. It can be seen from these figures that the agreement between the synthetic and observed vertical component displacement waveforms is again quite reasonable for all three explosions at these stations in the 3 to 25 km distance range. This suggests that the Mueller/Murphy model provides adequate explosion seismic source approximations for these three widely separated granitic source environments.

Comparisons of the simulated and observed vertical displacement waveforms for the 10/02/74, 9/02/69, 4/11/72 and 4/18/87 PNE events in limestone are presented in Figures 12–15, respectively. It can be seen with reference to these figures that for the first three of these explosions, the agreement between predicted and observed amplitude levels and dominant frequencies is quite reasonable, at least for the initial arrivals. However, the comparisons for the PNE event of 4/18/87 are less satisfactory in that the predicted amplitude levels are found to be 3 to 5 times larger than those observed at the three stations in the narrow distance range extending from 1.75 to 2.82 km. This is puzzling, particularly in view of the fact that the observed dominant frequency of the initial arrivals is well matched by the synthetics, which would not be expected if the actual source coupling was significantly different than predicted.

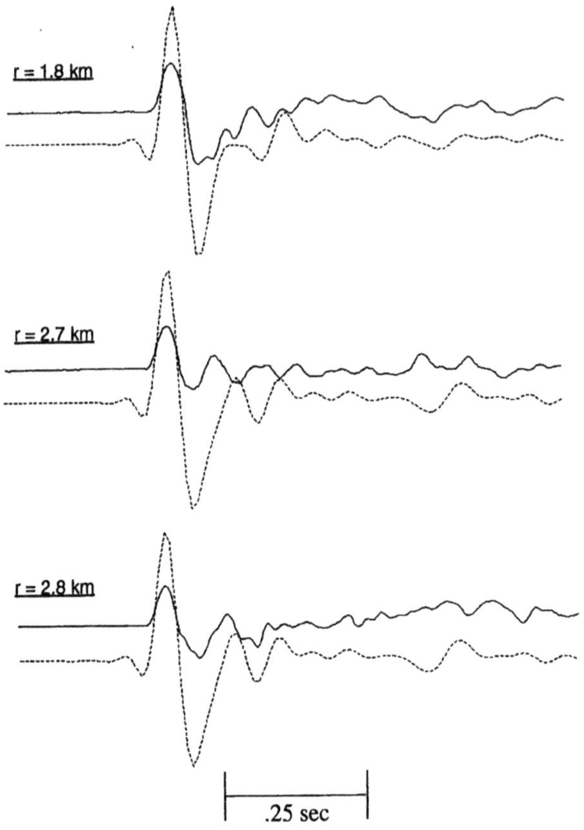

Figure 15
Comparison of observed (solid) and synthetic (dashed) near-regional, vertical component displacements for the PNE event of 4/18/87, which was a 3.2 kt explosion at a depth of 2055 m in limestone.

While it is the deepest of the four explosions in limestone, it is not significantly deeper than the 4/11/72 explosion (i.e., 2055 versus 1720 m) for which the amplitude levels of the observed and synthetic waveforms are in good agreement (cf. Figure 14). It may be that there is some unaccounted for local receiver site attenuation in this case, but there are no data currently available to test this hypothesis. Thus there is no obvious explanation for this discrepancy presently.

The final comparison is presented in Figure 16 which shows the observed and synthetic vertical component displacement waveforms for the 6/18/85 2.5 kt PNE event in sandstone. At an announced depth of 2859 m, this is the deepest known nuclear explosion conducted to date. Therefore, it is particularly significant that the synthetic waveforms predicted using the Mueller/Murphy source model provide such excellent fits to the observed amplitude levels and dominant frequencies at these three stations in the distance range extending from 1.41 to 6.00 km from this deeply overburied explosion.

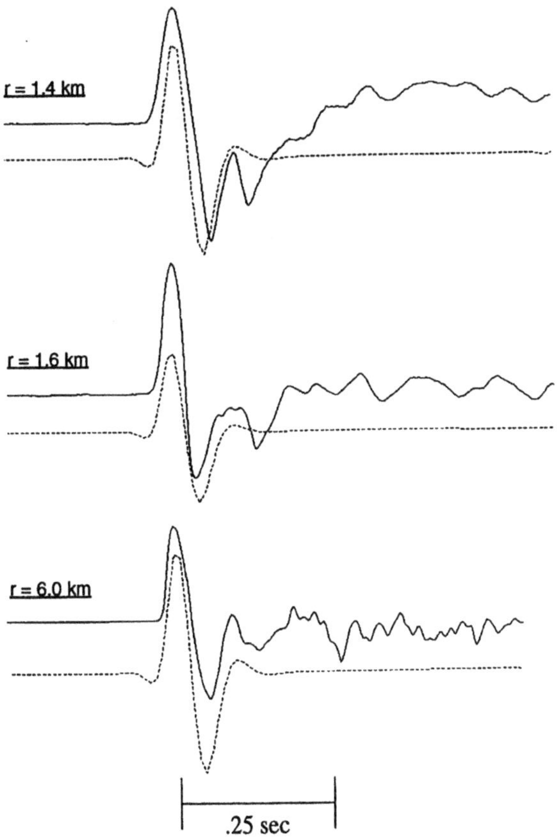

Figure 16
Comparison of observed (solid) and synthetic (dashed) near-regional, vertical component displacements for the PNE event of 6/18/85, which was a 2.5 kt explosion at a depth of 2859 m in sandstone.

In summary, considering the simplicity of the source and propagation path models employed, the overall agreement between the synthetic and observed waveforms shown in Figures 8–16 is quite remarkable and seems to confirm the fact that the Mueller/Murphy source approximation can generally account for the observed variations in seismic source coupling over the ranges of explosion yield, depth of burial and source media considered in this study. Moreover, although the data overlap between the PNE events represented in the near-regional and Borovoye regional data sets is not extensive enough to provide a definitive test of the statistical decomposition of the source and propagation path effects on the Borovoye data, the analyses have demonstrated that both data sets are generally consistent with the source yield and depth of burial scaling predicted by the Mueller/Murphy model, which supports the applicability of the proposed statistical model.

Summary and Conclusions

Our analyses of regional seismic data recorded from Soviet PNE events have provided considerably improved constraints on the range of variability of nuclear explosion seismic signals which could be encountered in global monitoring of the CTBT. In particular, statistical covariance analyses of the Borovoye PNE regional phase spectral data indicate that, with the exception of explosions in clay, the seismic source coupling in all the other sampled media (i.e., salt, sandstone, limestone, dolomite, tuff, gabbro, argillite) is approximately the same. Moreover, the observed frequency-dependent variations of all the different regional phase spectra with explosion yield and depth of burial have been shown to be generally consistent with the predictions of the Mueller/Murphy explosion source model for explosions in hard rock media. Results of theoretical simulation analyses of near-regional seismic data recorded from nine selected Soviet PNE tests have led to similar conclusions, in that both the dominant frequency content and absolute amplitude levels of most of the initial portions of the observed data were matched remarkably well, using combinations of the Mueller/Murphy source model and simple plane-layered approximations to the local subsurface geology at the various test locations. These results, taken together with the results of previous studies of the source characteristics of underground nuclear explosions conducted in water-saturated volcanic media at the Nevada Test Site (MURPHY, 1977; MURPHY *et al.*, 1989), lead us to conclude that the Mueller/Murphy seismic source model can provide a reasonable basis for estimating the expected variations in regional phase spectral composition over a wide range of explosive source conditions of potential interest in CTBT monitoring.

Acknowledgments

The authors thank M. E. Marshall for her assistance in previewing and processing the Borovoye data. This research was sponsored by the U.S. Air Force Phillips Laboratory under Contract F19628-95-C-0109.

REFERENCES

ADUSHKIN, V. V., and AN, V. A. (1990), *Seismic Observations and Monitoring of Underground Nuclear Explosions at Borovoye Geophysical Observatory*, Izv. Acad. Sci. USSR, Phys. Solid Earth, *12*, 1023–1031.

BOUCHON, M. (1981), *A Simple Method to Calculate Green's Functions for Elastic Layered Media*, Bull. Seismol. Soc. Am. *71*, 959–971.

FREUND, J. E., *Mathematical Statistics* (Prentice-Hall, Inc., Englewood Cliffs, N. J. 1962).

KIM, W.-Y., and EKSTRÖM, G. (1996), *Instrument Responses of Digital Seismographs at Borovoye, Kazakhstan, by Inversion of Transient Calibration Pulses*, Bull. Seismol. Soc. Am. *86*, 191.

MUELLER, R. A., and MURPHY, J. R. (1971), *Seismic Characteristics of Underground Nuclear Detonations. Part I. Seismic Spectrum Scaling*, Bull. Seismol. Soc. Am. *61*, 1975.

MURPHY, J. R. (1977), *Seismic Source functions and Magnitude Determinations for Underground Nuclear Detonations*, Bull. Seismol. Soc. Am. *67*, 135–158.

MURPHY, J. R., and O'BRIEN, L. J. (1977), *The Correlation of Peak Ground Acceleration Amplitude with Seismic Intensity and Other Physical Parameters*, Bull. Seismol. Soc. Am. *67*, 877–915.

MURPHY, J. R., BARKER, B. W., and O'DONNELL, A. (1989), *Network-averaged Teleseismic P-wave Spectra for Underground Explosions. Part I. Definitions and Examples*, Bull. Seismol. Soc. Am. *79*, 141–155.

MURPHY, J. R., and BARKER, B. W. (1994), *Application of Network-averaged Teleseismic P-wave Spectra to an Analysis of the Seismic Source Characteristics of Soviet PNE Explosions*, S-CUBED Report SSS-FR-94-14528.

MURPHY, J. R., SULTANOV, D. D., BARKER, B. W., KITOV, I. O., and MARSHALL, M. E. (1996), *Application of Soviet PNE Data to the Assessment of the Transportability of Regional Discriminants*, PL-TR-96-2290.

MURPHY, J. R., SULTANOV, D. D., BARKER, B. W., KITOV, I. O., and MARSHALL, M. E. (1997), *Further Analyses of Regional Seismic Data Recorded from the Soviet PNE Program: Implications with Respect to CTBT Monitoring*, PL-TR-97-2141.

SULTANOV, D. D., FERAPONTOVA, E. N., KAAZIK, P. B., KITOV, I. O., KUZNETSOV, O. P., MALYSHEV, YU. K., NEDOSHIVIN, N. I., OVTCHINNIKOV, V. M., RUBINSHTEIN, KH. D., and WILLEMSON, O. P. (1993), *Investigation of Seismic Efficiency of Soviet Peaceful Nuclear Explosions Conducted in Various Geologic Conditions*, ARPA Report under Contract MDA903-91-C-0151, Institute for Dynamics of the Geospheres, Russian Academy of Sciences.

SULTANOV, D. D., MURPHY, J. R., and RUBINSHTEIN, KH. D. (1999), *A Seismic Source Summary for Soviet Peaceful Nuclear Explosions*, Bull. Seismol. Soc. Am. *89*, 675–683.

(Received May 21, 1999, revised/accepted November 20, 1999)

Explosion Yield Estimation

Pure appl. geophys. 158 (2001) 2105–2121
0033–4553/01/112105–17 $ 1.50 + 0.20/0

Pure and Applied Geophysics

Seismic Source Characteristics of Nuclear Explosions in Water-filled Cavities

J. R. MURPHY,[1] D. D. SULTANOV,[2] N. RIMER[3] and B. W. BARKER[1]

Abstract—During the period from 1975 to 1979, the former Soviet Union conducted a series a six nuclear explosions in a water-filled cavity in salt which was created in 1968 by a tamped 27 kt explosion at a depth of 597 m at the Azgir test site at the north end of the Caspian Sea. Broadband, near-regional seismic data recorded from these tests have been processed and analyzed in an attempt to characterize the seismic source characteristics of these explosions and assess their relevance to the cavity decoupling evasion scenario. The results of these analyses indicate that the explosions in the water-filled cavity were not decoupled, but rather show evidence of enhanced seismic coupling with respect to that which would be expected from tamped explosions of the same yields in salt. Theoretical finite difference simulations of these tests have been conducted in which the complex, nonlinear interactions between the shock effects in both the water and surrounding salt medium have been explicitly modeled. The results of these simulations indicate that the most prominent yield dependent features of the observed seismic source functions can be largely explained by the dynamic interactions between the expanding and contracting steam bubbles generated by the explosions in water and the shock-wave reflections from the cavity wall. More specifically, it has been found that the shock-wave reflection from the cavity wall retards the expansion of the steam bubble in a yield dependent fashion relative to that expected in the open ocean, resulting in a smaller maximum bubble radius and a shorter bubble oscillation period.

Key words: Nuclear explosions, water-filled cavity, Azgir.

Introduction

The cavity decoupling evasion scenario continues to represent the greatest challenge to effective seismic monitoring of the Comprehensive Test-Ban Treaty (CTBT). However, despite the fact that the feasibility of this evasion concept was experimentally established over 30 years ago by the U.S. STERLING nuclear cavity decoupling test (SPRINGER *et al.*, 1968), a number of issues of importance with respect to seismic monitoring still remain unresolved. In our opinion, this is the case

[1] Maxwell Technologies, Inc., Reston Geophysics Office, 11800 Sunrise Valley Drive, Suite 1212, Reston, VA 20191, USA.
[2] Institute for Dynamics of the Geospheres, Russian Academy of Sciences, 38 Leninsky Prospect, Moscow, Russia 117334.
[3] Maxwell Technologies, Inc., Systems Division, 8888 Balboa Avenue, San Diego, CA 92123, USA.

because the available experimental database is too sparse and uncertain to provide a firm basis for resolving these questions and because the theoretical models being used to simulate the seismic source characteristics of such tests are strongly dependent on the still uncertain low pressure equations of state used to represent the response of real earth materials to explosive loading (MURPHY *et al.*, 1996). In particular, directly relevant seismic data from nuclear explosions in air-filled cavities in salt are currently available from just two events: the above referenced U.S. STERLING test and the Soviet A-3-2 test which was conducted at the Azgir test site on March 29, 1976 (ADUSHKIN *et al.*, 1992). However, it has recently become known that the Soviets also conducted six nuclear explosions of different yields in a single water-filled cavity at Azgir. Although the phenomenology associated with these nuclear explosions is clearly different in many respects from that associated with the primary evasion scenario of nuclear explosions in air-filled cavities, they do provide a unique database of seismic data recorded from repeated nuclear explosions in a cavity. Therefore, the study described in this paper has proceeded with the objective of analyzing the seismic source characteristics of these water-filled cavity tests in an attempt to assess their relevance with respect to the variables controlling the air-filled cavity decoupling evasion scenario, particularly those related to the dynamic response of the surrounding salt medium at low pressures.

Characteristics of the Azgir Water-filled Cavity Tests

The Soviet nuclear test site at Azgir is situated in the northwestern part of the Caspian Depression of the Russian Platform, near the border between the Astrakhan District and Kazakhstan. The Caspian Depression is characterized by thick sediments (up to 20 km) and intensive salt dome tectonics. The salt is generally covered by a 1–2 km thick layer of sediments, however the upper surfaces of the salt domes are encountered at depths of only 200–400 m in some locations, with occasional surface outcrops, such as that near the town of Azgir. One of these salt dome structures, known as the Great Azgir salt deposit, was used as a scientific test site to conduct 17 peaceful nuclear explosions (PNE) during the period from 1966 to 1979. This deposit consists of two domes, denoted the West and East Azgir salt domes, both of which were used in the testing program. The goals of this PNE test series conducted at Azgir were to address specific scientific problems and to develop new technologies for industrial applications of underground nuclear explosions.

The A-2 emplacement hole at Azgir is located in the northern part of the West dome at 47.9086°N, 47.9119°E and penetrates to a depth of 600 m with a diameter of 0.63 m. The stratigraphic sequence at this location consists of a 240 m layer of sediments, overlying a 56 m thick anhydrite cap layer which, in turn, overlies the massive salt formation which extends to the bottom of the hole and beyond. The hole was cased to a depth of 300 m, or to just below the anhydrite/salt interface. The nominal elastic properties of the salt medium, determined from laboratory analysis of samples recovered from emplacement hole A-2 indicate a compressional wave velocity of 4250 m/sec, a shear-wave velocity of 2400 m/sec and a density of 2.2 g/cm^3 (ADUSHKIN *et al.*, 1992). These values are quite comparable to those reported for the Tatum salt dome in Mississippi where the U.S. nuclear tests SALMON and STERLING were conducted. In fact, comparisons of free-field seismic data recorded from SALMON and tamped Azgir explosions suggest that the seismic source coupling is quite similar in these two salt domes (MURPHY and BARKER, 1994). These observations suggest that it should be possible to evaluate the seismic data recorded from the Azgir water-filled cavity tests using material models similar to those previously validated for the Tatum salt dome.

A tamped explosion with a yield of about 27 kt was conducted at a depth of 597 m in emplacement hole A-2 on 7/01/68. This explosion produced a stable, roughly spherical cavity that subsequently filled with water which flowed in from the pierced water table above the cap rock. A post-test, downhole survey indicated that this cavity had a maximum horizontal radius of 32.5 m and a total volume of 101,000 m^3. This total volume is equal to that which would be associated with a purely spherical cavity with a radius of 28.9 m. Six nuclear explosions with different yields were detonated in this water-filled cavity from 1975 to 1979. In each case, the emplacement hole was filled with water prior to the test and a special containment lock was established in the upper part of the hole to prevent water ejection. Slight variations of cavity shape and volume were observed after each explosion. For example, after the first 0.35 kt water-filled cavity explosion of 4/25/75, a volume increase of 3000 m^3, or about 3%, was measured. A detailed cavity survey was conducted following the fourth water-filled cavity test of 9/12/78 and it was found that the depth to the bottom of the cavity had decreased by about 3 m (from 595 to 592 m) due to accumulation of fragments of cables, casing and salt which had fallen into the cavity as a result of the repeated testing. However, the depth to the top of the cavity had also decreased by about 3 m (from 555 to 552 m) due to spallation of the cavity roof material, consequently the vertical extent of the cavity remained virtually unchanged. Thus, although the repeated testing damaged the cavity walls to some extent in this case, the Soviet experience at Azgir indicates that it is feasible to conduct multiple nuclear tests in a sufficiently large cavity in such a manner that cavity collapse or significant venting of material will not occur.

The source parameters of the seven nuclear explosions conducted in emplacement hole A-2 at Azgir are listed in Table 1 (SULTANOV *et al.*, 1999), where it can be seen that the yields of the water-filled cavity explosions varied over a factor of 50, ranging from 0.01 to 0.50 kt. Now a 28.9 m radius, air-filled cavity at a depth of 585 m in salt would be expected to decouple a 1.3 kt explosion to the same degree as that achieved for the U.S. STERLING explosion. It follows that if this cavity had been air-filled, all of the A-2 cavity tests would have been fully decoupled and the associated seismic signals would have been well below the teleseismic detection threshold. In fact,

Table 1

Source parameters for Azgir water-filled cavity tests (Emplacement coordinates: 47.9086°N, 47.9119°E)

Event	Date	Origin time (UT)	Depth (m)	Yield (kt)
A-2-1	07-01-68	04:02:00.5	597	27
A-2-2	04-25-75	05:00:00	583	0.35
A-2-3	10-14-77	06:59:59.10	588	0.10
A-2-4	10-30-77	06:59:59.07	586	0.01
A-2-5	09-12-78	04:59:58.49	584	0.08
A-2-6	11-30-78	07:59:58.5	586	0.06
A-2-7	01-10-79	08:00:00	580	0.50

however, SYKES (1996) reports that all six of these cavity tests were detected by one or more seismic arrays outside the U.S.S.R, and the two largest of these cavity explosions (i.e., A-2-2 and A-2-7) have been assigned m_b values in the 4.0 to 4.5 range. Since a fully tamped 1-kt explosion in salt at Azgir corresponds to an m_b value of around 4.5 (MURPHY and BARKER, 1994), it follows that these cavity tests were not decoupled and, in fact, seem to show enhanced coupling over some frequency bands, more consistent with what would be expected for explosions in water (EVERNDEN, 1970). Nevertheless, the seismic data recorded from this unique set of multiple cavity tests have the potential to provide valuable information regarding the dynamic response characteristics of the surrounding salt medium under repeated explosive loading, which is very relevant to assessments of the air-filled cavity decoupling evasion scenario.

Seismic Data Analysis

Seismic data from this series of explosions in Azgir emplacement hole A-2 were recorded at 28 stations located in the near-regional distance range extending from about 1 to 175 km. Both vertical and horizontal radial component data were recorded at stations located within 20 km of ground zero and full, three-component data sets were recorded at the more distant stations. Photographic recordings from the tamped explosion A2 and the first four of the water-filled cavity tests (i.e., A-2-2, A-2-3, A-2-4, A-2-5) have now been collected and carefully digitized at the Institute for Dynamics of the Geospheres using standardized procedures described by KITOV *et al.* (1994). Data for the other two cavity tests are not currently available. Most of these data were digitized at sampling rates of 200 samples/second or higher, which provides more than adequate resolution of the highest frequency components that can be reliably recovered from this recording medium. At a number of stations, data were recovered from most or all of these tests on a common instrument, thereby providing a basis for direct comparisons in which the propagation paths and recording parameters are held constant. These data were presented and discussed in detail by MURPHY *et al.* (1994)

and, therefore, in this paper we will focus exclusively on analyses of the recordings at three of the closest stations (i.e., 1.17, 1.71 and 7.8 km) for which the usable bandwidths of the data are wide enough to examine the most important seismic source characteristics of these tests.

The vertical component displacement data recorded at a distance of 1.17 km from the Azgir tamped and selected water-filled cavity tests are presented in Figure 1, where they have been arranged in order of decreasing yield from top to bottom. It is evident from this display that the source corner frequency of the tamped test is much lower than those of the cavity tests, consistent with the large differences in yield. It can also be seen that the character of the seismic signals recorded from the cavity tests changes as a function of yield, with the signals becoming progressively more complex as the yield decreases. This is particularly evident for the lowest yield A-2-4 recording which shows a prolonged ringing which is not evident in the recordings from the larger yield cavity shots. These qualitative observations are confirmed in Figure 2, which shows a comparison of the spectra computed from the signals in Figure 1, where the spectra have been smoothed to an effective frequency resolution of about 1 Hz. Note that because of the dynamic range limitations of the hand-digitized data, the spectrum for the tamped A-2-1 explosion is only considered to be reliable below about 5 Hz because of its low corner

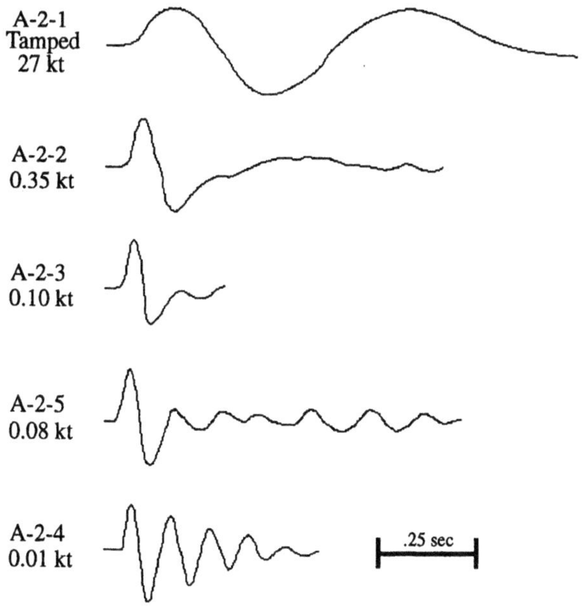

Figure 1
Comparison of vertical component seismic data recorded from the Azgir tamped and water-filled cavity explosions at a range of 1.17 km.

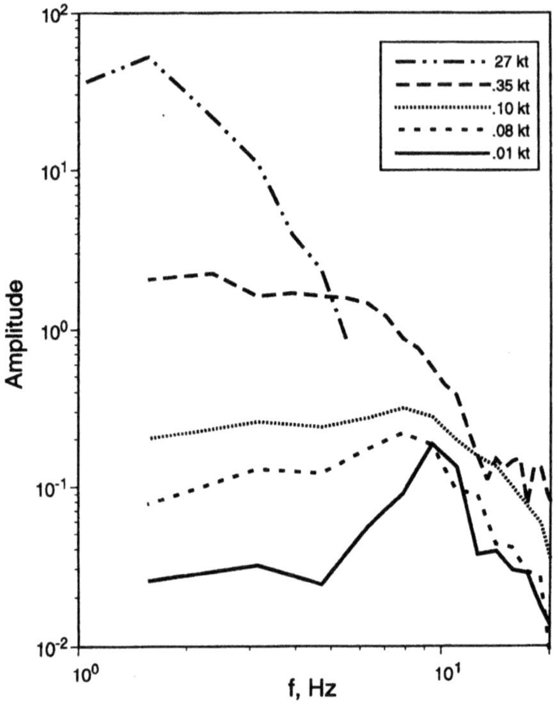

Figure 2
Comparison of displacement spectra computed from vertical component data recorded from the Azgir tamped and water-filled cavity tests at a range of 1.17 km.

frequency, while the cavity test spectra are estimated out to 20 Hz. Note also that the spectrum for the 0.01 kt cavity test exhibits a strong resonance peak at around 10 Hz, consistent with the previously noted oscillatory nature of the corresponding waveform in Figure 1. The spectra derived from the recordings of these four cavity tests at the 1.71 and 7.8 km stations are plotted in Figures 3 and 4, respectively. Comparing the spectra of Figures 2–4, it can be seen that the spectral shapes for any one explosion are quite similar at all three distances, which suggests that the propagation path effects are not very pronounced at these near-field distances.

The band-limited spectra corresponding to the tamped 27 kt A-2-1 event recordings at the three distances are shown in Figure 5 where they have been normalized to a common low frequency level for comparison purposes. Here, the low frequency ($f < 1$ Hz) levels have been approximated from the higher frequency data by determining the best-fitting Mueller/Murphy source function for the three individual spectra corresponding to this tamped explosion in salt. Once again, it can be seen that the three spectra are very similar in shape, suggesting that frequency-dependent propagation path effects may be of second-order importance in this case. If this were strictly true, the tamped explosion spectra at these three stations could be simply cube-

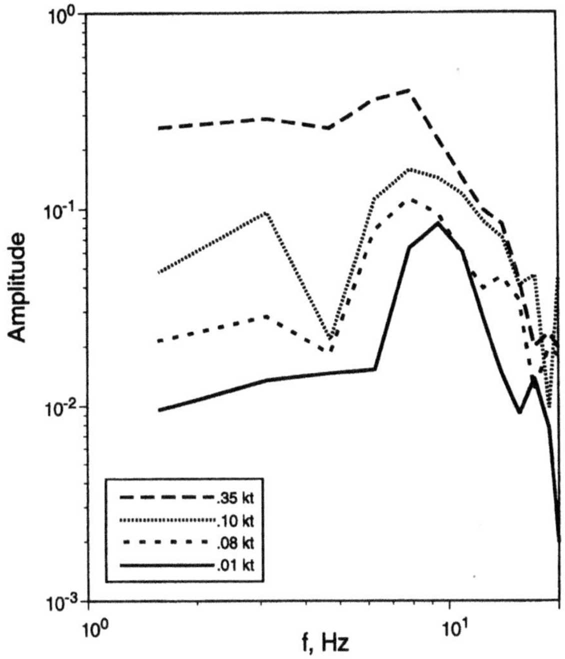

Figure 3
Comparison of displacement spectra computed from vertical component data recorded from the Azgir water-filled cavity tests at a range of 1.71 km.

root scaled to the yields of the cavity tests and used to compute spectral ratios which would then be equal to the frequency-dependent cavity/tamped source spectral ratios. The results of applying this approximation to the data of the four cavity tests are shown in Figure 6, where it can be seen that the cavity/tamped spectral ratios obtained for a given explosion at the three distances are reasonably consistent, suggesting that they may indeed represent useful approximations to the corresponding source spectral ratios. Figure 7 shows the results of logarithmically averaging these three estimates of the source spectral ratio for each cavity test. It can be seen from this figure that, to this degree of approximation, it can be concluded that the 0.35 kt A-2-2 water-filled cavity test shows enhanced coupling at low frequencies with respect to that expected from a tamped explosion of the same yield in salt, while the 0.10 kt A-2-3 and 0.08 kt A-2-5 cavity tests appear to be roughly comparable to tamped shots of those yields in salt over the entire frequency range of observation. The 0.01 kt A-2-4 cavity test spectral ratio, on the other hand, shows a pronounced resonance in the band around 10 Hz, where the overshoot of the low frequency level amounts to a factor of 5 or more. Consequently, the source spectral ratios corresponding to these four cavity tests show some pronounced yield dependencies that should correlate with differences in the seismic source generation mechanisms between these different explosions.

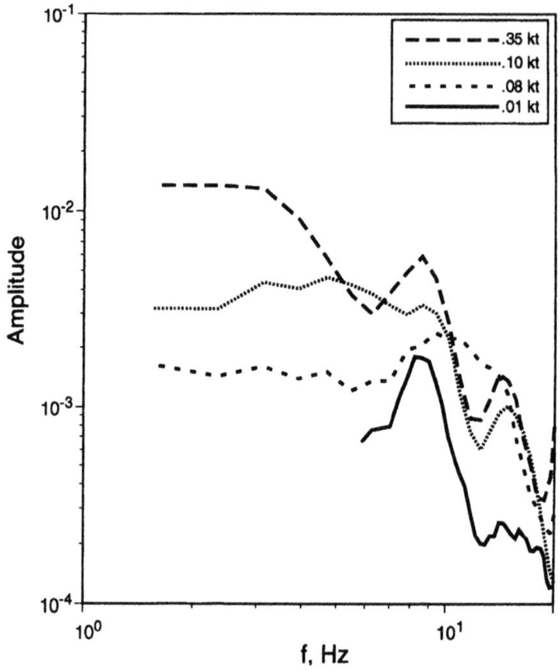

Figure 4
Comparison of displacement spectra computed from vertical component data recorded from the Azgir water-filled cavity tests at a range of 7.8 km.

Theoretical Simulation Analysis

The theoretical simulation of the seismic motions produced by nuclear explosions of different yields in the Azgir water-filled cavity is obviously complicated by the nonlinear interactions between the shock effects in both the water and the surrounding medium. In particular, the strong yield dependence of the seismic sources which was identified in Figure 7 suggests that it is necessary to explicitly account for the dynamic shock effects in the water-filled cavity. Consequently, in the following analysis we will focus on the results of a series of nonlinear finite difference simulations which explicitly account for such dynamic effects.

Before beginning the discussion of the specific finite difference calculations, it will be useful to briefly review the phenomenology associated with nuclear explosions in water. According to Pritchett (1971), near the burst point the internal energy imparted to the water by the explosion is sufficient that, upon expansion, the water will vaporize. For a 1 kt nuclear explosion, this vaporization will result in the generation of a steam bubble with an initial radius of about 3.7 m and an internal pressure of about 80 kilobars. In the open ocean, this high pressure bubble will

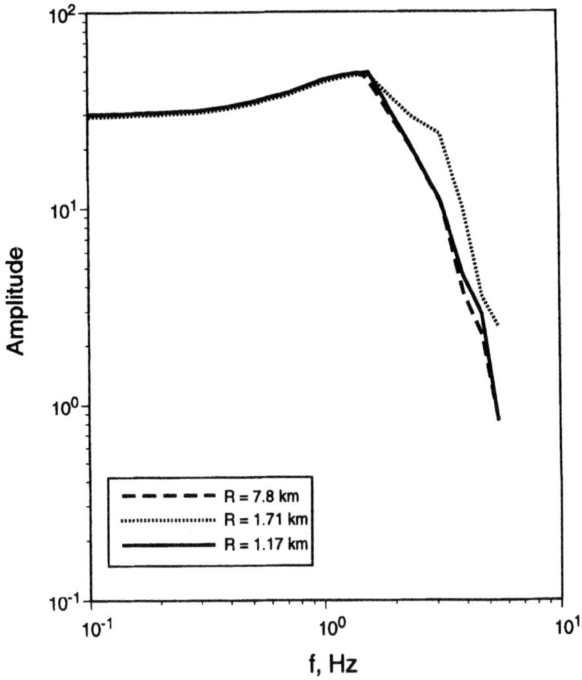

Figure 5
Comparison of amplitude-normalized displacement spectra corresponding to the vertical component recordings of the Azgir A-2-1 tamped 27 kt explosion at each of the three distances.

subsequently expand until the internal pressure decreases to slightly below overburden pressure (i.e., it overshoots the equilibrium point), whereupon it will rebound and collapse down to a minimum volume, where it is once again brought to a halt by the high internal pressure (COLE, 1948). At an overburden pressure of 60 bars, this initial expansion cycle results in a maximum bubble radius of about 37 m, or an expansion of bubble volume by about a factor of 1000. Once the bubble collapses to its minimum volume, the internal pressure is once again much higher than the overburden, and a new expansion cycle begins. This cycle of expansion and collapse would continue indefinitely in the absence of dissipation mechanisms. However, significant energy is in fact dissipated in each cycle due to Taylor instability and turbulence at the steam/water interface at bubble minimum and, as a result, only three cycles of expansion and collapse are observed from underwater nuclear sources (GLASSTONE and DOLAN, 1977). A pressure pulse is emitted at the end of each contraction phase, when the bubble is at minimum volume, and this constitutes the "bubble pulse" which is used to identify underwater nuclear explosions on hydroacoustic recordings. Of course, in the open ocean, the bubble also simultaneously rises to the surface due to its buoyancy, as it undergoes these expansion/contraction cycles.

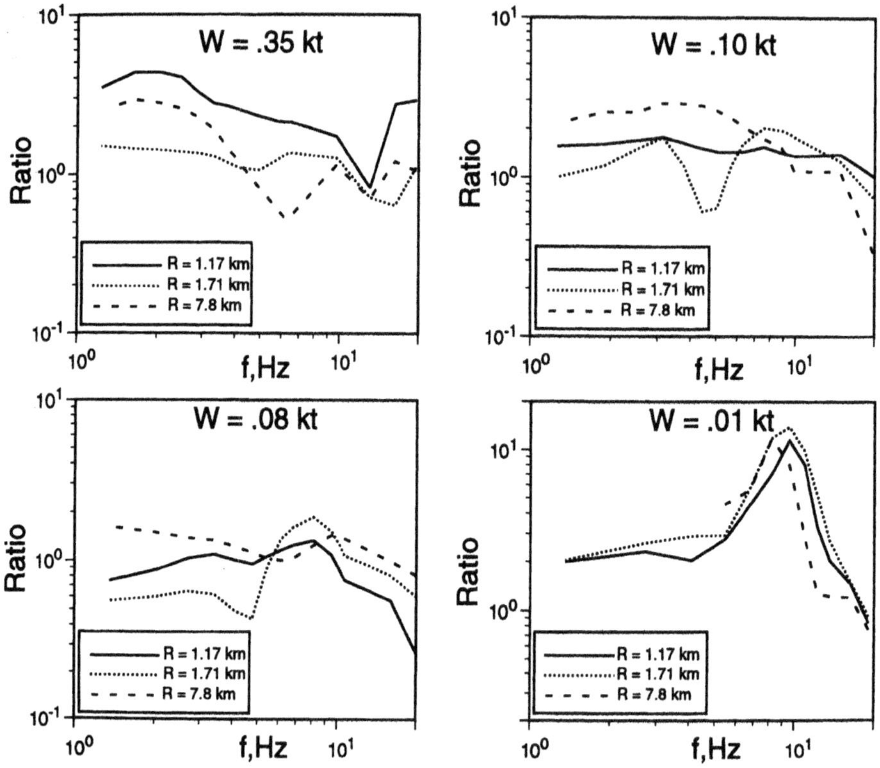

Figure 6
Individual cavity/tamped source spectral ratios estimated from the observed Azgir near-regional seismic data by cube-root scaling the observed tamped spectra to the yields of the water-filled cavity tests.

For nuclear explosions in a water-filled cavity, the simple steam bubble dynamics described above are complicated by the dynamic interactions with the cavity wall. These effects are schematically illustrated in Figure 8 where the interactions between the expanding gas bubble and the reflected shocks from the cavity wall are shown for reference purposes. We have attempted to theoretically model this complex process with a series of one-dimensional, nonlinear finite difference simulations of explosions in a purely spherical, 28.9 m radius, water-filled cavity in salt (STEVENS *et al.*, 1991; MURPHY *et al.*, 1997). Lithostatic pressure at the 597 m depth of the Azgir cavity is roughly 120 bars. For the purposes of the numerical simulations to be discussed below, it was assumed that the water-filled cavity was initially pressurized at the corresponding hydrostatic pressure of about 60 bars. The Lamé elastic solution for the stresses surrounding a pressurized spherical cavity was used to determine the initial stress distribution in the salt near the cavity. Parametric studies subsequently revealed that the results of the calculations to be discussed here were essentially unchanged when the initial cavity water pressure was reduced to 1 bar.

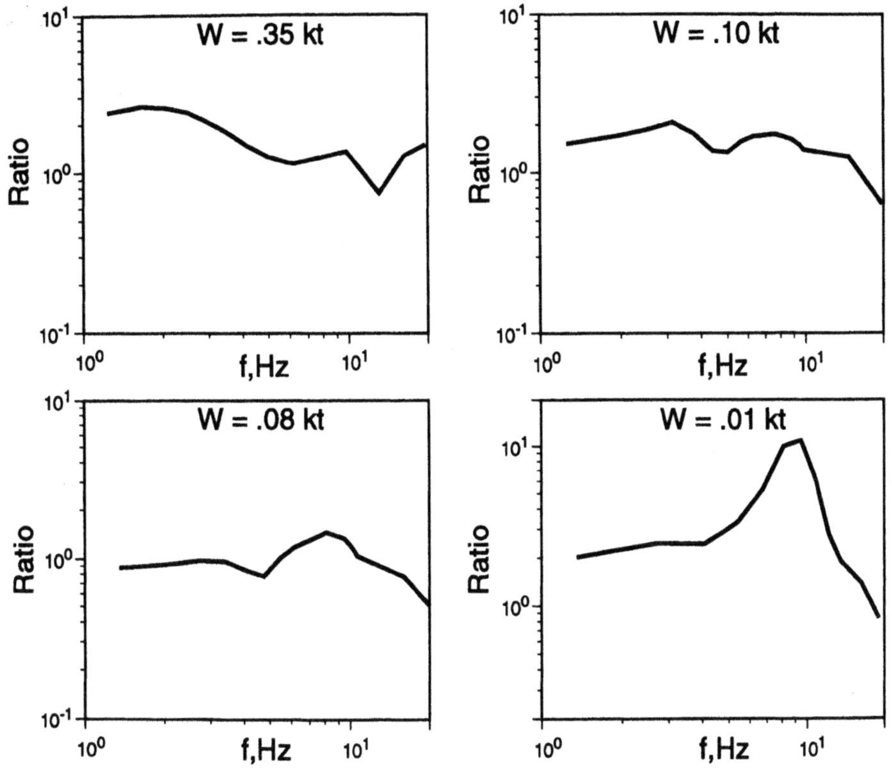

Figure 7
Average cavity/tamped source spectral ratios estimated from seismic data recorded from the Azgir tamped and selected water-filled cavity tests at three different near-regional stations (1.17, 1.71, 7.8 km).

The calculations were initialized with the device energy uniformly distributed in a water volume that gives an average initial pressure of 80 kilobars, sufficient to vaporize the required amount of water during the bubble expansion (MURPHY *et al.*, 1997). Parametric calculations indicate that a substantially smaller initial bubble results in approximately the same computed spectra while a significantly larger initial bubble (a lower initial pressure) does not model the steam bubble behavior well and greatly changes the results of interest. A detailed equation of state of water, AQUA (BAKER *et al.*, 1988), which includes the steam tables, and most other relevant features of water behavior, was used in the calculations.

Two very different strength models were used to describe the shear behavior of the surrounding salt. The first, to be referred to as Lab Salt, assumes that the Azgir site has a failure surface similar to that measured in the laboratory on small samples of dome salt from the SALMON nuclear event. It is approximated here as a parabola in stress difference Y versus mean stress P, with a stress difference in compression of 138 bars, at zero P, and a maximum value of Y in compression of 680 bars at $P \geq 600$ bars. This model provides the "high-strength bound" for the site. The second

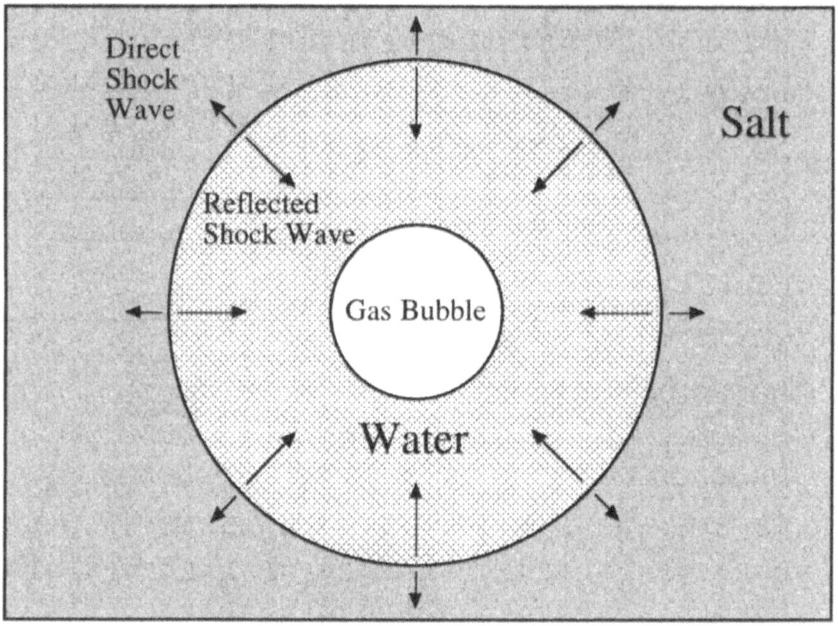

Figure 8
Schematic representation of shock-wave evolution associated with a nuclear explosion in a water-filled cavity.

strength model, generally known as the Cherry–Rimer (CR) model, was developed to simulate the particle velocity versus time ground-motion measurements made during the SALMON event (RIMER and CHERRY, 1982). This model assumes that the salt has a very low strength initially and that the strength increases greatly, to the lab strength, during the strain hardening which occurs as a result of the ground motion. This model, in general, results in inelastic behavior out to a far greater range than does the Lab Salt model. Both models have been used in a number of parametric studies of decoupling, such as those described by STEVENS *et al.* (1991). For the purposes of the present study it was found that, although the magnitudes of the seismic sources varied significantly between the two models, the resulting cavity/tamped source spectral ratios were relatively insensitive to the choice of model, therefore only the results from the simulations carried out using the CR model will be considered in the following discussion.

We will first consider the simulation of the lowest yield 0.01 kt A-2-4 cavity explosion. The sequence of events in this case is summarized in Figure 9 which shows snapshots of the computed pressure as a function of range in the cavity and surrounding salt medium at different instants of time during the shock-wave evolution. The dashed vertical lines on these snapshots denote the location of the

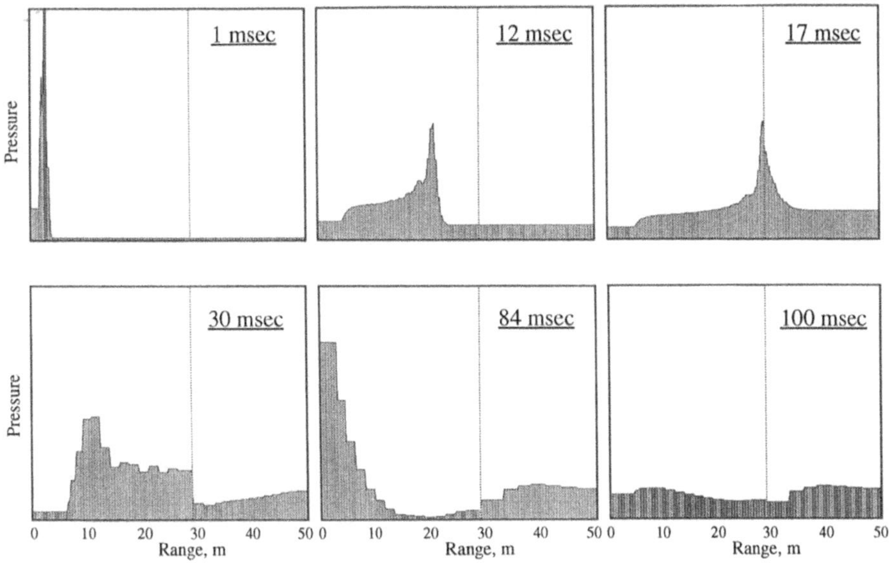

Figure 9
Snapshots of the computed pressure as function of range in the cavity and surrounding salt medium at selected times of the 0.01 kt Azgir A-2-4 simulation. The dashed vertical lines on these snapshots denote the location of the cavity wall.

cavity wall in this one-dimensional simulation. At this yield, the prompt vaporization of the surrounding water produces a steam bubble with an initial radius of about 0.8 m and an internal pressure of 80 kilobars. In Figure 9, the time-dependent radius of this steam bubble is represented by the plateaus of constant pressure extending out from the origin, which coincides with the center of the cavity. This gas bubble ultimately expands to a first maximum radius of 6.2 m in about 40 msec, which is significantly smaller than the maximum radius of 8 m that would be expected for such an explosion in an unbounded water medium. This is because the water borne shock wave, which reflects from the cavity wall at about 17 msec, arrives at the expanding bubble/water boundary at about 30 msec and retards further bubble expansion. As is indicated in Figure 9, the bubble then collapses to its first minimum radius of 3.7 m at about 80 msec, whereupon it begins a second cycle of expansion and collapse. Therefore, the period of the bubble oscillation in this case is about 80 msec, giving a characteristic frequency about 12 Hz, which is close to the 10 Hz resonant frequency observed in the corresponding spectral ratio for this test shown in Figure 7. As a point of reference, the same yield test under this overburden pressure in the open ocean would be expected to produce a steam bubble which would expand to its first maximum radius of 8 m in about 100 msec and collapse to its first minimum radius in about 200 msec, giving rise to a considerably lower bubble frequency of about 5 Hz.

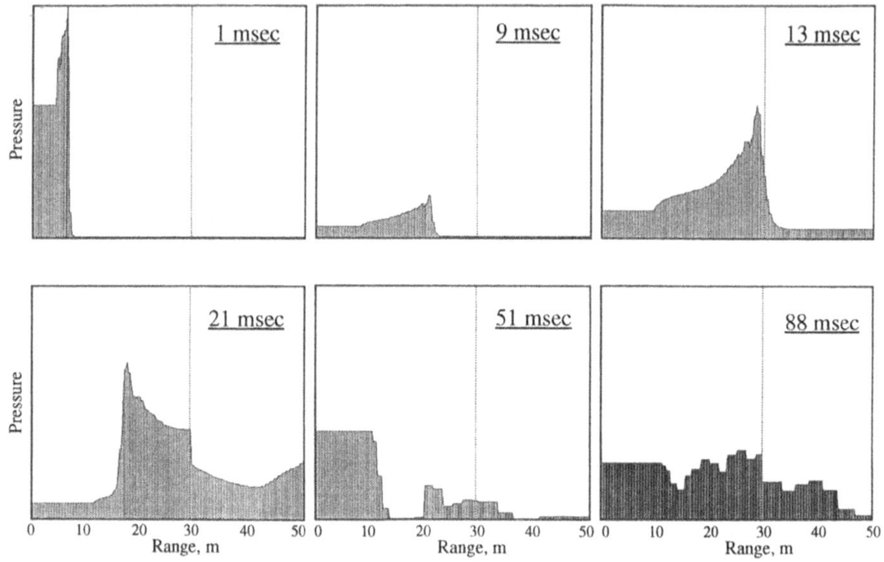

Figure 10
Snapshots of the computed pressure as a function of range in the cavity and surrounding salt medium at selected times of the 0.35 kt Azgir A-2-2 simulation. The dashed vertical lines on these snapshots denote the location of the cavity wall.

The corresponding theoretical simulation of the 0.35 kt A-2-2 explosion in that same cavity is summarized in Figure 10. In this case the initial vaporization radius is about 2.6 m and the computed first maximum expansion radius is only 11.7 m, as opposed to the broadly larger 26 m radius which would be expected in an unbounded water medium. This drastic reduction occurs because the shock reflection from the cavity wall arrives sooner (i.e., at around 20 msec in Fig. 10) and is stronger than for the previous 0.01 kt A-2-4 simulations and, therefore, has a stronger influence on terminating the bubble expansion phase. The bubble then collapses to its first minimum radius of 10.5 m in about 50 msec, after which the bubble oscillations damp out to almost nothing. Thus, for this case, there is only one weak expansion/contraction cycle and a less pronounced resonance peak would be expected, in qualitative agreement with the experimental results of Figure 7.

The theoretical cavity/tamped seismic source spectral ratios corresponding to these two simulations are shown in Figure 11, where they are compared with the corresponding average observed spectral ratios from Figure 7. Here, the theoretical results have been smoothed to approximate the frequency resolution of the observations for comparison purposes. It can be seen from this figure that the theoretical simulations provide a good qualitative description of the most pronounced observed differences in the seismic source characteristics of these two cavity tests. One notable discrepancy between theory and observation is that the

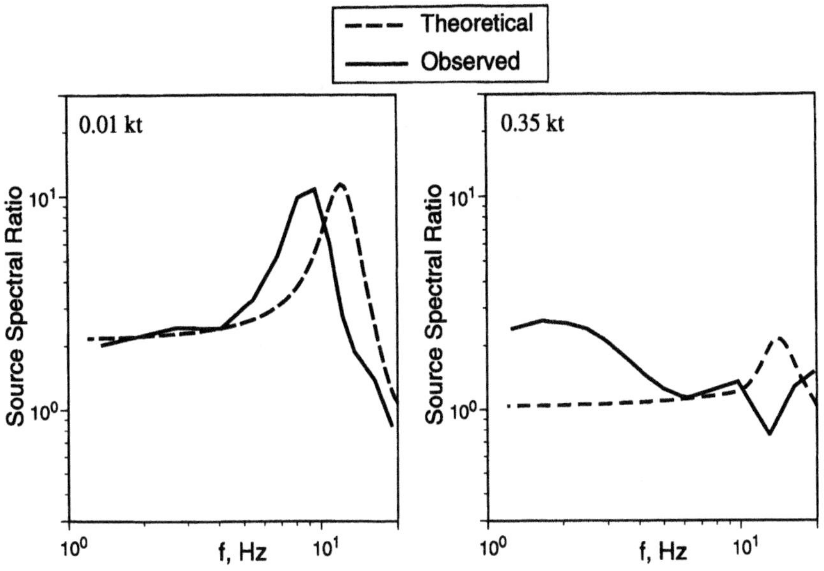

Figure 11
Comparison of observed and theoretical cavity/tamped source spectral ratios corresponding to 0.01 kt (left) and 0.35 kt (right) nuclear explosions in the Azgir water-filled cavity.

computed resonant frequency for the 0.01 kt simulation is somewhat too high. This is perhaps not surprising, given the facts that the yield is uncertain to some extent and the cavity is being approximated in the simulation as a perfect sphere, which it is not, and this might well affect the bubble period. That is, for the perfectly spherical cavity, a strong, coherent reflection is produced at the cavity wall and this reflected shock converges on the expanding gas bubble, effectively terminating its growth phase. In reality, the reflection from the actual, irregular cavity boundary could be expected to produce a less coherent reflected shock which would be less efficient at retarding the bubble growth, thereby leading to a larger bubble and longer associated bubble pulse period. Thus, given the limitations of the one-dimensional theoretical simulations described above, the degree of agreement shown in Figure 11 may be about as good as can be expected.

In any case, the evidence strongly suggests that variations in bubble pulse dynamics give rise to the most prominent yield dependent seismic source characteristics which have been identified in the data recorded from these Azgir water-filled cavity tests. While this is an interesting result, it indicates that the observed seismic characteristics are dominated by the water response rather than by the response characteristics of the surrounding salt medium. Thus, it does not appear that the data recorded from these cavity tests can provide new information that can be used to further constrain the salt equations of state to be used in theoretical simulations of the cavity decoupling evasion scenario.

Summary and Conclusion

The seismic source characteristics of a series of four nuclear explosions conducted in the same water-filled cavity at the Soviet Azgir test site have been thoroughly analyzed. The results of these analyses indicate that the seismic signals from these explosions were not decoupled in the manner which would be expected for comparable explosions in air-filled cavities and, in fact, indicate evidence of enhanced coupling with respect to tamped explosions of the same yields in salt, more consistent with what would be expected for explosions in water. The observed seismic source characteristics of these explosions have been systematically investigated through a detailed series of finite difference simulations in which the complex, nonlinear interactions between the shock effects in both the water and surrounding salt have been explicitly modeled. The results of these theoretical simulations indicate that the dynamic interactions between the expanding and contracting steam bubbles generated by the explosions in water and the shock-wave reflections from the cavity wall dominate the most prominent yield dependent features of the seismic source functions observed from these tests. Consequently, the seismic data observed from these tests do not appear to provide abundant new information which can be used to further constrain the low-pressure salt equation of state required for theoretical simulations of the cavity decoupling evasion scenario. However, these data do provide unique insight into the seismic effects of repeated nuclear explosions in a cavity over a range of energy release encompassing a factor of 50 in yield, and this insight may prove to be important in the evaluation of the feasibility of some proposed cavity decoupling scenarios.

Acknowledgments

The authors appreciatively thank J. L. Stevens for his assistance in carrying out the finite difference simulations. This research was sponsored by the U.S. Air Force Phillips Laboratory under Contract F19628-93-C-0126.

References

Adushkin, V. V., Sultanov, D. D., Kitov, I. O., and Kuznetsov, O. P. (1992), *Overview of the Experimental Data and Theoretical Simulations of Underground Nuclear Explosions Decoupled by Large Air Filled Cavities*, Rep. Russian Acad. Sci. *327*(1).

Baker, J., Peyton, S., Brown, F., and Champion, J. (1988), *Computational Equation of State – 1986 AQUA Water*, S-CUBED Technical Report SSS-TR-88-9680.

Cole, R. H., *Underwater Explosions* (Princeton University Press, Princeton, New Jersey 1948).

Evernden, J. F. (1970), *Magnitude versus Yield of Explosions*, J. Geophys. Res. *75*, 1028.

GLASSTONE and DOLAN, Effects of Nuclear Weapons (Government Printing Office, Washington, D. C. 1977).

KITOV, I. O., SULTANOV, D. D., ADUSHKIN, V. V., KOSTUCHENKO, V. N., KUZNETSOV, O. P., KAAZIK, P. B., NEDOSHIVIN, N. I., and RUBINSHTEIN, H. D. (1995), *Analysis of the Seismic Characteristics of U.S. and Russian Coupled and Cavity Decoupled Explosions in Salt*, Institute for the Dynamics of the Geospheres, Final Report 911412, August.

MURPHY, J. R., and BARKER, B. W. (1994), *Seismic Identification Analyses of Cavity Decoupled Nuclear and Chemical Explosions*, S-CUBED Technical Report SSS-TR-94-14399 (PL-TR-94-2036), ADA280947.

MURPHY, J. R., KITOV, I. O., STEVENS, J. L., SULTANOV, D. D., BARKER, B. W., RIMER, N., and FRIEDMAN, M. C. (1994), *Analyses of the Seismic Characteristics of U.S. and Russian Cavity Decoupled Explosions*, Rep. PL-TR-94-2295, Phillips Laboratory, ADA292881.

MURPHY, J. R., RIMER, N., and STEVENS, J. L. (1996), *Comments on Seismic Decoupling with Chemical and Nuclear Explosions in Salt by L. Glenn and P. Goldstein*, J. Geophys. Res. *101*, 346.

MURPHY, J. R., KITOV, I. O., RIMER, N., ADUSHKIN, V. V., and BARKER, B. W. (1997), *Seismic Characteristics of Cavity Decoupled Explosions in Limestone: An Analysis of Soviet High Explosive Test Data*, J. Geophys. Res. *102*, 27,393.

PRITCHETT, J. W. (1971), *An Evaluation of Various Theoretical Models for Underwater Explosion Bubble Pulsation*, Information Research Associates Report to ARPA IRA-TR-2-71, April.

RIMER, N., and CHERRY, J. T. (1982), *Ground-motion Predictions for the Grand Saline Experiment*, S-CUBED Topical Report to VELA Seismological Center, SSS-R-82-5673, VSC-TR-82-25, July.

SPRINGER, D., DENNY, M., HEALY, J., and MICKEY, W. (1968), *The Sterling Experiment: Decoupling of Seismic Waves by a Shot-generated Cavity*, J. Geophys. Res. *73*, 5995.

STEVENS, J. L., MURPHY, J. R., and RIMER, N. (1991), *Seismic Source Characteristics of Cavity Decoupled Explosions in Salt and Tuff*, Bull. Seismol. Soc. Am. *81*, 1272–1291.

SULTANOV, D. D., MURPHY, J. R., and RUBINSHTEIN, Kh. D. (1999), *A Seismic Source Summary for Soviet Peaceful Nuclear Explosions*, Bull. Seismol. Soc. Am. *89*, 675–683.

SYKES, L. R. (1996), *Re-evaluation of Evasion Possibilities for Conducting Nuclear Explosions in Underground Cavities in Former U.S.S.R*, Phillips Laboratory Report PL-TR-96-2073.

(Received May 21, 1991, accepted October 28, 1999)

Pure appl. geophys. 158 (2001) 2123–2171
0033–4553/01/112123–49 $ 1.50 + 0.20/0

Pure and Applied Geophysics

Application of Network-averaged Teleseismic *P*-wave Spectra to Seismic Yield Estimation of Underground Nuclear Explosions

J. R. MURPHY[1] and B. W. BARKER[1]

Abstract—A set of procedures is described for estimating network-averaged teleseismic *P*-wave spectra for underground nuclear explosions and for analytically inverting these spectra to obtain estimates of m_b/yield relations and individual yields for explosions at previously uncalibrated test sites. These procedures are then applied to the analyses of explosions at the former Soviet test sites at Shagan River, Degelen Mountain, Novaya Zemlya and Azgir, as well as at the French Sahara, U.S. Amchitka and Chinese Lop Nor test sites. It is demonstrated that the resulting seismic estimates of explosion yield and m_b/yield relations are remarkably consistent with a variety of other available information for a number of these test sites. These results lead us to conclude that the network-averaged teleseismic *P*-wave spectra provide considerably more diagnostic information regarding the explosion seismic source than do the corresponding narrowband magnitude measures such as m_b, M_s and $m_b(L_g)$, and, therefore, that they are to be preferred for applications to seismic yield estimation for explosions at previously uncalibrated test sites.

Key words: Nuclear explosions, yield estimation, *P*-wave spectra.

Introduction

One of the results of the intensive U.S. research program which was initiated to support monitoring of the Threshold Test-Ban Treaty (TTBT) was the development of a new capability to determine stable estimates of network-averaged teleseismic *P*-wave spectra for underground nuclear explosions and the subsequent development of procedures for analytically inverting these spectra to derive estimates of explosion yield and other source parameters (MURPHY *et al.*, 1989; MURPHY, 1989). This constituted an important advance in that previous seismic yield estimation procedures were based on empirical correlations between narrowband magnitude measures (i.e., m_b, M_s, $m_b(L_g)$) and explosion yield, and such measures are not well-suited for resolving possible trade-offs between source and propagation path effects at uncalibrated test sites. The broader band network-averaged *P*-wave spectra, on the other hand, provide additional frequency-dependent information which can be used to quantitatively evaluate such trade-offs and their possible effects on test site

[1] Maxwell Technologies, Inc., Reston Geophysics Office, 11800 Sunrise Valley Drive, Suite 1212, Reston, VA 20191, USA. E-mail: jrm@MAXwell.com

magnitude bias estimates. Seismic yield estimation procedures based on these network-averaged spectra were subsequently developed and applied to the estimation of m_b/yield relations for underground nuclear explosions at the Soviet test sites at Shagan River, Degelen Mountain, Novaya Zemlya and Azgir, the French Sahara and Mururoa test sites and the Chinese Lop Nor test site. The objectives of this paper are to document these procedures and summarize the principal results of their application to this wide variety of different nuclear test sites.

Overview of the Spectral Analysis Procedures

Before proceeding to a consideration of specific test sites, it is appropriate to review the definition of the network-averaged *P*-wave spectra and the algorithms that have been developed to determine seismic yield estimates from these spectra. As was indicated by MURPHY *et al.* (1989), the basic spectral data used in the analysis are obtained by bandpass-filtering the individual short-period teleseismic recordings through a comb of Gaussian filters spaced at intervals of 0.25 Hz. In order to maintain constant bandwidth, the filter Q values are taken to be linearly related to filter center frequency, f_c, with $Q = 12f_c$, producing frequency resolution of approximately 0.1 Hz and time resolution of approximately 2 seconds. The spectral

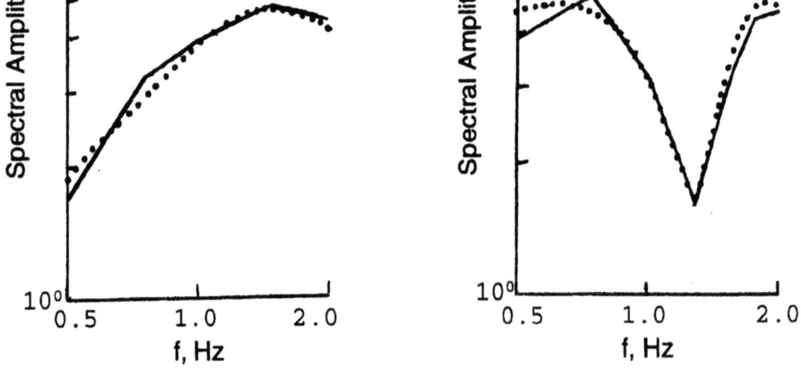

Figure 1
Comparison of normalized observed and best-fitting theoretical network-averaged *P*-wave spectra (left) and *pP* (right) for the NTS SLED explosion.

amplitude estimates are derived from the instrument-corrected, time-domain envelope functions produced by this bandpass operation. In particular, the *P*-wave spectrum is defined as the amplitude of the envelope maximum in each band in a 5-second window following *P* onset. A corresponding noise estimate is obtained by taking the average amplitude of the envelope maxima in each band over a 10-second window preceding the *P*-wave arrival time, and generally only those *P*-wave spectral amplitudes associated with signal-to-noise ratios of greater than 2 have been retained for subsequent analysis. For Nevada Test Site (NTS) explosions, this requirement limits the usable frequency band to about 0.5 to 2.5 Hz and, since NTS defines the calibration base used to estimate yields at other test sites, the m_b/yield analyses described in the following sections are confined to this same short-period band.

The network-averaging procedures applied to these single-station spectral amplitudes are patterned very closely after those commonly employed in the derivation of network-averaged m_b values for explosions. In particular, the logarithms of the spectral amplitudes are corrected for epicentral distance effects (i.e., in this case normalized to a reference epicentral distance of 60°) using the VEITH-

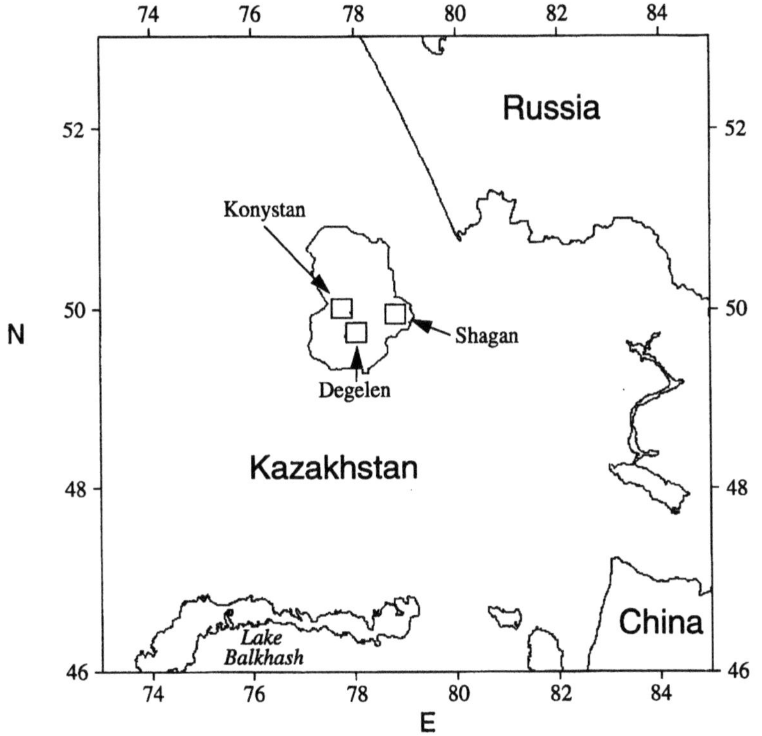

Figure 2
Map location of the Soviet Semipalatinsk test site showing the approximate outlines of the three areas used to conduct underground nuclear tests.

CLAWSON (1972) corrections which were shown by MURPHY *et al.* (1989) to be applicable over the entire short-period band extending from 0.5 to 2.5 Hz. These corrected spectral amplitudes in each frequency band for explosions at a given test site are then processed through the least-squares matrix factorization (LSMF) algorithm originally formulated by DOUGLAS (1966). In this approach, data recorded on a network of stations from explosions at a fixed test site are processed through a two-way linear regression model to simultaneously determine average station correction factors and station-corrected, network-averaged *P*-wave spectra under the constraint that the station correction factors at each frequency sum to zero (MURPHY *et al.*, 1989). The advantage of LSMF relative to direct averaging is that it provides a consistent framework for analyzing data on networks of stations for which the individual stations have recorded different numbers of the events under consideration. Roughly speaking, it provides, for each event, the best approximation to the network-averaged spectrum that would have been obtained if that event had been recorded by every station in the network. That is, any consistent frequency dependent station effects are accounted for in the estimation of the network-averaged values for explosions recorded by different subsets of the network stations.

In order to obtain a calibrated reference level for applications at uncalibrated test sites, the procedures described above were initially applied to teleseismic *P*-wave data

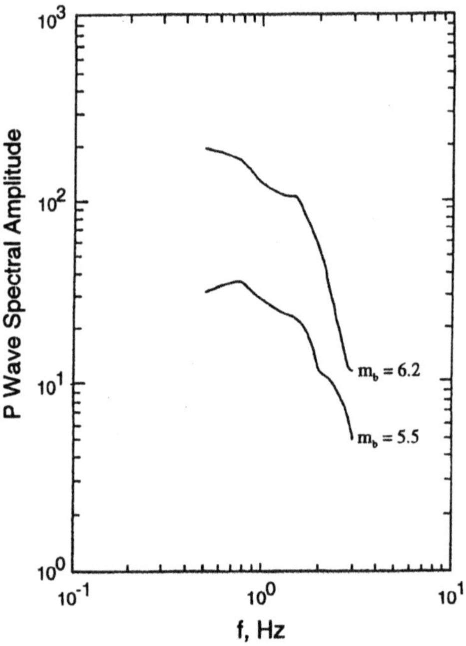

Figure 3
Comparison of predicted network-averaged *P*-wave spectra for nominal Shagan River explosions of $m_b = 6.2$ and $m_b = 5.5$.

recorded from a sample of 60 NTS explosions of known yield. These explosions were all detonated in good coupling media below the water table and encompass a yield range extending from about 50 to 1400 kt, with an associated range in m_b extending from 5.2 to 6.5. *P*-wave spectra were estimated from data recorded on a well-distributed network consisting of 50 teleseismic stations, each of which had recorded 5 or more of the selected explosions. The network-averaged spectra resulting from the LSMF analysis of these data were then used to derive interpolative scaling relations of the form:

$$m(f) = \alpha(f) + \beta(f)m_b \ , \tag{1}$$

where $m(f)$ denotes the logarithm of the network-averaged *P*-wave spectral amplitude at each frequency f, m_b denotes the corresponding network-averaged body-wave magnitude and $\alpha(f)$, $\beta(f)$ are frequency-dependent scaling coefficients which were estimated by performing separate least-squares regression analyses on the network-averaged spectral amplitude data in each frequency band. It should be noted that the linear approximation employed in equation (1) is not strictly applicable in situations where the seismic source function corner frequency migrates

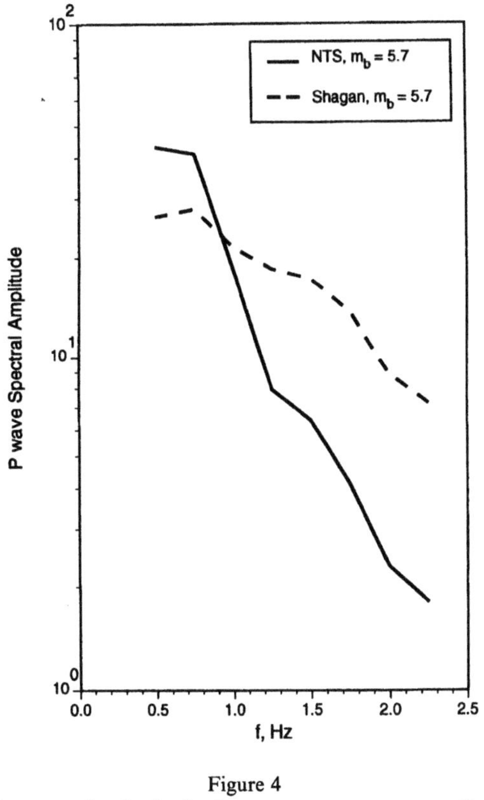

Figure 4
Comparison of network-averaged teleseismic *P*-wave spectra corresponding to nominal m_b = 5.7 explosions at the NTS and Shagan test sites.

through the short-period band as a function of explosion yield. However, it has been found that departures from linearity are small with respect to the typical data scatter (MURPHY *et al.*, 1989). Thus, while the inferred linear scaling relations may not be reliable for extrapolations outside the range of experience, they have been found to be adequate for interpolation purposes. In any case, given comparable scaling relations for any other uncalibrated test site, the test site m_b bias as a function of yield can be estimated using the following simple model. Let $C(f)$ denote the average NTS network-averaged spectrum evaluated at $m_b = 5.7$ (i.e., $W \approx 150$ kt) using equation (1) and $\tilde{\alpha}(f)$, $\tilde{\beta}(f)$ denote the best-fit scaling coefficients derived from data measured from explosions at the uncalibrated test site. Then, assuming that the differences in the average spectral composition of the P waves observed from these two test sites can be accounted for by the effects of a conventional, constant t^* anelastic attenuation operator (MURPHY, 1989), we wish to solve for the specific m_b and differential attenuation values $(\tilde{m}_b, \Delta t^*)$ for which

$$\left[10^{\tilde{\alpha}(f)+\tilde{\beta}(f)\tilde{m}_b}\right] e^{-\pi f \Delta t^*} = 10^{C(f)} \ . \tag{2}$$

Taking the logarithm of both sides and re-arranging gives

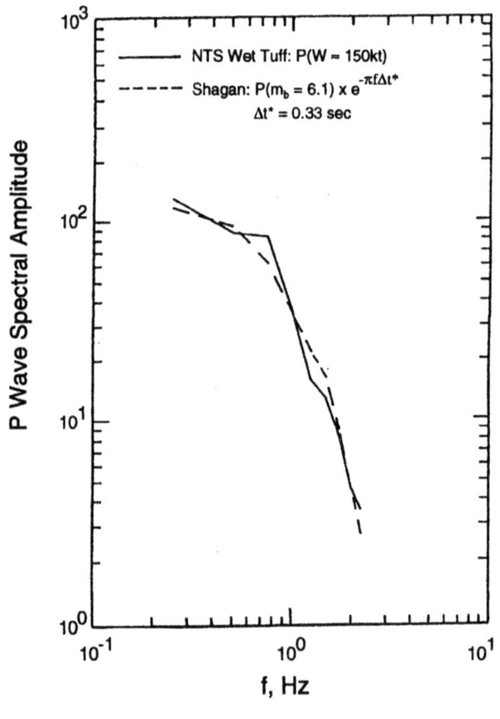

Figure 5
Comparison of the predicted network-averaged P-wave spectrum for an NTS explosion with $m_b = 5.7$ (i.e. $W \approx 150$ kt) with the scaled network-averaged P-wave spectrum for an equivalent explosion at the Shagan River testing area.

$$\frac{\tilde{\alpha}(f) - C(f)}{\pi f \log e} = -\frac{\tilde{\beta}(f)}{\pi f \log e}\tilde{m}_b + \Delta t^* \tag{3}$$

which is of the form $Y_i = mX_i + b$ with

$$Y_i = \frac{\tilde{\alpha}(f_i) - C(f_i)}{\pi f_i \log e} \quad X_i = \frac{\tilde{\beta}(f_i)}{\pi f_i \log e} \tag{4}$$

with $m = \tilde{m}_b$ and $b = \Delta t^*$. Thus, given the scaling laws which predict the network-averaged *P*-wave spectra as a function of m_b for explosions at an uncalibrated test site, $\tilde{m}_b$ and Δt^* can be found by an application of linear least-squares analysis to obtain an estimate of the test site magnitude bias with respect to NTS (i.e., $\Delta m_b = \tilde{m}_b - 5.7$) for explosions with yields of around 150 kt. Clearly, this same procedure can be repeated for a range of yield values to obtain an estimate of the average m_b/yield relation for the uncalibrated test site.

It is evident from the above discussion that the traditional m_b measure plays a central role in yield estimation and that it serves as a point of reference in the network-averaged *P*-wave spectral analyses. Therefore, it is appropriate to briefly summarize the origins of the m_b values which will be referred to throughout the remainder of the paper. One concern in the use of m_b for yield estimation purposes is

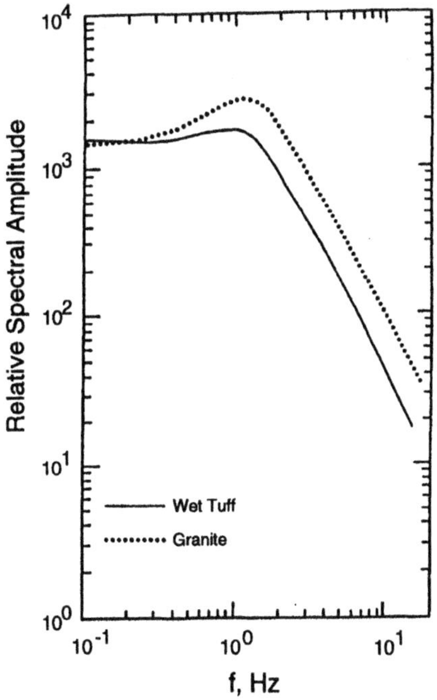

Figure 6
Comparison of Mueller/Murphy teleseismic *P*-wave source spectra for 150 kt explosions in wet tuff and granite emplacement media.

that the network-averaged values determined from different networks of stations might contain network bias which could affect the comparisons of different test sites. Fortunately, as will be illustrated below for the NTS and French Sahara test sites, it has been found that m_b values (as well as the corresponding *P*-wave spectra) determined from large, well-distributed networks of teleseismic stations, appear to be remarkably insensitive to the specific sample of stations employed in the averaging process. Therefore, for the purposes of the present analyses we have employed network-averaged m_b values determined by us through LSMF analyses of large samples of single station m_b values, which were carefully collected for each test site in conjunction with the Yield Estimation Experiment sponsored by the U.S. Air Force (MURPHY *et al.*, 1991). The yield estimation results based on these m_b values which are presented in the following sections have been found to be essentially identical to the corresponding results obtained using other carefully determined classified and unclassified m_b values for the same explosions, which has led us to conclude that these results are robust with respect to small variations in m_b determination procedures.

While the algorithm described above can be used to estimate an average m_b/yield relation for an uncalibrated test site, it provides no mechanism for estimating yields

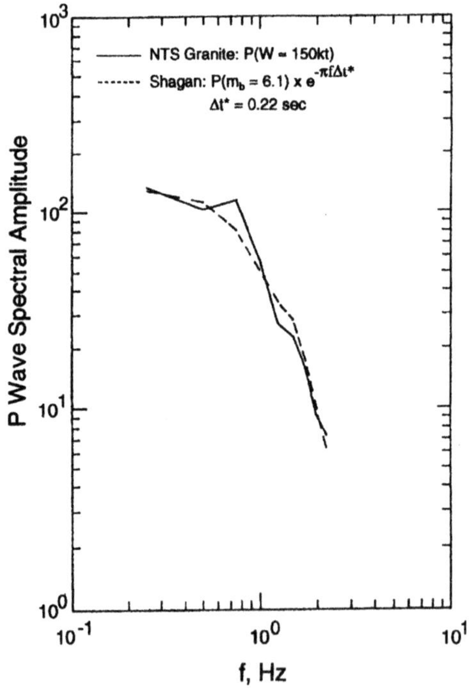

Figure 7
Comparison of the predicted network-averaged *P*-wave spectrum for an NTS explosion with a yield of about 150 kt in granite with the scaled network-averaged *P*-wave spectrum for an equivalent explosion at the Shagan River testing area.

of individual explosions from the observed network-averaged spectra. Consequently, an alternate, model-based yield estimation procedure has been developed which is based on the formal analytic inversion algorithm described by MURPHY (1989). This process entails correcting the observed spectrum for propagation effects, including both geometrical and anelastic attenuation, as well as free surface interaction effects, and then determining the yield of the Mueller/Murphy seismic source function (MUELLER and MURPHY, 1971; MURPHY, 1977) which provides the best average fit to the amplitude level and shape of the corrected spectrum. In order to estimate the teleseismic propagation operators needed to implement this procedure, MURPHY (1989) analyzed network-averaged *P*-wave spectra from a sample of 20 explosions with known yields and depths of burial which were detonated below the water table at NTS. The observed spectrum for each explosion was first divided by the Mueller/Murphy saturated tuff/rhyolite seismic source function corresponding to the associated yield and depth of burial values after which the resulting source-normalized spectra for the 20 explosions were then averaged to obtain an estimate of the frequency-dependent propagation path transfer function. This transfer function,

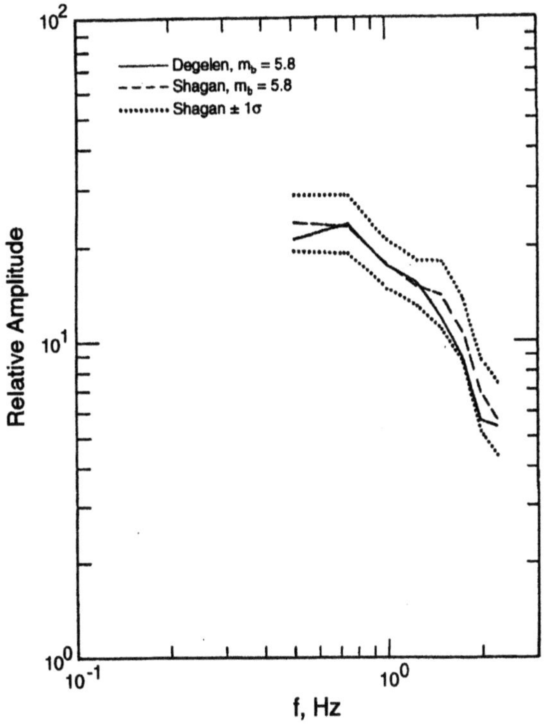

Figure 8
Comparison of network-averaged *P*-wave spectra corresponding to m_b = 5.8 explosions at the Degelen Mountain and Shagan River test sites.

denoted $T(f)$ was then decomposed into a product of elastic and anelastic propagation terms in the form

$$T(f) = G \cdot e^{-\pi f t^*} , \tag{5}$$

where G denotes the frequency-independent effect associated with elastic P-wave propagation to the reference epicentral distance of 60°, which is assumed to be independent of the test site location, and t^* is the anelastic attenuation parameter which will, in general, be different for different test sites. The analysis of the selected NTS data indicated that the value of t^* applicable to that site is approximately 0.75 seconds (MURPHY, 1989). This serves as the calibrated reference value that is modified for application to other test sites using the Δt^* values with respect to NTS obtained from the corresponding solutions to equation (3).

Once the observed network-averaged spectrum has been corrected for propagation path effects, it can be analyzed using an interactive software module which has been developed to provide automated estimates of explosion yield and other source parameters. The first step in this process is to determine the yield of the Mueller/Murphy seismic source function which provides the best average fit to the absolute

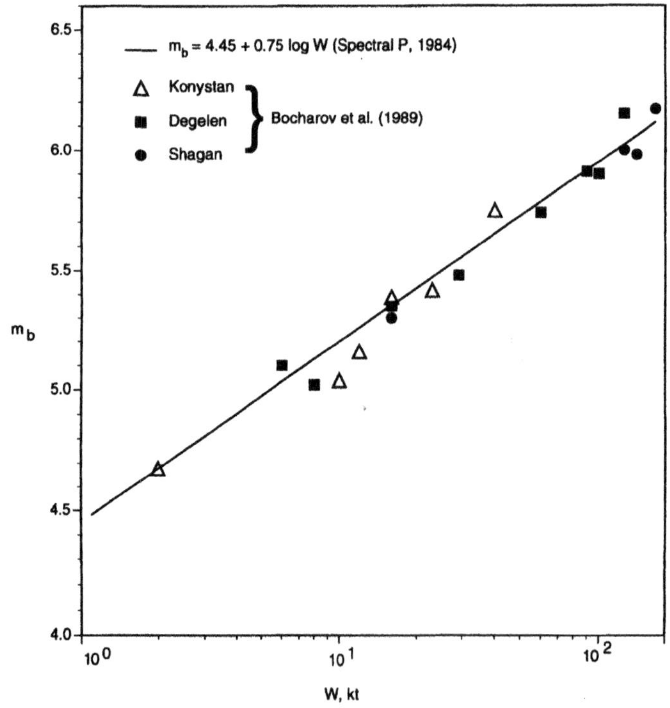

Figure 9

Comparison of the Semipalatinsk test site m_b/yield relation inferred from statistical analyses of network-averaged P-wave spectra (solid line) with selected m_b/yield values for that test site which were subsequently published by BOCHAROV et al. (1989).

amplitude level and shape of the propagation path corrected spectrum over the frequency band 0.5 to 2.5 Hz. This best-fitting source function is then divided out of the corrected spectrum and the resulting frequency-dependent function is automatically processed to obtain an estimate of the characteristics of the surface reflected *pP* phase. This is accomplished by fitting the corrected and source normalized observed spectrum to a simple *pP* model, based on the elastic approximation in which the effect of *pP* on the normalized observed spectrum is represented as

$$\left|\frac{P(\omega)+pP(\omega)}{P(\omega)}\right| = \left|1 - Ae^{i\omega t_0}\right| = \sqrt{1+A^2-2A\cos\omega t_0}\ , \qquad (6)$$

where A denotes the ratio of *pP* to *P*-wave amplitude, t_0 denotes the delay time of *pP* with respect to *P* and it is assumed that *pP* has the same time dependence as *P*. Although equation (6) represents the effects of interference between the direct *P* and *pP* arrivals on the normalized *P*-wave spectrum, it depends only on the *pP* parameters and, consequently, best-fitting solutions to (6) will be referred to simply as "*pP* models" in the following discussions. The approximate automatic algorithm proceeds by finding the frequency (f_0) corresponding to the minimum amplitude

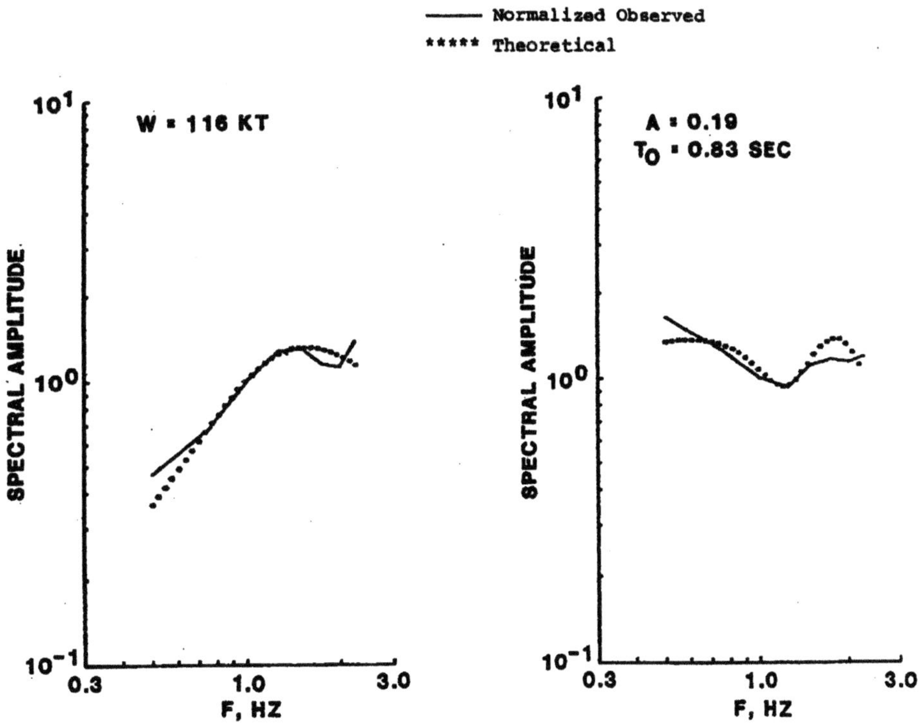

Figure 10
Comparison of normalized observed and best-fitting theoretical network-averaged *P*-wave spectra (left) and *pP* (right) for the Soviet JVE explosion of 9/14/88.

(A_{min}) of the normalized observed spectrum in the 0.5 to 2.5 Hz band. It then finds the maximum spectral amplitudes in the bands $f < f_0$ (denoted $A_{max(1)}$) and $f > f_0$ (denoted $A_{max(2)}$) and computes

$$A_{max} = \frac{A_{max(1)} + A_{max(2)}}{2} \ . \tag{7}$$

Initial estimates of the pP parameters are then computed as

$$A = \frac{A_{max} - A_{min}}{A_{max} + A_{min}} \qquad t_0 = \frac{1}{f_0} \ . \tag{8}$$

Now since the spectral data are discretely sampled, the estimated frequency of minimum amplitude, f_0, may not coincide with the true minimum and, consequently, a search is conducted over the frequency range $f_0 - 0.15$ to $f_0 + 0.15$ Hz. For each value of f_0 in this range, the estimated amplitude ratio is varied over the range $A - 0.10$ to $A + 0.10$ and the final, optimum values of f_0 and A are taken to be those which minimize the RMS difference between the logarithms of the normalized

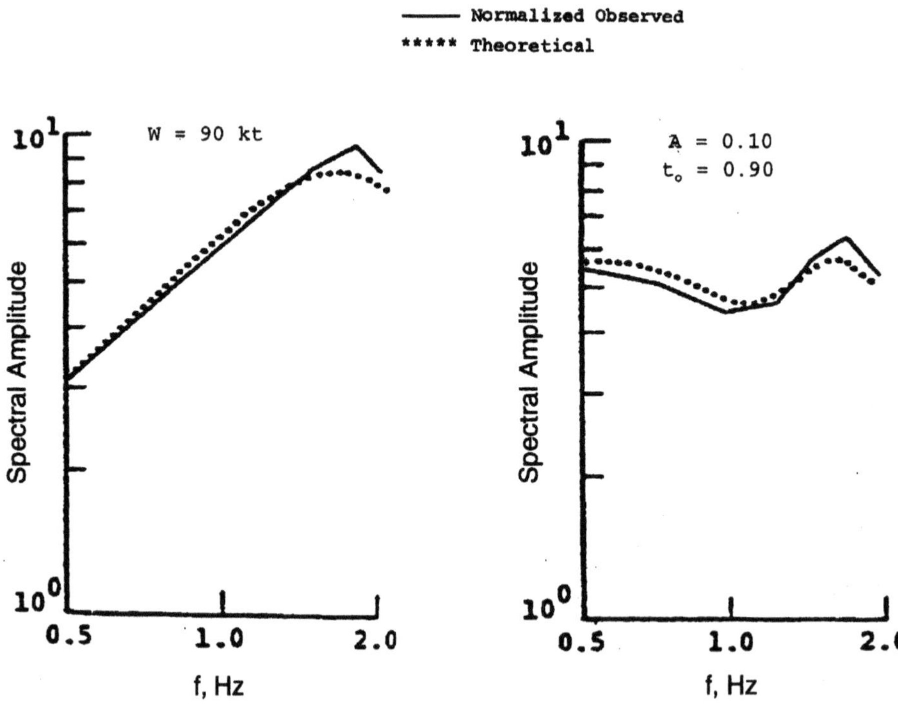

Figure 11
Comparison of normalized observed and best-fitting theoretical network-averaged P-wave spectra (left) and pP (right) for the Degelen Mountain explosion of 2/26/67.

observed and predicted spectral amplitudes. The final step in the automatic processing provided by the software module is that this optimum *pP* frequency-dependent operator is divided out of the propagation path corrected observed spectrum and the result is again used to determine the yield of the seismic source function which provides the best fit to the corrected data. At this point, the system provides the operator with the capability to interactively modify the automatically determined yield and *pP* parameters in order to explore the effects of these variables on the derived solution and, if necessary, refine it.

An example of the results of the above process is shown in Figure 1 for the $m_b = 5.9$ SLED explosion at NTS. The left side of this figure shows the final fit and associated yield estimate corresponding to the displayed model fit to the propagation corrected observed network-averaged spectrum for SLED which has been divided by the best-fitting *pP* model shown on the right side of this figure. Note that, although the data for this explosion are described remarkably well by the simple model described above, the inferred free surface reflection coefficient for *pP* (i.e., $A = 0.5$) is significantly smaller than the value of about 0.8 which would be predicted by simple linear, elastic theory. This is a common observation for nuclear explosions detonated at normal containment depths, and has been interpreted by a number of investigators as an indication that nonlinear surface interaction phenomena such as spall perturb the surface reflected phase. Nevertheless, as is indicated in Figure 1, the simple model of equation (5) has been found to account quite well for observed spectral modulation effects in most cases and, consequently, we will continue to refer to this

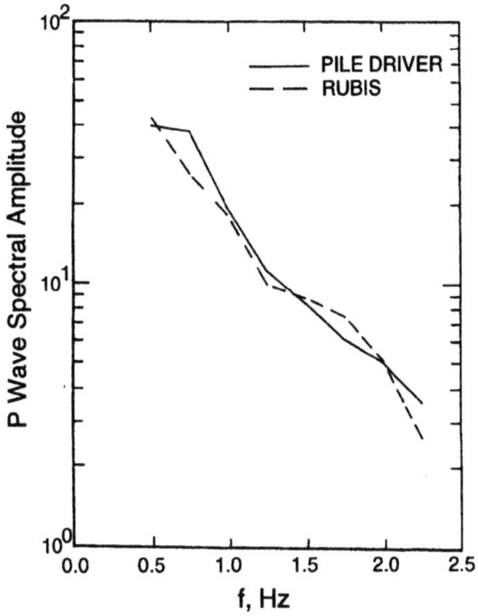

Figure 12
Comparison of network-averaged *P*-wave spectra for the NTS PILE DRIVER and French Sahara RUBIS explosions of comparable yield in granite.

arrival as *pP* and to model its effect on the *P*-wave spectra using a "quasi-linear" model with a reduced, effective free surface reflection coefficient (MURPHY, 1989).

The above example illustrates some of the advantages of using *P*-wave spectra rather than the traditional narrowband m_b, M_s and $m_b(L_g)$ magnitude measures for yield estimation purposes. That is, while the different effects of explosion yield, anelastic attenuation and *pP* on the observed data cannot be resolved from narrowband amplitude measurements, their characteristic frequency-dependent effects on the corresponding *P*-wave spectra can be explicitly modeled and corrected for to obtain spectral yield estimates which account for the effects of these variables. This is not to say, of course, that the spectral analysis procedures are completely immune from possible trade-offs between the different effects. Thus, for example, it is certainly conceivable that both the frequency-dependent source coupling and anelastic attenuation at a previously uncalibrated test site differ from those of the NTS reference in such a manner that they coincidentally offset each other over the short-period band. However, in the following analyses we will generally prefer the

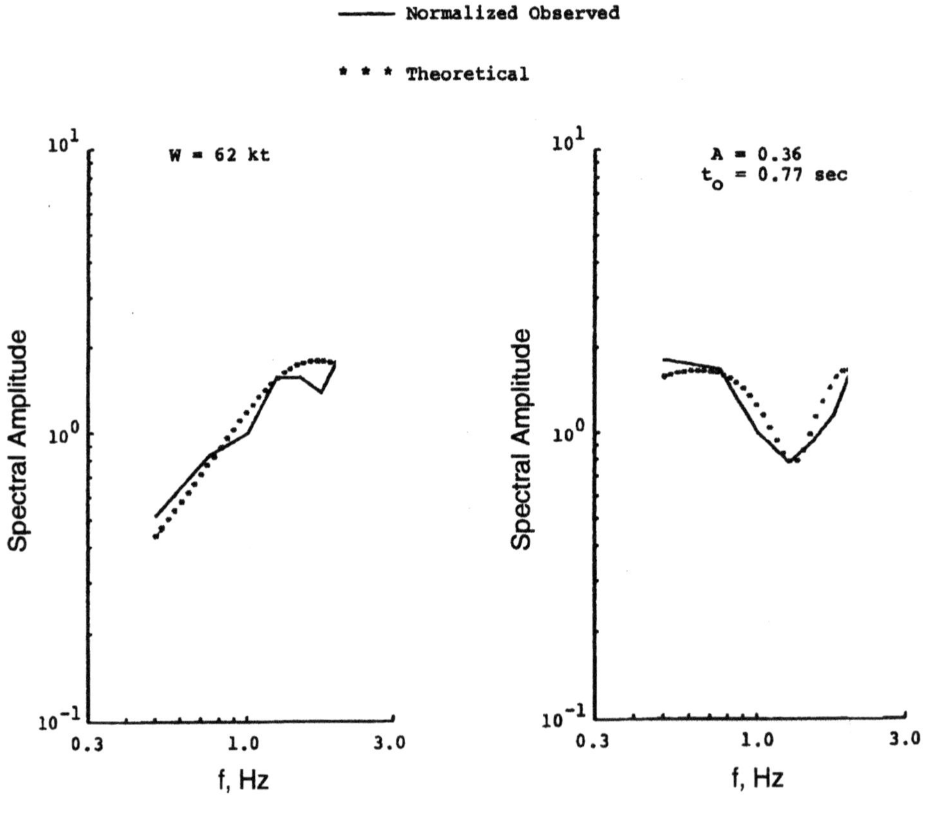

Figure 13
Comparison of normalized observed and best-fitting theoretical network-averaged *P*-wave spectra (left) and *pP* (right) for the 62 kt NTS explosion PILE DRIVER.

simplest explanations which are consistent with the observations. Thus, for example, if the network-averaged *P*-wave spectra corresponding to explosions of the same m_b value at two different test sites are essentially the same with regard to amplitude level and shape over the entire short-period band, this will be interpreted as an indication that both the explosion source coupling and anelastic attenuation at the two sites must be very similar. As will be seen in the following sections, this interpretation has generally proven to be correct in applications to test sites where significant explosion yield calibration data are available.

In the following sections the spectral analysis procedures described above will be further illustrated through applications to data recorded from underground explosions conducted at a wide variety of different nuclear test sites.

Soviet Semipalatinsk Test Site

The map location of the former Soviet Union (FSU) Semipalatinsk test site in Central Asia is shown in Figure 2, where it is indicated that three separate areas of

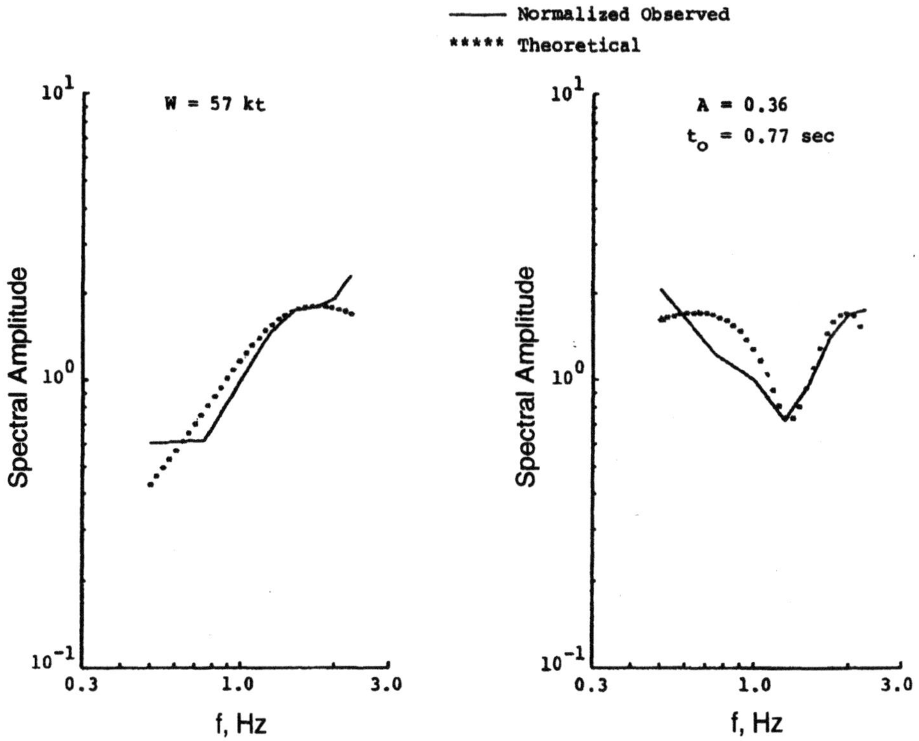

Figure 14
Comparison of normalized observed and best-fitting theoretical network-averaged *P*-wave spectra (left) and *pP* (right) for the French Sahara explosion RUBIS.

the site (i.e., Shagan River, Degelen Mountain and Konystan) were used to conduct underground nuclear tests. The initial focus of the TTBT research program was on the Shagan River testing area, which was the location where the largest and most controversial tests were conducted after the U.S. and FSU agreed to abide by the yield limit of 150 kt specified in the treaty. For purposes of the network-averaged *P*-wave spectral analysis, a sample of short-period, teleseismic data recorded from 38 Shagan River explosions at stations of a 30-station network was assembled and analyzed. The explosions encompassed a range of m_b extending from about 5.0 to 6.2 and only stations that recorded 5 or more of the explosions were used in the analysis. These data were processed in the identical fashion as the NTS data discussed above to obtain network-averaged *P*-wave spectral estimates corresponding to each of the 38 selected explosions. As with the NTS data, the resulting Shagan spectral amplitudes were then statistically analyzed to derive interpolative scaling relations as a function of m_b in the form of equation (1). These derived scaling relations were found to be reasonable and generally consistent with what would be expected on the

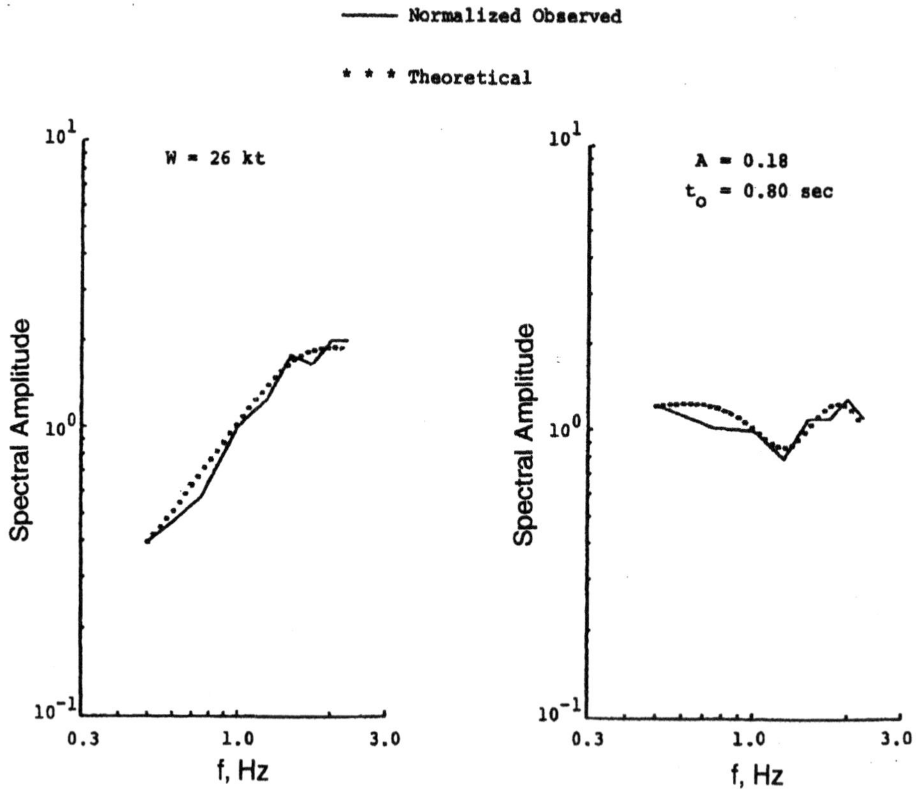

Figure 15
Comparison of normalized observed and best-fitting theoretical network-averaged *P*-wave spectra (left) and *pP* (right) for the French Sahara explosion BERYL.

basis of simple seismic source theory. This observation is illustrated in Figure 3, which shows a comparison of the network-averaged *P*-wave spectra predicted by these scaling relations for nominal Shagan River explosions corresponding to m_b values of 5.5 and 6.2. It can be seen from this example that the predicted spectra show the expected frequency-dependent scaling with source size, scaling directly with m_b at around 1 Hz and increasing more slowly with m_b at higher frequencies.

Given the frequency-dependent m_b spectral scaling relations for explosions at both the NTS and Shagan River test sites, it is possible to directly compare the network-averaged *P*-wave spectra which would be expected on average from explosions of the same m_b at the two test sites. Such a comparison is shown in Figure 4 for an m_b value of 5.7, which corresponds to a yield of about 150 kt at NTS. It can be seen from this figure that the spectral amplitude levels for the two test sites are comparable for frequencies of about 1 Hz, consistent with their common m_b value. However, it is evident that the overall shapes of the two spectra are dramatically different, with the Shagan River spectrum showing lower amplitudes than the NTS spectrum at frequencies below 1 Hz, and progressively higher

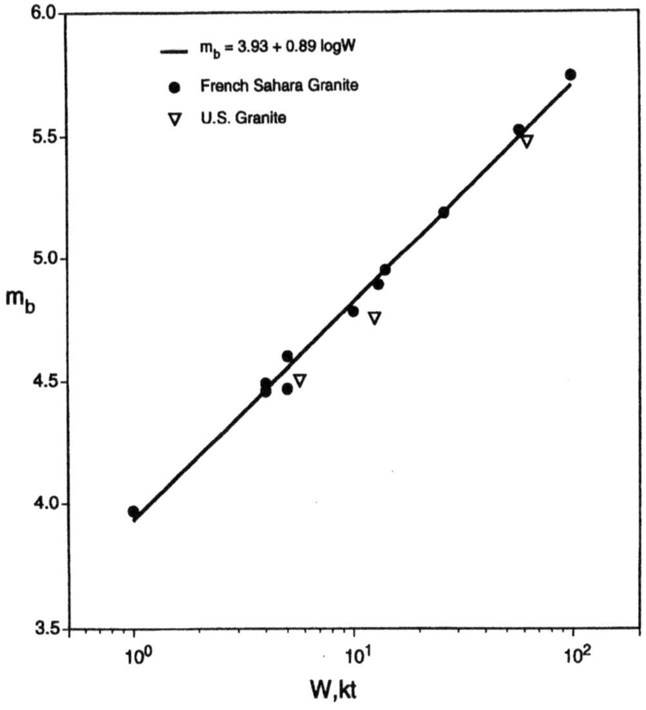

Figure 16
Comparison of the m_b/yield relation inferred from the network-averaged *P*-wave spectral analysis of the French Sahara explosions with the observed m_b/yield values for the three U.S. explosions in granite.

amplitudes at frequencies above 1 Hz, consistent with a significant anelastic attenuation bias between these two test sites.

In an attempt to quantify this apparent test site bias, the inversion algorithm of equation (3) was solved to find the model parameters which best reconcile the average observed NTS and Shagan River explosion spectra. The results of this inversion indicate that a Shagan River explosion with an m_b value of 6.1, together with an associated Δt^* value of 0.33 seconds, provides the best fit to the observed network-averaged P-wave spectrum corresponding to a 150 kt explosion in a saturated tuff/rhyolite emplacement medium below the water table at NTS. The degree to which this model describes the observed spectral differences between these two test sites is indicated in Figure 5, where it can be seen that this combination of $\tilde{m}_b$, Δt^* values reconciles the two network-averaged P-wave spectra to well within the data uncertainties. Thus, these inversion results indicate a magnitude bias of about 0.4 units m_b between the Shagan River and NTS m_b/yield relations at a yield of around 150 kt. Certainly, the above analysis is based on the assumption that the seismic source coupling at Shagan River is comparable to that below the water table at NTS, and, given the differences in the known subsurface geology at the two sites, it is not necessarily obvious that this is the case. That is, the source media at explosion

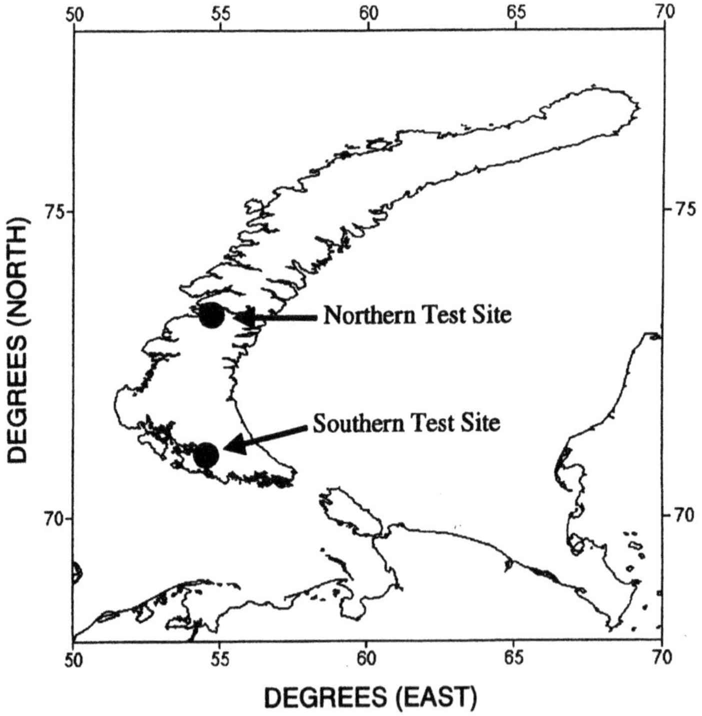

Figure 17
Map locations of the two principal underground nuclear testing areas on Novaya Zemlya.

emplacement depths at the Shagan River site are known to be hard rocks which would appear to be more geologically similar to granite than to the saturated tuff/rhyolite media characteristic of the calibration explosion database at NTS. Thus, it is conceivable that there might be frequency-dependent differences in the *P*-wave source characteristics for explosions of the same yield at the two test sites that could contaminate the estimate of test site bias. One way to test for this possibility would be to directly compare network-averaged *P*-wave spectra from explosions at Shagan River with those from explosions in granite at NTS. Unfortunately, the available sample of NTS granite data (i.e., HARDHAT and PILE DRIVER) is too small to be used as a basis for deriving empirical scaling laws as a function of m_b for direct use in the inversion algorithm. However, the Mueller/Murphy source model does provide approximate seismic source representations for explosions in both wet tuff and granite media (MUELLER and MURPHY, 1971; MURPHY, 1977, 1996), and these can be used to assess the potential effects of differences in seismic source coupling. The estimated teleseismic *P*-wave source functions predicted by this model for explosions with yields of 150 kt in these two media are compared in Figure 6, where it can be seen that the granite source function exhibits somewhat greater overshoot and a higher corner frequency than the corresponding wet tuff source function. In order to

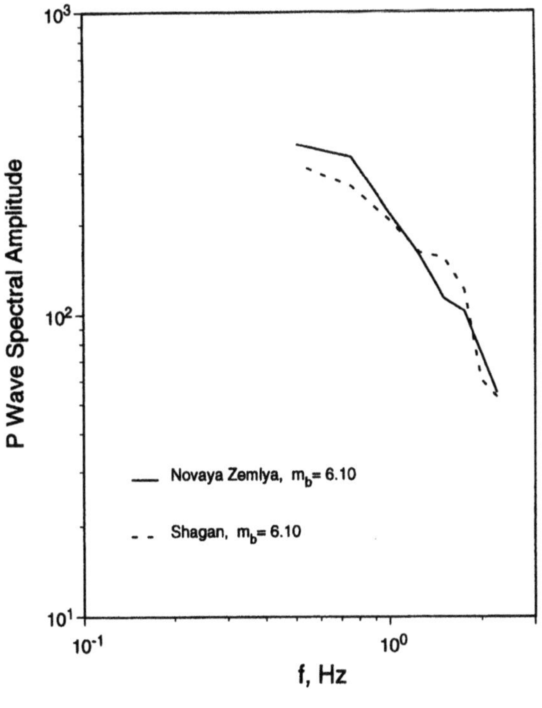

Figure 18
Comparison of network-averaged *P*-wave spectra corresponding to m_b = 6.1 explosions at the Novaya Zemlya and Semipalatinsk test sites.

approximately account for these estimated source-coupling differences, the 150 kt NTS spectrum of Figure 5 was multiplied by the frequency-dependent ratio of the granite to wet tuff source functions shown in Figure 6 to obtain an estimate of the corresponding network-averaged *P*-wave spectrum for a 150 kt explosion in NTS granite. Using this NTS granite spectrum as a reference, best-fitting Shagan River $\tilde{m}_b$ and Δt^* values were redetermined using equation (3), giving $\tilde{m}_b = 6.1$, $\Delta t^* = 0.22$ seconds. That is, the $\tilde{m}_b$ estimate remained unchanged, while the Δt^* decreased to account for the added high frequency energy in the granite source function. The model fit corresponding to this granite solution is shown in Figure 7 where it can be seen that it is quite comparable to that of Figure 5 over the entire short-period band. Therefore, the estimate of test site m_b bias between Shagan River and NTS is fairly insensitive to assumptions concerning the relative seismic source coupling characteristics at the two sites, at least over the range of source variability generally associated with the differences between the saturated volcanic and granite source media at NTS. Consequently, the inversion analysis described above was repeated using NTS network-averaged spectra corresponding to yields of 50 and 300 kt, and it

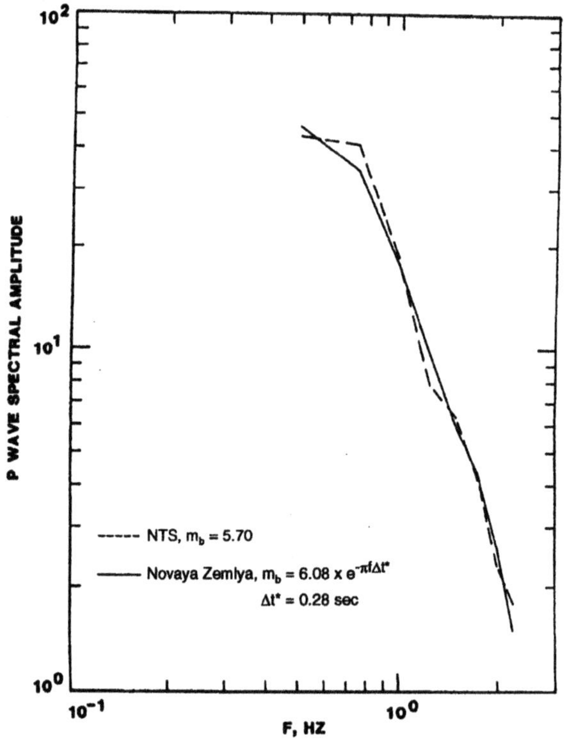

Figure 19
Comparison of the network-averaged *P*-wave spectrum corresponding to a 150 kt NTS explosion (i.e. $m_b = 5.7$) with the scaled network-averaged *P*-wave spectrum for an equivalent explosion at the Novaya Zemlya test site.

was found that the derived test site bias estimate with respect to the NTS m_b/yield relation is essentially independent of yield and consistent with an inferred Shagan River m_b/yield relation of the form

$$m_b = 4.45 + 0.75 \log W \ . \tag{9}$$

The above analysis was next extended to the Degelen Mountain testing area of the Semipalatinsk site through an analysis of data recorded from a sample of 9 Degelen explosions in the m_b range extending from about 5.0 to 6.0. These data were recorded on essentially the same network of stations used in the corresponding analysis of the Shagan River testing area. Once again, network-averaged *P*-wave spectra were estimated for each of the 9 explosions and these spectral amplitude data were used to derive interpolative scaling relations of the form of equation (1) for explosions at Degelen Mountain. The network-averaged teleseismic *P*-wave spectrum predicted for an $m_b = 5.8$ Degelen Mountain explosion using these derived scaling relations is shown in Figure 8, where it is compared with the corresponding Shagan River spectrum predicted for this same value of m_b. It can be seen from this comparison

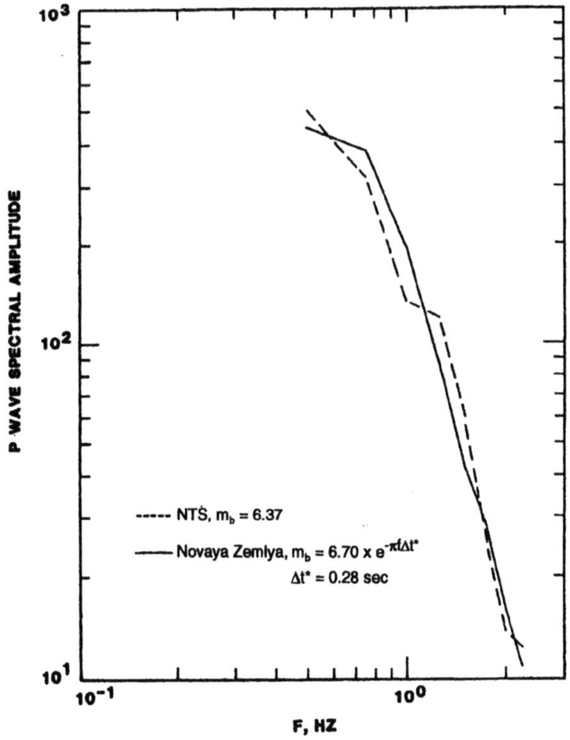

Figure 20
Comparison of the network-averaged *P*-wave spectrum corresponding to a 1000 kt NTS explosion (i.e., $m_b = 6.37$) with the scaled network-averaged *P*-wave spectrum for an equivalent explosion at the Novaya Zemlya test site.

that these two spectral estimates are essentially identical over the entire short-period band. In particular, the estimated Degelen spectrum falls within the plus and minus one standard error of estimate bounds on the Shagan River average, which are shown as dotted lines on this figure. Thus, these *P*-wave spectral data are consistent with the hypothesis that the seismic source coupling and test site bias with respect to NTS are comparable for these two areas on the Semipalatinsk test site. On the basis of these results, the m_b/yield relation given by equation (9) was proposed for application to all underground nuclear explosions at the Semipalatinsk test site. This inference, which was originally formulated in 1984 before any yield values had been released by the FSU, later proved to be remarkably consistent with the calibration information which was subsequently published by BOCHAROV *et al.* (1989). This fact is illustrated in Figure 9 where it can be seen that the m_b/yield relation inferred from the comparisons of network-averaged *P*-wave spectra provides an excellent fit to these subsequently published yields for the three testing areas of the Semipalatinsk test site. It can also be seen from this figure that the m_b/yield values for the Shagan River and Konystan explosions in hard rock are very consistent with those for the Degelen explosions in granite, which suggests that the granite seismic source coupling model is applicable at all three of these testing areas.

As with the NTS analysis described previously, the statistical analysis results summarized above also provide a basis for direct modeling of the observed network-averaged spectra for individual explosions at the Semipalatinsk test site. For

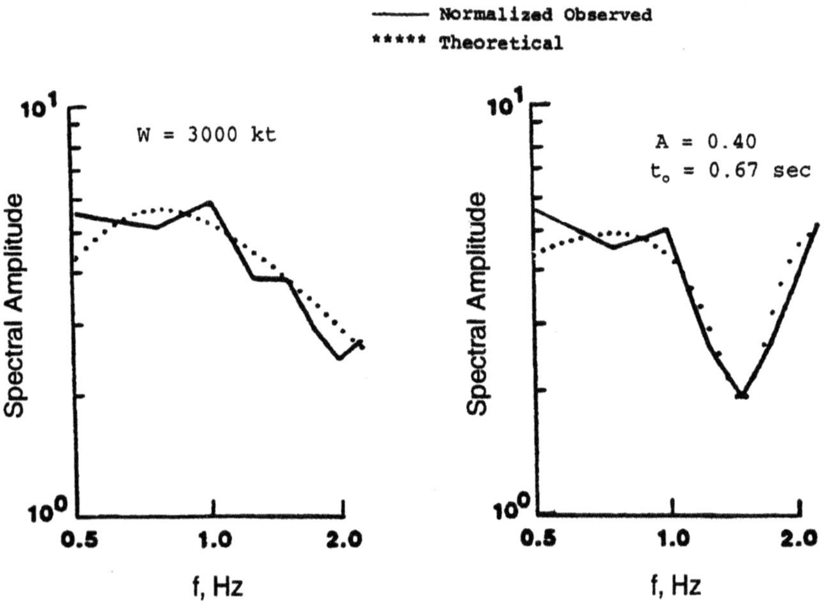

Figure 21
Comparison of normalized observed and best-fitting theoretical network-averaged *P*-wave spectra (left) and *pP* (right) for the southern Novaya Zemlya explosion of 10/27/73.

example, the observed spectrum for the Soviet Joint Verification Experiment (JVE) of 9/14/88 in granite was modeled using the Mueller/Murphy granite seismic source model together with the teleseismic *P*-wave propagation operator determined from the NTS analysis, modified to be consistent with the Δt^* value between NTS and Semipalatinsk determined above (i.e., $\Delta t^* = 0.22$ seconds). The resulting model fit to the observed JVE spectrum is shown in Figure 10, where it can be seen that the agreement between predicted and observed is very good over the short-period band for a source consistent with a yield of 116 kt, which agrees remarkably well with the independent hydrodynamic/seismic range of 110–120 kt quoted by SYKES and EKSTRÖM (1989) for this explosion. A similar comparison of predicted and observed spectra for the $m_b = 6.0$ Degelen Mountain explosion of 2/26/67 is shown in Figure 11. In this case, excellent agreement is obtained for a granite seismic source corresponding to a yield value of 90 kt, which is consistent with the yield range of 20–150 kt reported by BOCHAROV *et al.* (1989) for this explosion and also with the average Semipalatinsk m_b/yield relation of equation (9). Thus, systematic comparisons of network-averaged teleseismic *P*-wave spectra from explosions at the NTS and Semipalatinsk test sites led to a capability to estimate seismic yields for

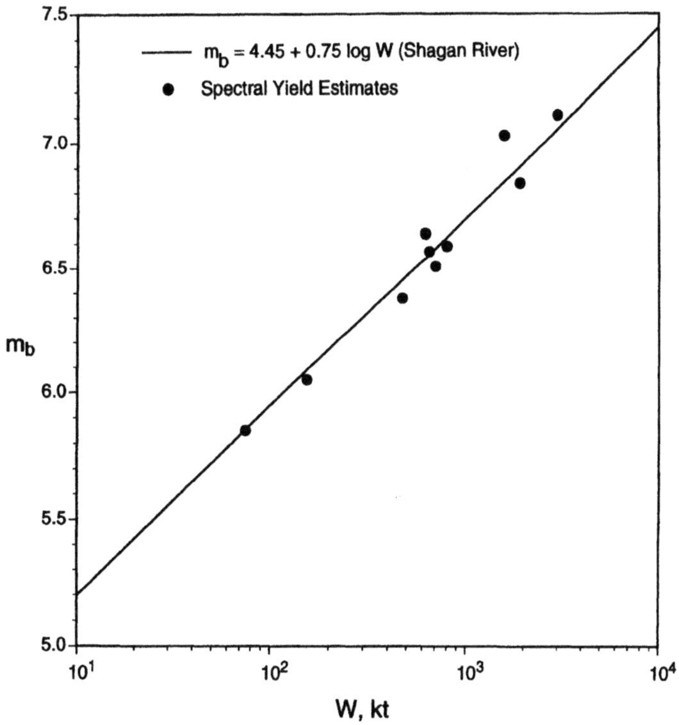

Figure 22
Comparison of spectral yield estimates for selected Novaya Zemlya explosions with the nominal Shagan River m_b/yield relation.

Semipalatinsk explosions which proved to be remarkably consistent with the yield information subsequently published by the FSU. This validation of the derived spectral analysis methodology provided the motivation to conduct similar analyses for other nuclear test sites.

French Sahara Test Site

Between November, 1961 and February, 1966, the French conducted a total of 13 underground nuclear tests in granite at a site in the Hoggar Mountains located in the central Sahara Desert region of southern Algeria (24°N, 5°E). Reported m_b/yield data from these tests proved to be very consistent with the limited U.S. granite data set, as well as with data observed from well-coupled explosions below the water table at NTS. However, despite this apparent consistency, controversy arose over the years concerning the proper interpretation of these observational results. One element of this controversy centered around questions regarding differences in the physical properties of the French and U.S. granites and their possible influence on seismic

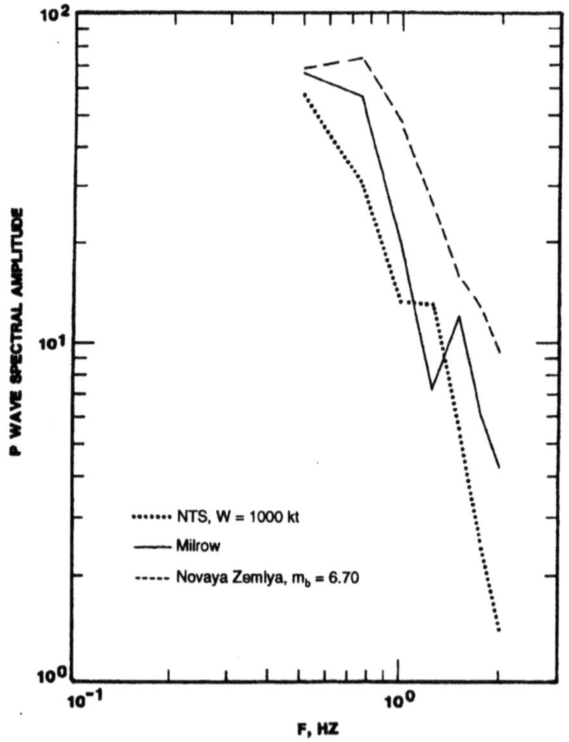

Figure 23
Comparison of network-averaged *P*-wave spectra corresponding to megaton level explosions at the NTS, Amchitka and Novaya Zemlya test sites.

coupling efficiency. In particular, the fact that the scaled cavity radii reported by the French were found to be significantly smaller than those observed from U.S. explosions in granite (HEUZE, 1983) led some to speculate that the corresponding seismic source functions might be quite different at low frequencies (BACHE, 1982). Of course, in order for this hypothesis to be consistent with the observational data, any such differences in seismic coupling would have to be effectively canceled by offsetting differences in upper mantle attenuation beneath the two test sites, at least in the narrow frequency band around 1 Hz, and no convincing evidence of such attenuation differences has ever been presented. Moreover, as was noted previously, narrowband m_b data are poorly suited to evaluating such trade-offs between source and propagation path effects.

In an attempt to resolve this controversy, short-period, teleseimic *P*-wave data recorded from the 11 largest of these French Sahara explosions on a network of 35 selected stations were processed and analyzed to estimate network-averaged *P*-wave spectra for each event. These explosions encompass a range of m_b extending from about 4.0 to 5.7, and each of the selected stations recorded 3 or more of the 11 explosions. The first step in the analysis of the resulting network-averaged *P*-wave spectra was to directly compare them with corresponding spectra from U.S. granite

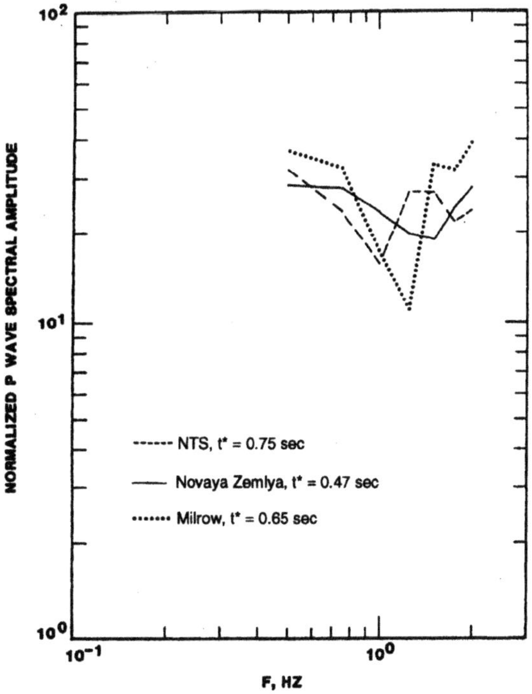

Figure 24
Comparison of source (Mueller/Murphy) and path (t^*) normalized network-averaged *P*-wave spectra corresponding to megaton level explosions at the NTS, Amchitka and Novaya Zemlya test sites.

explosions of approximately the same yield. Such a comparison is presented in Figure 12 for the 58 kt French Sahara RUBIS explosion of 10/20/63 and the 62 kt U.S. PILE DRIVER explosion in granite. It can be seen from this figure that these two network-averaged spectra are virtually identical over the entire short-period band extending from 0.50 to 2.25 Hz. This is quite a remarkable observation given that these two explosions were detonated at sites over 10,000 km apart and were recorded on very different networks of stations. More specifically, it supports our hypothesis that network-averaged *P*-wave spectra estimated from data recorded on large, well-distributed networks of teleseismic stations are essentially unbiased. In any case, the comparison shown in Figure 12 strongly supports the hypothesis that both the explosion source coupling and upper mantle attenuation (t^*) at the French Sahara test site must be very similar to those for NTS explosions in granite, at least over the short-period band of interest here.

It follows from the above conclusion that if we have a quantitative understanding of the observed network-averaged teleseismic *P*-wave spectra for NTS explosions in granite, then it should be directly applicable to the analysis of the corresponding observed spectra from French Sahara explosions. The extent to which the NTS granite spectra can be described by the model consisting of the Mueller/Murphy granite seismic source function (which was inferred from near-regional seismic

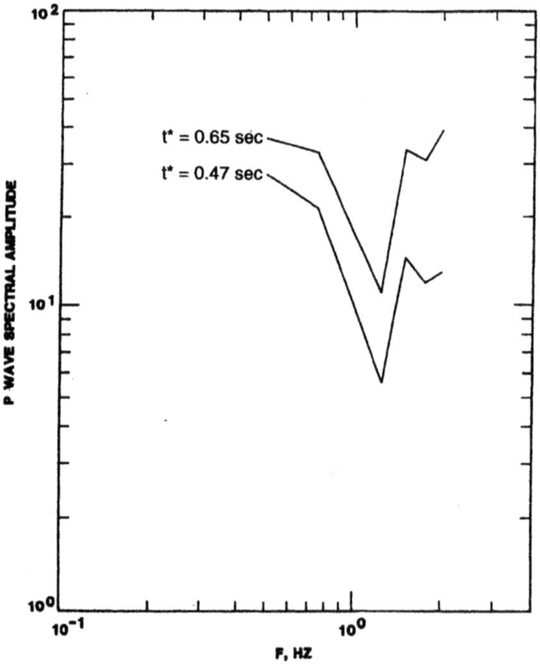

Figure 25
Comparison of source-normalized Milrow spectra which have been corrected for path attenuation effects using the best-fitting Novaya Zemlya (0.47 sec) and Amchitka (0.65 sec) t^* values.

recordings of the U.S. SHOAL explosion) and the elastic propagation transfer function and associated anelastic t^* value inferred from the analysis of data recorded from explosions in tuff/rhyolite below the water table at NTS is illustrated for the PILE DRIVER explosion in Figure 13. It can be seen from this figure that the simple model provides a remarkably good fit to the observed PILE DRIVER spectrum for the announced yield value of 62 kt, confirming the validity of the Mueller/Murphy granite/wet tuff source spectral ratio discussed previously in connection with Figure 6. This exact model was applied to the observed network-averaged *P*-wave spectrum for the French Sahara RUBIS explosion, and the final, best-fitting results are given in Figure 14, where it can be seen that excellent agreement is obtained for a Mueller/Murphy granite source corresponding to a yield of 57 kt, which is essentially identical to the announced yield of 58 kt for this explosion. Therefore, the deterministic NTS granite model explains the observed network-averaged, teleseismic *P*-wave spectrum for this French Sahara explosion to well within the data

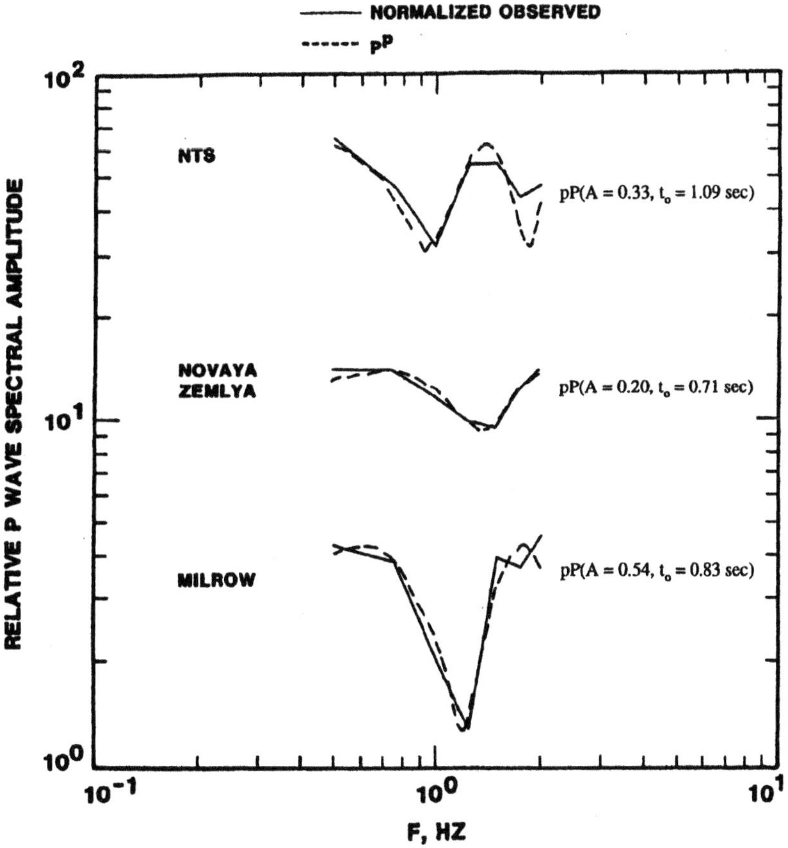

Figure 26
Comparison of source and path normalized network-averaged *P*-wave spectra from Figure 24 with spectra predicted by best-fitting "*pP*" models.

uncertainty. Taken together with the Semipalatinsk granite results described in the previous section, the comparisons of Figures 13 and 14 suggest that the proposed quantitative deterministic model has now been validated for explosions in granite, at least to the extent that the granite source environments at these three widely separated test sites are representative of the range of coupling to be expected for explosions in this medium.

Similar model inversions of the observed network-averaged spectra were conducted for each of the other 10 selected French Sahara explosions. A second example, for the $m_b = 5.2$ BERYL explosion of 5/01/62 is shown in Figure 15, where it can be seen that the NTS granite model once again provides an excellent fit over the entire short-period band for an inferred yield value of 26 kt. The m_b values for all the analyzed French Sahara explosions are plotted as a function of their corresponding spectral yield estimates in Figure 16. Also shown on this figure is the least-squares m_b/yield relation determined from these French Sahara data and the observed m_b/yield values for the three U.S. explosions in granite (i.e., HARDHAT, SHOAL and PILE DRIVER). As would be expected from the above discussion, this

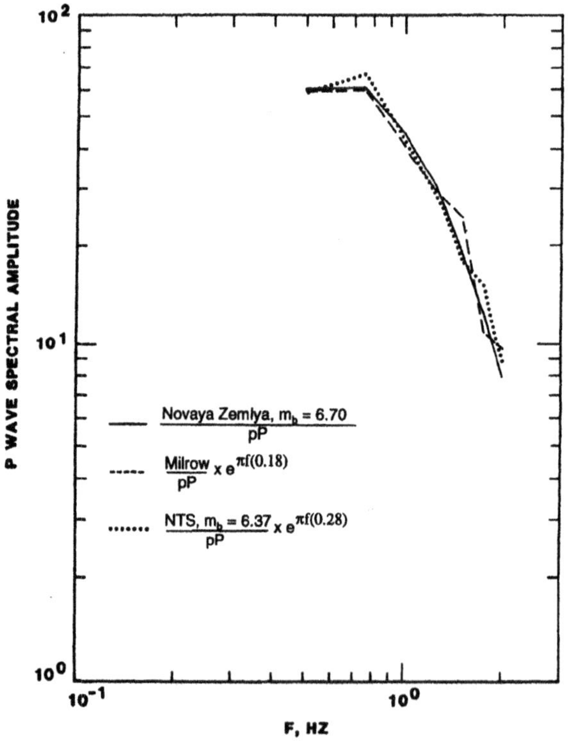

Figure 27
Comparison of pP and Δt^* corrected network-averaged P-wave spectra corresponding to megaton level explosions at the NTS, Amchitka and Novaya Zemlya test sites.

inferred French Sahara m_b/yield relation is seen to be very consistent with the observed m_b/yield values for U.S. explosions in granite.

Soviet Novaya Zemlya and U.S. Amchitka Test Site

Novaya Zemlya is located about 500 km north of the Arctic Circle and is composed of a North Island and a South Island which are divided by the Matochkin Shah Strait. The largest of the Soviet underground nuclear tests were conducted at two testing areas on the South Island, the map locations of which are shown in Figure 17. The southern test site (Krasino) is located on the southwestern margin of the South Island at about 70.9°N, 53.9° near the coastline of the Barents Sea, while the northern test site (Matochkin Shah) is located on the southern coast of the Matochkin Shah Strait at about 73.4°N, 54.8°E.

A sample consisting of short-period teleseismic *P*-wave data recorded from 20 Novaya Zemlya underground nuclear tests on a network of 50 stations has been processed and analyzed in an attempt to characterize the seismic source coupling at this test site. These selected explosions encompass a range of m_b extending from about 5.5 to 7.0, and each of the stations employed in the analysis recorded data from 4 or more of the explosions. Of these 20 tests, 17 were carried out at the northern Matochkin Shah testing area, while only 3 were conducted at the southern Krasino testing area. However, the teleseismic *P*-wave data recorded from the

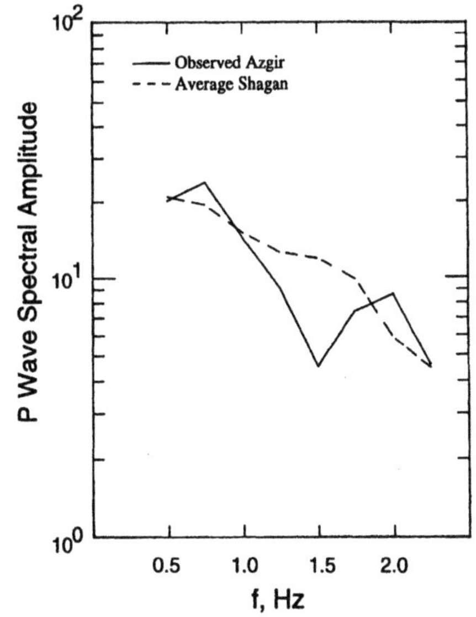

Figure 28
Comparison of network-averaged *P*-wave spectrum for the Soviet Azgir explosion of 1/17/79 with the average Shagan River explosion spectrum corresponding to the same value of m_b.

explosions at these two testing areas were found to be quite consistent, at least on a network average basis, and so no distinction between these two test areas is made in the following analysis. It should also be noted at the outset that information recently published by the RUSSIAN FEDERATION (1996) indicates that all 17 of the Matochkin Shah tests considered in this study were multiple explosions. However, this was not known at the time the analysis was originally conducted and, consequently, they are analyzed as equivalent single explosions in the following discussion. This clearly constitutes an added source of uncertainty which needs to be considered in evaluating the results of the analysis, but it will not be possible to quantify this uncertainty until more detailed information becomes available regarding the characteristics of these multiple tests.

As in the analyses of explosions at other test sites, the *P*-wave data recorded from this selected sample of Novaya Zemlya explosions were processed to obtain network-averaged *P*-wave spectral estimates for each event and the resulting network-averaged spectral amplitudes were then used to derive linear scaling laws versus m_b in the form of equation (1). Figure 18 shows a comparison of the average Novaya

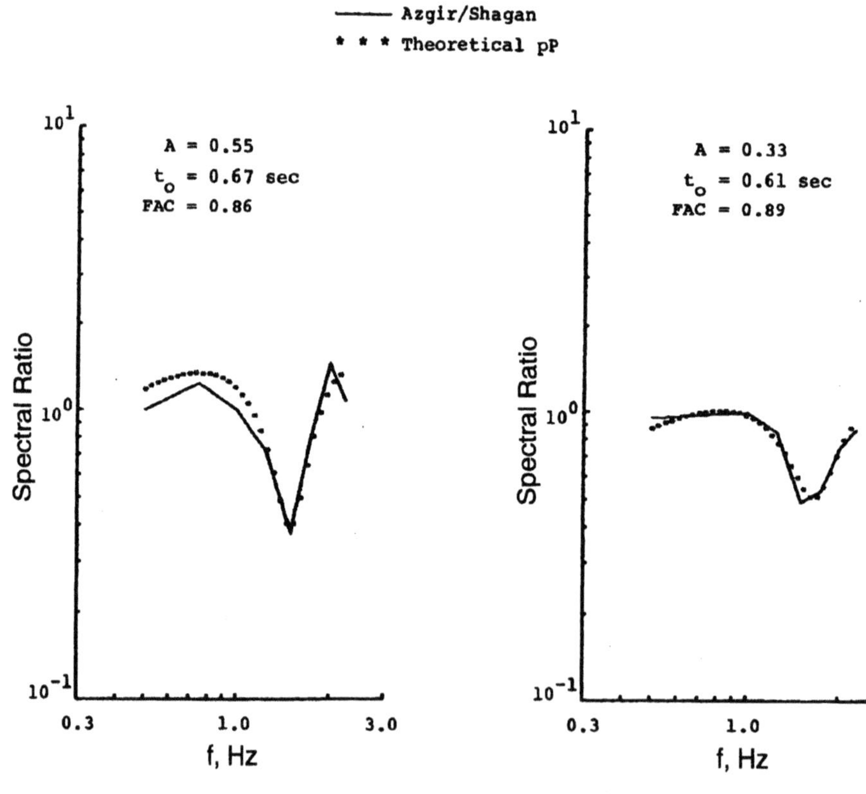

Figure 29
Comparison of observed Azgir/Shagan spectral ratios for the Soviet Azgir explosions of 1/17/79 (left) and 7/1/68 (right) with the corresponding best-fitting theoretical *pP* models.

Zemlya spectrum predicted by these scaling laws for an m_b value of 6.1 with the corresponding Shagan River spectrum, which was previously associated with a yield of about 150 kt at that test site. It can be seen that these two spectra are very similar over the entire short-period band, which suggests that the seismic source coupling and upper mantle attenuation must be quite similar at these two test sites. Not surprisingly then, when the Novaya Zemlya scaling laws are used to estimate the $\tilde{m}_b$ and Δt^* values which best reconcile the Novaya Zemlya spectra with the spectrum corresponding to a 150 kt explosion below the water table at NTS according to equation (3), we find $\tilde{m}_b = 6.08$ and $\Delta t^* = 0.28$ seconds, which are very comparable to the previously derived Shagan River estimates of $\tilde{m}_b = 6.10$ and $\Delta = 0.33$ seconds. The degree to which this model reconciles the average observed spectra from the two test sites at a yield of around 150 kt is illustrated in Figure 19, where it can be seen that it explains the average observed spectral differences to well within the data uncertainties.

As was noted previously, the largest historical Soviet underground nuclear explosions were conducted at the Novaya Zemlya test site, and there has also been

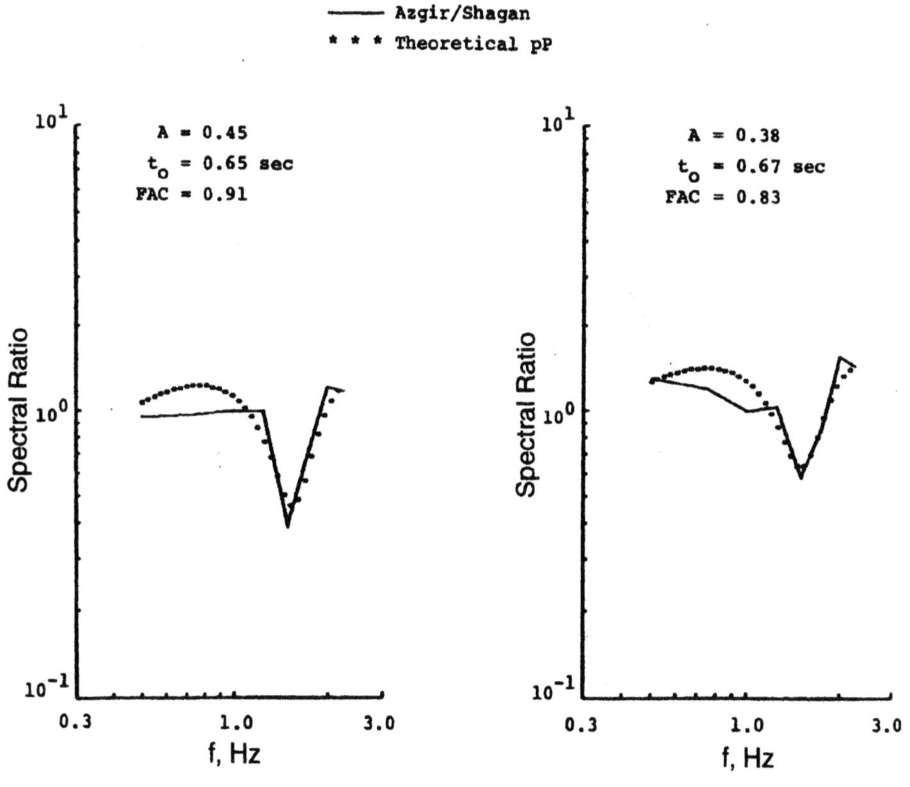

Figure 30
Comparison of observed Azgir/Shagan spectral ratios for the Soviet Azgir explosions of 12/22/71 (left) and 7/14/79 (right) with the corresponding best-fitting theoretical *pP* models.

considerable interest in yield estimation for these larger explosions. For this reason, the linear inversion analysis was repeated for a yield of 1000 kt (i.e., $m_b \approx 6.37$ at NTS) and the results are shown in Figure 20 where it can be seen that an $\tilde{m}_b$ value of 6.70 at Novaya Zemlya, together with the same Δt^* value of 0.28 seconds found previously at 150 kt, provides the best fit to the average 1000 kt NTS spectrum. Thus, the derived m_b estimates for explosions with yields of 150 and 1000 kt at Novaya Zemlya are in complete agreement with the inferred Shagan River m_b/yield relation given by equation (9).

On the basis of the above results, the individual network-averaged P-wave spectra for numerous larger Novaya Zemlya explosions were analyzed using the deterministic model, employing a Δt^* value of 0.28 seconds with respect to the NTS reference. An example of the results of this analysis is shown in Figure 21 for the multi-megaton Krasino explosion of 10/27/73. It can be seen that in this case an excellent fit is obtained using a Mueller/Murphy seismic source function corresponding to a yield of 3000 kt. The model-based spectral yield estimates for all these selected Novaya

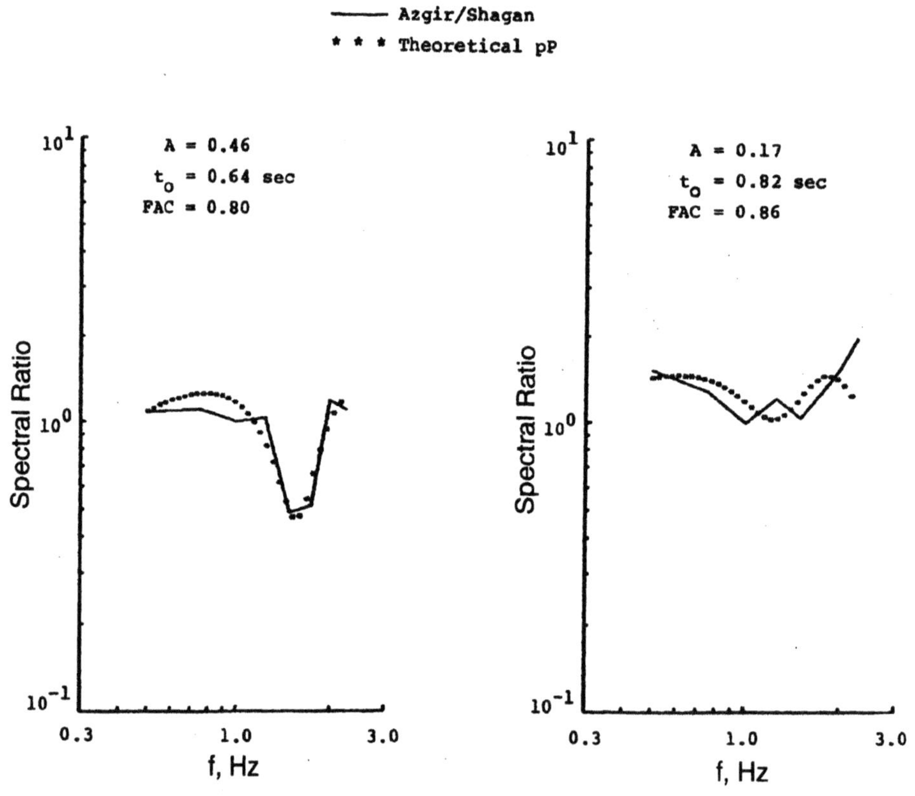

Figure 31
Comparison of observed Azgir/Shagan spectral ratios for the Soviet Azgir explosions of 7/29/76 (left) and 10/24/79 (right) with the corresponding best-fitting theoretical pP models.

Zemlya explosions are plotted versus their corresponding m_b values in Figure 22, where the inferred Shagan River m_b/yield relation of equation (9) is also shown for reference purposes. It can be seen that these Novaya Zemlya spectral yield estimates are very consistent with the average Shagan River m_b/yield relation, which implies that the same seismic yield estimation methodology is applicable at these two former Soviet test sites.

Although the results of the preceding analysis provide for an internally consistent interpretation of the Semipalatinsk and Novaya Zemlya data, new information regarding the yearly aggregate yields of Novaya Zemlya explosions has recently been published by the RUSSIAN FEDERATION (1996) which raises questions regarding the accuracy of our inferred m_b/yield relation for that test site. That is, although the yields of individual explosions are not given and the data appear to be internally inconsistent to some extent, the published aggregate yield information can be interpreted in such a manner as to suggest that the Semipalatinsk m_b/yield relation of equation (9) may underestimate the yields of the Novaya Zemlya tests by as much as a factor of two. Given that the network-averaged *P*-wave spectra corresponding to explosions of the same m_b at these two test sites are observed to be so similar over the entire short-period band (cf., Figure 18), this can only be true if the seismic source coupling, or the combination of source coupling and anelastic attenuation are significantly different at these two test sites. That is, as was noted previously in the discussion of the spectral analysis procedures, either the seismic source coupling at

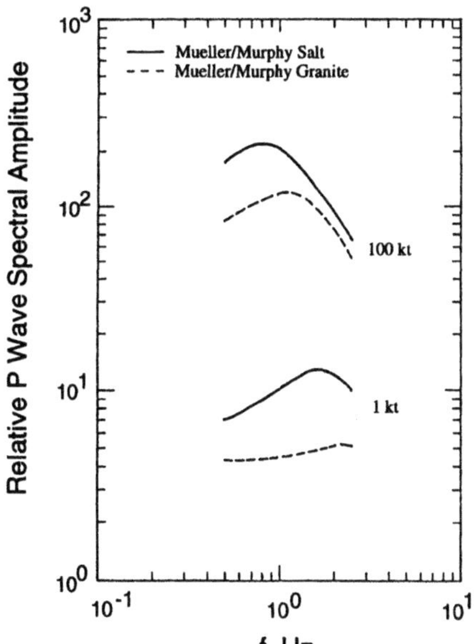

Figure 32
Comparison of estimated Mueller/Murphy teleseismic *P*-wave seismic source functions for explosions in salt and granite with yields of 1 and 100 kt.

Novaya Zemlya is lower than that at Semipalatinsk by a constant factor which is independent of frequency over the entire short-period band, or both the source coupling and anelastic attenuation differ in frequency-dependent manners such that they coincidentally combine to produce a frequency-independent offset in the teleseismic *P*-wave spectra. While neither of these possibilities can be definitively ruled out for Novaya Zemlya at this time, such frequency-independent offsets have never been observed at any other yield calibrated test site. An alternate hypothesis could be that the previously noted multiple explosion configurations of many of the analyzed Novaya Zemlya tests might be biasing the spectral analysis results or m_b values in some fashion. Again, however, it would be surprising if any such interference effects were to be independent of frequency over the entire short-period band. A definitive resolution of these questions must be deferred until more precise information is released regarding the yields and configurations of the individual Novaya Zemlya tests.

An alternate yield estimation methodology which has been proposed for Novaya Zemlya involves direct comparisons with seismic data recorded from the 3 explosions at the U.S. Amchitka test site. Two of them (MILROW and CANNIKIN) had yields

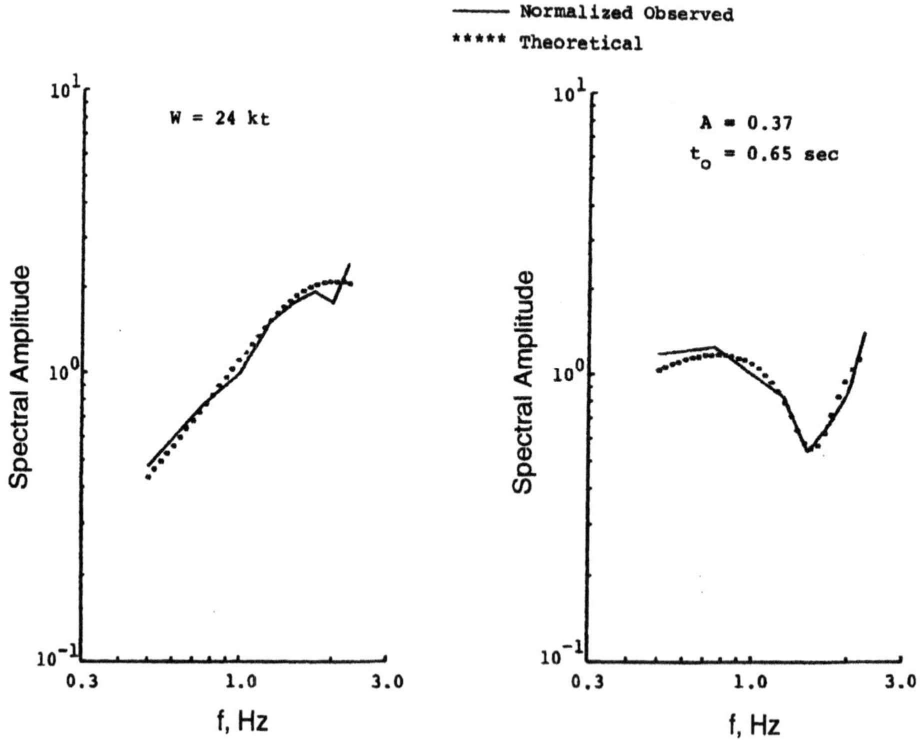

Figure 33
Comparison of normalized observed and best-fitting theoretical network-averaged *P*-wave spectra (left) and *pP* (right) for the Soviet Azgir explosion of 7/01/68.

in the megaton range of interest at Novaya Zemly and, some investigators have suggested that the upper mantle attenuation beneath these two island test sites should be comparable. If true, this would suggest that the yields of the larger Novaya Zemlya explosions could be estimated through direct comparisons with observed Amchitka seismic data, thereby eliminating the uncertainty associated with the estimation of the attenuation bias correction.

In an attempt to evaluate this hypothesis, short-period seismic data recorded from the megaton level MILROW explosion on a network of 30 teleseismic stations were processed and analyzed to obtain an estimate of the network-averaged *P*-wave spectrum for that event. The resulting MILROW spectrum is shown in Figure 23 where it is compared with the corresponding network-averaged spectra expected from megaton level explosions at NTS and Novaya Zemlya. It can be seen from this figure that the differences between the spectra characteristic of the three test sites are strongly frequency-dependent. The nature of these differences is easier to visualize once the spectra have been normalized to approximately correct for source and propagation path effects. For this reason, the network-averaged *P*-wave spectra of Figure 23 were divided by the Mueller/Murphy seismic source function corresponding to a 1000 kt explosion and multiplied by the nominal teleseismic attenuation

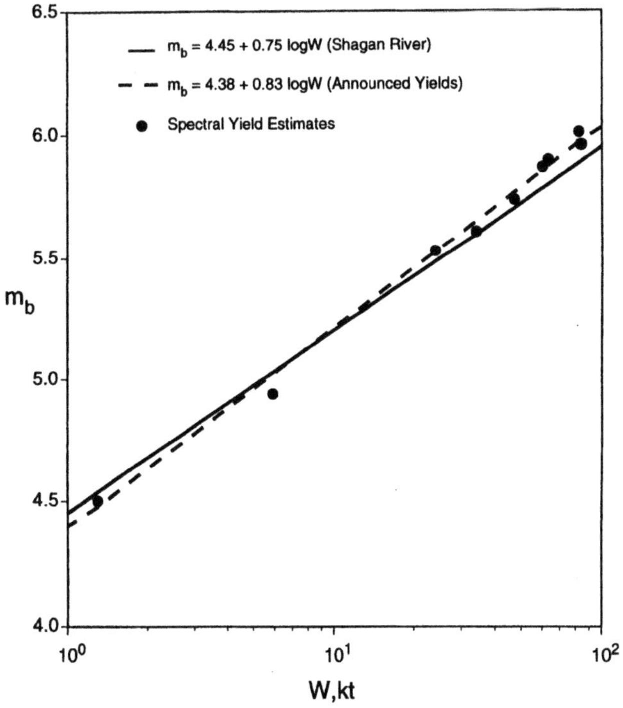

Figure 34
Comparison of spectral yield estimates for selected Azgir explosions with the least-squares fit to the announced Russian yields (dashed) and the nominal Shagan River m_b/yield relation (solid).

factor, $e^{\pi f t^*}$, where t^* was chosen to approximately eliminate any remaining systematic decrease of spectral amplitude with increasing frequency. The results of applying this normalization to the spectra of Figure 23 are shown in Figure 24, together with the associated estimated values of t^*. It can be seen that both the characteristic frequencies and amplitudes of the remaining spectral modulations are quite different for the three sites, being decidedly more pronounced for MILROW than for average megaton level explosions at either NTS or Novaya Zemlya. Furthermore, the inferred t^* value for MILROW is seen to be intermediate between the values estimated for the other two test sites. This difference is illustrated more explicitly in Figure 25, where the results of applying both the Novaya Zemlya and best-fitting t^* corrections to the source normalized MILROW spectrum are compared. It is clear from this comparison that the t^* values for Amchitka and

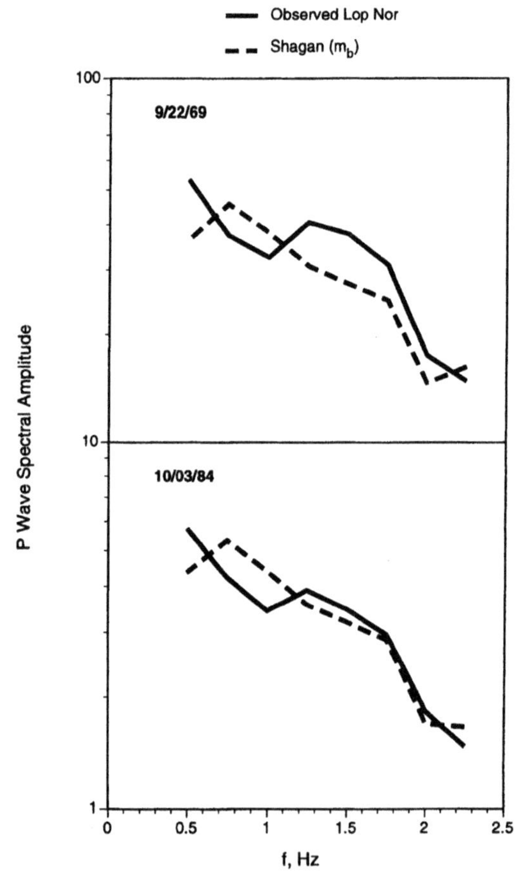

Figure 35

Comparison of observed network-averaged *P*-wave spectra for the Lop Nor explosions of 9/22/69 (m_b = 5.0, top) and 10/03/84 (m_b = 5.1, bottom) with corresponding average Shagan River spectra for explosions having the same m_b values.

Novaya Zemlya cannot be the same unless there are significant differences in the seismic source functions characteristic of these two test sites.

The observed interference patterns in the normalized spectra of Figure 24 were again modeled using the "quasi-linear" *pP* representation of equation (6). The degree to which this simple *pP* model can explain the observed spectral modulations is illustrated in Figure 26 where the inferred relative amplitudes (A) and delay times (t_0) for *pP* are listed for each site. It can be seen that these results indicate that the *pP*/*P* amplitude ratio is larger for MILROW than for average megaton level explosions at the other two test sites. Figure 27 shows the final results of correcting the three network-averaged *P*-wave spectra of Figure 23 for effects due to these inferred differences in *pP* and anelastic attenuation with respect to Novaya Zemlya. It can be seen that the combination of these two corrections brings the three spectra into

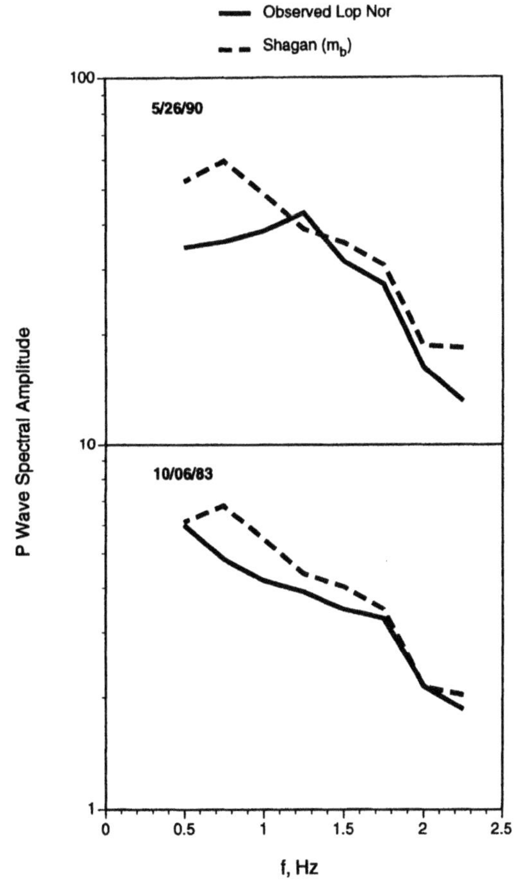

Figure 36
Comparison of obsesrved network-averaged *P*-wave spectra for the Lop Nor explosions of 5/26/90 (m_b = 5.3, top) and 10/06/83 (m_b = 5.4, bottom) with corresponding average Shagan River spectra for explosions having the same m_b values.

remarkably close agreement, which seems to support our previous conclusion that a 1000 kt explosion at Novaya Zemlya correlates with an m_b value of about 6.7. A corollary to this is that Amchitka does not constitute a good analog site for yield estimation at Novaya Zemlya.

Soviet Azgir Test Site

During the course of their underground nuclear testing program, the FSU conducted a significant number of nuclear explosions at sites away from their principal testing areas at Semipalatinsk and Novaya Zemlya. A number of these explosions were detonated in salt, notably at a site near Azgir, which is located just

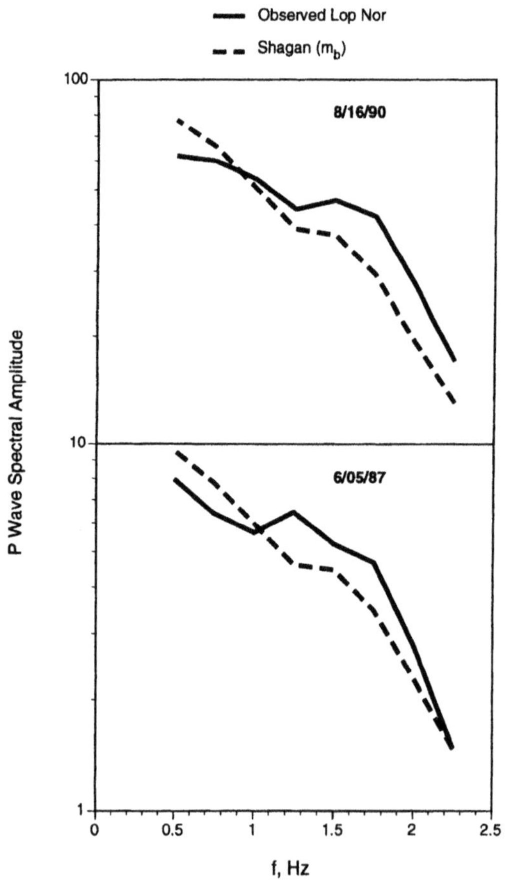

Figure 37
Comparison of observed network-averaged *P*-wave spectra for the Lop Nor explosions of 8/16/90 (m_b = 6.1, top) and 6/05/87 (m_b = 6.2, bottom) with corresponding average Shagan River spectra for explosions having the same m_b values.

north of the Caspian Sea. Since salt is the primary medium of concern with regard to the cavity decoupling evasion scenario, accurate determination of the m_b values for fully coupled explosions in salt with yields in the 1–10 kt range plays an important role in assessments of general seismic monitoring capability (OFFICE OF TECHNOLOGY ASSESSMENT, 1988). Consequently, teleseismic *P*-wave data from a sample of Azgir explosions in salt were assembled and analyzed in an attempt to derive an independent assessment of the seismic source coupling associated with such explosions.

The analysis of the seismic source characteristics of the Azgir salt medium focused on short-period teleseismic *P*-wave data recorded from 10 Azgir explosions on a network of 27 stations. These selected explosions, all of which were detonated within 10 km of 47.9°N, 48.0°E, encompass a range in m_b extending from about 4.5 to 6.0, and each of the stations employed in the analysis recorded data from three or more of the explosions. Once again, these Azgir data were processed through a series of narrowband filters, and the resulting spectral amplitudes in each band were then statistically analyzed to obtain frequency-dependent station correction factors and network-averaged *P*-wave spectra corresponding to each of the 10 explosions.

The first step in the analysis of the derived Azgir network-averaged *P*-wave spectra was to compare them directly with estimated Shagan River spectra corresponding to the associated m_b values. Such a comparison is shown in Figure 28

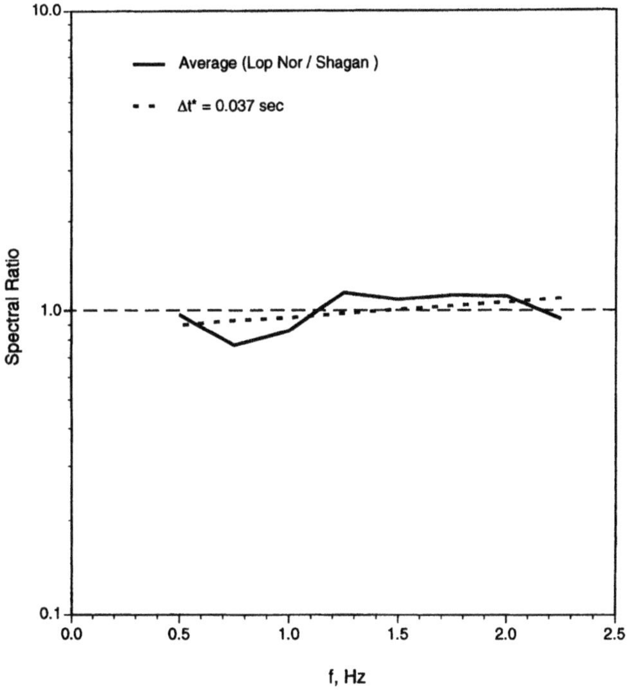

Figure 38
Average Lop Nor/Shagan *P*-wave spectral ratios computed for six Lop Nor explosions with m_b > 5.0.

for the $m_b = 5.96$ Azgir explosion of 1/17/79. It can be seen from this figure that these two spectra show quite similar average trends of decrease in spectral amplitude level with increasing frequency, despite the fact that the Azgir spectrum shows stronger modulation effects, which are presumably associated with interference from secondary arrivals. This observation is typical and is interpreted as an indication that *pP* effects are substantially stronger for these Azgir explosions than for the corresponding average Shagan events for which such *pP* effects appear to be essentially negligible over the frequency band of interest here. Ratios of observed Azgir to predicted Shagan spectra for six selected Azgir events are shown in Figures 29–31, where they are compared with the theoretical *pP*/*P* spectral ratios which provide the best fits to these data. It can be seen from these comparisons that the simple theoretical model fits the observations remarkably well, which suggests that the Azgir spectra are essentially identical to the Shagan spectra corresponding to the same m_b values, except for an enhanced *pP* effect at Azgir. There is, however, evidence of some small, constant amplitude offset between the two, as indicated by the quantity labeled FAC listed on each of these figures, which denotes the constant factor by which the theoretical *pP*/*P* ratio must be multiplied to bring it into overall

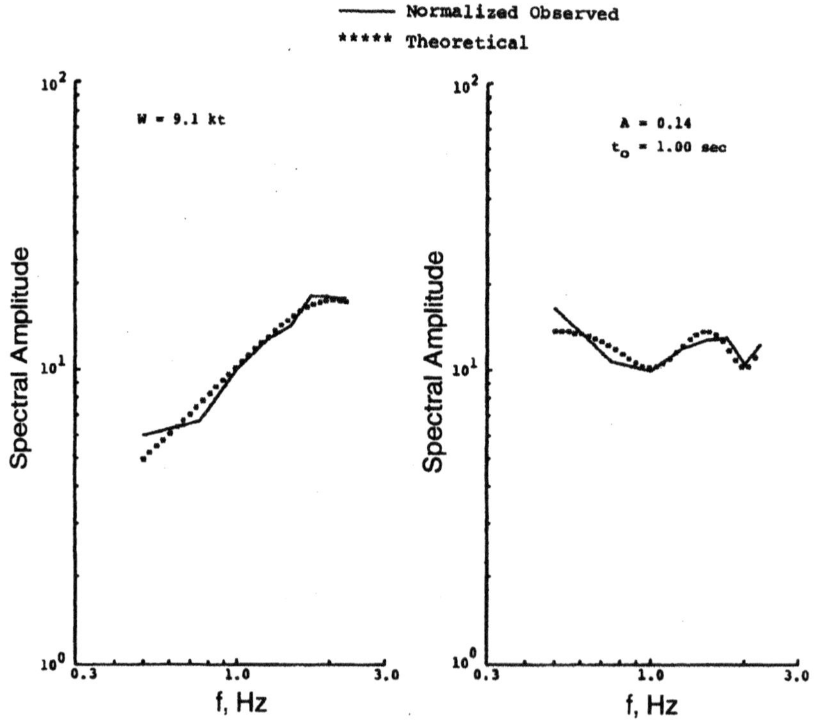

Figure 39
Comparison of normalized observed and best-fitting theoretical network-averaged *P*-wave spectra (left) and *pP* (right) for the Lop Nor explosion of 10/03/84.

agreement with the observations. For this data set FAC averages to a value of about 0.85, which suggests that, on average, *pP* effects at Azgir produce a slight increase in m_b over that which might be expected from an explosion of comparable yield at Shagan. We will return to this point later in conjunction with the discussion of the inferred m_b/yield relation for explosions at Azgir. In any case, the absence of any systematic trends with frequency in Figures 29–31 strongly suggests that both the explosion source coupling and upper mantle attenuation at Azgir must be quite comparable to that at Shagan.

This apparent consistency between the explosion source coupling at Azgir and Shagan is somewhat surprising in view of the fact that a number of previous investigators have concluded that explosions in salt are expected to couple significantly better than explosions in granite. Thus, for example, WERTH and HERBST (1963) inferred seismic source functions (i.e., reduced displacement potentials) from free-field recordings of the U.S. HARDHAT explosion in granite and GNOME explosion in salt and concluded that explosions in salt couple significantly better in the frequency band characteristic of m_b. Similarly, the original Mueller/Murphy source models for explosions in salt (based on SALMON data) and granite

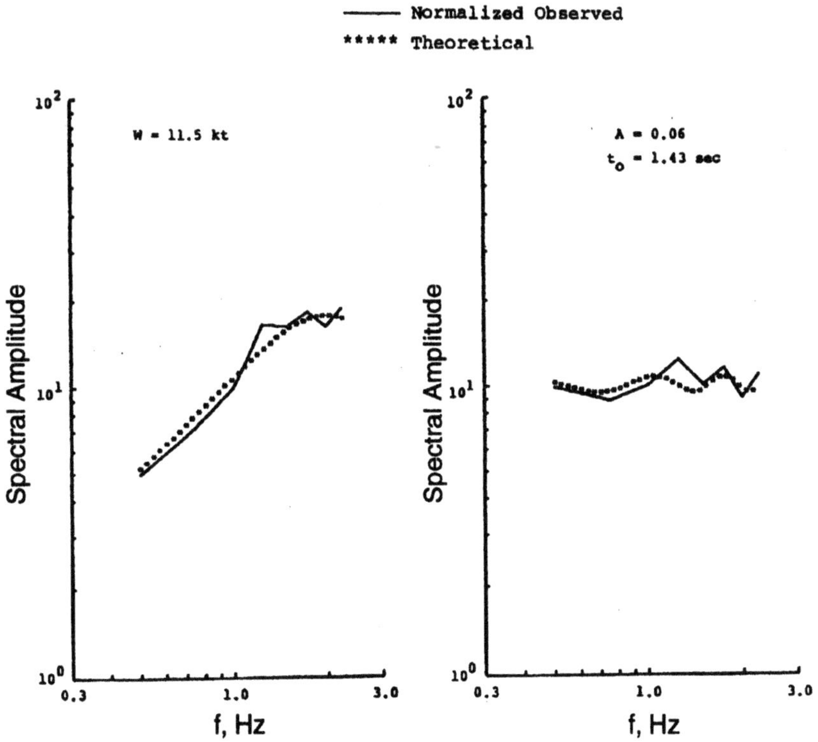

Figure 40
Comparison of normalized observed and best-fitting theoretical network-averaged *P*-wave spectra (left) and *pP* (right) for the Lop Nor explosion of 5/26/90.

(MUELLER and MURPHY, 1971) predict significant yield-dependent differences in coupling efficiency between the two media, as is illustrated for yields of 1 and 100 kt in Figure 32. On the other hand, it seems clear that the spectral ratio data shown in Figures 29–31 are inconsistent with any large frequency and yield-dependent differences of the kind displayed in Figure 32. Moreover, given the consistency of the granite results shown in the preceding sections, the likely explanation for the discrepancy is that the salt source functions estimated in these previous studies were overestimates. This explanation is qualitatively consistent with the more recent hypotheses of DENNY and GOODMAN (1990) and others to the effect that the nonlinear regime for U.S. explosions in salt, such as SALMON, is considerably larger than previously thought and that failure to recognize this fact led to overestimates of the salt seismic source function in these earlier studies.

In any case, given the empirical results summarized in Figures 29–31, it has been assumed for purposes of this analysis that the Mueller/Murphy granite source function and propagation path transfer function which were used to successfully match the observed Shagan River JVE spectrum (cf., Fig. 10) are directly applicable

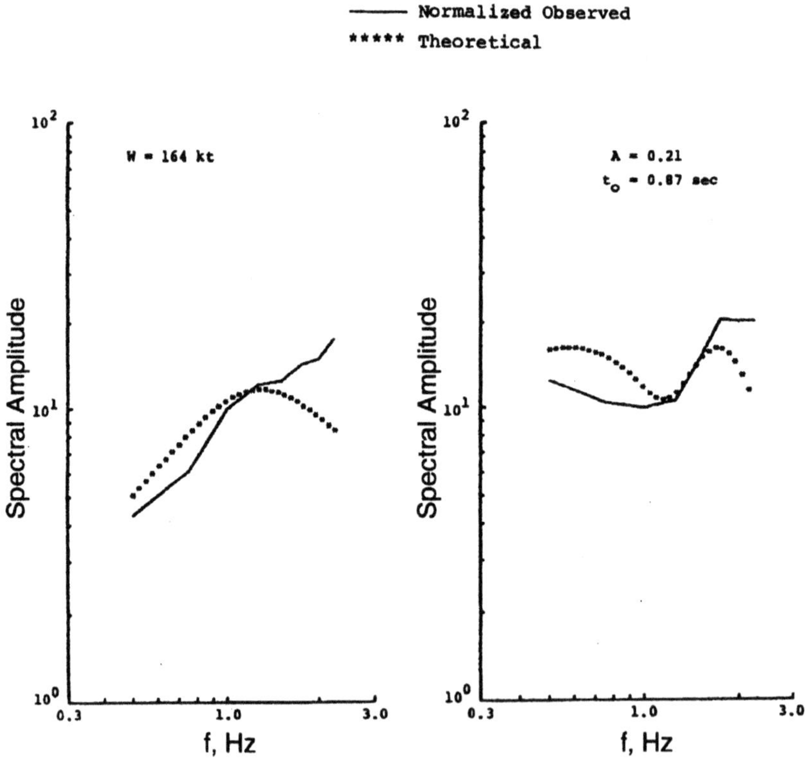

Figure 41
Comparison of normalized observed and best-fitting theoretical network-averaged *P*-wave spectra (left) and *pP* (right) for the Lop Nor explosion of 8/16/90.

at Azgir. Consequently, these models were employed to estimate seismic yields from the individual network-averaged *P*-wave spectra corresponding to each of the 10 selected Azgir explosions. The results of applying this inversion procedure to the observed network-averaged *P-wave* spectrum for the Azgir explosion of 7/1/68 are shown in Figure 33, where it can be seen that excellent agreement is obtained for a seismic source function corresponding to a yield of 24 kt. This is in remarkably good agreement with the announced yield of 27 kt for this test (SULTANOV *et al.*, 1999) and, therefore, provides strong evidence to support our previous inference that the seismic source coupling and upper mantle attenuation at Azgir must be essentially the same as those at the Shagan River test site.

The model-based spectral yield estimates for all 10 of the selected Azgir explosions are plotted versus their corresponding m_b values in Figure 34, where the least-squares fit to the announced Russian yields and the nominal Shagan River m_b/yield relation are also shown for reference purposes. It can be seen from this figure that the m_b/yield relation derived using the announced yields is in very good agreement with the spectral yield estimates, confirming the applicability of the

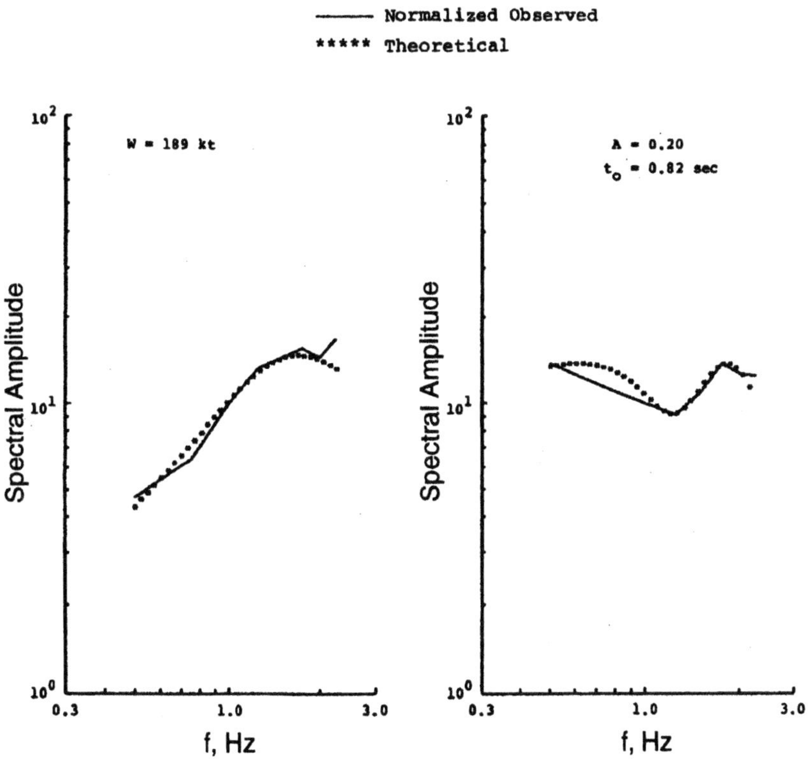

Figure 42
Comparison of normalized observed and best-fitting modified theoretical network-averaged *P*-wave spectra (left) and *pP* (right) for the Lop Nor explosion of 8/16/90.

selected modeling procedure. These spectral yield estimates are also seen to be generally quite consistent with the Shagan River m_b/yield relation, although there is a tendency for them to be somewhat low with respect to this relation for the larger events by an average of about 30%. This is consistent with the Azgir/Shagan spectral ratio results shown previously in Figures 29–31 and may be related to differences in the *pP* characteristics at Azgir relative to those at the other test sites. Additional data from explosions in salt must be analyzed in order to determine whether these relatively small differences are consistent enough to justify the formulation of a separate m_b/yield relation for explosions in this emplacement medium. In any case, it is concluded on the basis of the results presented above that the seismic source coupling in the salt domes at Azgir is very comparable to that in granite.

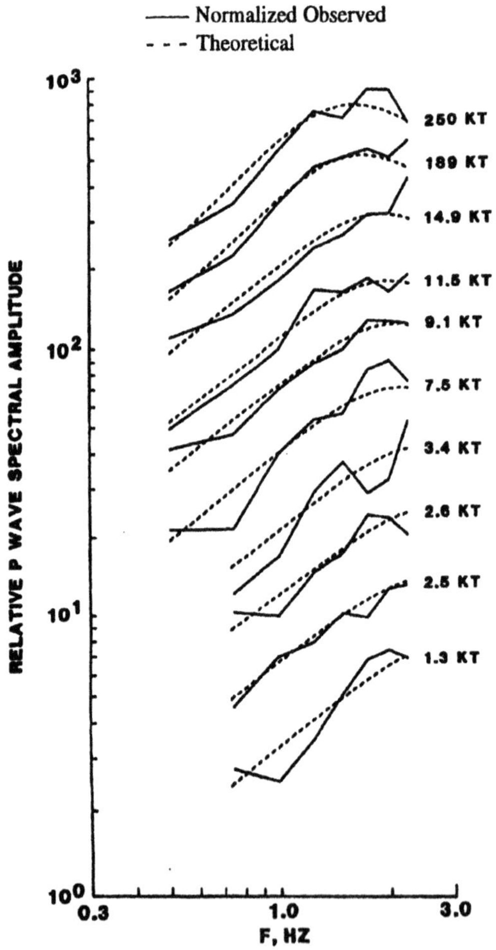

Figure 43
Composite summary of the final model fits to the attenuation and *pP* corrected observed network-averaged *P*-wave spectra corresponding to the 10 selected Lop Nor explosions.

Chinese Lop Nor Test Site

Since 1969, the Chinese have conducted a number of underground nuclear tests at a site in northwestern China (41.5°, 88.5°E) adjacent to their atmospheric nuclear test grounds at Lop Nor. In an attempt to characterize the seismic source coupling at this test site, a sample of short-period teleseismic *P*-wave data recorded from ten of these explosions on a network of 26 stations was collected and analyzed. The selected explosions encompass a range in m_b extending from about 4.5 to 6.2, and each of the stations used in the analysis recorded data from three or more of the explosions. As for the other test sites discussed above, the observed short-period data were processed and analyzed to obtain estimates of the network-averaged *P*-wave spectra corresponding to each of the ten explosions.

As with the Azgir spectra discussed in the preceding section, the first step in the analysis of the observed Lop Nor network-averaged *P*-wave spectra was to directly compare them with average Shagan River spectra corresponding to the individual event m_b values. Such average spectral estimates are available for Shagan River explosions in the magnitude range from about $5.0 < m_b < 6.2$, and the comparisons for the six Lop Nor explosions associated with m_b values in this range are shown in

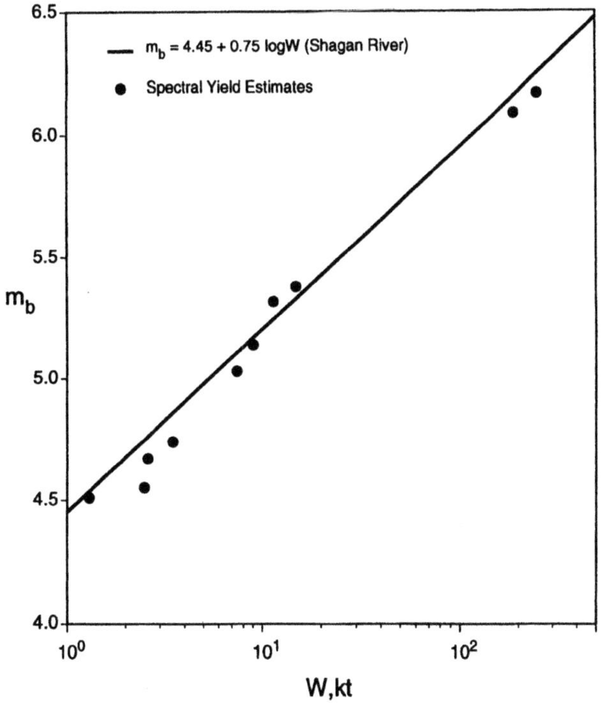

Figure 44
Comparison of spectral yield estimates for selected Lop Nor explosions with the nominal Shagan River m_b/yield relation.

Figures 35–37. It can be seen from these figures that, allowing for the normal variability between events, the Lop Nor spectra are quite comparable to those for Shagan River explosions of the same m_b over the entire short-period band between 0.5 and 2.25 Hz. The only indications of any consistent frequency-dependent differences are for the two largest explosions of Figure 37, where it appears that the Lop Nor spectra are slightly enhanced at high frequency with respect to the corresponding Shagan River spectra. We will return to this point later in our discussion of the spectral yield estimates for these two Lop Nor events. The degree of similarity in the spectra from these two test sites is brought out more explicitly in Figure 38, where the average of the individual ratios of observed Lop Nor to estimated Shagan River spectra computed over the six explosions of Figures 35–37 is plotted as a function of frequency over the short-period band extending from 0.50 to 2.25 Hz. It can be seen from this average ratio that there is no evidence of any significant trend with frequency that might be indicative of differences in source coupling or anelastic attenuation between these two sites. This conclusion is quantitatively confirmed by the fact that the best fit Δt^* value of 0.037 sec shown on this figure is found to not be statistically and significantly different from zero.

On the basis of the test site comparison results shown in Figure 38, spectral yield estimation analyses of the Lop Nor explosions have been carried out using the same Mueller/Murphy source function model and propagation path transfer function which were used in the previous analysis of the Shagan River spectra. The results of applying this inversion procedure to the observed network-averaged spectra from the Lop Nor $m_b = 5.1$ explosion of 10/03/84 and the $m_b = 5.3$ explosion of 5/26/90 are shown in Figures 39 and 40, respectively, where it can be seen that the selected analytical model provides very precise fits to the observed spectra for seismic source functions corresponding to the listed yield values. A similar comparison for the larger $m_b = 6.1$ Lop Nor explosion of 8/16/90 is shown in Figure 41, where it can be seen that the model fit is less satisfactory in this case, showing discrepancies suggestive of an unaccounted for shift in source function corner frequency. With regard to the seismic source function, the two factors which can affect the source corner frequency in the Mueller/Murphy model are depth of burial and source medium velocity (MURPHY, 1989). Thus, for example, an increase in the source medium velocity of about 30% between the smaller and larger Lop Nor explosions would produce a corner frequency shift comparable to that observed for the two large Lop Nor explosions of Figure 37. This fact is illustrated in Figure 42 for the Lop Nor explosion of 8/16/90. It can be seen from this comparison that the postulated increase in velocity results in a significantly improved model fit over that shown in Figure 41, to one which is comparable to those shown for the lower yield explosions in Figures 39 and 40. Although there is currently no direct evidence to confirm such a variation in velocity, it does not seem unreasonable that the medium velocity at depths corresponding to 200 kt explosions (~700 m) might be significantly higher than the medium velocity at depths corresponding to 10 kt explosions (~250 m) at some test sites. In any case, the evidence strongly suggests that a shift in corner frequency, in addition to that normally expected from the variation in yield, is required to reconcile the observed changes in the observed Lop Nor spectra for explosions with yields of around 10 and 200 kt. It should be noted, however, that, as with the *pP* corrections, perturbations of the source medium velocity such as that proposed above have very little effect on the final yield estimates. Thus, although the data fit of Figure 42 is significantly improved over that shown in Figure 41, the corresponding yield estimates differ by only about 15%. That is, the spectral yield estimates are relatively insensitive to such variations in the modeling assumptions.

The final results of the spectral yield estimation analyses for the ten Lop Nor explosions are concisely summarized in Figure 43, where the attenuation and *pP* corrected observed spectra are compared with the theoretical spectra corresponding to the annotated best-fitting yield values. It can be seen that the simple theoretical model accounts remarkably well for variations in the observed short-period *P*-wave spectra over the entire yield range extending from about 1 to 250 kt. To put this observation in quantitative terms, the 95 percent confidence interval about the average misfit to these data is only about a factor of 1.3. The inferred yield values are plotted versus their corresponding m_b values in Figure 44, where the nominal Shagan River m_b/yield relation is again shown for reference purposes. It can be seen from this figure that the m_b/yield relation corresponding to these Lop Nor spectral yield estimates is quite consistent with the results of the previous Shagan River analysis, which suggests that the same seismic yield estimation procedures are applicable at these two test sites.

Summary and Conclusion

The investigations summarized in this paper have focused on the application of network-averaged teleseismic *P*-wave spectra to seismic yield estimation of underground nuclear explosions at a wide variety of different test sites. Results of these analyses indicate that such spectral data can be analytically inverted using simple models to obtain estimations of explosion yield and m_b/yield relations which have generally proven to be remarkably consistent with the detailed yield data which have subsequently become available for a number of these test sites. This methodology was initially applied to explosions at the former Soviet Semipalatinsk test site, and it has been demonstrated that the resulting m_b/yield relation, which was originally derived in 1984, proved to be completely compatible with both the results of the Soviet Joint Verification Experiment of 1988 and with the yield calibration information for that site which was later published by BOCHAROV *et al.* (1989). This success prompted the subsequent applications of the same methodology to explosions at the French Sahara, Chinese Lop Nor, U.S. Amchitka and former Soviet Novaya Zemlya and Azgir test sites which have been briefly summarized in the preceding sections. The resulting *P*-wave spectral yield estimates were shown to be in good agreement with the accurate explosion yield values which have subsequently been published for individual explosions at the French Sahara, former Soviet Azgir and U.S. Amchitka test sites. Moreover, it was also demonstrated that the observed network-averaged *P*-wave spectra for a number of currently uncalibrated explosions at the former Soviet Novaya Zemlya and Chinese Lop Nor test sites can be accounted for with high precision by the predictions of the proposed simple model. While these results are encouraging, it should be noted that they do not necessarily imply that this spectral yield estimation methodology is universally applicable, in that it is always possible that underground nuclear tests could be conducted in some previously untested medium for which the seismic source coupling is significantly different from any of those encountered in the past. In fact, as was noted previously, there are some recently published aggregate yield data for explosions at the former Soviet Novaya Zemlya test site which may indicate that the seismic source coupling there is significantly different from that encountered at any of the other test sites considered in this paper, in a manner which is independent of frequency over the entire short-period frequency band. Clearly, any such dramatic departures from previous experience would present serious problems for any currently available seismic yield estimation methodology. However, we feel that we have demonstrated that the short-period teleseismic *P*-wave spectral data and associated analysis procedures provide considerably more diagnostic information regarding the explosion seismic source than do the corresponding narrowband magnitude measures such as m_b, M_s and $m_b(L_g)$, and, therefore, are to be preferred for applications to seismic yield estimation for explosions at previously uncalibrated test sites.

Acknowledgements

The authors thank M.E. Marshall for her assistance in previewing and processing the seismic data used in the various analyses described in this paper. We would also like to thank Paul Richards and Bob Blandford for their careful reviews which led to significant improvement of the manuscript. This work was sponsored by the U.S. Air Force Phillips Laboratory under contract PL-TR-92-2076.

References

Bache, T. C. (1982), *Estimating the Yield of Underground Nuclear Explosions,* Bull. Seismol. Soc. Am. *72,* S131–S168.

Bocharov, V. S., Zelentsov, S. A., and Mikhailov, V. N. (1989), *Characteristics of 96 Underground Nuclear Explosions at the Semipalatinsk Test Site,* Atomic Energy (Atomnaya Energiya) *67,* 3.

Denny, M. D. and Goodman, D. M. (1990), *A Case Study of the Seismic Source Function: Salmon and Sterling Reevaluated,* J. Geophys. Res. *95,* 19, 705–19, 724.

DOUGLAS, A. (1966), *A Special Purpose Least-squares Program*, AWRE Report No. 0-54/66, Atomic Weapons Research Establishment, U.K. Aldermaston, Berkshire.

HEUZE, F. E. (1983), *A Review of Geomechanics Data from French Nuclear Explosions in the Hoggar Granite, with Some Comparisons to Tests in U.S. Granite*, Lawrence Livermore National Laboratory, UCID-19812.

MUELLER, R. A. and MURPHY, J. R. (1971), *Seismic Characteristics of Underground Nuclear Detonations, I: Seismic Spectrum Scaling*, Bull. Seismol. Soc. Am. *61*, 1675–1692.

MURPHY, J. R. (1977), *Seismic Source Functions and Magnitude Determinations for Underground Nuclear Detonations*, Bull. Seismol. Soc. Am. *67*, 135–158.

MURPHY, J. R., BARKER, B. W. and O'DONNELL, A. (1989), *Network-averaged Teleseismic P-wave Spectra for Underground Explosions, I: Definitions and Examples*, Bull. Seismol. Soc. Am. *79*, 1–15.

MURPHY, J. R. (1989), *Network-averaged Teleseismic P-wave Spectra for Underground Explosions, II: Source Characteristics of Pahute Mesa Explosions*, Bull. Seismol. Soc. Am. *79*, 16–31.

MURPHY, J. R., STEVENS, J. L., O'NEILL, D. C., BARKER, B. W., MCLAUGHLIN, K. L., and MARSHALL M. E. (1991), *Development of a Comprehensive Seismic Yield Estimation System for Underground Nuclear Explosions*, Annual Technical Report to Phillips Laboratory, PL-TR-91-2161.

MURPHY, J. R. (1996), *Types of seismic events and their source descriptions*, In *Monitoring a Comprehensive Test-Ban Treaty, Proceedings of the NATO Advanced Study Institute*, Kluwer Academic Publishers, pp. 225–245.

OFFICE OF TECHNOLOGY ASSESSMENT (1988), *Seismic Verification of Nuclear Testing Treaties*, OTA-ISC-361, May.

RUSSIAN FEDERATION MINISTRIES OF ATOMIC ENERGY AND DEFENSE (1996), *USSR Nuclear Weapons Test and Peaceful Nuclear Explosions, 1949 through 1990*, Russian Federation Nuclear Center - VNIIEF, ISBN 5-85165-062-1.

SULTANOV, D. D., MURPHY, J. R., and RUBINSTEIN, Kh. D. (1999), *A Seismic Source Summary for Soviet Peaceful Nuclear Explosions*, Bull. Seismol. Soc. Am. *89*, 575–584.

SYKES, L. L. and EKSTRÖM, G. (1989), *Comparison of Seismic and Hydrodynamic Yield Determinations for the Soviet Joint Verification Experiment of 1988*, Proc. Nat. Acad. Sci. USA *86*, 3456–3460.

VEITH, K. F. and CLAWSON, G. E. (1972), *Magnitude from Short-period P-wave Data*, Bull. Seismol. Soc. Am. *62*, 435–452.

(Received July 21, 1999, revised December 20, 1999, accepted February 14, 2000)

Pure appl. geophys. 158 (2001) 2173–2216
0033–4553/01/112173–44 $ 1.50 + 0.20/0

Pure and Applied Geophysics

Effects of Source RDP Models and Near-source Propagation: Implication for Seismic Yield Estimation

CHANDAN K. SAIKIA,[1,2] D. V. HELMBERGER,[2] R. J. STEAD[3] and B. B. WOODS[1]

Abstract—It has proven difficult to uniquely untangle the source and propagation effects on the observed seismic data from underground nuclear explosions, even when large quantities of near-source, broadband data are available for analysis. This leads to uncertainties in our ability to quantify the nuclear seismic source function and, consequently the accuracy of seismic yield estimates for underground explosions. Extensive deterministic modeling analyses of the seismic data recorded from underground explosions at a variety of test sites have been conducted over the years and the results of these studies suggest that variations in the seismic source characteristics between test sites may be contributing to the observed differences in the magnitude/yield relations applicable at those sites. This contributes to our uncertainty in the determination of seismic yield estimates for explosions at previously uncalibrated test sites. In this paper we review issues involving the relationship of Nevada Test Site (NTS) source scaling laws to those at other sites. The Joint Verification Experiment (JVE) indicates that a magnitude (m_b) bias (δm_b) exists between the Semipalatinsk test site (STS) in the former Soviet Union (FSU) and the Nevada test site (NTS) in the United States. Generally this δm_b is attributed to differential attenuation in the upper-mantle beneath the two test sites. This assumption results in rather large estimates of yield for large m_b tunnel shots at Novaya Zemlya. A re-examination of the US testing experiments suggests that this δm_b bias can partly be explained by anomalous NTS (Pahute) source characteristics. This interpretation is based on the modeling of US events at a number of test sites. Using a modified Haskell source description, we investigated the influence of the source Reduced Displacement Potential (RDP) parameters ψ_∞, K and B by fitting short- and long-period data simultaneously, including the near-field body and surface waves. In general, estimates of B and K are based on the initial P-wave pulse, which various numerical analyses show to be least affected by variations in near-source path effects. The corner-frequency parameter K is 20% lower at NTS (Pahute) than at other sites, implying larger effective source radii. The overshoot parameter B appears to be low at NTS (although variable) relative to other sites and is probably due to variations in source conditions. For a low B, the near-field data require a higher value of ψ_∞ to match the long-period M_S and short-period m_b observations. This flexibility in modeling proves useful in comparing released FSU yields against predictions based on m_b and M_S.

Key words: RDP, NTS, scaling relationships, synthetics, Gasbuggy, Rulison etc.

[1] URS Group, Inc., 566 El Dorado Street, Pasadena CA 91101, USA.
E-mail: chandan_saikia@urscorp.com
[2] Seismological Laboratory, California Institute of Technology, Pasadena CA 91125, USA
[3] SAIC, Center for Monitoring Research, 1300N 17th Street, Arlington VA 22209, USA

Introduction

The seismic source level and yield for underground nuclear explosions have now been extensively documented for both the United States (DOUGLAS and MARSHALL, 1995) and the former Soviet Union (DOUGLAS and MARSHALL, 1995; SULTANOV *et al.*, 1999). Source models and scaling relationships for underground explosions have been studied for several decades with an emphasis on comparing teleseismic data sets consisting of short-period m_b's and long-period surface wave M_S's. Classical amplitude measures of the *P* wave such as m_b can only account for factors including attenuation, *pP* (depth phase) and initial source time function in a crude fashion. Similar measures on surface waves, M_S, suffer from their own complicating factors such as effective attenuation, contamination by tectonic release and source coupling. It has been known for some time that these two magnitude-based yield estimators display distinct regional behavior, presumably caused by differences in the above factors. Recent broadband studies of earthquakes also indicate considerable complexity and non-uniformity of source time functions; modern source descriptions are mostly expressed in terms of seismic moment and slip distribution on fault surfaces, especially for larger earthquakes ($M_W \geq 6$). Small events cannot be seen teleseismically and estimating teleseismic M_S, in particular, becomes difficult because of poor signal-to-noise ratios (SNR). However, at regional distances they are observed with good SNR, although the two source types, earthquake and explosions, show distinctly different spectral contents.

Figure 1a displays m_b:M_S measurements taken from a sample of NTS (Nevada Test Site, US) and SRS (Shagan River Site in the former Soviet Union, FSU) explosions (MARSHALL *et al.*, 1979). The regression lines for these two populations indicate distinctly different trends. The separation of the populations is well established at larger yields, as demonstrated in many studies (SYKES and CIFUENTES, 1984; GIVEN and MELLMAN, 1986). Whatever scatter, most significant for the SRS events, is observed is partly caused by random effects and partly by deterministic factors that can be modeled and corrected. We have plotted the m_b vs M_S relation for a set of 10 explosions from the Lop-Nor test site in China, shown by solid hexagons, which appear to follow the m_b vs M_S data from the NTS. A least-squares fit to this Chinese data set yields a m_b:M_S relationship given by $m_b = 0.98\ M_S + 3.06$. A preliminary attempt to determine which scaling relation is appropriate for the Indian and Pakistani nuclear test sites proved difficult, as the m_b vs M_S for the recent nuclear tests in these two countries do not plot distinctly with any one relation in Figure 1b. However, following the NTS relationship, SAIKIA *et al.* (1999) found that a yield of 10 kt is consistent with the regional Rayleigh waves observed at station NIL (Nilore, Pakistan); this estimate was also inferred by MURPHY (1998). The yields of the Pakistani explosions are slightly smaller. Figure 1b compares the m_b:M_S for a set of peaceful nuclear explosions (PNE's) from the United States and the FSU, and for a set of off-NTS explosions, for example CANNIKIN, FAULTLESS, GASBUGGY,

LONGSHOT, MILROW and RULISON, and others for which there is considerable variation in the relationship of m_b:M_S.

The strategy of this paper is to review the near-field and far-field results from the US testing experience and explain the above anomalies in terms of differences in the

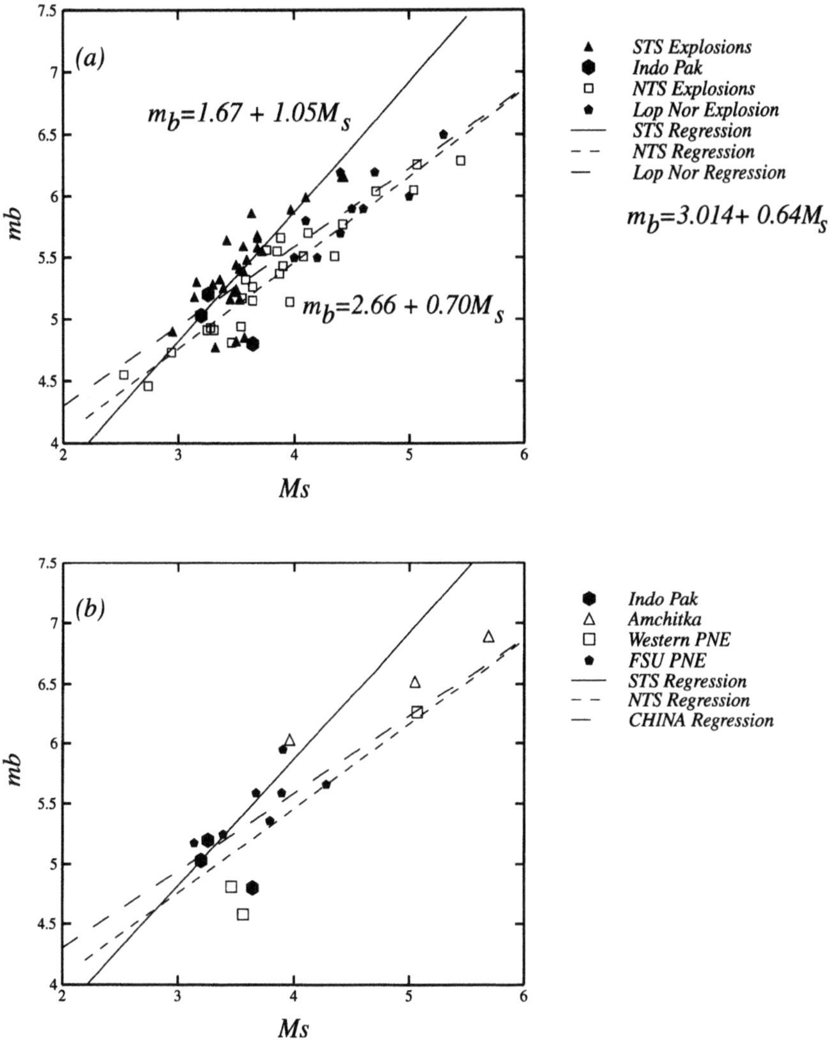

Figure 1

(a) Plots of m_b vs M_S regression curves for Shagan River, USSR (solid triangles), and NTS explosions (after MARSHALL *et al.*, 1979). Also plotted are the m_b vs M_S regression curve for the Chinese explosions at Lop Nor test site and their corresponding data points (solid hexagons). The corresponding m_b vs M_S data points for the Indian and Pakistani explosions are also plotted (large solid circles). (b) Similar regression curves as in (a) shown for Soviet PNEs and the off-NTS explosions CANNIKIN, FAULTLESS, GASBUGGY, LONGSHOT, MILROW and RULISON (indicated by the first letters of their names).

source excitation and path effects, which include upper-mantle Q (i.e., t^*), to help facilitate the applicability of the m_b:M_S-yield relationship at other test sites.

Numerous papers have discussed why the m_b:M_S values from the Amchitka shots differ from those at Pahute Mesa. In one of the earliest studies VON SEGGERN (1978) investigated differences between the BOXCAR (Pahute Mesa) and MILROW (Amchitka) shots which had nearly the same yield. The m_b for BOXCAR is 0.3 units smaller than that for MILROW, while the M_S for the BOXCAR is approximately 0.5 units larger. VON SEGGERN (1978) concluded that since the teleseismic P waves from NTS are lower in frequency than those from Amchitka, the attenuation under NTS is stronger, thus making for the t^* and m_b difference. Similarly, the JVE explosions involving the same shot sizes in the US and FSU indicated a similar m_b offset believed to be caused by the difference in the attenuation in the upper-mantle between the two test sites. However, the M_S's from the Shagan river events, which are the least contaminated by tectonic release (as indicated by the absence of Love waves), are roughly 0.3 to 0.5 units less than those for NTS shots near the 150 kt testing limits (STEVENS, 1986); these results are for the same events studied by SYKES and CIFUENTES (1984). However, for events which have strong-tectonic release, the Rayleigh waves may be reduced in amplitude and phase shifted and the Love waves can still be small. It is at least well known for the Pahute Mesa events that the tectonic release is primarily caused by a strike-slip mechanism and, therefore, the average M_S estimates from a network should be less influenced by the source radiation. In addition, the tectonic release radiation in the Pahute Mesa events has no identifiable influence on the short-period P waves (WALLACE et al., 1986). These discrepancies between the various test sites suggest that the long-period source levels measured by M_S relative to the short-period levels measured by m_b are site-related, and thus these differences can be used to establish working broadband source models.

A similar conclusion was reached by BURDICK et al. (1984), LAY et al. (1984a) and HARTZELL et al. (1983), based on the near-field vs far-field observations for the Amchitka shots and comparing them to the NTS shots. Following their approach, we assume a modified Haskell source representation containing three variable parameters and which is given by

$$\psi(\tau) = \psi_\infty \left[1 - e^{-K\tau}\left[1 + K\tau + \frac{(K\tau)^2}{2} - B(K\tau)^3\right]\right]$$

where the RDP is defined by $\psi(\tau)/R$, τ is the reduced time, where $\tau = t - R/\alpha$ (α being the velocity) and R is the distance between the source and receiver. The study by MUELLER and MURPHY (1971) established the basic scaling laws relating the constant K and ψ_∞ to yield. The parameter K is directly related to the corner frequency, and depends on source strength such that the larger the event, the longer is the source duration, or

$$K = C_1 \frac{h^{0.42}}{W^{0.33}} ,$$

where W is the yield in kt, h is the depth in meters, and C_1 is a constant determined by near-field modeling. Assuming the overshoot parameter $B = 1$, HARTZELL *et al.* (1983) find that at Amchitka $C_1 = 4.7$, while at Pahute Mesa $C_1 = 4.0$. These results suggest that the effective elastic cavity radius is larger for the softer rocks at Pahute by about 18% relative to other test sites, resulting in smaller corner frequencies. Coupling seems to be about 30% more efficient at Amchitka, as reported by the above studies.

The overshoot parameter B provides the means to uncouple the short-period signals, used for the m_b measurements, from the static long-period level which controls the M_S measurements. In short, allowing B to vary provides an extra degree of freedom in modeling recorded data. Figure 2 shows RDPs for the Amchitka modeling which will be discussed later.

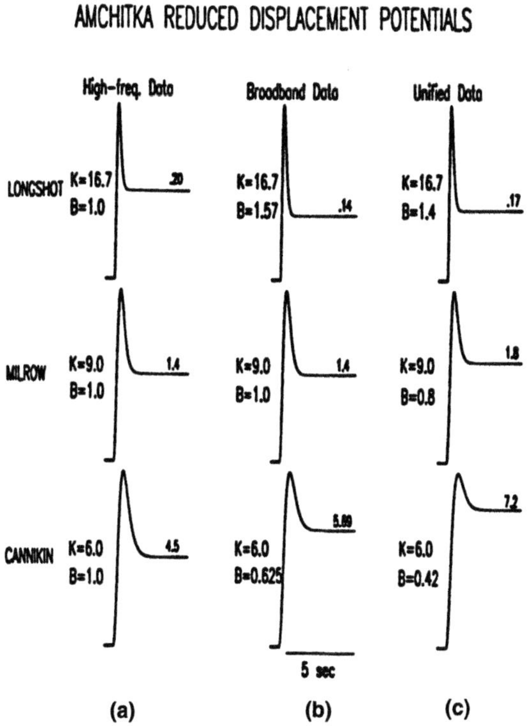

Figure 2

Reduced displacement potentials (RDP) (a) based on short period signals: (b) from broadband data: fixing $B = 1$ for MILROW, and adjusting the B's of LONGSHOT and CANNIKIN to match Rayleigh wave differentials; (c) from broadband data: utilizing long-period P waves, the phase pS, and Rayleigh waves. (Figure modified from LAY *et al.*, 1984.)

The problem of determining magnitude-yield scaling at low yields has several aspects which differ significantly from those of higher yields, specifically when the details of event emplacement and the available data type vary significantly. In this paper, we discuss the detailed issues related to the relationship of the NTS scaling behavior to that at the other sites. It is demonstrated that the basic explosion source model is independent of site. We find that in addition to a 30% weaker ψ_∞ value and a 20% weaker K value at NTS than the other sites, the overshoot parameter B also appears to be lower than unity at NTS and more variable, probably due to the variation in source conditions. At other US sites B is larger and more compatible with the Amchitka results, but shows a strong depth dependence for which it decreases with depth. The evidence that the nature of coupling and the shape of RDP varies from site to site was found in FAULTLESS data, with the event being strongly enhanced in the short-period energy relative to the long-period source level. The off-test-site explosions GASBUGGY and RULISON provide strong support for the site dependence of M_S and m_b yield scaling. Applying the non-NTS scaling laws to the Novaya Zemlya shots, we find rather modest yields with no events as large as CANNIKIN.

Scaling Relationship at Amchitka

The data set for Amchitka consists of near-in strong-motion records and teleseismic seismograms containing body and surface waves. The right panel in Figure 3 compares the long-period records of the two Amchitka events MILROW and CANNIKIN at DUG (Dugway, Utah). Since the paths are nearly identical, any difference in these seismograms from the two events must be source related. In earlier efforts BURDICK *et al.* (1984) established the linkage between the near-in data (see left panel in Fig. 3) and the teleseismic short-period amplitudes. Their source models, while fitting the short-period data very well, do not fit the relative amplitude of Rayleigh waves shown in the right panel. They predict CANNIKIN/MILROW long-period amplitude ratios of about 3, assuming $B = 1$, whereas the observed ratios are near 6. The Amchitka events have negligible Love waves, therefore we would not expect any tectonic-release effects on the Rayleigh waves unless this station is located at a Love-wave minimum (TOKSÖZ and KEHRER, 1972). The ratio of the observed long-period P waves between the two events is about 2.8 which is not consistent with the short-period prediction. The observed ratio for the surface waves corresponds directly to the difference in yield of 5:1 or the difference in surface wave magnitudes of 5.7:4.9 (LIEBERMANN and BASHAM, 1971; WILLIS *et al.*, 1972). The M_S for LONGSHOT is 3.9 and these surface-wave magnitude estimates fit the well-known equation

$$M_S = \log(Y) + 2$$

which is an often quoted result. The RDPs can be adjusted to fit surface waves by allowing B to vary with depth as

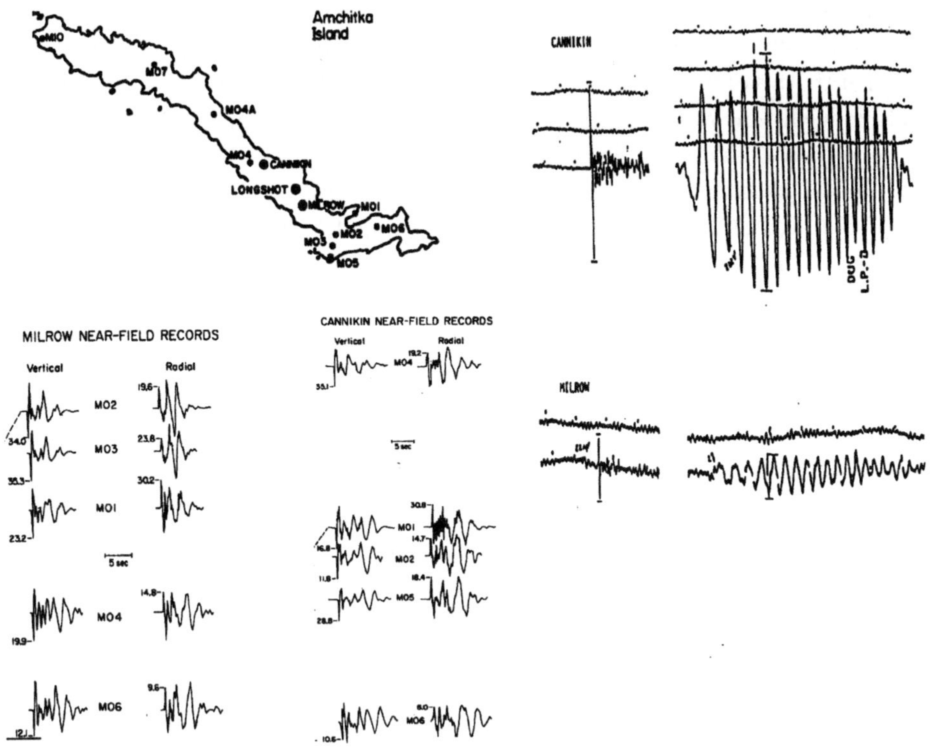

Figure 3
Shown in the left is Amchitcka Island and records from events CANNIKIN and MILROW. Vertical and radial component velocity seismograms are presented along with the slant distances from event to station (modified from BURDICK *et al.*, 1984). To the right is shown long-period vertical *P* and Rayleigh waves for CANNIKIN and MILROW at station DUG (Dugway, Utah). For MILROW the body wave is larger but for CANNIKIN the surface wave is larger.

$$\log B = 0.057 - 0.9701 \log h$$

(LAY *et al.*, 1984a,b). Since the source parameters are dependent on yield and depth, they also obtained the following relationship

$$\psi_\infty = 1.69 \times 10^8 \frac{W^{0.98}}{h^{0.33}}$$

which produces a long-period static displacement. We will review some of these efforts and modify the final scaling law (unified data, Fig. 2) to better match the absolute amplitudes of the CANNIKIN event.

Modeling of Near-in Seismograms Using Varying Crustal Structures at Amchitka

Six good data sets of strong-motion seismograms of the type displayed in the left panel of Figure 3 are available for MILROW and CANNIKIN for modeling the effects of wave propagation and the source RDP model. While the source time-function model controls the *P* waves, the crustal model completely determines the relative arrivals of the various phases such as *P*, *pP* and the Rayleigh waves; however both influence the relative amplitude and frequency content of the Rayleigh waves.

The crustal model for Amchitka was determined in detail from refraction and drill-hole measurements (ENGDAHL, 1972). By analyzing the MILROW near-in data, BURDICK *et al.*, (1984) adjusted Engdahl's model, allowing a *P*-wave gradient in the top 3 km of the crust in order to fit the move-out of the *pP* wave relative to *P*, and also incorporated an *S*-wave gradient to fit the travel time and shape of the Rayleigh waves. The crustal model so obtained for MILROW was further modified for the CANNIKIN blast, although the modifications were very slight. Geologic constraints on the structure are also available in a report by ORPHAL *et al.* (1970) wherein geologic cross sections from the Amchitka blasts CANNIKIN, MILROW and LONGSHOT to various near-field stations are given. Density, and *P* and *S* velocities are available in the well-log information (PERRET, 1973; PERRET and BREDING, 1972) and can be used to define sharp-boundary, gradient and random media models for the Amchitka structure (Fig. 4). The effects of these structures can be quite profound, even on the near-field waveforms. In the following

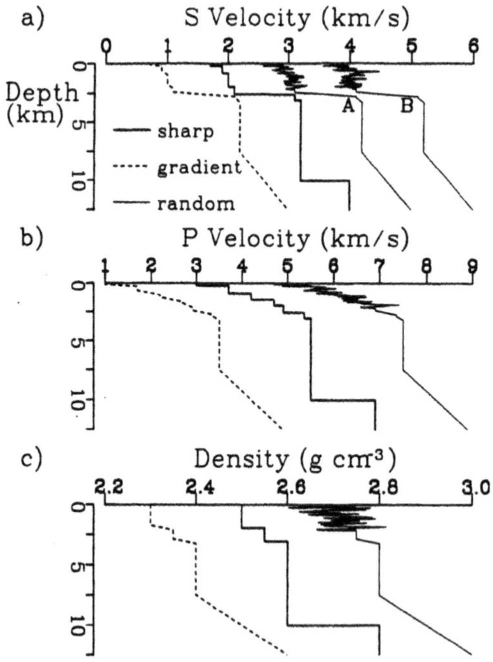

Figure 4
S- and *P*-wave profiles of velocity structures and density profiles used in the examination of various crustal models: sharp, gradational (gradient) and random media. The profiles of sharp boundaries are located correctly along the vertical axis and the others are shifted for clarity.

paragraphs we will present results from a canonical study demonstrating the relative importance of such structures using finite-difference simulations (STEAD and HELMBERGER, 1988).

Figure 5 compares observed and synthetic waveforms computed for various models at stations $M01$ and $M06$ at 8.0 and 11.5 km respectively for MILROW,

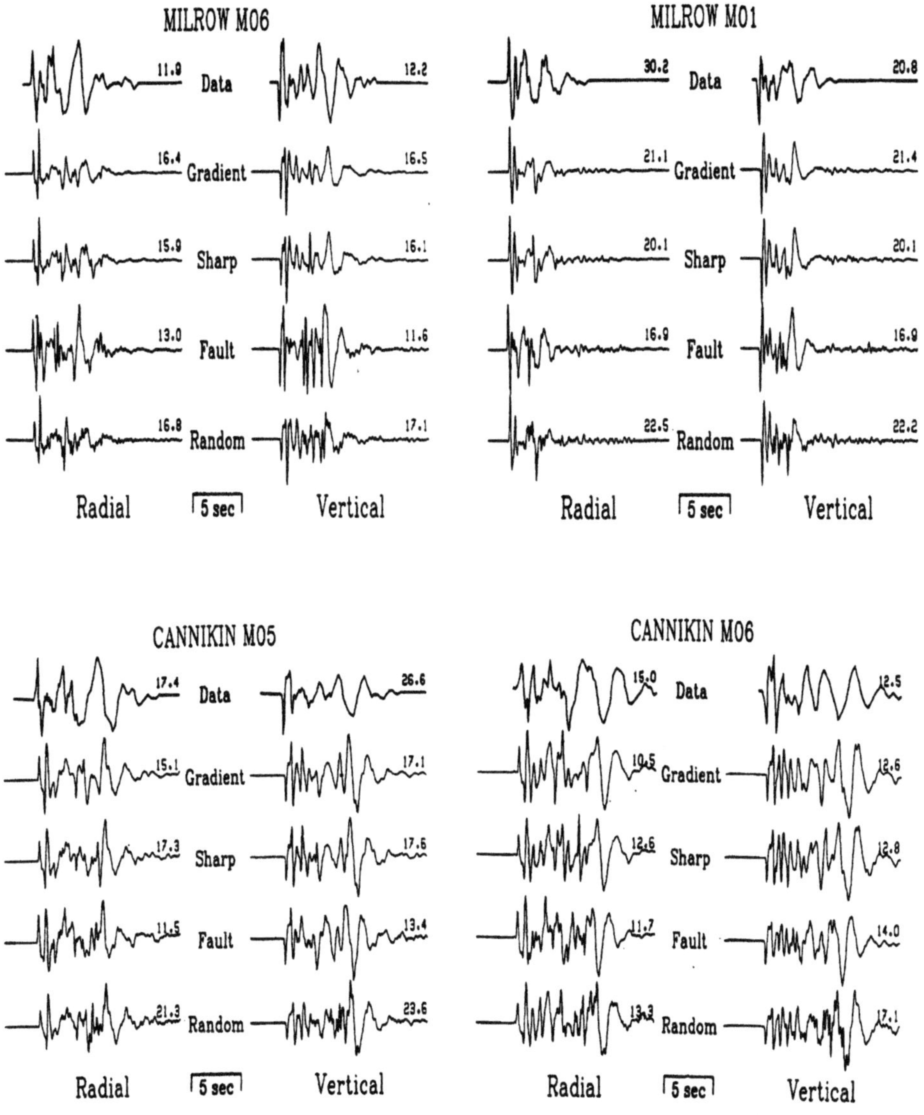

Figure 5

Comparison of data to four models: sharp boundary, gradient, random and a fault model. Top panel shows comparison at stations $M01$ ($r = 8$ km) and $M06$ ($r = 11.5$ km) for MILROW and bottom panel shows the same at stations $M05$ ($r = 15.8$ km) and $M06$ ($r = 18.7$ km) for CANNIKIN. Amplitudes are given in cm/s.

and *M*05 and *M*06 at 15.8 and 18.7 km respectively for CANNIKIN so that we can evaluate which model yields the best agreement with the recorded data. We also have included an additional velocity model; we call it the faulted model, which is approximated from the data presented in ORPHAL *et al.* (1970) and is based on the structure between the blast and station *M*05. The velocity in this model includes a fault near the shots and is the same as the sharp-boundary model below the source. At *M*01 for MILROW the first arrival is well modeled by both the sharp- and gradient-boundary models and the long-period part of the signal is better fit by the faulted model. Station *M*06 yields the same result, with the faulted model fitting the data better. For CANNIKIN the gradient model is the best fitting one according to the relative amplitude of the Rayleigh wave and its waveform shape. From this we can infer that the effect of the fault on the larger, long-period and more distant source CANNIKIN is not as strong as that for MILROW. Figure 6 presents a comparison of finite-difference simulations for the effect of a fault with 1 km offset on synthetic seismograms computed at ranges of 7, 8, 10 and 12 km (heavy lines) with those from a flat-layered model (thin lines, taken from VIDALE and HELMBERGER, 1987). Compared to the flat-layered case, Rayleigh waves in the faulted model are larger in amplitude and arrive earlier. The waves arrive earlier because the faulted model has raised the faster material to the surface and their amplitudes are larger because the layers that dip down away from the source tend to convert body waves to surface waves. While the surface waves and other phases have these variations in the travel times and amplitudes, the *P*-wave portion where the measure of m_b and source RDP parameters are made show no significant change, thus they are probably least influenced by the near-source path effect and probably more reliable for source estimation.

Effect of RDP Sources on Synthetics

The above modeling of the MILROW and CANNIKIN data suggests that structure has a significant effect on shaping the near-field observations and their amplitudes. However, the synthetics are based on the source RDP of HELMBERGER and HADLEY (1981). There are other RDP sources also (VON SEGGERN and BLANDFORD, 1972; MUELLER and MURPHY, 1971) which are frequently used in modeling explosions (Appendix A). In Figure 7 we compare the various RDP representations; the far-field displacement time functions are shown at the top which indicate little difference between them. The RDP functions are plotted next, and the level of permanent offset, ψ_∞, is 1.4×10^{11} for the Helmberger and Hadley, and von Seggern and Blandford sources, and 2.4×10^{11} for the Mueller and Murphy source. At the bottom are shown the displacement spectra which are also similar except that the Mueller and Murphy source has a higher long-period level. Figure 8 compares data and synthetics generated at station M01 for MILROW and M05 for CANNIKIN for these three RDPs, where the RDP parameters for the three are given in Table 1. There

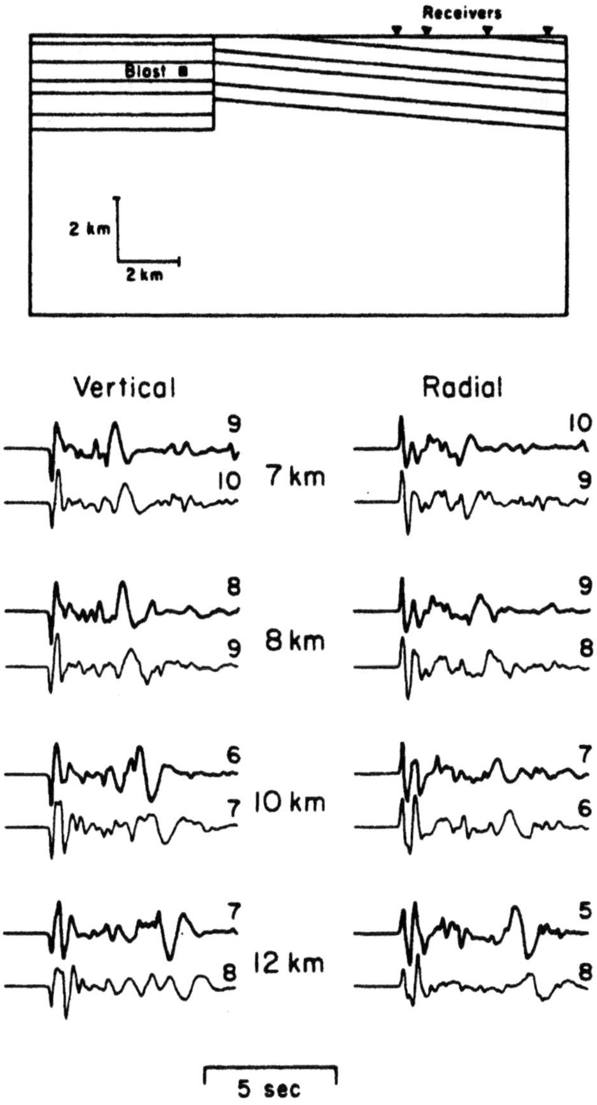

Figure 6
Effect of a fault with 1 km offset. Synthetic seismograms computed using finite difference scheme at 7, 8, 10 and 12 km are compared with those computed using a flat-layered model. Note that while the later phases are affected in arrival times and amplitudes, the initial onset remains almost unaffected and so the m_b estimate using the onset waveform is stable and can be used to investigate the parametric influence of the overshoot parameter B in the source RDPs for explosions.

is little difference between the three RDP sources for MILROW; amplitudes are within 5% of those for the data; there is a slight preference for the Mueller and Murphy source for CANNIKIN in terms of waveform fit. The amplitude of the RDP seismograms for this shot are within 35% of those of the data in all cases, and within 20% in every case

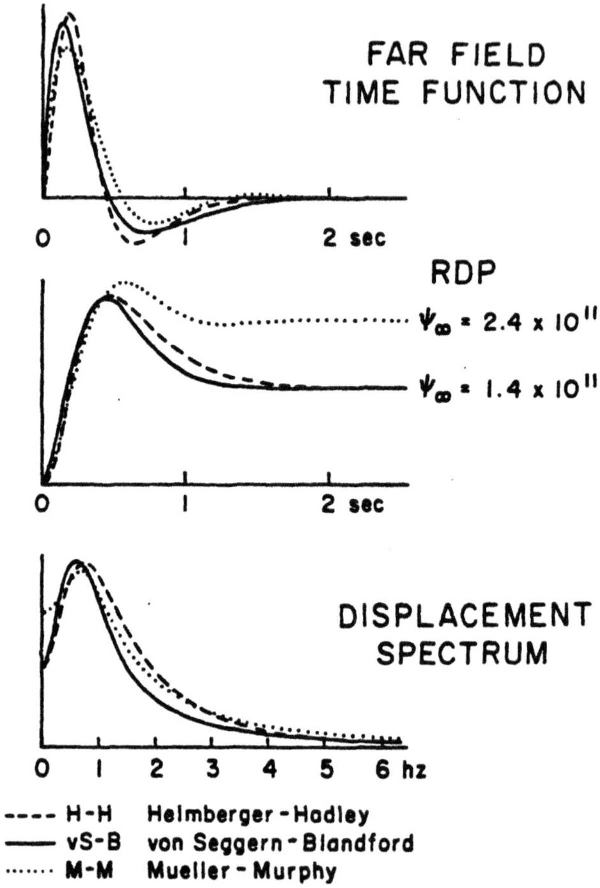

Figure 7
Comparison of various RDP source functions (Helmberger and Hadley, Mueller and Murphy, and von Seggern and Blandford). The top figure shows the far-field response, i.e., the time derivative of RDP. The center figure shows the RDPs of the three sources and the bottom figure shows their respective displacement spectra.

but one. In the following section we discuss how *B* and *K* values needed to be adjusted in fitting the observed data, using only the Helmberger-Hadley source RDP model as the waveforms are not significantly sensitive to the type of source RDP chosen.

Analysis of Source RDP Parameters B and K

The RDP source parameters can be varied to better fit the near-field data. Figure 9 (upper panel) compares observed and synthetic waveforms computed at M01 for CANNIKIN for a suite of source models consisting of various *B* and *K* value combinations; the most stable portion of the observed-synthetic matches is in the initial few seconds of the records. The synthetic consists of the diving *P* wave

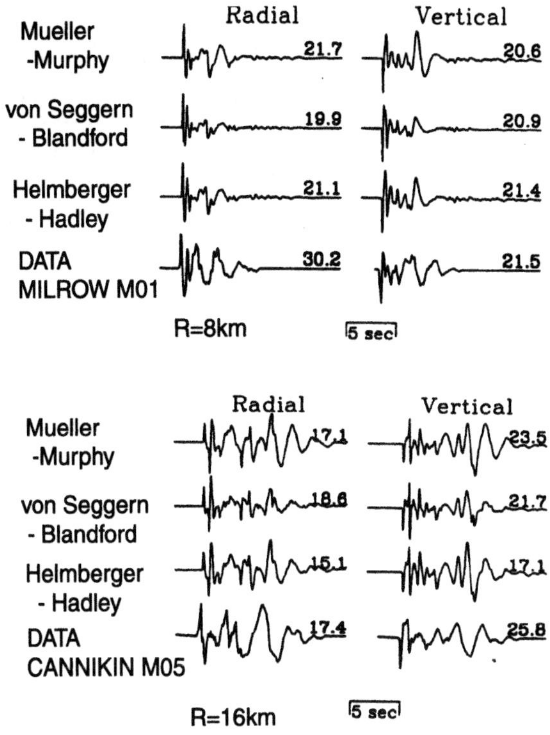

Figure 8
Comparison of the Mueller-Murphy, von Seggern-Blandford and Helmberger-Hadley source RDPs used with the synthetics generated from the gradient model. The top panel shows data recorded at $M01$ from MILROW and the bottom panel shows data at $M05$ from CANNIKIN. The Mueller-Murphy source seems to fit the data marginally better than the yield results from the other two sources.

followed by the opposite polarity pulse, *pP*. However, since *pP* cancels out a portion of the *P* wave, it becomes difficult to resolve all three parameters, *K*, *B* and ψ_∞, because ψ_∞ trades-off with *B* as shown in the lower panel of Figure 9. For example, a source function with a ψ_∞ of 7.3 and *B* of 0.5 fits data using a value 6 for *K* just as well as a source function with a ψ_∞ of 4.3 and *B* of unity. By holding *B* at some fixed value, ψ_∞ can be estimated for a given event and compared with teleseismic short-period data for large events to establish a realistic t^* (attenuation) operator for *P* waves. BURDICK *et al.* (1984) used *B* equal to unity and estimated a t^* of 0.9 for *P* waves, as shown in Figure 2a. LAY *et al.* (1984a) developed a scaling law based on broadband data using the intercorrelation technique, which allows a determination of source strength. They also fixed *B* to unity for MILROW and adjusted it for LONGSHOT and CANNIKIN to match the Rayleigh waves (Fig. 2, LAY *et al.*, 1984b). In this paper we demonstrate that a further modification to this source scaling relation is possible using three types of data, namely the long-period *P* waves,

Table 1

The RDP parameters for the three types of sources

Factor	MILROW	CANNIKIN
	HELMBERGER and HADLEY (1981) source	
K (s^{-1})	9.0	6.0
B	1.0	0.625
ψ_∞ (cm^3)	1.0×10^{11}	5.69×10^{11}
	VON SEGGERN and BLANDFORD (1972) source	
K′ (s^{-1})	9.0	6.0
B′	1.0	0.625
ψ_∞ (cm^3)	1.0×10^{11}	5.69×10^{11}
	MUELLER and MURPHY (1971) source	
Yield (Kt)	1000	5000
h (m)	1125	1725
V_p (km s^{-1})	4.2	4.7
V_s (km s^{-1})	2.0	2.0
ρ (g cm^{-3})	2.5	2.5
A/A_{cal}	2.0	2.0
Comp. factor	0.8	0.8
Prop. factor	2.0	2.0

See Appendix-A for detailed explanation of factors.

the phase *pS* and direct modeling of the Rayleigh waves, provided that the absolute long-period level can be established (see, Fig. 2c).

In general, the ratio of short-period *P*-wave amplitude to the long-period Rayleigh wave amplitude (SPZ/LPZ, measured on the vertical seismograms) depends mostly on the values of t^* (Fig. 10). This makes the long-period *P* waves less useful for inferring source information as the interaction of *pP* with *P* becomes somewhat less sensitive to the variation of the *B* value, thus a wide range of values can fit the data. The average, estimated from 28 measurements of SPZ/LPZ ratios, is 0.65, which is consistent with a t^* of about unity, in agreement with BURDICK *et al.* (1984).

The teleseismic *pS* phase is the most promising for inferring source information, since these waveforms are not subjected to any strong interference such as in the case of the *P-pP* waveforms BURGER *et al.* (1986a). However, there are complications, such as the *S-P* conversion at the Moho beneath the receiver, which can affect the *pS* phase. The late arriving so-called SV-coupled PL waves can also cause problems for this phase. Other complications can occur at the source, caused by tectonic release and possible spall interaction. These problems are discussed at length by COHEE and LAY (1988) in their analysis of *pS* waves from some Novaya Zemlya events.

To further investigate the trade-off between RDP source parameters and attenuation (t^*), the *pS* waveforms from CANNIKIN were collected at 21 worldwide stations and the *pS* amplitudes were corrected to a distance of (Δ) of

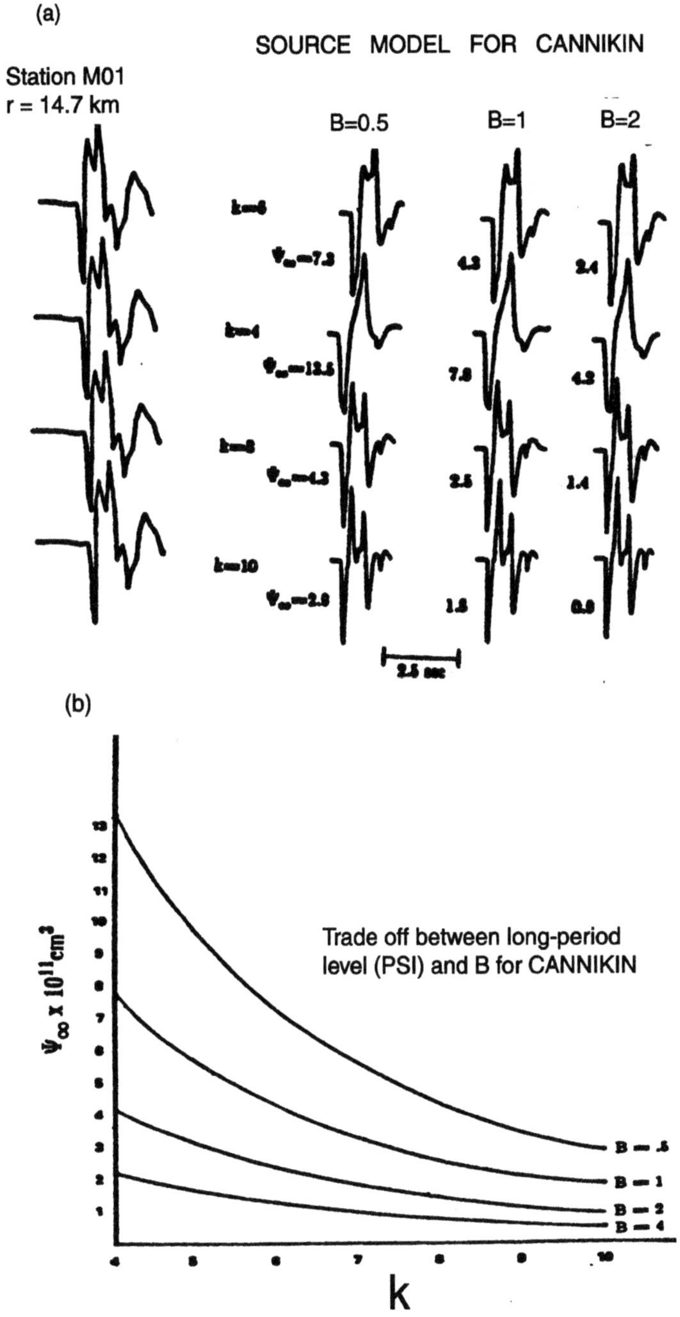

Figure 9
A suite of source models for event CANNIKIN (top panel). The ψ_∞ determined from the first swing in the waveform is printed to the left of each synthetic. The bottom panel shows the trade-off of ψ_∞ and B for CANNIKIN for a specific K.

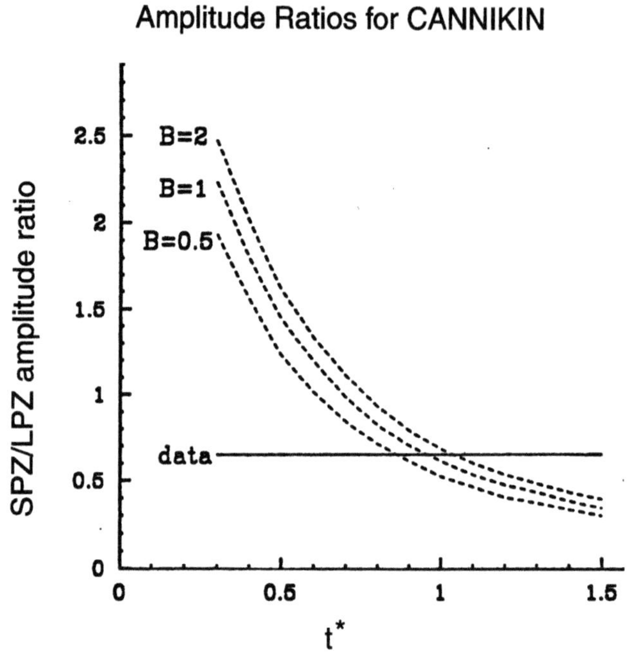

Figure 10
Amplitude ratios of short-period vertical (SPZ) to long-period vertical (LPZ) for the event CANNIKIN. The average observed ratio using 28 LPZ records is 0.65, indicated by the solid line.

50° (Table 2). Figure 11 (taken from BURDICK *et al.*, 1984) shows waveforms for a subset of 10 stations and three synthetics for different values of t^* using the source model ($K = 6$, $B = 1$), and the velocity structure from the near-field observations, assuming that the elastic conversion of *P*-to-*S* wave is −0.405 for a ray parameter of 0.12 s/km, a value consistent with those in modeling of the *pS* phase for earthquakes (LANGSTON and HELMBERGER, 1975). The actual reflection point occurs outside the region of strongest spall (a distance of roughly 1.5 km for CANNIKIN) because of the change in ray parameters (0.07 for *pP* and 0.12 for *pS*). Yet another difficulty is in the estimation of t^*. However, assuming that t^* for *S* is 4 times that of t^* for *P*, an expected value for t^* of *S* is 3.6, which is supported by many earthquake studies. Under this assumption, it becomes possible to determine the appropriate combination of *B* and ψ_∞ which satisfy the *pS* waveforms and their amplitudes. Table 3 gives the synthetic *pS* amplitude showing the trade-off between *B* and t^* for combinations of $K = 1$ and ψ_∞ values. Clearly, a wide range of combinations can fit the recorded amplitudes. If we assume t^* to be between 3.0 and 3.5, it is clear that a low value for *B* is preferred to match the average amplitude of 2083 μm (Table 2). In particular, a *B* value of 0.5 yields a ψ_∞value of 7.3×10^{11} cm^3, remarkably similar to the estimate reported by TOKSÖZ and KEHRER (1972) which is 7.2×10^{11} cm^3. Thus, we set the level of CANNIKIN to be 7.3×10^{13} cm^3 and adjusted the other

Table 2

CANNIKIN pS *amplitudes* (μ)

Station	Radial SV	Vertical SV	Tangential SH	Radial SV corrected to 50°	Vertical SV corrected to 50°	Tangential SH corrected to 50°
ALQ	1402	526	584	1473	559	610
BKS	3543	2421	1854	3302	2261	1727
BLA	879	493	290	1092	610	356
CHG	1428	450	282	1799	559	356
COR	3655	1298	866	3150	1118	737
CTA	1367	188	737	1905	254	1016
DAV	3145	904	1430	3658	1041	1651
DUG	1463*	998	—	1422*	965	—
KLB	800	343	180	1092	457	254
KBS	312**	389	389	305**	381	381
LON	1537	721	968	1321	610	838
MAT	3046	1092	899	2438	864	711
NDI	1265	472	175	1702	635	229
NOR	996	—	653	965	—	635
PMG	2108	996	747	2591	1219	914
POO	1872	589	627	2896	914	965
QUE	1222	445	145	1727	635	203
RAB	4227	874	1415	4775	991	1600
SHA	1847	673	1201	2210	813	1473
SHK	3241	1041	638	2743	889	533
TUC	1120	1275	747	1168	1321	787
Average				2083	855	799
Standard deviation				1069	436	485
Number of samples				21	20	20

* Measured on EW component. ** Emergent onset.

events accordingly to obtain the RDPs given in Figure 2c. If we follow MUELLER and MURPHY (1971), namely that yield, Y, is proportional to effective cavity volume, we expect

$$\psi_\infty = C_1 \frac{Y}{h^{0.27}} \quad K = C_2 \frac{h^{0.42}}{Y^{0.33}} \ .$$

The two scaling constants, C_1 and C_2 can then be estimated by regression, assuming the yields of LONGSHOT, MILROW and CANNIKIN. The results are $C_1 = 12.3 \times 10^8$ and $C_2 = 4.7$. The parameter B is dependent of depth and is given by

$$\log B = -0.05 - 1.3 \log h$$

which is slightly stronger than proposed by LAY *et al.* (1984), namely

$$\log B = 0.06 - 0.97 \log h \ .$$

As will be discussed in a later section, the above formulae prove quite effective in explaining the GASBUGGY-RULISON observations.

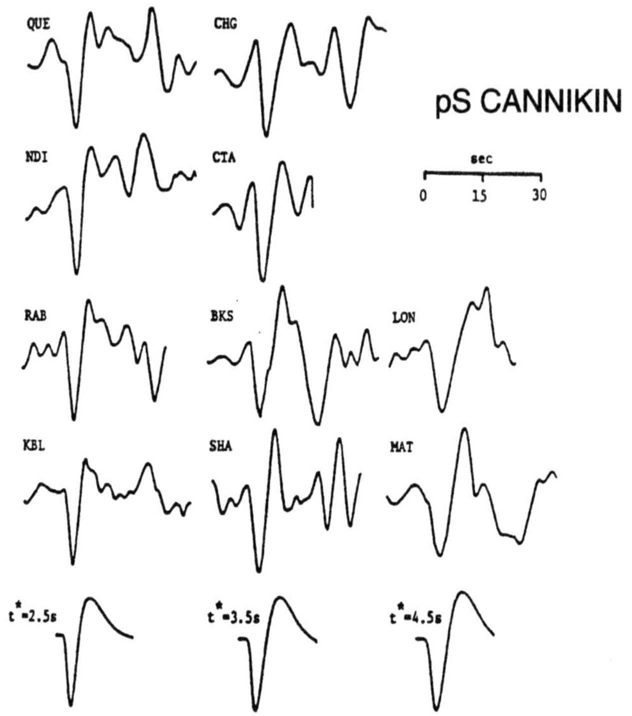

Figure 11

Comparison of observed long-period *pS* waveforms and synthetics using different values of t^*_β for event CANNIKIN. The source model and velocity structure are obtained from near-field modeling (taken from BURDICK *et al.*, 1984).

Source Scaling Relationship at NTS

The data set at NTS is substantially larger than at Amchitka, but again consists mostly of collections of strong-motion and teleseismic seismograms. The teleseismic estimate of long-period ψ_∞ for the NTS events are less certain than the estimates for the Amchitka events for several reasons. First, the *F* factor, or tectonic release, is markedly higher, in general, than at Amchitka, making it more difficult to correct for the surface waves. Secondly, the geologic structure is more complex in the source region, so that individual shots sample different source media (MURPHY, 1989). ψ_∞ estimates determined from seismic moments obtained from Rayleigh wave modeling (STEVENS, 1986), along with the ψ_∞ estimates from the intercorrelation technique (LAY, 1985), are given in Table 4. Note that ψ_∞ estimated from m_b is consistently smaller than ψ_∞ estimated from M_S; the ratio is about 3 to 1, although there is a considerable scatter. The depth effect on *B* is not particularly obvious in this data set. Also, the LP(SV)/SP(P) ratio is smaller although the event CHESHIRE is deeper than ESTUARY, which does not support the *B* Vs. depth dependence, discussed

Table 3

Synthetic pS *amplitudes* (μ) (K = 6)

t^*_β	B = 0.5, $\psi_\infty = 7.3 \times 10^{11}$ cm^3	B = 1, $\psi_\infty = 4.3 \times 10^{11}$ cm^3	B = 2, $\psi_\infty = 2.4 \times 10^{11}$ cm^3	B = 3, $\psi_\infty = 1.7 \times 10^{11}$ cm^3
1.5	4395	3324	2784	2652
2.0	3416	2503	2030	1904
2.5	2759	1965	1553	1442
3.0	2278	1591	1226	1127
3.5	1920	1320	996	904
4.0	1643	1190	823	741

Table 4

Strength estimates from surface waves and body waves

Events	Date	ψ_∞ from M_S (cm^3)	ψ_∞ from m_b (cm^3)	Depth (km)	Ratio
Scotch	05/23/67	9.1×10^{10}	1.9×10^{10}	0.98	4.8
Stinger	03/22/68	4.8×10^{10}	1.5×10^{10}	0.67	3.2
Sled	08/29/68	5.3×10^{10}	3.0×10^{10}	0.73	1.8
Almendro	06/06/73	1.9×10^{11}	6.8×10^{10}	1.06	2.8
Tybo	05/14/75	1.0×10^{11}	4.6×10^{10}	0.77	2.2
Stilton	06/03/75	4.8×10^{10}	2.4×10^{10}	0.73	2.0
Mast	06/16/75	1.2×10^{11}	4.9×10^{10}	0.91	2.4
Cheshire	02/14/76	2.0×10^{11}	3.5×10^{10}	1.17	5.7
Estuary	03/09/76	2.0×10^{11}	3.5×10^{10}	0.87	5.7
Pool	03/17/76	4.3×10^{10}	4.4×10^{10}	0.88	1.0

earlier in connection with the Amchitka data. The Rayleigh wave strength given by STEVENS (1986) is in general agreement with those given by GIVEN and MELLMAN (1986) so that a factor of about three appears appropriate in the ψ_∞ off-sets between short- and long-period estimates, with considerable scatter. The phase *pS* also has been used to estimate the source excitation. However, this proves difficult because of the tectonic release and the contamination produced by interference. The orientation of the best double-couple mechanism fitting the SH reveals mostly a strike-slip mechanism with SH maximums near the WWSSN stations OXF and FLO. These stations are naturally rotated allowing good estimates of ψ_∞ from the EW motions as displayed in Figure 12. A plot of their amplitudes versus yields also has proved quite accurate. In general, they correlate well with carefully corrected Rayleigh wave studies (GIVEN and MELLMAN, 1986). Consequently, there appears to be considerable variation in estimates of source strength at 1 sec vs long-period ($T \geq 10$ sec), which cannot be explained by a strike-slip tectonic release contribution.

A plot of averaged *P*-wave amplitudes vs *pS* at OXF is displayed in Figure 13. The curve displays the Amchitka experience assuming LAY *et al.* (1984) param-

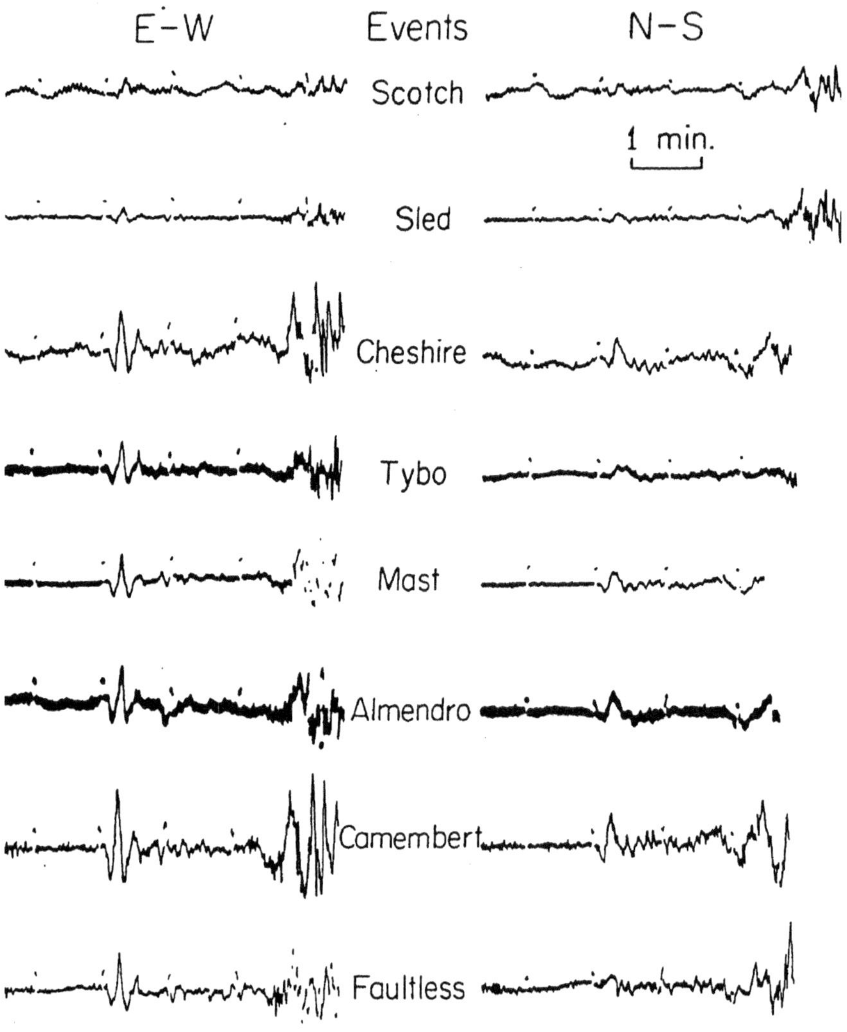

Figure 12
Shear-wave observations at OXF where the east-west component is primarily *SV* (*pS*). The *SH* motions are the strongest along this azimuth and *SV* should be the least contaminated by tectonic release.

eters. The effect of changing B with ψ_∞, or M_S fixed can be seen in Figure 14. Thus there is a clear tendency for deep events to display lower ratios as expected if overshoot decreases with burial depth. Given the complexity of the geology of Pahute Mesa, it is difficult to confirm this. There is, also, an abundance of free-field data at distances of less than a few kilometers; most of these data are for small events with yields less than 15 kt. The few measurements that do exist for large events are mostly inconsistent; this apparently is caused by local tectonic release effects (MURPHY, 1991). The strong-motion records at distances beyond a

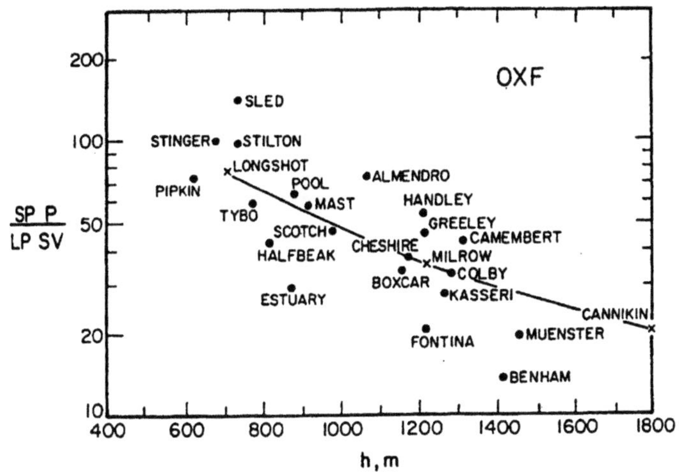

Figure 13
Ratios of (globally averaged short-period P waves (LAY, 1985)/long-period pS amplitudes recorded at OXF) for Pahute Mesa events plotted against depth.

few kilometers are vastly more complicated at NTS than at Amchitka, presumably because of the complex local geology (BARKER *et al.*, 1991). However, the near-field motions of the type discussed earlier for Amchitka can still be modeled using the structure derived by BARKER *et al.* (1991). Figures 15a–e show complete seismograms for the best recorded events. The columns on the left are appropriate for the $B = 1$ assumption used earlier in the extensive intercorrelation study (LAY *et al.*, 1984b). The synthetics on the right correspond to using a slightly different scaling law than BARKER *et al.* (1991), which allows for adjustments in ψ_∞ and K to fit the estimated long-period source strengths assuming $B = 0.1$. All synthetic seismograms were computed using the frequency-wavenumber integration method (SAIKIA, 1994) and the velocity model given in Table 5. The quality of fits is not as good as in the Amchitka case, but overall the fits assuming $B = 0.1$ are nearly as good as those for $B = 1$. Thus, it appears that allowing B to vary by about a factor of 10 trades-off with variations in ψ_∞ by about a factor of three. Experiments conducted at still smaller B's begins to indicate problems in fitting the extended P wave.

Following the results of strong-motion modeling assuming $B = 1$ (HARTZELL *et al.*, 1983), we used the parameterization given earlier in equations to obtain

$$C_1 = 20.1 \times 10^8 \quad C_2 = 4 \ ,$$

where C_1 has been increased by a factor of 3.

If we use the corner frequency scaling relative to Amchitka, we obtain the synthetic comparisons given in the column on the right in Figures 15(a–e). These begin to produce excessive short-period energy at the nearest stations and do not fit the acceleration data (HELMBERGER *et al.*, 1991). Therefore, the corner frequency difference between the two sites appears real.

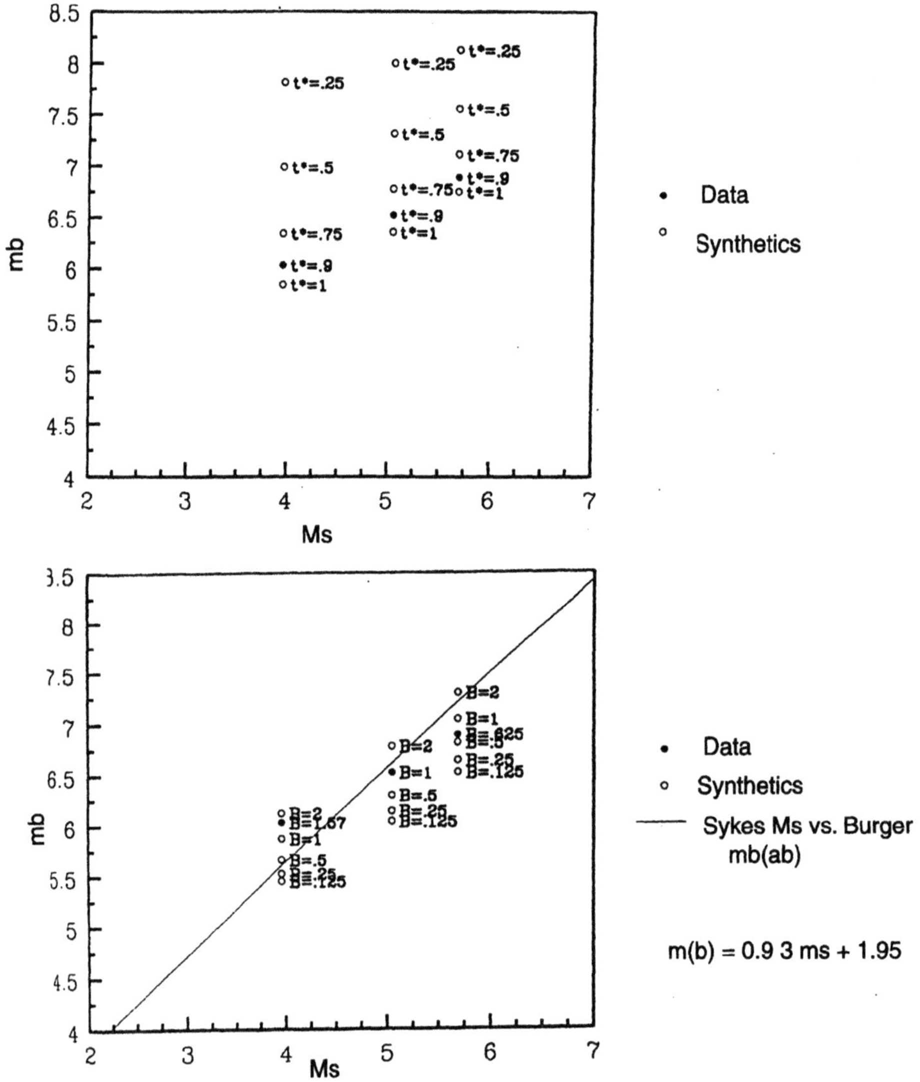

Figure 14
(a) Effects of varying t^* on three Amchitka source RDPs (LONGSHOT, MILROW and CANNIKIN, LAY *et al.*, 1984); source RDP parameters are held constant. Solid dots represent actual m_b:M_S values. (b) Effects of varying the overshoot parameter B on the same shot, but other parameters t^* are held constant. The line shows the regression $m_b = 1.95 + 0.93M_S$, derived form M_S values of SYKES and WIGGINS (1985) and $m_b(ab)$ values of BURGER *et al.* (1986). The lower slope of the regression implies a different B vs. yield relationship than that found for Amchitka.

Off-test-site Events

The local and teleseismic data from various off-site tests have been extensively studied in attempts to explain the scatter from standard $m_b(M_S)$ vs Y curves. For example, LRSM stations were established at FAULTLESS, RULISON, and GASBUGGY epicenters to measure the relative attenuation in order to explain δm_b's with respect to δt_α differences. These results generally do not appear to explain the anomalies (DER *et al.*, 1980). For example, the attenuation at the RULISON site is nearly the same as at GASBUGGY, and attenuation at FAULTLESS is the same as it is at NTS. Thus, anomalies in m_b remain unexplained. On the other hand, M_S values prove effective in estimating these relative yields (YACOUB, 1983).

RULISON and GASBUGGY, a Test of Source Scaling

The two off-NTS nuclear blasts, RULISON and GASBUGGY, provide particularly useful test cases for determining the transportability of different methods of estimating yield. These two nuclear events were detonated 304 km apart in nearby sedimentary basins in New Mexico and Colorado. Although the announced yield of RULISON was 40 kt and that of GASBUGGY was 29 kt, GASBUGGY has a larger m_b. A thorough analysis of this discrepancy was presented by MURPHY and ARCHAMBEAU (1986). Although the amplitude differences between the two events are of the order that might be expected from a difference in t_α^* of 0.4 sec between the two events, their analysis of the P-wave displacement spectra of the two events indicated that the effective t_α^* for the two events are essentially the same, in agreement with DER *et al.* (1980) results.

MURPHY and ARCHAMBEAU (1986) present evidence that the RULISON-GASBUGGY anomaly is principally caused by tectonic release associated with RULISON. They suggest that this release of tectonic energy was oriented in such a fashion as to destructively interfere with and reduce the amplitudes of the teleseismic short-period P waves for RULISON. The magnitude of the proposed release is such that the amplitude of the tectonic P waves is roughly one half of the amplitude of the explosion P waves. Although the tectonic release is a possible explanation, in Appendix B we demonstrate that the tectonic release contributions in these events are not significant to explain the discrepancy. Another interpretation is that this reduction in RULISON's amplitude is due to a decrease in the B factor as suggested by the above scaling relations for Amchitka.

Using the Amchitka scaling relations (7) and (10) of LAY *et al.* (1984a), the predicted values for GASBUGGY ($h = 1.29$ km, $Y = 29$ kt) are $B = 0.89$, $K = 31.3$, and $\psi_\infty = 4.31 \times 10^9$ cm^3, or $M_0 = 2.35 \times 10^{22}$ dyne-cm. For RULISON ($h = 2.57$ km, $Y = 40$ kt), the predicted values are $B = 0.46$, $K = 37.7$, and $\psi_\infty = 4.71 \times 10^9$ cm^3, or $M_0 = 2.94 \times 10^{22}$ dyne-cm. The predicted values for these events using our new Amchitka scaling relations are $B = 0.64$, $K = 31.3$, and

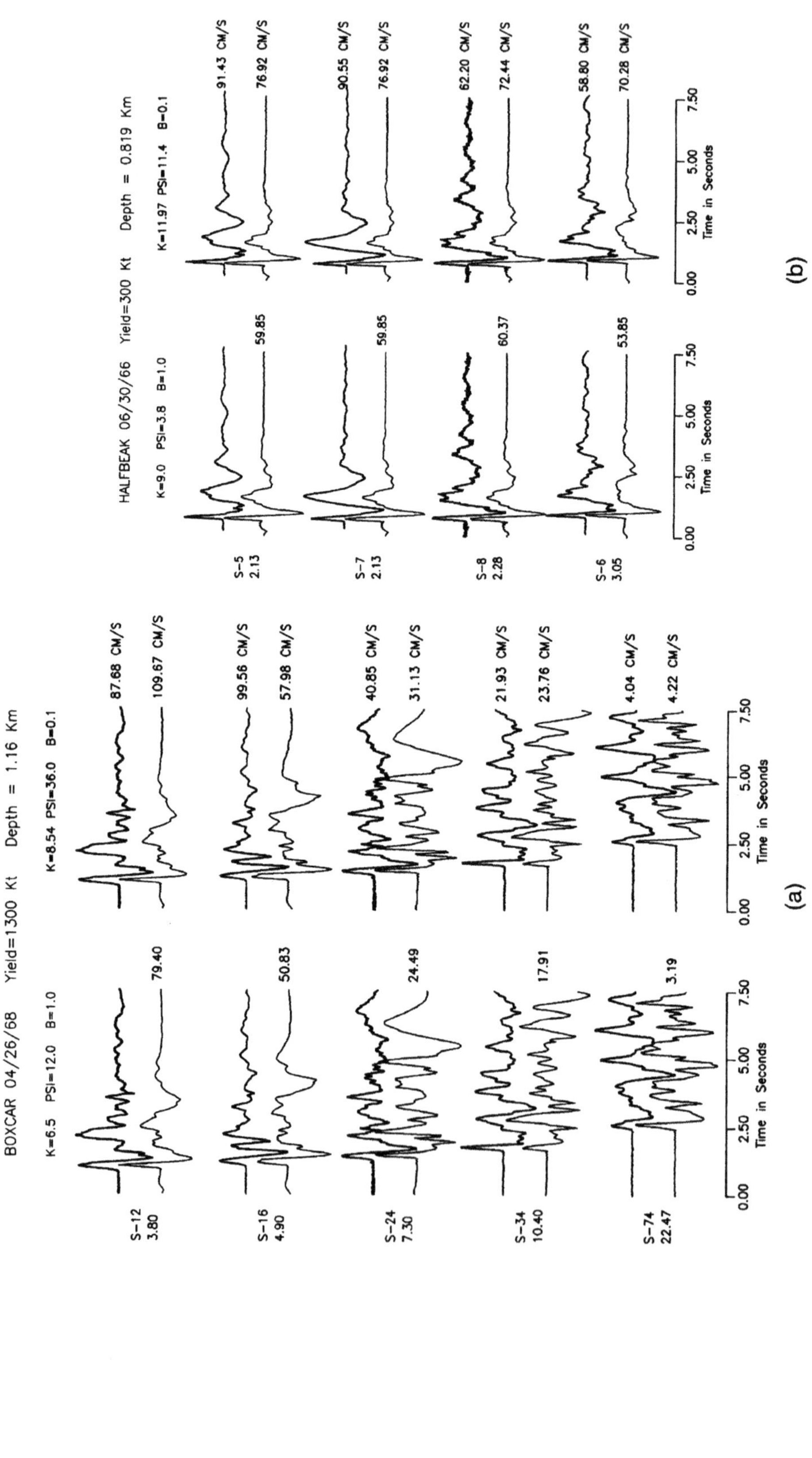
BOXCAR 04/26/68 Yield=1300 Kt Depth = 1.16 Km
K=6.5 PSI=12.0 B=1.0
K=8.54 PSI=36.0 B=0.1
S-12 3.80
79.40
87.68 CM/S
109.67 CM/S
S-16 4.90
50.83
99.56 CM/S
57.98 CM/S
S-24 7.30
24.49
40.85 CM/S
31.13 CM/S
S-34 10.40
17.91
21.93 CM/S
23.76 CM/S
S-74 22.47
3.19
4.04 CM/S
4.22 CM/S
0.00 2.50 5.00 7.50
Time in Seconds
(a)
HALFBEAK 06/30/66 Yield=300 Kt Depth = 0.819 Km
K=9.0 PSI=3.8 B=1.0
K=11.97 PSI=11.4 B=0.1
S-5 2.13
59.85
91.43 CM/S
76.92 CM/S
S-7 2.13
59.85
90.55 CM/S
76.92 CM/S
S-8 2.28
60.37
62.20 CM/S
72.44 CM/S
S-6 3.05
53.85
58.80 CM/S
70.28 CM/S
0.00 2.50 5.00 7.50
Time in Seconds
(b)

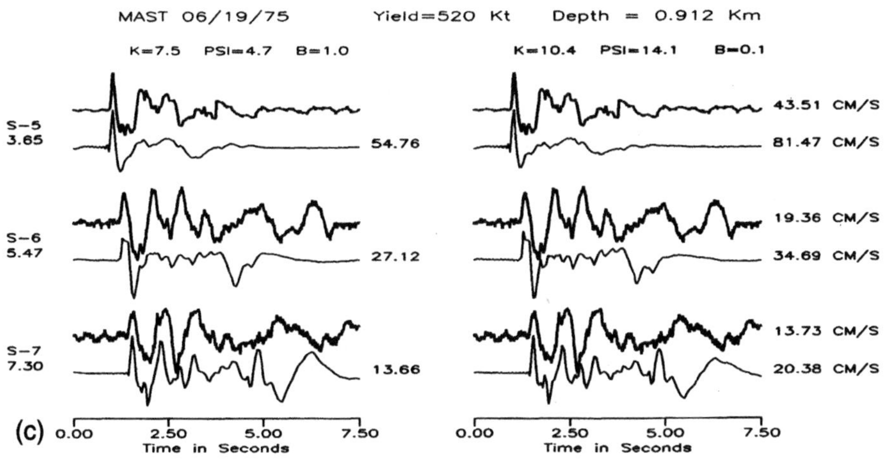

Figure 15c
Near-in observations (top bold traces) and predictions (bottom traces) for event MAST. The station names and distances are printed on the left; amplitudes are printed on the right, and values of K, ψ_∞, and B are printed above each column.

$\psi_\infty = 5.16 \times 10^9$ cm^3 or $M_0 = 2.81 \times 10^{22}$ dyne-cm for GASBUGGY and $B = 0.26$, $K = 37.7$, and $\psi_\infty = 5.91 \times 10^9$ cm^3, or $M_0 = 3.74 \times 10^{22}$ dyne-cm (Fig. 16). The moments are close to those predicted by surface-wave modeling (BURDICK *et al.*, 1984). They report $\psi_\infty = 3.6 \times 10^9$ and 3.3×10^9 cm^3 for GASBUGGY and RULISON respectively, with a strong tectonic bias factor associated with RULISON in agreement with MURPHY and ARCHAMBEAU (1986). Synthetic seismograms calculated using these source function parameters and the crustal structures given by Murphy and Archambeau predict that the m_b for GASBUGGY is 0.23 units higher than that for RULISON. This difference is primarily caused by the small B value for RULISON due to its greater source depth.

Although the Amchitka scaling laws favorably predict the differential δm_b between GASBUGGY and RULISON, an examination of the recorded RDPs of GASBUGGY (Fig. 16) shows that the prediction is not perfect. Only one of the observed RDPs exhibits a significant overshoot. The mean overshoot is 15%, corresponding to a B of 0.32. A K value approximately equal to 27 with a

◄

Figure 15a
Near-in observations (top bold traces) and predictions (bottom traces) for event BOXCAR. The station names and distances are printed on the left; amplitudes are printed on the right, and values of K, ψ_∞, and B are printed above each column.

Figure 15b
Near-in observations (top bold traces) and predictions (bottom traces) for event HALFBEAK. The station names and distances are printed on the left; amplitudes are printed on the right, and values of K, ψ_∞, and B are printed above each column.

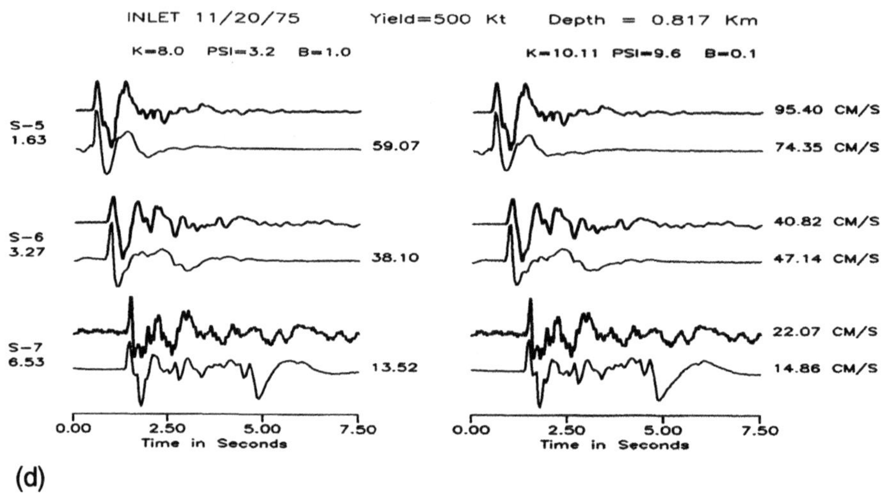

Figure 15d
Near-in observations (top bold traces) and predictions (bottom traces) for event INLET. The station names and distances are printed on the left; amplitudes are printed on the right, and values of K, ψ_∞, and B are printed above each column.

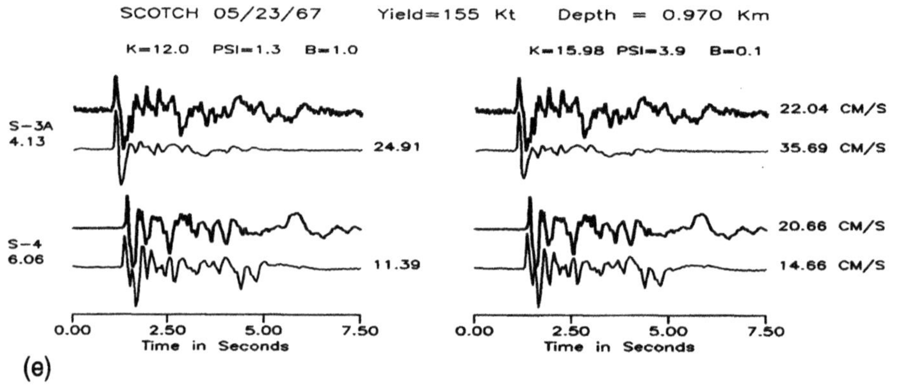

Figure 15e
Near-in observations (top bold traces) and predictions (bottom traces) for event SCOTCH. The station names and distances are printed on the left; amplitudes are printed on the right, and values of K, ψ_∞, and B are printed above each column.

$\psi_\infty = 5.9 \times 10^9$ cm^3 was needed to model the rise time (see the bottom trace). These observations of the direct RDP are rather unique in that gas well boreholes were available for subsurface observations (PERRET, 1982). They were all taken at a slant range of 0.5 km, at several azimuths well outside the nonlinear zone. The average RDP at the bottom of Figure 16 displays a higher ψ_∞ and a lower overshoot than

Table 5

Pahute Mesa velocity structure model

Layer	α, km/s	β, km/s	ρ, g/cm^3	Thickness, km
1	2.30	1.35	1.90	0.36
2	2.80	1.50	2.00	0.40
3	3.30	1.50	2.25	0.70
4	4.00	1.90	2.30	0.70
5	4.60	2.00	2.40	0.75
6	5.30	2.50	2.50	0.80
7	5.50	2.95	2.70	2.25
8	6.10	3.50	3.00	10.00
9	7.00	4.00	3.00	10.00

Model from HARZTELL *et al.* (1983).

predicted by the scaling laws from Amchitka, but move in line with the B behavior at NTS.

The moments obtained from modeling the surface waves (BURDICK *et al.*, 1989) are 2.0×10^{22} and 2.1×10^{22} dyne-cm, respectively for GASBUGGY and RULISON, which are somewhat low compared to both the observed and predicted RDP. These discrepancies are probably within modeling errors, although evidence for undershoot in this experiment is strong.

FAULTLESS and ALMENDRO, a Further Test of Scaling Laws

Since the P waveforms from Pahute Mesa events have been studied in great detail by many investigators, it is natural to compare data from the FAULTLESS ($m_b = 6.4$) event with a Pahute Mesa event (ALMENDRO, $m_b = 6.1$) having a comparable yield (Figure 12). Both blasts were detonated in similar environments at a depth of 1.1 km. ALMENDRO ($m_b = 6.1$) was located roughly 150 km to the south. The pairs of seismograms are from the same teleseismic WWSSN stations, where it appears that the FAULTLESS seismograms are anomalously long period. This suggests a higher t_α^* for FAULTLESS, but several detailed attenuation experiments indicate comparable values of t_α^* at these two sites. Long-period information pertaining to ALMENDRO and FAULTLESS is indicative of similar M_Ss for the two events with relatively low tectonic release (BURDICK *et al.*, 1984). Figure 12 also shows similar looking pS phases for both events. Therefore, it is appropriate to assume the same ψ_∞ for each event and determine a B_h (indicating a depth dependence) differential so that the difference in amplitude (m_b) between these two events can be explained. One way to accomplish this is to treat ALMENDRO as a known master event and fix its source. Although the interaction with the free surface above the sources is a well-known complex issue, the process can be idealized by comparing the waveforms of ALMENDRO (Fig. 17) with all the other Pahute

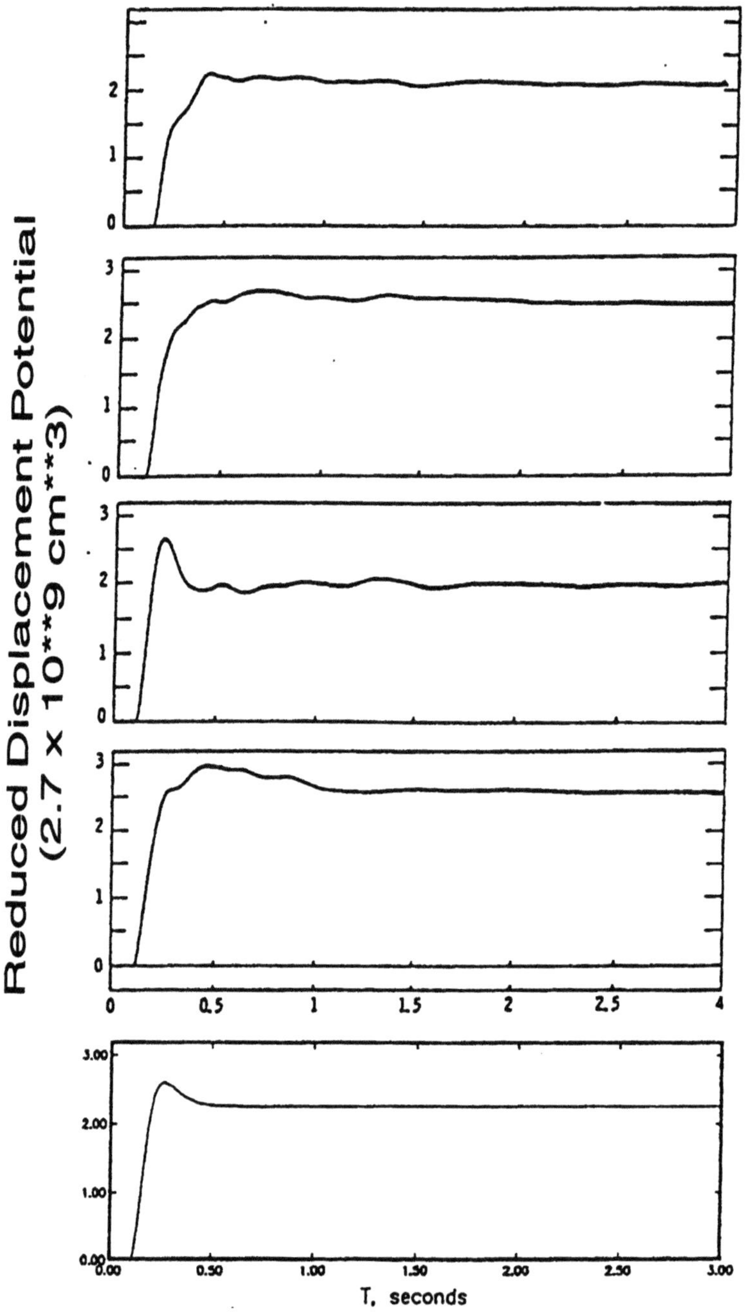

Figure 16
The upper four traces display the observed RDPs obtained from the GASBUGGY explosion, after PERRET (1982). The bottom trace displays a scaled RDP.

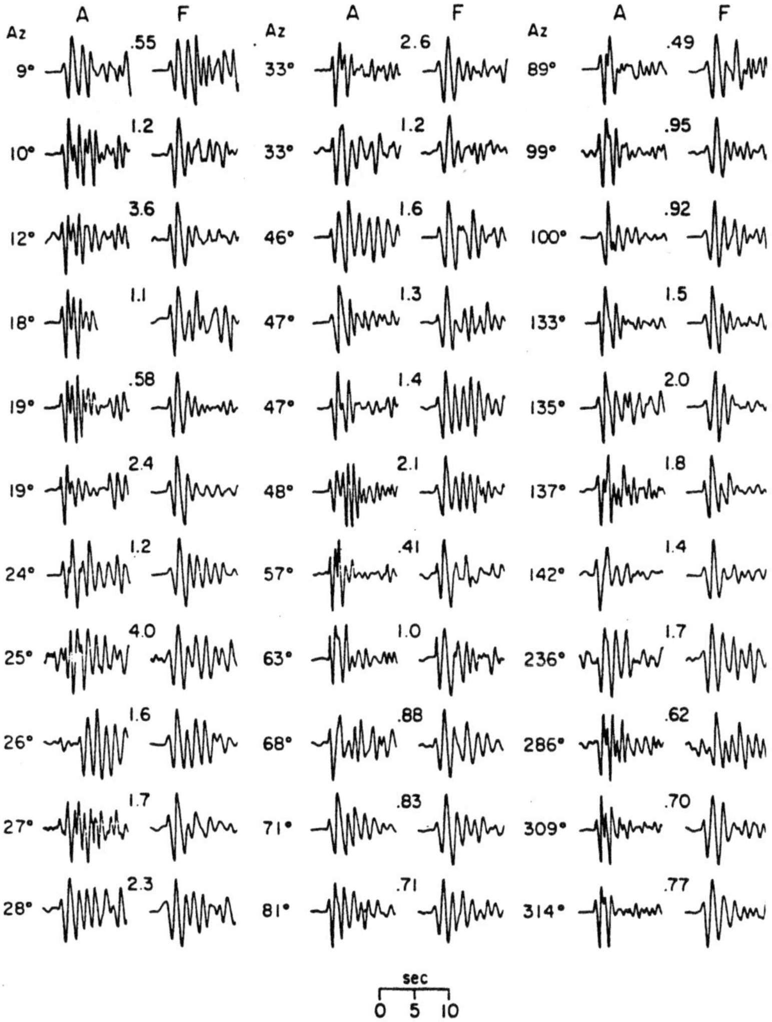

Figure 17
A comparison of seismograms of observed WWSSN P waveforms for the two explosions ALMENDRO and FAULTLESS as a function of azimuth (same station). The numbers indicate the ratios of peak amplitudes (FAULTLESS/ALMENDRO).

events (LAY, 1985). He obtained a time separation of 1 s, $pP\text{-}P = 1$, and amplitude ratio, pP/P, of 1 for ALMENDRO.

Assuming the effective source function of ALMENDRO as known, the FAULTLESS source can be determined by convolving the FAULTLESS records with the ALMENDRO source for different B values. A grid search of B values indicated a B value near a factor of two greater than ALMENDRO, which produced an excellent match to the data, yielding a differential m_b (δm_b) value of 0.3 between the two events (BURDICK *et al.*, 1984). A less involved estimate can be obtained from Figure 14 with respect to δm_b vs M_S.

Qualitatively it appears that the Pahute Mesa events have a depth dependence similar to the Amchitka events but which starts with smaller B values. However, given the complexity in geology and the other reasons for source variations, it may not be easy to demonstrate this effect conclusively. The contrast between the PNE's RULISON and GASBUGGY, fired in similar rock types but at greatly different depths, appears to be the best test case available, followed by the ALMENDRO-FAULTLESS comparison.

Scaling Relations at Soviet Sites

In this section we discuss some of the yield equations put forward for sites outside the United States before the Soviets released their yield data. Given the obvious disparity in m_b vs M_S values discussed earlier, Figure 1 leads to a great deal of speculation about the relative size of the Soviet yields relative to the US shots. The use of m_b and the US-scaling laws suggests verification problems (ALEWINE and BACHE, 1983). On the other hand, using M_S suggests compliance (SYKES and WIGGINS, 1986). They argued that using events with only small Love waves demonstrates that Amchitka M_S's and yield-laws could be used to estimate yields, namely

$$M_S = 0.97 \log Y + 2.16$$

for Novaya Zemlya (NZ, SYKES and WIGGINS, 1986) and

$$M_S = 0.95 \log Y + 2.16$$

for Semipalatinsk test site (STS, SYKES and CIFUENTES, 1984), Y being the yield in kt. Their argument was strengthened at the STS by a detailed study by STEVENS (1986) in which the author compared log-moments of the 10 largest NTS explosions (Pahute, $m_b \approx 5.3$–5.8) and the 10 largest STS ($m_b \approx 5.8$–6.2) events fired after 1976 following the 150 kt threshold compliance by the Threshold Test-Ban Treaty. Based on these data, BURDICK *et al.* (1989) established an M_S differential *pf* 0.36 (M_S (Pahute) – M_S (STS)) between the two test sites, with an average seismic moment (M_0) of 7.58×10^{22} dyne-cm for the STS events vs an average M_0 of 1.74×10^{23} dyne-cm for the NTS events. Several factors, including tectonic strain release, differences in body-wave attenuation, or an unusual source medium could account for these differences. Correction for tectonic strain release can increase the seismic moment for some events, but the STS explosions still do not become consistent with the standard models of a granite-like source medium (STEVENS, 1986). Thus, he argues against the tectonic release and suggests "one possible explanation for large body waves and small surface waves from East Kazakhstan explosions would be the presence of a source medium at the test site that results in a highly peaked source spectrum with a large amplitude near 1 Hz and an unusually small long-period level". Clearly, this feature could be caused by a medium dependent B factor, as discussed earlier. Studies like this lead to the establishment of δm_b-bias for Soviet sites. This δm_b-bias for Soviet sites is usually attributed to attenuation (MURPHY, 1998), but could also be produced by an increase of B which would explain the δm_b bias simultaneously (Fig. 14). The alternative explanation is that the δm_b bias is caused by a destructive interference of seismic waves with tectonic release not always associated with strong Love waves, thus requiring larger yields at STS. While this procedure proved effective at STS as demonstrated by the JVE experiments, the M_S vs yield relation predicts a yield of 2800 kt for the largest NZ event (Oct. 27, 1973, $m_b = 6.98$, $M_S = 5.5$), a value somewhat lower than the value that would be obtained from the m_b^{ISC} vs yield curve of ADUSHKIN (1996, see his Fig. 6); a value near 4000 kt is more appropriate. He fits the NZ yields with a break in the magnitude scale at 150 kt as follows

$$m_b = 0.76 \log Y + 4.32 \quad y \leq 150\,\text{kt},$$

$$m_b = 0.48 \log Y + 5.10 \quad y \geq 150\,\text{kt} \ .$$

A similar break occurs in the few points from the experiments at Amchitka which can also be interpreted as caused by a change in B values for the larger events.

It appears that tectonic release systematically reduced the explosion-generated Rayleigh waves at Novaya Zemlya. This same feature occurred for the pS studies conducted by BURGER *et al.* (1986a,b) and COHEE and LAY (1988). Unfortunately this is a difficult feature to correct for, given the ambiguous Love-wave patterns. For the southern Novaya Zemlya (SNZ) explosions the patterns can be fit with a strike-slip or thrust mechanism producing different SV patterns, which interfere with the explosion-generated pS, yielding different estimates of ψ_∞ based on the mechanism. For example, for the 27th October, 1973 event a strike-slip mechanism would require a ψ_∞ of 3.5×10^{11} cm^3, while a thrust solution would require a ψ_∞ of 7.3×10^{11} cm^3 to fit the observations, which is larger by about a factor of 2.1. A more detailed study of this event by COHEE and LAY (1988) produced a value of 3.8×10^{11}, which proves to be quite accurate (Adushkin's results) assuming the Amchitka relationship.

Another effort aimed at yield estimation was conducted by BURGER *et al.* (1986) applying the waveform intercorrelation method. While this method proves effective in ranking events in order of yield and making relative comparisons, it suffers from a lack of baseline ψ_∞ yield scaling information and t^* uncertainty. They argue that the Amchitka scaling law of LAY *et al.* (1984b), with $B = 1$ and a $t^* = 0.5$, produces reasonable synthetic waveform shapes. Their results are displayed in Table 6. If they assumed $t^* = 0.65$, all the yields would be larger by 1.9 whereas a

Table 6

Various parameters for some selected Novaya Zemlya and Amchitka events

Date	O.T.	Lat(°N)	Lon(°E)	m_b	*Ya* (kt)	*Yb* (kt)	*pP-P*(sec)	*SP/LP*
			Northern Novaya Zemlya					
10.27.66	05:57:58	73.44	54.75	6.50	770	600	0.74	1.2
10.14.70	05:59:57	73.31	55.15	6.88	2100	1714	0.65	1.7
09.27.71	05:59:55	73.39	55.10	6.63	770	973	0.73	1.3
09.12.73	06:59:54	73.30	55.16	6.98	2700	2824	0.71	1.7
08.29.74	09:59:58	73.37	55.09	6.53	870	629	0.71	1.0
08.23.75	08:59:58	73.37	54.64	6.37	550	604	0.62	1.3
10.21.75	11:59:57	73.35	55.08	6.5	700	554	0.69	1.3
			Southern Novaya Zemlya					
10.27.73	06:59:57	70.78	54.18	7.14	3200	3886	0.70	0.68
11.02.74	04:59:57	70.82	54.18	6.90	1600	1624	0.58	1.6
10.18.75	08:59:56	70.84	53.69	6.68	1400	1166	0.55	1.1
			Amchitka (CANNIKIN)					
11.06.71					< 5000 (*Ya*+)			0.41

Ya are yield estimates taken from DAHLMAN O., and H. ISRAELSON (1977). *Monitoring Underground Nuclear Explosions*, Elsevier, New York.
Yb are yield estimates taken from Table 1 of BURGER *et al.* (1986a).
SP/LP are short-period to long-period amplitude ratios, taken from BURDICK and HELMBERGER, 1979.
pP-P values taken from Table 6 of BURGER *et al.* (1986a).

$t^* = 0.35$ would decrease the yield estimates by a factor of 0.44. A similar estimate of this trade-off can be seen again in Figure 14, assuming the ψ_∞ and yield are directly related to M_S. The depth determinations also are displayed in Table 6 as (*pP-P*). Note that the variation for the tunneled shots (NZ) is small, while the drilled-shots (SNZ) events behave more like the US-Pahute experience, where large events are fired at a deeper depth than small events. The ratio of short-period *P* to long-period *P* is also given, taken from BURDICK and HELMBERGER (1979), except for the CANNIKIN event which has been increased slightly based on more data (BURGER *et al.*, 1986). Only the 10/27/93 event has a similarly low ratio. Note that the two largest NNZ events (10/14/70 and 9/12/73) have the largest ratios – indicative of a large *B* on over-shoot, as displayed earlier in Figure 10. However, there is also a trade-off with decreasing depth that drives the ratio up as discussed by BURDICK and HELMBERGER (1979). At any rate, the ratios seem to be bracketed quite nicely by t^*'s from shot-to-shot, in the range of 0.5 to 0.65, for a variety of *B*'s. While the *pP-P* times and t^* values displayed here and at other test-sites are quite similar to those determined from spectral analysis (MURPHY *et al.*, 1992) the *pP*-amplitude estimates are considerably larger. For example, the amplitudes *pP* for Pahute events are typically low between 0.3 and 0.6 (MURPHY, 1989) while intercorrelation yields values near 1. It appears that some of our changes in source description are caused by the complex surface interactions. Essentially one can fix the source behavior and adjust the *pP* amplitude to satisfy data assuming simple elastic behavior (large *pP*'s) and change the *B* factor. A study of near-in Soviet shot data (NNZ) of the type considered at Amchitka could help establish the *B*-factor dependents since the depths are relatively fixed for the tunnel shots. However modeling broadband data either in the time domain or spectral domain helps to narrow the ranges of yield estimation. Unfortunately shots in uncalibrated regions will prove difficult to estimate, especially without controls on the crustal structure and attenuation.

Conclusions

While the contribution from the tectonic release is often considered as one possible explanation for the variation in Rayleigh waves which may cause M_S to vary, in this study we have shown that shallow station structure and lateral velocity variations can also significantly effect the ground motions, based on the near-in data recorded from the Amchitka blast MILROW. The velocity structure is at least as important as the choice of explosion source time function even for the near-field velocity seismograms. A velocity structure model with gradational boundaries near-in the upper crust produces better agreement with the data than those simulated using sharp boundaries. We found that effects of various types of source functions, for example those of HELMBERGER and HADLEY (1981), MUELLER and MURPHY (1971), and VON-SEGGERN and BLANDFORD (1971), are marginally different from one another, although the Mueller-Murphy source yielded better agreement to the near-in data, particularly that recorded from the CANNIKIN blast. The amplitude for the MILROW observations are within 5% of those predicted for all three RDP source types.

After reviewing the broadband data from US explosions, including near-in strong motions, teleseismic short periods (m_b) and teleseismic long periods (M_S), we conclude that NTS (Pahute) may have unique source properties relative to the other test sites. In particular the ratio of (M_S/m_b) is anomalously large, which can be explained kinematically by the absence of overshoot, i.e., the Rayleigh waves are large. Other sites such as Amchitka and the PNEs, RULISON and GASBUGGY, indicate substantial overshoot. Furthermore, this overshoot seems to be a function of source depth, decreasing with depth. This feature leads to a break-over in m_b:yield curves near 150 kt which appears to fit the NZ experience. The break-over is less clear in the NTS data because of the lack of overshoot for shallow events. The long-period to short-period *P*-wave data for the Soviet events support a larger overshoot (our interpretation) or a combination of δt^* bias plus source dependent *pP* interaction (MURPHY, 1998). To resolve this issue requires near-in observations at the various former Soviet Union test sites.

Appendix A

Mueller Murphy and von Seggern Explosion Sources

In this section we present the RDP explosion source functions of VON SEGGERN and BLANDFORD (1972) and MUELLER and MURPHY (1971) which are also frequently used in the study of explosions and which we have employed to demonstrate that crustal structure has more effect in determining the amplitudes and shaping the waveforms rather than the type of RDP used. We start with the von Seggern-Blandford RDP source which is given by

$$\psi(t) = \psi_\infty[1 - \mathrm{e}^{-K't}(1 + K't - B'(K't)^2)] \ ,$$

where ψ_∞ is the source strength, K prime and B prime are corner frequency and overshoot parameters similar to K and B in the HELMBERGER and HADLEY (H-H) source.

The RDP source of MUELLER and MURPHY (1971) can be expressed most easily as a convolution (BARKET *et al.*, 1985) as

$$\psi(t) = \frac{r_{el}V_p^2}{4\mu}P(t) * F(t) \ ,$$

where the * indicates convolution and $P(t)$ and $F(t)$ are as follows

$$P(t) = \{(P_{os} - P_{oc})\mathrm{e}^{-\alpha t} + P_{oc}\}H(t) \quad F(t) = \frac{\sin(bt)e^{at}}{\beta b}$$

where P_{os} is dynamic (peak shock) cavity pressure, P_{oc} is static (steady-state) cavity pressure and r_{el} and r_e are elastic and cavity radii respectively, and are given by

$$P_{oc} = \frac{4}{3}C_3\mu\frac{r_e^3}{r_{el}^3} \quad P_{os} = 1.5\,\rho g h \quad R_{el} = C_1\frac{W^{0.33}}{(h/100)^{0.42}} \quad r_e = C_2\frac{W^{0.29}}{(h/100)^{0.11}}$$

and

$$\alpha = C_4\omega_0 = \frac{C_4 r_{el}}{V_p} \quad \beta = \frac{\lambda + 2\mu}{4\mu} \quad a = \frac{-\omega_0}{2\beta} \quad b = \frac{\omega_0}{\beta}(\beta - 0.25)^{0.5} \ .$$

The constants C_1 through C_4 are calibration constants dependent on the source medium (STEAD, 1989). We have shown a comparison of these three sources in Figure 7; the top graph shows the time derivative of the RDP, which is equivalent to the far-field displacement, the middle graph shows the RDPs of the sources where the long-period level ψ_∞ is indicated and the bottom shows the amplitude spectra of the three far-field displacement time functions.

Table B-1

GASBUGGY Gutenberg-Richter magnitudes WCC measurements, 21°–100°

Station	Delta(°)	Azimuth	$A(\mu$, 0-$p)$	T(sec)	A/T	log(A/T)	B	*Mag.*	*Resid.*
BLA	21.4	80.	198.5	1.1	177.7	2.25	3.14	5.39	−0.27
OGD	25.7	70.	70.7	1.2	57.0	1.76	3.43	5.19	−0.07
CMC	31.5	354.	4.1	1.2	3.5	0.54	3.70	4.24	0.88
COL	37.0	333.	74.3	1.2	63.4	1.80	3.50	5.30	−0.18
SJG	40.5	106.	9.1	1.2	7.7	0.89	3.45	4.34	0.78
CAR	44.7	115.	54.4	0.8	69.0	1.84	3.64	5.48	−0.36
KtG	54.4	24.	13.6	1.3	10.6	1.03	3.80	4.83	0.29
NNA	56.2	143.	8.3	1.6	5.3	0.72	3.80	4.52	0.60
ARE	62.7	141.	26.7	1.0	27.8	1.44	3.93	5.38	−0.26
ESK*	67.9	37.	18.0	1.0	18.0	1.26	4.00	5.26	−0.14
LOR	76.2	41.	31.5	1.2	26.7	1.43	3.90	5.33	−0.21
TRI	82.3	37.	11.3	1.3	8.5	0.93	3.93	4.86	0.26
IST	87.1	32.	29.6	1.5	20.4	1.31	4.01	5.32	−0.20

Number of Stations = 13, Mean magnitude 5.12, s.d. 0.29.

Table B-2

GASBUGGY Veith-Clawson magnitudes WCC measurements, 21°–100°

Station	Delta(°)	Azimuth	$A(\mu$, p-$p)$	T(sec)	A/T	log(A/T)	B	*Mag.*	*Resid.*
BLA	21.4	80.	335.7	1.1	300.4	2.48	2.82	5.30	−0.28
OGD	25.7	70.	112.4	1.2	90.6	1.96	3.22	5.17	−0.15
CMC	354.0		9.1	1.2	7.8	0.89	3.37	4.26	0.76
COL	37.0	333.	113.5	1.2	96.9	1.99	3.34	5.33	−0.30
SJG	40.5	106.	19.2	1.2	16.3	1.21	3.32	4.53	0.49
CAR	44.7	115.	90.7	0.8	115.0	2.06	3.34	5.40	−0.38
KtG	54.4	24.	27.3	1.3	21.3	1.33	3.39	4.72	0.30
NNA	56.2	143.	16.5	1.6	10.5	1.02	3.40	4.42	0.60
ARE	62.7	141.	38.2	1.0	39.9	1.60	3.45	5.05	−0.03
ESK*	67.9	37.	35.0	1.0	35.0	1.54	3.48	5.02	0.00
LOR	76.2	41.	50.6	1.2	43.0	1.63	3.53	5.16	−0.14
TRI	82.3	37.	20.6	1.3	15.6	1.19	3.57	4.76	0.26
IST	87.1	32.	52.6	1.5	36.2	1.56	3.66	5.22	−0.20

Number of Stations = 13, Mean magnitude 5.02, s.d. 0.27.

Appendix B

Analysis of GASBUGGY and RULISON Data

In this we present a detailed discussion on the amplitudes of the short-period *P*-wave observations (m_b) and relative strengths of the regional surface waves recorded from GASBUGGY and RULISON with an emphasis towards establishing the presence of tectonic-release effects that may accompany these explosions.

Table B-3

RULISON Gutenberg-Richter magnitudes WCC measurements, 21°–100°

Sation	Delta (°)	Azimuth	$A(\mu$, 0-$p)$	T(sec)	A/T	$\log(A/T)$	B	*Mag.*	*Resid.*
SCP	23.0	77.	41.9	0.8	53.4	1.73	3.30	5.03	−0.01
OGD	25.4	75.	30.8	0.9	34.5	1.54	3.46	5.00	0.02
COL	34.4	331.	18.5	1.1	16.2	1.21	3.70	4.91	0.11
ARE*	65.2	141.	17.0	1.0	17.0	1.23	4.00	5.23	−0.21
ESK*	66.0	37.	7.1	0.6	10.9	1.04	4.00	5.04	−0.02
LOR*	74.5	41.	10.4	1.1	9.0	0.96	3.80	4.76	0.26
MAT	83.0	312.	16.4	1.1	15.4	1.19	4.00	5.18	−0.16

Number of Stations = 7, Mean magnitude 5.02, s.d. 0.15.

Table B-4

RULISON Veith-Clawson magnitudes WCC measurements, 21°–100°

Station	Delta (°)	Azimuth	$A(\mu, p\text{-}p)$	T(sec)	A/T	$\log(A/T)$	B	*Mag.*	*Resid.*
SCP	23.0	77.	124.0	0.8	157.8	2.20	2.94	5.14	−0.22
OGD	25.4	75.	54.1	0.9	60.6	1.78	3.19	4.97	−0.06
COL	34.4	331.	36.5	1.1	32.0	1.50	3.35	4.85	0.06
ARE*	65.2	141.	30.0	1.0	30.0	1.48	3.46	4.94	−0.02
ESK*	66.0	37.	14.2	0.6	21.8	1.34	3.46	4.80	0.11
LOR*	74.5	41.	18.7	1.1	16.3	1.21	3.52	4.73	0.19
MAT	83.0	312.	26.1	1.1	24.5	1.39	3.58	4.97	−0.06

Number of Stations = 7, Mean magnitude 4.91, s.d. 0.13.

Table B-5

Summary of m_b's from stations recording both RULISON and GASBUGGY

Station	GASBUGGY m_b	RULISON m_b	Δm_b (G-R)
TUC	5.13	5.01	0.11
GSC	5.67	5.13	0.54
BKS	5.31	5.38	−0.07
ATL	4.51	4.63	−0.12
OGD	5.19	5.00	0.19
COL	5.30	4.91	0.39
ARE	5.38	5.23	0.15
ESK	5.26	5.04	0.22
LOR	5.33	4.76	0.57
TRI	4.86	*	
CAR	5.48	*	
NNA	4.52	*	

* At or below noise level.

Amplitude of Short-period P Waves (m_b)

We recalculated the ISC magnitude using the log(A/T) values reported in the ISC bulletin for RULISON and GASBUGGY and m_b for these events were estimated to be 5.07 and 5.09, respectively. For comparison with other measurements we also measured WWSSN short-period data for both, using the Gutenberg-Richter (GR) and Veith-Clawson (VC) m_b correction scales. To avoid problems with mantle triplications, we used observations from 21° and beyond (Tables B-1 to B-4) to estimate m_b. To derive an accurate picture, we estimated m_b's for the two events using the stations that recorded both the events; hoping to minimize differences due to station structure or mantle paths. Table B-5 compares the GR m_b's for which the waveforms were analyzed. GASBUGGY had higher m_b estimates at every station with the exception of two regional stations BKS and ATL. At several stations, GASBUGGY signals could be detected, but not those of RULISON. Using the most stringent test, that is using only the teleseismic stations which recorded both events, namely OGD, COL, ARE, ESK and LOR, thus eliminating the problem of amplitude variation due to the mantle triplication, and minimizing the path and station differences between the two events, we found the GR m_b to be 5.29 for GASBUGGY which is 0.3 magnitude units larger than that of RULISON. The VC correction tables, which include many WWSSN stations at regional and near-regional distances, also showed a δm_b of 0.29 units of magnitude, which is in agreement with MURPHY and ARCHAMBEAU (1986).

Modeling of Regional Surface Waves

We collected WWSSN records from North American stations at epicentral distances less than 15°, hand-digitized and then rotated them. Seismograms from 10 stations were collected for RULISON and 8 for GASBUGGY (Figs. B-1 and B-2). The EW component at GSC for GASBUGGY had an incorrectly reported gain, but was corrected by the time of RULISON blast.

The clearest indication of the presence of the non-explosive energy on long-period seismograms is the presence of Love-wave motions. Regional seismograms for GASBUGGY show no evidence of Love waves with the tangential surface-wave energy remaining near or at noise level. However, RULISON shows significant Love-wave energy particularly at stations DAL, JCT and GO. The stations DUG, ALQ, and GSC show enhanced Love wave, but TUC has a low level. Stations BKS, COR and LON are too low in amplitude where Love waves also cannot be detected on the tangential waveforms.

To investigate whether surface-wave modeling can resolve the RULISON/GASBUGGY m_b anomaly, we generated fundamental-mode surface-wave synthetic seismograms as the higher modes did not significantly contribute to the principal features (Figs. B-3 and B-4). We established a tectonic release model by combining the results of Murphy and Archambeau (1986) and Yacoub (1984) for RULISON which has a 45° dip-slip double-couple source striking approximately at 142°. The

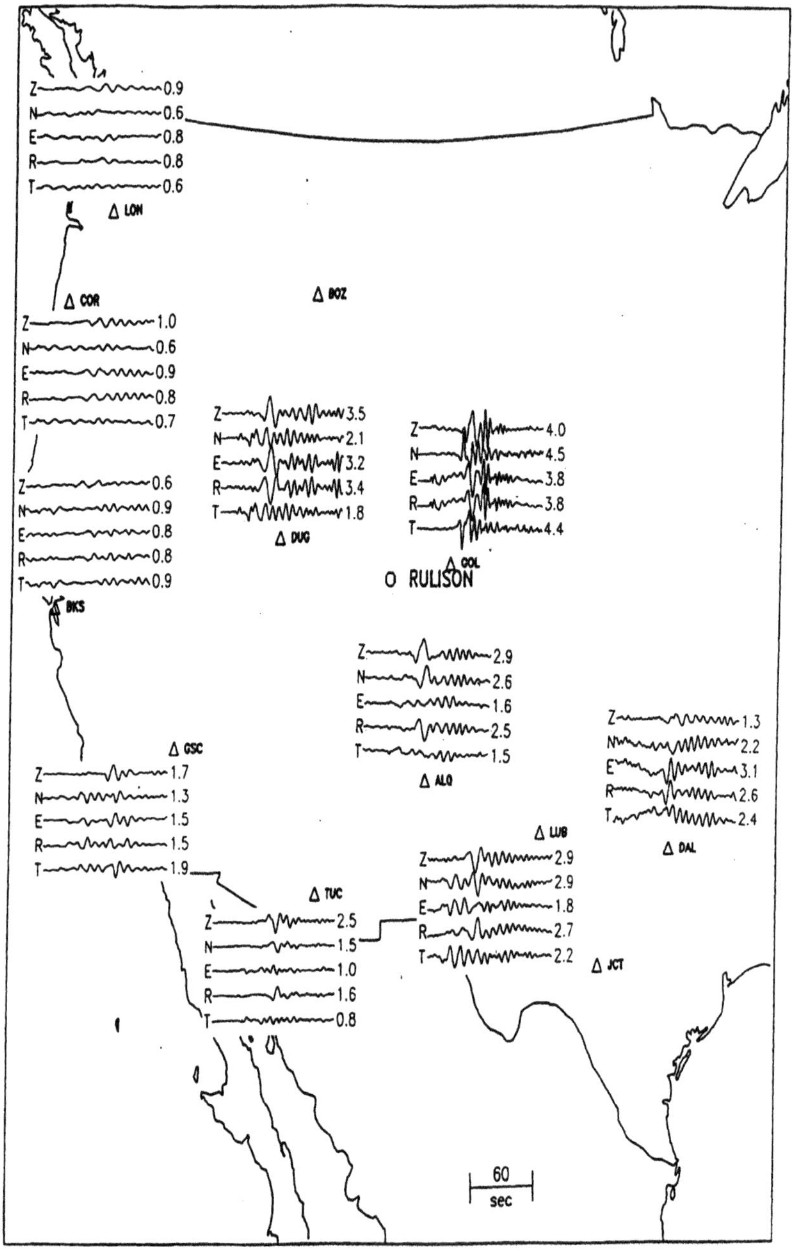

B-1

Figure B-1
Examples of regional WWSSN long-period surface waves for RULISON, plotted next to the map locations of recording stations; the peak amplitudes are given in microns. Note the strong tangential motions at GOL, DAL and JCT.

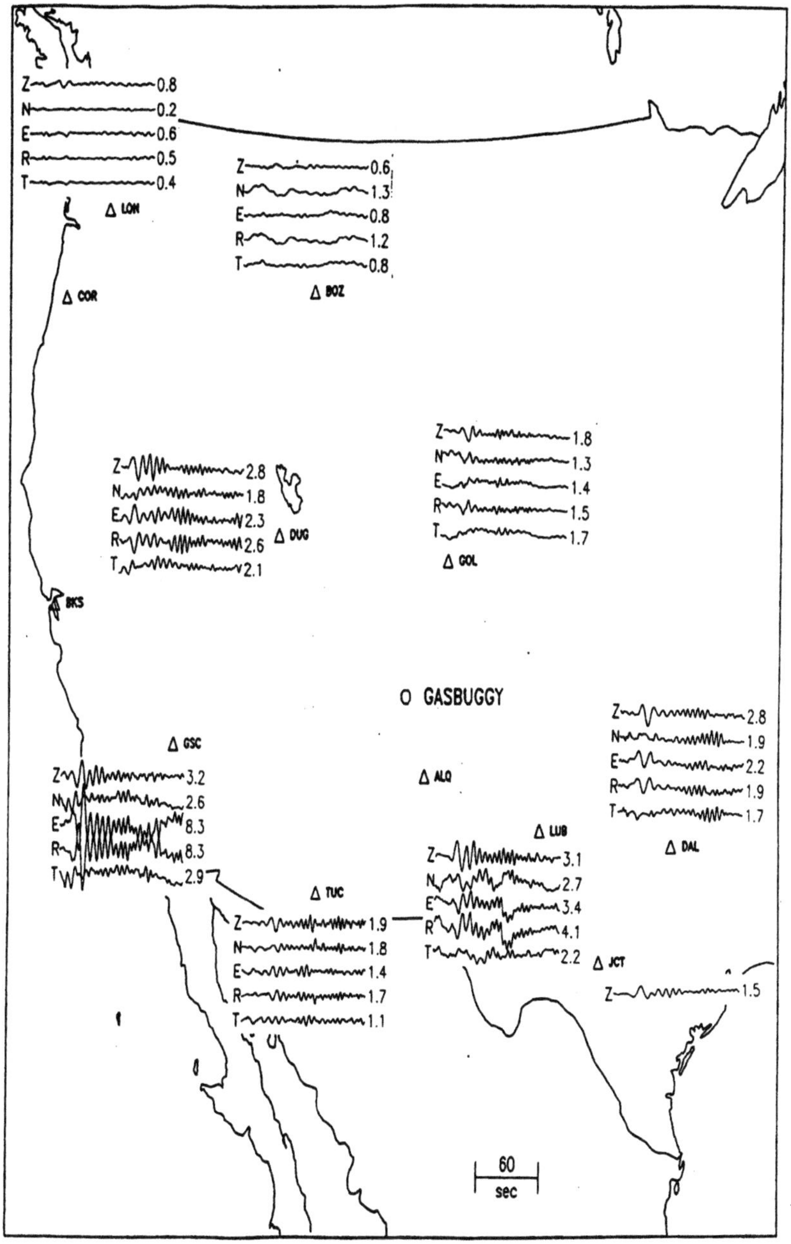

B-2

Figure B-2
Examples of regional WWSSN long-period surface waves for GASBUGGY, plotted next to the map locations of recording stations; the peak amplitudes are given in microns. Note that the tangential motions are generally weak.

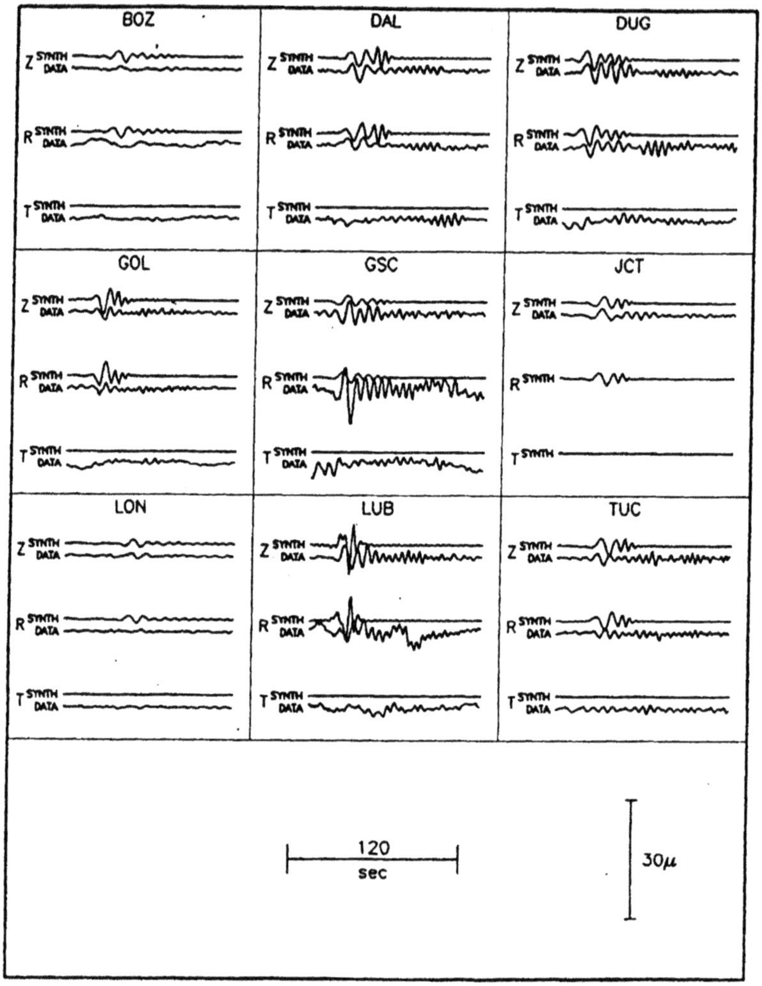

Figure B-3
Synthetic modeling of GASBUGGY. All traces are plotted at the same vertical scale. Synthetic seismograms are calculated using an explosion M_0 of 1.91×10^{22} dyne-cm and no tectonic-release. Only the fundamental-mode contributions are included.

seismic moment (M_0) of the explosion component of the two events, was estimated by fitting the amplitudes of the vertical synthetic seismograms vs data.

We have used available crustal structures from nearby stations and source regions (MURPHY and ARCHAMBEAU, 1986; STEVENS, 1986). The final explosion moment determined by amplitude fitting is 1.99×10^{22} dyne-cm for GASBUGGY and 2.05×10^{22} dyne-cm for RULISON. The tectonic release moment for RULISON is 8.52×10^{21} dyne-cm. This implies an *F-factor* (the strength of the double-couple/the strength of the explosion) of 0.42. With these moment estimates, the ψ_∞ estimates for RULISON (density = 2.35 gm/cm^3; $\alpha = 4.6$ km/s) and GASBUGGY

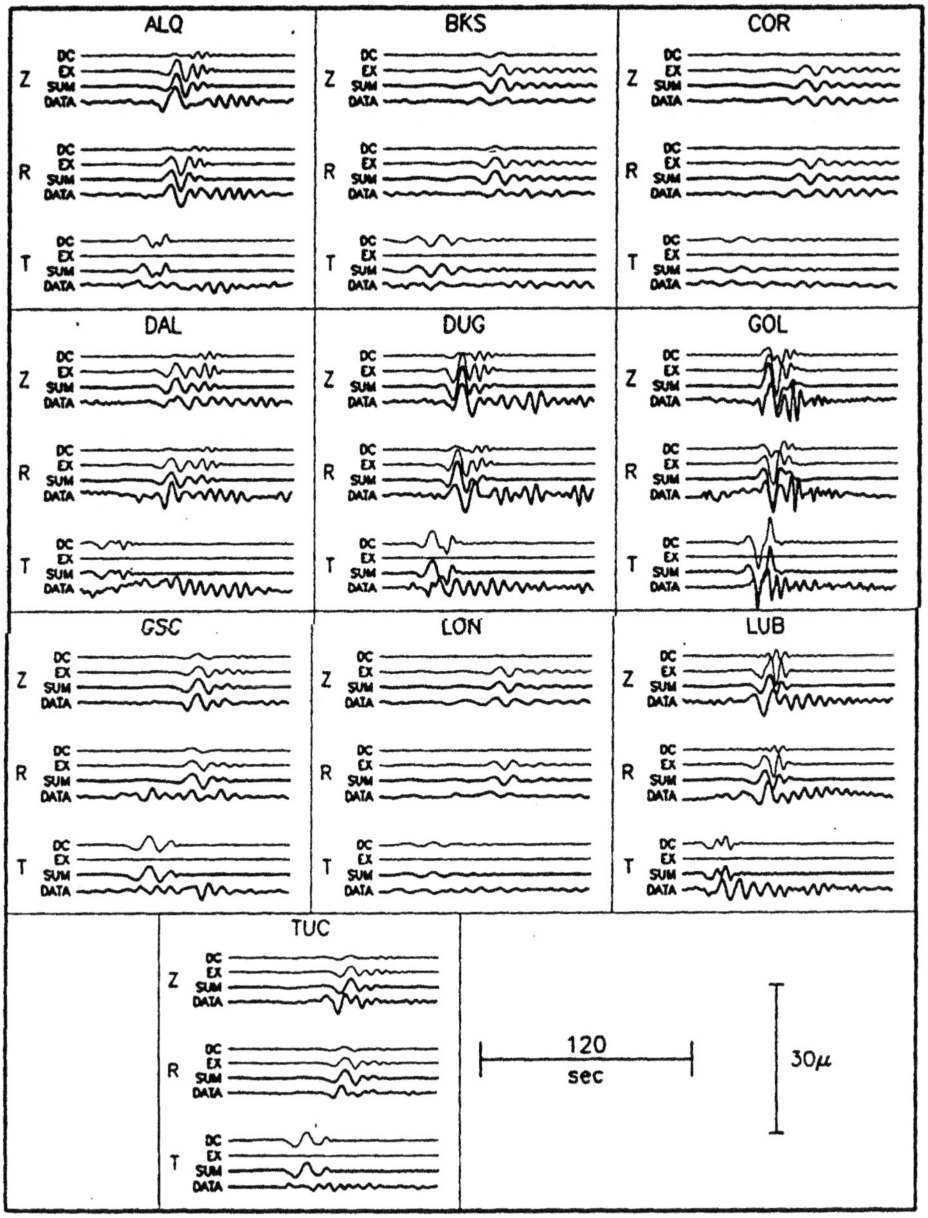

Figure B-4

Synthetic modeling of RULISON. All traces are plotted at the same vertical scale. Synthetic seismograms are calculated using an explosion M_0 of 2.05×10^{22} dyne-cm and a double-couple (tectonic release) M_0 of 8.52×10^{21} dyne-cm. The explosion contributions labeled as EX and the double-couple contribution as DC. The final synthetic is the sum of the two (SUM).

(density = 2.58 gm/cm^3; $\alpha = 4.1$ km/s) are 3280 m^3 and 3650 m^3, respectively. Note that ψ_∞ for GASBUGGY is 1.11 higher than that for RULISON even though the yield of GASSBUGY is smaller than that of RULISION (0.725 times of the RULISON yield). Note also that this value of GASBUGGY's ψ_∞ is 0.61 times the value estimated based on the Mueller-Murphy source and the average observed free-field value from Figure 16.

Acknowledgment

The authors would like to thank John R. Murphy, Maxwell Federal Technologies, Thorne Lay, University of California, Santa Cruz and an anonymous reviewer for their comments during the review of the original manuscript, which helped significantly in enhancing the manuscript. Special thanks to L. J. Burdick and David G. Harkrider for their interest, contributions and fruitful discussions during the various phases of this study.

References

ADUSHKIN, V. V., *Monitoring of underground nuclear tests by seismic stations in the former Soviet Union (FSU)*. In *Monitoring a Comprehensive Test Ban Treaty* (eds Husebye, E. S., and Dainty, A. M.) (Kluwer Academic Publishers 1996) pp. 35–52.

ALEWINE, R. W., and BACHE, T. (1983), *Monitoring a Threshold Test Ban Treaty*, AGU special section on verification.

BARKER, J. S., HARTZELL, S. H., BURDICK, L. J., and HELMBERGER, D. V. (1991), *Analysis of Local Seismic Waveforms from Underground Nuclear Explosions: Forward Modeling for Effective Surface Functions at Pahute Mesa*, J. Geophys. Res. *96*, 10,129–10,143.

BARKER, J. S., BURDICK, L. J., and WALLACE, T. C. (1985), *Analysis of Near-field Seismic Waveforms from Underground Nuclear Explosions*, Scientific Report #1, AFGL-TR-85-0321, Woodward-Clyde Consultants, Pasadena, California.

BURDICK, L. J., and HELMBERGER, D. V. (1979), *Time Functions Appropriate for Nuclear Explosions*, Bull. Seismol. Soc. Am. *69*, 957–973.

BURDICK, L. J., WALLACE, T., and LAY, T. (1984), *Modeling in the Near-field and Teleseismic Observations from the Amchitka Test Site*, J. Geophys. Res. *89*, 4373–4388.

BURDICK, L. J., GARNER, E. J., MCLAREN, J. P., WOODS, B. B., HELMBERGER, D. V., and HARKRIDER, D. G. (1989), *Uncertainty in Yield Scaling at Low Yield*, Final Report to AFTAC under Contract No. F08606-89-C-0024.

BURGER, R. W., LAY, T. X., and BURDICK, L. J. (1986a), *Examining the Relative Yield of Novaya Zemlya tests by Waveform Intercorrelation*, Geophys. J.R. Astr. Soc. *87*, 775–800.

BURGER, R. W., LAY, T. X., WALLACE, T. C., and BURDICK, L. J., (1986b), *Evidence of Tectonic Release in Long-period S Waves from Underground Nuclear Explosions at the Novaya Zemlya Test Sites*, Bull. Seismol. Soc. Am. *76*, 733–755.

COHEE, B. P., and LAY, T., (1988), *Modeling Teleseismic SV Waves from Underground Explosions with Tectonic Release: Results for Southern Novaya Zemlya*, Bull. Seismol. Soc. Am. *78*, 1158–1178.

DAHLMAN, O., and ISRAELSON (1977), *Monitoring Underground Nuclear Explosions* (Elsevier & New York).

DER, Z. A., MCELFRESH, T. W., and O'DONNELL, A. (1980), *Results of the SDCS Experiment*, SDAC-TR-80-4, Teledyne-Geotech, Alexandria, Virginia.

DOUGLAS, A., and MARSHALL, P. D., *Seismic source size and yield for nuclear explosions*. In *Monitoring a Comprehensive Test Ban Treaty* (eds Husebye, E. S., and Dainty, A. M.) (Kluwer Academic Publishers 1995) pp. 309–353.

ENGDAHL, E. (1972), *Seismic Effects of the MILROW and CANNIKIN Nuclear Explosions*, Bull. Seismol. Soc. Am. *62*, 1411–1423.

GIVEN, J. W., and MELLMAN, G. R. (1986), *Estimating Explosion and Tectonic Release Source Parameters of Underground Nuclear Explosions from Rayleigh and Love Wave Observations*, Technical Report, DARPA, ARPA order No. 5308.

HARTZELL, S. H., BURDICK, L. J., and LAY, T. (1983), *Effective Source Functions for Pahute Mesa Nuclear Tests*, Final Technical Report WCCP-R-83-3, Woodward-Clyde Consultants, Pasadena, California.

HELMBERGER, D. V. (1985), *Long-period S Waves and Yield Estimation*, 7th Annual DARPA Seismic Research Symposium.

HELMBERGER, D. V., BURDICK, L. J., and STEAD, R. (1991), *Modeling Near-field NTS and Amchitka*. In *Geophysical Monograph 65: Explosion Source Phenomenology* (eds Steven R. Taylor, Howard J. Patton, and Paul G. Richards) (American Geophysical Union, 2000 Florida Avenue, NW, Washington, DC 20009) pp. 35–45.

HELMBERGER, D. V., and HADLEY, D. M. (1981), *Seismic Source Functions and Attenuation from Local and Teleseismic Observations of the NTS Events JORUM and HANDLEY*, Bull. Seismol. Soc. Am. *71*, 51–67.

LANGSTON, C. A., and HELMBERGER, D. V. (1975), *A Procedure for Modeling Shallow Dislocation Sources*, Geophys. J.R. Astr. Soc. *42*, 117–130.

LAY, T. (1985), *Estimating Explosion Yield by Analytical Waveform Comparison*, Geophys. J.R. Astr. Soc. *82*, 1–31.

LAY, T., HELMBERGER, D. V., and HARKRIDER, D. G. (1984a), *Source Models and Yield-scaling Relations for Underground Nuclear Explosions at Amchitka Island*, Bull. Seismol. Soc. Am. *74*, 843–862.

LAY, T., BURDICK, L. J., and HELMBERGER, D. V. (1984b), *Estimating the Yields of the Amchitka Tests by Waveform Intercorrelation*, Geophys. J.R. Astr. Soc. *78*, 181–208.

LIEBERMANN, R. C., and BASHAM, P. W. (1971), *Excitation of Surface Waves by the Aleutian Underground Explosion MILROW (October 2, 1969)*, J. Geophys. Res. *76*, 4030–4034.

MARSHALL, P. D., SPRINGER, D. L., and RODEAN, H. C. (1979), *Magnitude Corrections for Attenuation in the Upper Mantle*, Geophys. J.R. Astr. Soc. *28*, 431–458.

MUELLER, R. A., and MURPHY, J. R. (1971), *Seismic Characteristics of Underground Nuclear Detonations: Part I, Seismic Scaling Law of Underground Detonations*, Bull. Seismol. Soc. Am. *61*, 1675–1692.

MURPHY, J. R., *Types of seismic events and their source description*. In *Monitoring a Comprehensive Test-Ban Treaty* (Kluwer Academic Publishers, Netherland 1998) pp. 225–245.

MURPHY, J. R., (1998), *Seismic Yield Estimation for the Indian and Pakistani Nuclear Tests of May, 1998*, 20th Ann. Seismic Res. Symp. on monitoring a CTBT, Santa Fe, New Mexico.

MURPHY, J. R. *Free-field seismic observations from underground nuclear explosions*. In *Geophysical Monograph 65: Explosion Source Phenomenology* (eds Steven R. Taylor, Howard J. Patton, and Paul G. Richards) (American Geophysical Union, 2000 Florida Avenue, NW, Washington, DC 20009, 1991) pp. 25–34.

MURPHY, J. R. (1989), *Network Averaged Teleseismic P-wave Spectra for Underground Explosions Part 2: Source Characteristics of Pahute Mesa Explosions*, Bull. Seismol. Soc. Am. *79*, 156–171.

MURPHY, J. R. (1977), *Seismic Source Functions and Magnitude Determinations for Underground Nuclear Detonations*, Bull. Seismol. Soc. Am. *67*, 135–158.

MURPHY, J. R., and ARCHAMBEAU, C. B. (1986), *Variability in Explosion Body Wave Magnitudes: An Analysis of the RULISON/GASBUGGY Anomaly*, Bull. Seismol. Soc. Am. *76*, 1087–1113.

MURPHY, J. R., BARKER, B. W., JENAB, J. N., and MARSHALL, M. E. (1992), *Yield Estimate Research*, Presented at the 14th Ann. PL/DARPA Seismic Res. Symp., pp. 304–310.

ORPHAL, D. L., SPIKER, C. T., WEST, L. R., and WRONSKI, M. D. (1970), *Analysis of Seismic Data, MILROW Event*, Report NVO-1163-209, Environmental Research Corporation, Alexandria, Virginia, pp. 81.

PERRET, W. (1982), *The Physics of Non-isotropic Source Effects, Close-in Seismic Source Parameters*, DARPA/AFOSR Symposium (GSD-8203), Las Vegas, Nevada.

PERRET, W. (1973), *Ground Motion in the Vicinity of the CANNIKIN Nuclear Explosion*, Report SLA-73, 0043, Sandia Laboratories, Albuquerque, New Mexico, pp. 79.

PERRET, W., and BREDING, D. (1972), *Ground Motion in the Vicinity of an Underground Nuclear Explosion in the Aleutian Islands: MILROW Event*, Report SL-RR-72 0668, Sandia Laboratories, Albuquerque, New Mexico, pp. 84.

SAIKIA, C. K. (1994), *Modified Frequency-wavenumber Algorithm for Regional Seismograms Using Filon's Quadrature – Modeling of Lg Waves in Eastern North America*, Geophys. J. Int. *118*, 142–158.

SAIKIA, C. K., ZHU, L., WOODS, B. B., and THIO, H. K. (1999), *Path calibration and source characterization in and around India.* In 21st Seismic Res. Symp. *Technologies for Monitoring the CTBT*, Las Vegas, pp. 243–253.

STEAD, R. J. (1989), *Finite Differences and a Coupled Analytic Technique with Applications to Explosions and Earthquakes*, Ph.D. Dissertation, California Institute of Technology, Pasadena, CA, pp. 337.

STEAD, R. J., and HELMBERGER, D. V. (1988), *Numerical-analytical Interfacing in Two Dimensions with Application to Modeling NTS Seismograms*, J Pure appl. geophys. *128*, 101–193.

STEVENS, J. L. (1986), *Estimation of Scalar Moments from Explosion-generated Surface Waves*, Bull. Seismol. Soc. Am. *76*, 123–151.

SULTANOV, D. D., MURPHY, J. R., and RUBINSTEIN, K. D. (1999), *A Seismic Source Summary for Soviet Peaceful Nuclear Explosions*, Bull. Seismol. Soc. Am. *89*, 640–647.

SYKES, L. R., and CIFUENTES, I. L. (1984), *Yields of Soviet Underground Nuclear Explosions from Seismic Surface Waves: Compliance with the Threshold Test-Ban Treaty*, Proceedings of the National Academy of Sciences, *81*, March 1984, 1922–1925.

SYKES, L. R., and WIGGINS, G. C. (1986), *Yields of Soviet Underground Nuclear Explosions at Novaya Zemlya, 1964–1976 from Seismic Body and Surface Waves*, Proc. National Academy of Sciences USA, *83*, 201–2056.

TOKSÖZ, M. N., and KEHRER, H. H. (1972), *Tectonic Strain-release Characteristics of CANNIKIN*, Bull. Seismol. Soc. Am. *62*, 1425–1438.

VIDALE, J. E., and HELMBERGER, D. V., *Path effects in strong motion seismology.* In *Methods of Computational Physics* (ed. Bruce Bolt) (Academic Press, New York 1987) pp. 267–319.

VON SEGGERN, D., and BLANDFORD, R. (1972), *Source Time Functions and Spectra for Underground Nuclear Explosions*, Geophys. J. *31*, 8–97.

VON SEGGERN, D. H. (1978), *Intersite Magnitude-yield Bias Exemplified by the Underground Nuclear Explosion MILROW, BOXCAR and HANDLY*, Technical Report SDAC-TR-77-4, Teledyne Geotech, Alexandria, Virginia.

WALLACE, T. C., HELMBERGER, D. V., and LAY, T. (1986), *Reply to Comments by A. Douglas, Y. B. Young and N. S. Lyman and a Note on the Revised Moments for Pahute Mesa Tectonic Release*, Bull. Seismol. Soc. Am. *76*, 313–318.

WILLIS, D. E., GEORGE, G. D., POETZL, K. G., SALTZER, C. E., SHAKAL, A. F., TORFIN, R. D., WOODZIK, T. L., and WOOLSIN, C. (1972), *Seismological Aspects of the CANNIKIN Nuclear Explosion*, Bull. Seismol. Soc. Am. *62*, 1377–1395.

YACOUB, N. K. (1983), *Instantaneous Amplitudes: A New Method to Measure Seismic Magnitude*, Bull. Seismol. Soc. Am. *73*, 1345–1355.

(Received October 20, 1999, revised May 18, 2000, accepted May 26, 2000)

Pure appl. geophys. 158 (2001) 2217–2226
0033–4553/01/112217–10 $ 1.50 + 0.20/0

Pure and Applied Geophysics

Yield Estimation for Semipalatinsk Underground Nuclear Explosions Using Seismic Surface-wave Observations at Near-regional Distances

V. V. ADUSHKIN[1]

Abstract—A statistical procedure is described for estimating the yields of underground nuclear tests at the former Soviet Semipalatinsk test site using the peak amplitudes of short-period surface waves observed at near-regional distances (Δ < 150 km) from these explosions. This methodology is then applied to data recorded from a large sample of the Semipalatinsk explosions, including the Soviet JVE explosion of September 14, 1988, and it is demonstrated that it provides seismic estimates of explosion yield which are typically within 20% of the yields determined for these same explosions using more accurate, non-seismic techniques based on near-source observations.

Key words: Yield estimation, semipalatinsk, nuclear explosions.

Introduction

Underground nuclear tests conducted at the former Soviet Semipalatinsk test site (STS) in Central Asia were routinely monitored by a network of near-regional (Δ < 150 km) seismic stations. The locations of these stations with respect to the Balapan and Degelen testing areas of the STS are shown in Figure 1, where it can be seen that the network consisted of a *N-S* profile of nine permanent stations located within the STS (i.e., stations 1–9), supplemented by an additional seven temporary stations situated in various inhabited areas around the test site. Three-component seismic sensors were installed at these sites, either on hard rock outcrops or on specially prepared concrete pads. The frequency responses of these seismic systems were essentially flat to velocity from about 0.2–0.5 to 20 Hz, with a corresponding dynamic range of 40–50 dB. The data were recorded on analog magnetic tape and, at some points, on photographic paper. Also shown on Figure 1 are the surface traces of a number of significant faults in the region, including the Kolba-Tchingiz, Tchinrauz and Shagan faults which intersect the Balapan testing area. The heterogeneous zones associated with these deep faults are observed to be tens to

[1] Institute For Dynamics of the Geospheres, Russian Academy of Sciences, Russia. E-mail: adushkin@idg1.chph.ras.ru

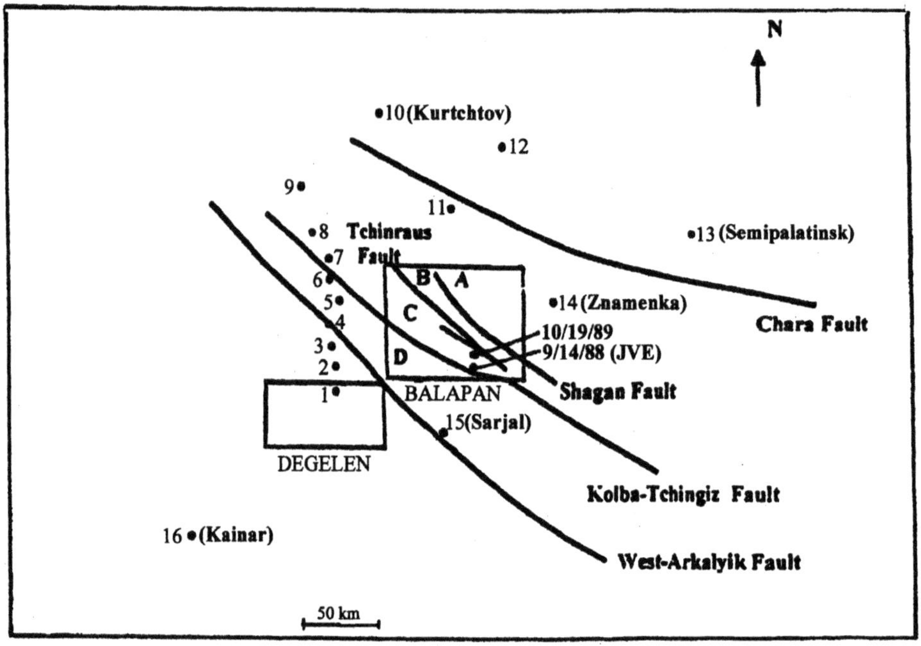

Figure 1
Map locations of the seismic stations of the near-regional Semipalatinsk network and deep regional faults which intersect the test site. The letters A, B, C, D denote Balapan source regions bounded by the regional faults.

hundreds of meters wide and, as will be discussed in more detail below, it has been found that they disturb the axial symmetry and propagation efficiency of the regional seismic waves produced by underground explosions at the Balapan testing area. This paper provides a brief summary of the methodology which was developed to estimate yields of STS explosions using the seismic signals recorded on this near-regional network of stations (ADUSHKIN *et al.*, 1990).

Characterization of the Observed Near-regional Seismic Data

A profile of typical vertical displacement waveforms recorded by network stations at various distances from the epicenter of the 85-kt Balapan explosion of October 19, 1989 (cf. Fig. 1) is presented in Figure 2. It can be seen that at distances beyond 10 km, the longitudinal (*P*), shear (*S*) and surface (*R*) waves are well separated in time, which permits one to consistently identify specific wave groups and to determine their characteristic parameters. Furthermore, at these near-regional distances, the signal-to-noise ratios are generally observed to be very good, and there is no difficulty in detecting the onsets of the different major seismic phases.

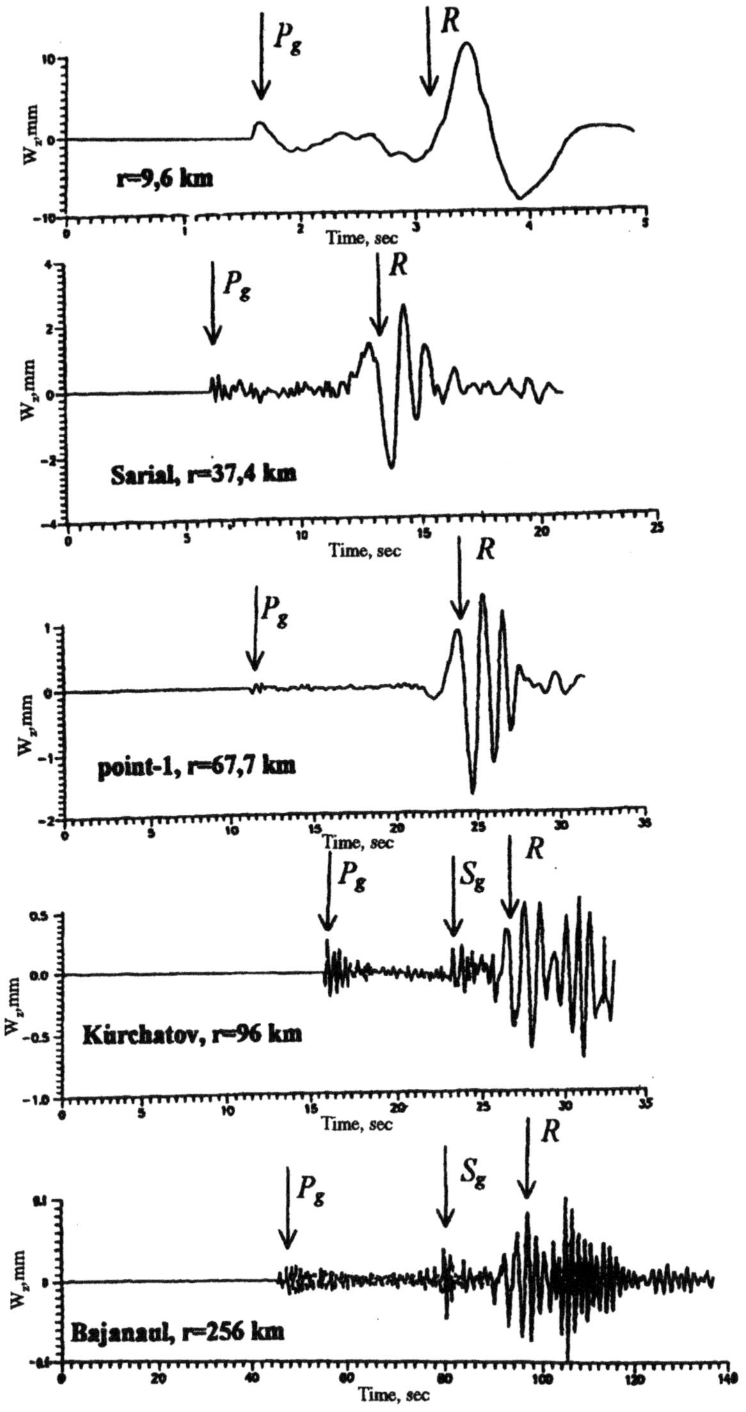

Figure 2
Profile of vertical displacement waveforms recorded by network stations at various distances from the 85-kt Balapan explosion of October 19, 1989.

However, the observed signals are complex, containing many individual seismic arrivals associated with the complex geologic structures and layering along the propagation paths.

The travel-time curves which have been inferred for the principal observed crustal and upper mantle seismic phases in this region are shown in Figure 3. These curves indicate that the P_g phase refracted in the upper crustal layers is the first arrival at distances less than about 150 km, where it is associated with apparent velocities in the range extending from 5.3 to 6.6 km/sec. At distances greater than about 200 km, the P_n arrival corresponding to the critical refraction from the Moho becomes the first arrival, and it is associated with apparent velocities in the range extending from 7.7 to 9.1 km/sec. The corresponding shear phases, S_g and S_n, have similar origins to P_g and P_n and are characterized by apparent velocities of about 3.1 km/sec for S_g and 3.9 to 5.5 km/sec for S_n. Beyond about 100 km, the S_n phase is observed to have the largest amplitude in the body-wave group. The last principal arrival considered is the Rayleigh surface wave, *R*, which has a dominant group velocity of about 2.75 km/sec.

It can be seen from the representative waveforms shown in Figure 2 that the surface-wave displacement amplitudes are several times larger and that their predominant periods are 4 to 10 times longer than those of the body waves measured at the same distances. It follows that the surface waves should be less affected by inhomogeneities along the propagation path and by varying properties of the source emplacement media than are the corresponding body waves. Moreover, because the seismic network used during the periods of intensive testing in the 1960s and 1970s measured only low frequency displacements, it was natural to focus the yield estimation research on the larger amplitude surface wave data. More recently, broader band equipment has been employed in the network, and yield estimation procedures based on higher frequency body waves and the L_g phase have also been investigated (KOSTYUCHENKO *et al.*, 1998).

Development of Surface Wave-yield Estimation Methodology

As the first step in the formulation of a seismic yield estimation procedure based on *R*-wave observations, the variations of the amplitudes and periods of the *R*-wave as functions of yield and distance were analyzed in detail. Analyses of the recorded waveform data indicated that the duration and waveform shape of the *R* wave at a fixed distance are essentially independent of yield, emplacement medium and propagation path. However, both the duration and number of cycles of motion in the *R* wave increase systematically with increasing distance, such that the duration $t \sim r^{0.5}$. Oscillations in the *R* wave are observed to be approximately sinusoidal and, consequently, are characterized by the apparent period of peak displacement, *T*. It was found that the dependence of *T* on yield is very weak ($T \sim q^{0.05}$) over the range

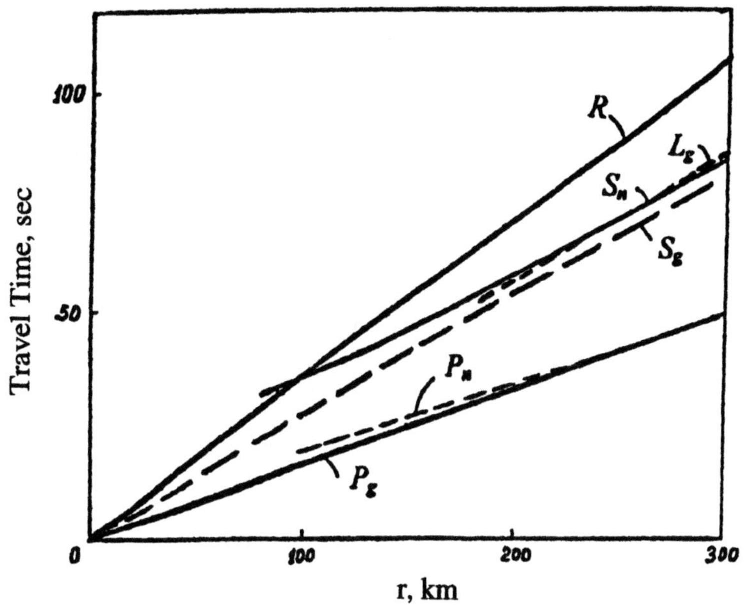

Figure 3
Average travel-time curves for the various regional seismic phases in the Semipalatinsk region.

investigated (i.e., 0.02 to 126 kt), and that T increases with distance according to the power-law relation $T \sim r^{0.23}$, with the observed periods being very similar on the vertical and radial components of motion.

The variations of the R-wave peak amplitudes with yield and distance were also studied in detail. Figure 4 shows plots of observed peak vertical (W_z) and radial (W_x) R-wave amplitudes as a function of distance for a Balapan explosion with a yield of 32 kt. These amplitudes were measured from seismograms recorded at stations located in a very narrow azimuth window from the event (i.e., less than 10°) in order to reduce variations due to propagation path differences and possible azimuthal asymmetries in the seismic radiation pattern. Similar relationships were constructed for an additional 15 explosions with yields in the 1 to 100 kt range, and these were combined to infer an average distance decay rate of $r^{1.4}$ for R waves in this region.

Given this average decay rate for the region, the dependence of the R wave amplitudes on yield was estimated by analyzing the distance-normalized, observed amplitudes, $W \cdot r^{1.4}$, as a function of yield, q, at each station. Figure 5 displays such a relationship for a station lying in the distance range of 80–100 km from the selected sample of explosions. In this case, the seismograms were all recorded from Balapan explosions detonated in similar emplacement media within a 10 km by 20 km area bounded by two of the regional faults of Figure 1. As can be seen from Figure 5, the yield dependence of the distance-normalized R-wave amplitudes at this station can be described with good accuracy by a simple power law, $W \cdot r^{1.4} \sim q^m$, with $m = 0.88$.

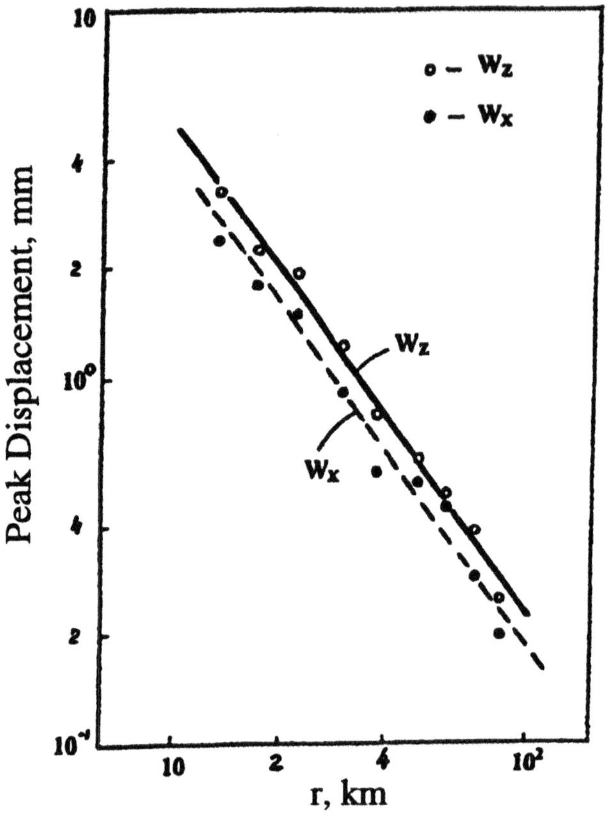

Figure 4
Peak displacements of radial (W_x) and vertical (W_z) components of surface waves as a function of distance.

Similar results were obtained for the other stations, with values of the exponent m falling in the range 0.85 to 0.92 and averaging value 0.9. Thus, the R-wave amplitude dependence on yield and distance at a given station can be represented as

$$W = Aq^{0.9}r^{-1.4} \tag{1}$$

where A is a station-dependent amplitude constant. Using equation (1), best-fitting A values were estimated for each station by computing logarithmic averages of the individual A estimates determined using the known values of W, q and r corresponding to seismograms recorded from about 100 STS explosions. Table 1 lists the resulting average A values determined for stations 1–16 of Figure 1. It can be seen that these estimated coefficients, A_i, vary by more than a factor of 4 between stations and, consequently, they were not averaged to obtain a single scaling relationship, but were retained as individual station coefficients which account for the influence of local geology beneath the stations and other unmodeled propagation path variables on the observed R-wave amplitudes at each station.

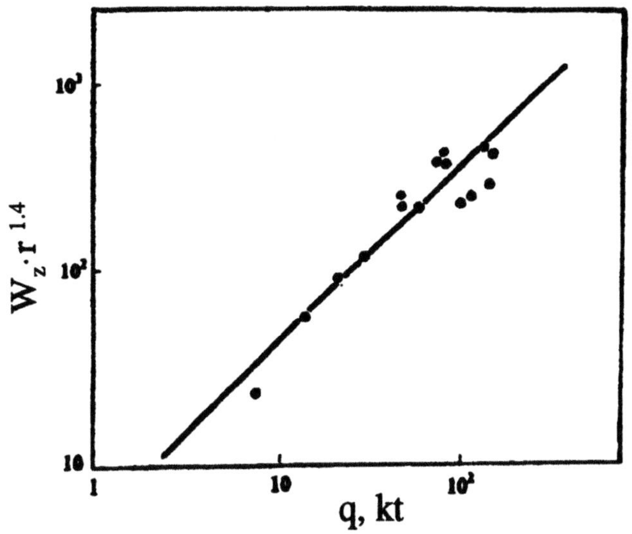

Figure 5
Dependence of $W \cdot r^{1.4}$ on explosion energy for one of the seismic stations.

Table 1

Average station Rayleigh wave amplitude coefficients for explosions at the Semipalatinsk Test Site

Station	1	2	3	4	5	6	7	8	9	10	11	12	13	14	15	16
$\bar{A}_i$	8.3	8.6	5.1	4.3	3.4	8.3	7.4	7.1	3.0	6.3	4.6	4.0	4.3	8.3	12.6	10.0

A further significant reduction in variance was obtained when each station coefficient, A_i, was estimated separately for each of the four Balapan source regions (i.e., A, B, C, D) shown bounded by the regional fault traces in Figure 1. Thus, the final coefficients depend on both station i and source region j (j = A, B, C, D) and are denoted as A_{ij}. This subdivision by source region was found to be statistically significant in that the observed variation of the log A_{ij} values was reduced by 25% with respect to that of the corresponding log A_i values. Several thousand station recordings from explosions within the four Balapan source regions were processed to obtain estimates of log A_{ij}, with separate coefficients for the z and x components at each station. The resulting station coefficients A_{ij} were found to vary from 3 to 12.6, with variations of as much as a factor of 3.5 across all stations for explosions within a fixed source region, and variations of as much as a factor of 2 across all source regions at a fixed station. However, the source region effect was not observed to be the same at all stations, and, in fact, the average coefficients at some stations showed variations as small as 15% across the four source regions. In general, these coefficients account for unmodeled variations in the physical properties of the source emplacement medium, the propagation paths from the sources to the receivers and the local geology beneath each station.

Using equation (1), together with the estimated station coefficients A_{ij} and the associated standard deviation of those estimates, σ_{ij}, the following procedure has been formulated for estimating explosion yields from the observed near-regional surface-wave amplitudes. First, separate yield estimates are determined for each of the stations which recorded an explosion using equation (1) rewritten in the form

$$\log q_i^{x,z} = \frac{(\log W_i^{x,z} - \log A_{ij}^{x,z} - 1.4 \log r_i)}{0.9} \tag{2}$$

where $q_i^{x,z}$ are the yield estimates determined from the x and z component R-wave displacement amplitudes observed at station i, $W_i^{x,z}$, $A_{ij}^{x,z}$ are the station coefficients at station i for an explosion in source region j, and r_i is the distance from the explosion to station i. Since the accuracy of these individual yield estimates depends upon the uncertainty in the corresponding average A_{ij} determinations, the final yield estimate is obtained as a weighted average of the individual yield estimates in the form

$$\overline{\log q} = \frac{\sum_i a_{ij}^x \log q_{ij}^x + \sum_i a_{ij}^z \log q_{ij}^z}{\sum_i (a_{ij}^x + a_{ij}^z)} \tag{3}$$

where $a_{ij} = 1/\sigma_{ij}^2$. That is, inverse variance weighting is applied under the assumption that the individual yield estimates are independent. The corresponding standard deviation in this average yield estimate is then given as

$$\sigma = \sqrt{\frac{\sum_i a_{ij}^x \left(\log q_{ij}^x - \overline{\log q}\right)^2 + \sum_i a_{ij}^z \left(\log q_{ij}^z - \overline{\log q}\right)^2}{(N-1)\sum_i \left(a_{ij}^x + a_{ij}^z\right)}} \tag{4}$$

where N denotes the total number of recordings from a given explosion.

The average value of σ_{ij} associated with the estimation of the station coefficients A_{ij}, using data measured from explosions of known yield, corresponds to a multiplicative factor of about 13%, which, under the assumption that the amplitude values are log-normally distributed, implies a yield estimation uncertainty at the 2σ level of around 10% for determinations based on 10 R-wave amplitude measurements from this sample of explosions. Thus, this proposed seismic yield estimation methodology based on near-regional surface wave amplitude is very precise for explosions at the STS.

Sample Applications of the Yield Estimation Methodology

In order to evaluate the actual achievable accuracy of the proposed yield estimation methodology, seismic yield estimates were determined using R-wave amplitudes measured from a sample of STS explosions for which accurate yields have been determined using non-seismic methods. This sample of events included 34 explosions in boreholes at Balapan with yields ranging from 1.4 to 175 kt and 13 explosions in tunnels at Degelen with yields ranging from 1.1 to 79 kt. The ratio of the seismic yield estimates, $\bar{q}$, to the actual yields, q_0, for the 34 Balapan explosions are plotted as a function of q_0 in Figure 6. It can be seen from this figure that, although the observed seismic yield estimation accuracy for these events is quite good, it is somewhat lower than the nominal 10% accuracy referenced above, being on the order of 15%. This discrepancy is probably due to a number of factors, including the fact that some of these seismic yield estimates were based on data from fewer than 10 stations, and the fact that the accuracy of the non-seismic yield estimates for some of these borehole explosions is uncertain. The corresponding seismic yield estimates for the selected sample of Degelen explosions, which were determined using the average station amplitude coefficients of Table 1, were found to be more accurate, on the order of 9%, which is very consistent with the nominal estimate based on the uncertainty of the station coefficients.

The yield estimation procedure was also applied to seismic data recorded from the September 14, 1988 Soviet JVE explosion at Balapan (cf., Fig. 1). In this case, *R*-wave amplitude data were available from all 16 of the stations of Figure 1, and these were analyzed to obtain a seismic yield estimate of 122.6 ± 7.1 kt. This yield estimate agrees very well with the independent hydrodynamic, seismic yield range of 110–120 kt cited by SYKES and EKSTRÖM (1989), as well as with other reported

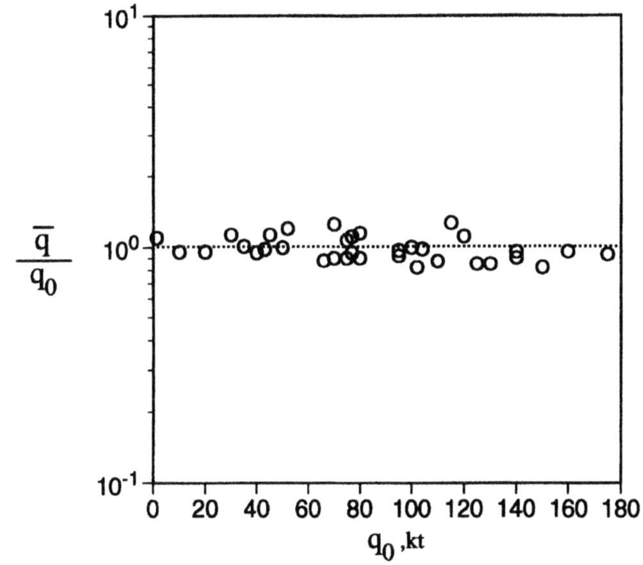

Figure 6
Ratios of estimated ($\bar{q}$) to true (q_0) yields as a function of q_0 for a sample of 34 Balapan explosions.

hydrodynamic and "CORRTEX" yield estimates (MINATOM, 1996) for this explosion.

Summary and Conclusions

The investigations which have been summarized in this paper were directed toward the development of a seismic yield estimation methodology for explosions at the Semipalatinsk test site, based on near-regional surface wave observations. Scaling relations were formulated to define the dependence of the observed short-period Rayleigh wave peak amplitudes on yield and distance, and detailed statistical analyses were conducted to derive individual station coefficients which depend upon explosion location within the test site. Using the measured uncertainties in these station terms, an inverse variance weighting procedure was defined for combining the individual yield estimates from the various stations into a statistically optimum seismic estimate of explosion yield. This methodology was then applied to data recorded from a large sample of Semipalatinsk explosions, including the Soviet JVE test of September 14, 1988, and it was demonstrated that it can provide highly accurate estimates of explosion yield for events at this test site.

REFERENCES

ADUSHKIN, V. V., KULIKOV, V. I., SAFONOV, F. F., and SOLOMONOV, A. A. (1990), *Procedure for Determination of a Underground Explosion Yield from Parameters of Seismic Waves at Near-regional Distances*, Izvestiya ANSSSR (Notes of AS USSR), Physics of the Earth *12*, 21–30.

KOSTYUCHENKO, V. N., RODIONOV, V. N., and SULTANOV, D. D. (1998), *Seismic waves from underground nuclear explosions*. In *Peaceful Nuclear Explosion – III*. (Vienna, Intern. Atomic Energy Agency 1998) pp. 447–461.

RUSSIAN FEDERATION MINISTRIES OF ATOMIC ENERGY AND DEFENSE (MINATOM) (1996), *USSR Nuclear Weapons Test and Peaceful Nuclear Explosions, 1949 through 1990*, Russian Federation Nuclear Center – VNIIEF, ISBD 5-85165-062-1.

SYKES, L. L. and EKSTRÖM, G. (1989), *Comparison of Seismic and Hydrodynamic Yield Determinations for the Soviet Joint Verification Experiment of 1998*, Proc. Nat. Acad. Sci. USA *86*, 3456–3460.

(Received June 5, 1999, revised January 19, 2000, accepted January 31, 2000)

Pure appl. geophys. 158 (2001) 2227–2251
0033–4553/01/112227–25 $ 1.50 + 0.20/0

Pure and Applied Geophysics

Yield Estimation from Surface-wave Amplitudes

JEFFRY L. STEVENS[1] and JOHN R. MURPHY[1]

Abstract — Surface-wave amplitudes from explosion sources show less variation for a given event than body wave amplitudes, so it is natural to expect that yield estimates derived from surface waves will be more accurate than yield estimates derived from body waves. However, yield estimation from surface waves is complicated by the presence of tectonic strain release, which acts like one or more earthquake sources superimposed on top of the explosion. Moment-tensor inversion can be used to remove the tectonic component of the surface waves, however moment-tensor inversion for shallow sources is inherently non-unique so the explosion isotropic moment cannot be determined with the necessary accuracy by this means. Explosions on an island or near a mountain slope can exhibit anomalous surface waves similar to those caused by tectonic strain release. These complications cause yield estimates derived from surface waves to be less accurate than yield estimates from body waves recorded on a well-calibrated network with good coverage. Surface-wave amplitudes can be expressed as a surface-wave magnitude M_s, which is defined as the logarithm of the amplitude plus a distance correction, or as a path corrected spectral magnitude, log M_0', which is derived from the surface-wave spectrum. We derive relations for M_s vs. yield and log M_0' vs. yield for a large data set and estimate the accuracy of these estimates.

Key words: Surface wave, explosion, yield, magnitude, moment, moment tensor.

Introduction

The Threshold Test-Ban Treaty (TTBT) between the United States and the Soviet Union, which went into effect in 1976, limited the yield of underground nuclear tests to 150 kilotons. Because of this, accurate estimation of explosion yields from seismic data became very important, and a great deal of research was performed to identify all of the factors that caused variations in seismic amplitudes. This caused renewed interest in estimating explosion yields from surface waves. These yield estimates were then used as a consistency check on other phases, or combined with the other estimates to obtain a unified yield (RINGDAL *et al.*, 1992; MURPHY, 1993) using all measurements at the same time.

Surface waves have some advantages and disadvantages for yield estimation compared to other phases. Surface-wave amplitudes exhibit less scatter than other phases, with a network standard deviation in M_s typically about half the standard deviation for m_b (BACHE, 1982). Also, since surface waves are usually measured at periods of about 20 seconds, they are much less sensitive to small-scale variations in

[1] Science Applications International Corporation. E-mail: Jeffry.L.Stevens@saic.com

the earth than short-period regional and teleseismic phases. Since the measured period is considerably longer than the explosion source duration, this also means that surface waves are insensitive to the explosion source function. That is, while surface-wave amplitudes will vary in different materials with source coupling, they are not affected by details of the source spectrum which are flat at periods longer than a few seconds. On the negative side, surface waves cannot be measured to as small a magnitude as m_b. This is particularly true for explosions where m_b is 1–2 magnitude units higher than M_s. Also, while surface waves are less sensitive to variations in attenuation and scatter than m_b, they are subject to variations due to tectonic strain release and near-source structure that have little effect on m_b. The sources and magnitude of these variations are discussed later in this paper.

Because surface waves are long-period measurements, surface-wave amplitudes are approximately proportional to explosion yield, and the slope of the magnitude yield curve is close to one. That is, we can write $M_s = \log Y + C$, where Y is the yield in kilotons (KT) and C is a constant that may depend on source medium, but is independent of explosion yield. For example, BACHE (1982) using data from MARSHALL *et al.* (1979), found $M_s = \log Y + 2.05 \pm 0.21$. There are a number of other similar relations in the literature. MURPHY (1977) found $M_s = 0.84 \log Y + 2.14$ and $M_s = 1.33 \log Y + 1.20$ for explosions with yield less than and greater than 100 kilotons, respectively. MARSHALL *et al.* (1979) found $M_s = 0.97 \log Y + 2.16$ for explosions in salt and granite, and $M_s = 1.06 \log Y + 1.88$ for all explosions in their data set including poorly coupling events above the water table. In general, however, these relations have been derived as best fits to data including events in different materials with varying depths, and are also consistent with unit slope for explosions in similar materials at comparable depth.

Since an idealized explosion is spherically symmetric, and when embedded in a plane layered structure is cylindrically symmetric, such an explosion should generate no Love waves and should generate Rayleigh waves with the same amplitude in all directions. It is well known, however, that explosions generate Love waves, which can be quite large in some cases, and that the Rayleigh waves not only have a radiation pattern, but can be reversed, as if from an implosive source, in some cases (e.g., TOKSÖZ and KEHRER, 1972). Figure 1 shows an example of Rayleigh and Love waves recorded at the same station for two explosions at the Soviet Semipalatinsk test site. The first set of seismograms shows a normal Rayleigh wave and a small Love wave; the second shows a reversed Rayleigh wave and a Love wave comparable in amplitude to the Rayleigh wave. This anomalous radiation can be explained by superposition of one or more earthquake-like sources on top of the explosion source. Explanations for this extra source include tectonic strain release (ARCHAMBEAU, 1972; STEVENS, 1982; DAY *et al.* 1987; HARKRIDER *et al.*, 1994), earthquake triggering (AKI and TSAI, 1972), and passive block motion (SALVADO and MINSTER, 1980; MASSE, 1981). VIECELLI (1973) suggested that spall could be a strong generator of surface waves, and a possible mechanism for reversed Rayleigh waves. However

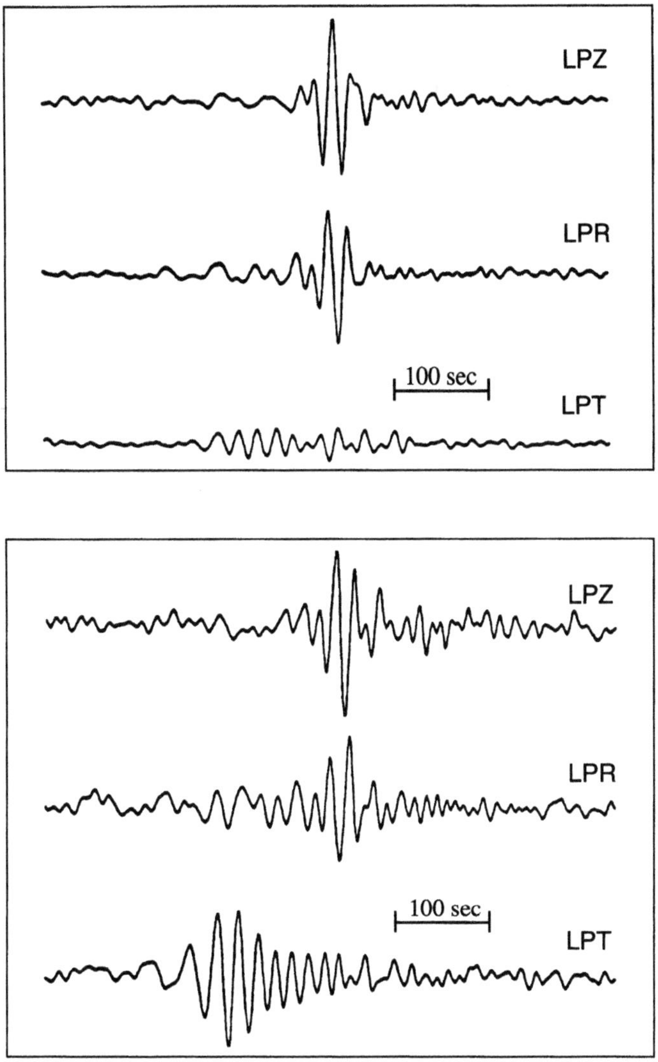

Figure 1
Long-period vertical, radial, and transverse components of displacement recorded at the same station from two explosions at the Semipalatinsk test site. The first event (top) has a normal Rayleigh wave and a small Love wave on the transverse component. The seismogram from the second event (bottom) shows a Love wave comparable in amplitude to the Rayleigh wave, and the Rayleigh wave phase shifted, approximately reversed, with respect to the normal Rayleigh wave.

DAY *et al.* (1983), showed conclusively that spall cannot contribute significantly to surface waves at periods longer than a few seconds. An initially puzzling aspect of the observations of Rayleigh wave reversals was an apparent time delay of the anomalous Rayleigh waves compared to normal Rayleigh waves from nearby locations (RYGG, 1979; HELLE and RYGG, 1984; HERRIN and GOFORTH, 1986). However, this time delay proved to be easily explained by interference between the explosion and double-couple source functions (DAY and STEVENS, 1986; DAY *et al.*, 1987; STEVENS *et al.*, 1991).

All of the feasible mechanisms for explaining the anomalous surface waves have the characteristic that they can be described at the periods of interest (greater than about 10 seconds) by superposition of one or more double-couple sources, or an equivalent moment-tensor source, on top of a point explosion source. Moment-tensor inversion has been used to separate the isotropic and nonisotropic parts of the source so that the explosion yield could be determined from the isotropic source alone. A fundamental problem with moment-tensor inversion for shallow sources is that the solution is non-unique and some additional constraint is required. GIVEN and MELLMAN (1986) performed moment-tensor inversions for 37 Shagan River (Russia) explosions and 47 Nevada Test Site (NTS) explosions. They constrained the Shagan River explosions to a 45 degree thrust mechanism, and the NTS explosions to a strike-slip mechanism, and the inversion then determined the strike of the double-couple source and the isotropic and double-couple moments for each event. They determined these constraints by plotting the function $0.9 \log M_i - m_b$ as a function of F factor (the ratio of nonisotropic to isotropic moment) for the isotropic moment M_i derived for different source constraints, and finding the source constraint that minimized the slope of this function. The assumption here is that isotropic moment is proportional to yield Y and m_b is proportional to $0.9 \log Y$ so that the difference $0.9 \log M_i - m_b$ should be constant if the constraint is correct and consistent for a test site. It is important to recognize, however, that selecting different mechanisms for the two test sites creates a very significant bias in the results as the thrust fault assumption increases M_i relative to the strike-slip assumption.

EKSTRÖM and RICHARDS (1994) found that the constraint of a constant mechanism at each site was too strong and led to unrealistic results in some cases. They used announced yields, and yields derived from m_b and Lg as additional constraints and performed moment-tensor inversions on a larger set of Shagan River explosions, and were then able to determine both the dip-slip and strike-slip components. They found solutions more consistent with Shagan River geology, and determined a moment yield relation of $\log M_i = \log Y + 14.31$ where Y is yield in kilotons and M_i is isotropic moment in Newton-meters. However, they also found that because of the variability in the source mechanism they were unable to determine isotropic moments that gave satisfactory results without detailed knowledge about each explosion.

An assumption used in most moment-tensor inversions is that the nonisotropic source can be described by a single double-couple source. However, this is not necessarily the case. DAY *et al.* (1987), for example, performed axisymmetric calculations of an explosion in a uniformly prestressed medium and showed that Rayleigh wave reversals could be obtained with shear stress levels of 100 bars or greater. Uniformly prestressed means that the M_{xx} and M_{yy} moment-tensor components are equal. This source produces reduced or reversed Rayleigh waves, but no Love waves, making moment-tensor inversion for the isotropic source even more difficult to accomplish.

In addition to problems with tectonic release, STEVENS *et al.* (1993) found that a long-period amplitude reduction can occur if the explosion is on an island or near a mountain slope. If there is a free surface such as a cliff or a sharp mountain boundary, it allows stress to be relaxed in the horizontal direction, leading to a reduction in the amplitude of generated surface waves. This happens because the distortion caused by the explosion goes into a permanent displacement of the slope rather than being radiated away by surface waves. This effect appears to have been a contributor to anomalous surface waves from Novaya Zemlya explosions.

Surface-wave Magnitudes

Yield estimates derived from surface waves are calculated by measuring a surface-wave magnitude and then using a magnitude yield relation to estimate the yield. "Magnitude" is used in a generalized sense here to mean a set of surface-wave measurements with corrections applied to give a measure of source strength, and therefore includes isotropic moment derived from moment-tensor inversion as well as time domain and spectral magnitudes. We consider three types of magnitudes: time domain M_s, path-corrected spectral magnitude $\log M_0'$, and isotropic moment M_i. We also compare results for network-averaged and maximum likelihood magnitudes.

The surface-wave magnitude M_s corrects for distance and is a measure of the size of the source, or more accurately, the part of the source that generates surface waves. The IASPEI formula for M_s is given by

$$M_s = \log \frac{A}{T} + 1.66 \log \Delta + 0.3 \ , \tag{1}$$

where A is the zero-to-peak amplitude in nanometers, Δ is the distance in degrees and T is the measured period in seconds. M_s magnitudes reported by the U.S. Geological Survey (USGS) are made on the vertical component and are measured between periods of 18 and 22 seconds. A problem with the IASPEI formula is that the distance correction becomes increasingly inaccurate at short distances. An improved M_s formula with better distance independence was derived by REZAPOUR and PEARCE (1998) and is based on the theoretical functional form for propagating surface waves (SATO, 1967), with constant factors derived using the entire ISC M_s data set:

$$M_s = \log \frac{A}{T} + \frac{1}{3}\log(\Delta) + \frac{1}{2}\log(\sin(\Delta)) + 0.0046\Delta + 2.370 \ . \tag{2}$$

The $1/3 \log(\Delta)$ term is an Airy phase approximation for the effect of time domain dispersion. This formula has recently been adopted by the International Data Center (IDC), replacing the IASPEI M_s formula.

The path-corrected spectral magnitude $\log M_0'$ is defined as the logarithm of the spectral amplitude divided by the Green's function for an explosion-generated surface wave for the source to receiver path:

$$M_0' = \left| u_z(\omega, r, \theta) \Big/ \frac{S_1^{x}(\omega, h_x) S_2(\omega) \exp[-\gamma_p(\omega) r + i(\varphi_0 - \omega r / c_p(\omega))]}{\sqrt{a_e \sin(r/a_e)}} \right| , \qquad (3)$$

where u_z is the measured spectral amplitude on the vertical component, S_1^{x} is a function that depends on the properties of the source region, S_2 is a function that depends on the receiver region, r is the source to receiver distance, a_e is the radius of the earth, and c_p and γ_p are the phase velocity and attenuation coefficients averaged over the source-to-receiver path. Definitions of the functions are given in Appendix A. For an explosion source with no tectonic release, $M_0' = 3\frac{\beta^2}{\alpha^2} M_i$ where M_i is the explosion isotropic moment. M_0' is defined this way so that the source region excitation function is not explicitly dependent on local material properties at shallow source depths. The functional form of the path-corrected spectral magnitude is similar to the Rezapour and Pearce magnitude, equation (2), except for omission of the Airy phase dispersion correction which is not needed in the frequency domain. Equation (3) has the advantage that it can be regionalized, calculating the earth structure dependent quantities using different structures in different regions. STEVENS and MCLAUGHLIN (2001) discuss this in more detail.

Moment-tensor inversion is a technique for determining the explosion isotropic moment M_i by removing the contaminating nonisotropic source which generates surface waves that are superimposed on top of the explosion-generated surface waves. Although moment-tensor inversion is widely used in earthquake studies, explosions are a special case because the solution becomes nonunique for shallow sources. Details of the techniques for moment-tensor inversion for shallow explosion sources are given by GIVEN and MELLMAN (1986) and EKSTRÖM and RICHARDS (1994), and are summarized in Appendix A. In brief, moment-tensor inversion requires an additional constraint on the nonisotropic source, and this constraint has a very significant effect on the solution. The constraint usually takes the form of assuming that the nonisotropic source can be described by a double-couple source, and then fixing the strike, dip, or rake, or some combination of these parameters. The magnitude of the error that can result from the incorrect choice of this constraint is shown in Figure 2. This figure shows the error in log isotropic moment that results when tectonic release of general rake λ and dip δ are interpreted under the assumption that the tectonic release corresponds to a strike-slip mechanism. The error in moment was calculated for $F = 0.3$, where F is the ratio of double-couple to isotropic moment, and is as large as ± 0.2 magnitude units.

The isotropic moment inferred from moment-tensor inversion will be equal to the average path-corrected spectral magnitude (log of equation (3)) plus a term that depends on the magnitude and orientation of the nonisotropic source. A strike-slip

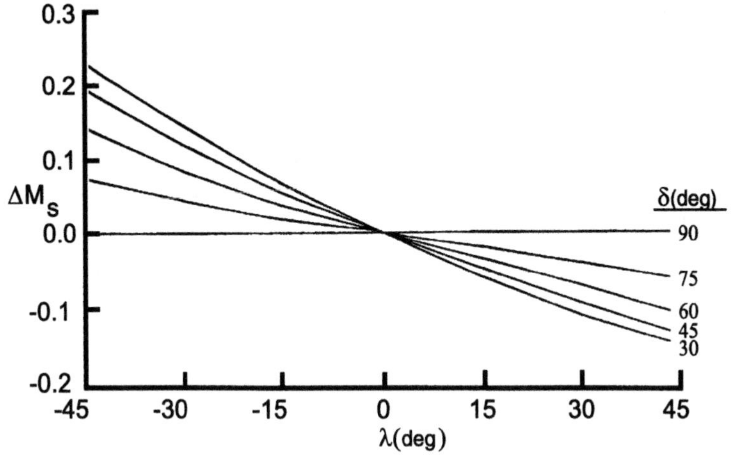

Figure 2
Error in log isotropic moment (labeled ΔM_s) for an event with $F = 0.3$ as a function of rake, for several values of dip, when the tectonic release is interpreted under a strike-slip assumption (rake $\lambda = 0°$, dip $\delta = 90°$). F is the ratio of double-couple to isotropic moment.

constraint has a minimal effect on the result, because the nonisotropic source increases the surface waves over a 180 degree range and decreases the surface waves over a 180 degree range, so if coverage is good, then the isotropic moment will be very close to the average value of the path-corrected spectral magnitudes. A 45-degree dip-slip (thrust) mechanism, on the other hand, decreases the surface-wave amplitudes at all azimuths, so moment-tensor inversion with this constraint increases the isotropic moment. Conversely, a normal fault increases the surface-wave amplitudes at all azimuths, so moment-tensor inversion with this constraint decreases the isotropic moment. A vertical dip-slip mechanism is never used as a constraint because this source generates very small surface waves at shallow depths. A more complicated situation is a complex linear vector dipole (CLVD) source, which is equivalent to two 45-degree thrust faults at right angles to each other. The CLVD source is cylindrically symmetric and therefore generates no Love waves and no azimuthal variations in Rayleigh waves. Consequently, it cannot be removed by moment-tensor inversion, but it can affect the surface-wave amplitudes as strongly as a thrust mechanism.

Surface-wave Magnitude Measurements

Surface-wave magnitudes, M_s and path-corrected spectral magnitudes log M_0' were measured from a large historical data set of surface waves from underground nuclear tests. M_s measurements were made using the techniques now in the operational system at the International Data Center (see STEVENS and MCLAUGHLIN, 2001), and using the Rezapour and Pearce definition of M_s. Noise levels were measured for all waveforms where a surface wave could not be identified. These noise levels were used as upper bounds for calculation of maximum likelihood (MLE) magnitudes, and maximum likelihood magnitudes and station corrections were derived for all events. Table 1 lists the events, network-averaged and MLE magnitudes, and known or estimated yields for each event. The purpose of maximum likelihood magnitudes is to compensate for the bias that results when lower amplitude measurements are lost in noise and only higher amplitude measurements are used. However, MLE magnitudes tend to be smaller than network-averaged magnitudes even for well-recorded events, so we treat them here as separate magnitude measures.

Table 1 also lists isotropic moments from the moment-tensor inversion results of GIVEN and MELLMAN (1986). EKSTRÖM and RICHARDS (1994) give a more extensive list of moment-tensor inversion results for the Shagan River test site, however their moments were normalized to moments derived from regional phases, and are therefore somewhat higher than the Given and Mellman moments, which were derived using the surface-wave path corrections of STEVENS (1986).

Magnitude Yield Relations

Yields have been released for only a few of the events listed in Table 1. The yields for U.S. tests are from DOE (1994), and for Russian tests from MIKHAILOV (1996). Yields for the three Amchitka explosions and for Handley are approximate. The three earliest Shagan River explosions have announced yields. We have also included estimated yields derived from m_b and m_bLg measurements as derived by RINGDAL *et al.* (1992) for the remaining Shagan River explosions. For each magnitude type, we have derived a linear data fit with slope 1 of the form:

$$M = \log Y + C \ . \tag{4}$$

Table 2 lists the value and standard deviation of the constant C for each magnitude for all events with known yields, and for each test site. Figures 3–7 show the different magnitudes plotted vs. yield for all of the data.

Figure 8 shows the difference between log isotropic moment and log yield plotted as a function of F factor for all of the events with both isotropic moments and yields or yield estimates. The scatter is comparable to or greater than the uncertainty due to the variation in source mechanism shown in Figure 2.

We can also compare the moment estimates with theoretical and empirical models of explosion source functions. The Mueller Murphy source model (MUELLER and MURPHY, 1971; MURPHY and MUELLER, 1971) is an empirical fit to the explosion source function that was derived from near-field records of explosions in a variety of materials. Depth dependence is included explicitly in the Mueller Murphy source, and for explosions at standard containment depth ($\sim$122 m/kt$^{1/3}$) this model predicts moment proportional to $Y^{0.76}$. We also compare with two numerical models of explosions in tuff, salt and granite (STEVENS and DAY, 1985). These calculations were

Table 1

Surface-wave magnitude measurements for 207 nuclear explosions

Testsite	Date	Event Name	m_b	M_s Ave	M_s MLE	Stdev	Log M_0' Ave	Log M_0' MLE	Stdev	Nsig	Nnoise	Log M_i	Yield
AHAGGAR	1963293	rubis	5.62	4.26	4.16	0.25	16.18	16.00	0.22	2	0		52
AHAGGAR	1965058	saphir	5.88	4.64	4.60	0.16	16.43	16.35	0.16	3	0		120
AMCHITKA	1965302	longshot	6.03	3.95	3.76	0.12	15.62	15.44	0.06	8	5		80
AMCHITKA	1969275	milrow	6.52	5.00	4.94	0.06	16.56	16.49	0.06	23	3		≈ 1000
AMCHITKA	1971310	cannikin	6.89	5.70	5.68	0.07	17.31	17.26	0.06	26	2		< 5000
CLIMAX	1966153	piledriver	5.56	3.99	3.99	0.14	15.97	15.86	0.06	10	1		62
DEGELEN	1967057	deg26feb67	6.00	3.80	3.64	0.25	15.85	15.74	0.01	2	0		
DEGELEN	1972345	deg10dec72	5.60	4.42	4.39	0.07	16.14	16.07	0.05	7	0		
DEGELEN	1977302	deg29oct77	5.60	3.96	3.74	0.11	15.75	15.47	0.09	4	4		
DEGELEN	1978085	deg26mar78	5.61	4.07	3.61	0.10	15.92	15.45	0.09	3	6		
DEGELEN	1978112	deg22apr78	5.28	3.36	3.17	0.04	15.40	15.06	0.08	3	5		
DEGELEN	1978209	deg28jul78	5.66	3.36	3.15	0.04	15.32	15.01	0.04	2	6		
DEGELEN	1979151	deg31may79	5.24	3.35	3.04	0.11	15.21	14.83	0.07	2	9		
DEGELEN	1980143	deg22may80	5.49	3.26	3.12	0.10	15.10	14.92	0.07	4	8		
DEGELEN	1980213	deg31jul80	5.30	3.70	3.56	0.13	15.50	15.35	0.09	5	6		
DEGELEN	1987157	deg06jun87	5.30	2.07	2.02	0.25	14.66	14.52	0.04	2	24		
DEGELEN	1987198	deg17jul87	5.80	3.77	3.33	0.22	15.73	15.33	0.30	4	17		
LOPNOR	1996160	lop08jun96	5.69	4.05	4.03	0.09	15.72	15.75	0.04	4	1		
MURUROA	1977078	mur19mar77	5.73	4.13	4.04	0.16	15.91	15.88	0.18	3	2		
MURUROA	1977328	mur24nov77	5.80	4.16	3.95	0.10	15.77	15.67	0.20	3	3		
MURUROA	1978334	mur30nov78	5.80	4.37	3.96	0.17	15.86	15.62	0.08	4	5		
MURUROA	1979206	mur25jul79	6.03	4.05	3.94	0.04	15.79	15.65	0.13	5	3		
MURUROA	1980338	mur03dec80	5.58	3.84	3.52	0.12	15.59	15.24	0.08	2	7		
MURUROA	1982206	mur25jul82	5.60	3.88	3.79	0.06	15.61	15.54	0.08	6	8		
MURUROA	1983109	mur19apr83	5.60	4.05	3.94	0.08	15.86	15.75	0.10	8	8		
MURUROA	1983145	mur25may83	5.90	3.97	3.82	0.08	15.74	15.64	0.11	8	7		
MURUROA	1984133	mur12may84	5.70	3.87	3.56	0.10	15.61	15.37	0.13	4	11		
MURUROA	1984307	mur02nov84	5.70	4.23	3.71	0.11	15.93	15.47	0.11	2	12		
MURUROA	1984341	mur06dec84	5.60	3.99	3.80	0.11	15.68	15.50	0.10	7	5		
MURUROA	1985330	mur26nov85	5.80	4.04	3.95	0.06	15.93	15.65	0.18	5	10		
MURUROA	1987140	mur20may87	5.60	3.89	3.64	0.09	15.52	15.34	0.07	4	7		

Table 1

continued

Testsite	Date	Event Name	m_b	M_s Ave	M_s MLE	Stdev	Log M_0' Ave	Log M_0' MLE	Stdev	Nsig	Nnoise	Log M_i	Yield
NN_ZEMLYA	1967294	nnz21oct67	5.99	3.82	3.69	0.19	15.49	15.33	0.07	7	6		
NN_ZEMLYA	1968312	nnz07nov68	6.11	4.17	4.09	0.22	16.06	15.97	0.13	8	0		
NN_ZEMLYA	1970287	nnz14oct70	6.77	5.00	4.94	0.07	16.56	16.49	0.05	28	4		
NN_ZEMLYA	1973255	nnz12sep73	6.96	5.37	5.29	0.10	16.88	16.77	0.11	10	3		
NN_ZEMLYA	1974241	nnz29aug74	6.54	4.94	4.87	0.08	16.54	16.43	0.05	24	2		
NN_ZEMLYA	1975235	nnz23aug75	6.55	4.85	4.76	0.10	16.27	16.14	0.09	18	4		
NN_ZEMLYA	1976273	nnz29sep76	5.77	3.63	3.41	0.14	15.34	15.10	0.07	6	2		
NN_ZEMLYA	1976294	nnz20oct76	4.89	3.40	3.25	0.01	15.02	14.72	0.04	2	2		
NN_ZEMLYA	1977244	nnz01sep77	5.71	4.22	4.27	0.17	15.40	15.33	0.12	5	0		
NN_ZEMLYA	1978222	nnz10aug78	6.04	3.94	3.64	0.08	15.54	15.13	0.07	6	9		
NN_ZEMLYA	1978270	nnz27sep78	5.68	4.16	4.09	0.05	15.80	15.74	0.06	9	1		
NN_ZEMLYA	1979267	nnz24sep79	5.80	4.24	4.14	0.07	15.75	15.63	0.06	7	2		
NN_ZEMLYA	1979291	nnz18oct79	5.85	4.21	3.79	0.13	15.37	15.05	0.06	2	7		
NN_ZEMLYA	1980285	nnz11oct80	5.80	3.89	3.76	0.11	15.52	15.36	0.06	7	6		
NN_ZEMLYA	1981274	nnz01oct81	5.91	4.13	4.10	0.09	15.75	15.67	0.07	9	4		
NN_ZEMLYA	1982284	nnz11oct82	5.52	3.82	3.66	0.13	15.38	15.22	0.06	9	6		
NN_ZEMLYA	1983230	nnz18aug83	5.84	3.97	3.86	0.05	15.65	15.54	0.10	2	16		
NN_ZEMLYA	1983268	nnz25sep83	5.71	3.82	3.52	0.09	15.31	15.10	0.08	6	12		
NN_ZEMLYA	1984299	nnz25oct84	5.80	3.99	3.92	0.06	15.76	15.65	0.06	10	6		
NN_ZEMLYA	1987214	nnz02aug87	5.80	3.93	3.81	0.09	15.61	15.42	0.08	11	11		
NN_ZEMLYA	1988128	nnz07may88	5.60	4.01	3.89	0.08	15.51	15.45	0.05	14	4		
NN_ZEMLYA	1988339	nnz04dec88	5.90	4.06	3.93	0.08	15.71	15.59	0.05	12	4		
NN_ZEMLYA	1990297	nnz24oct90	5.40	3.99	3.86	0.08	15.61	15.44	0.10	14	5		
PAHUTE	1966181	halfbeak	6.10	4.70	4.58	0.25	16.70	16.49	0.06	2	0		365
PAHUTE	1967143	scotch	5.70	4.55	4.50	0.08	16.47	16.34	0.07	12	1	16.46	155
PAHUTE	1968082	stinger	5.60	4.23	4.14	0.10	16.15	16.04	0.04	10	2		
PAHUTE	1968117	boxcar	6.30	5.36	5.32	0.06	17.07	16.94	0.10	11	2		1300
PAHUTE	1968242	sled	5.90	4.28	4.23	0.09	16.19	16.11	0.04	12	2		
PAHUTE	1968354	benham	6.30	5.47	5.35	0.07	17.10	16.95	0.07	8	3		1150
PAHUTE	1969127	purse	5.80	4.48	4.43	0.07	16.34	16.24	0.04	18	1		
PAHUTE	1970085	handley	6.50	5.26	5.22	0.06	16.94	16.80	0.07	9	3		> 1000

PAHUTE	1973157	almendro	6.10	4.88	4.72	0.08	16.59	16.49	0.07	11	4		
PAHUTE	1975134	tybo	6.00	4.71	4.61	0.06	16.49	16.42	0.03	11	4		
PAHUTE	1975154	stilton	5.90	4.26	4.18	0.07	16.23	16.09	0.07	10	3		
PAHUTE	1975170	mast	6.10	4.71	4.55	0.09	16.50	16.39	0.06	10	2		
PAHUTE	1976045	cheshire	6.00	4.87	4.68	0.13	16.67	16.47	0.06	11	5		
PAHUTE	1976069	estuary	6.00	4.93	4.67	0.10	16.61	16.41	0.04	5	2		
PAHUTE	1976077	pool	6.10	4.56	4.45	0.08	16.38	16.20	0.04	7	2		
PAHUTE	1978101	backbeach	5.50	4.14	3.92	0.08	16.05	15.76	0.09	4	6		
PAHUTE	1978243	panir	5.60	4.00	3.93	0.09	15.86	15.74	0.07	13	3		
PAHUTE	1978350	farm	5.50	4.17	4.07	0.07	15.94	15.79	0.08	8	4		
PAHUTE	1979162	pepato	5.50	4.15	4.10	0.06	16.08	16.03	0.04	22	2		
PAHUTE	1979269	sheepshead	5.60	4.07	4.01	0.06	15.95	15.86	0.03	17	4		
PAHUTE	1980117	colwick	5.40	4.00	3.93	0.11	15.90	15.79	0.06	7	2		
PAHUTE	1980164	kash	5.60	4.24	4.11	0.06	16.11	15.97	0.06	16	8		
PAHUTE	1980207	tafi	5.50	4.29	4.20	0.05	16.11	16.01	0.04	15	3		
PAHUTE	1981157	harzer	5.50	4.00	3.62	0.17	15.60	15.48	0.07	4	3		
PAHUTE	1987108	delamar	5.50	4.06	3.97	0.05	15.89	15.62	0.09	8	6		
PAHUTE	1987120	hardin	5.50	4.08	3.95	0.11	16.08	15.76	0.16	9	12		
PAHUTE	1987267	lockney	5.70	4.22	4.15	0.06	16.01	15.95	0.05	13	1		
PAHUTE	1988230	kearsarge	5.50	4.05	3.87	0.08	15.85	15.61	0.10	7	9		
SHAGAN	1969334	sha30nov69	6.00	3.78	3.68	0.08	15.89	15.74	0.07	4	2		125
SHAGAN	1972307	sha02nov72	6.14	3.87	3.80	0.09	15.73	15.64	0.07	11	1		165
SHAGAN	1972345	sha10dec72	6.00	4.42	4.39	0.07	16.14	16.07	0.06	7	0		140
SHAGAN	1973204	sha23jul73	6.18	4.12	4.07	0.08	15.85	15.71	0.05	10	2		212
SHAGAN	1973348	sha14dec73	5.82	3.82	3.76	0.10	15.95	15.88	0.04	5	0		80
SHAGAN	1976186	sha04jul76	5.81	3.80	3.82	0.12	15.67	15.65	0.15	5	0		65
SHAGAN	1976342	sha07dec76	5.90	4.11	3.84	0.08	15.72	15.39	0.13	2	2		54
SHAGAN	1977149	sha29may77	5.77	3.26	3.11	0.25	15.39	15.16	0.01	2	2		44
SHAGAN	1977180	sha29jun77	5.22	3.16	3.01	0.25	14.96	14.74	0.22	2	2		9
SHAGAN	1977248	sha05sep77	5.74	3.90	3.81	0.06	15.59	15.50	0.09	4	2		78
SHAGAN	1977302	sha29oct77	5.54	3.97	3.74	0.10	15.76	15.47	0.09	4	4		50
SHAGAN	1978162	sha11jun78	5.86	4.16	4.11	0.10	15.93	15.86	0.06	11	1		58
SHAGAN	1978186	sha05jul78	5.83	3.53	3.39	0.12	15.33	15.11	0.10	4	4		57
SHAGAN	1978241	sha29aug78	5.95	3.65	3.48	0.12	15.59	15.44	0.06	5	4	15.91	119

Table 1

continued

Testsite	Date	Event Name	m_b	M_s Ave	M_s MLE	Stdev	Log M_0' Ave	Log M_0' MLE	Stdev	Nsig	Nnoise	Log M_i	Yield
SHAGAN	1978258	sha15sep78	5.99	4.00	3.83	0.09	15.75	15.62	0.07	7	3	16.05	81
SHAGAN	1978308	sha04nov78	5.56	3.82	3.65	0.13	15.45	15.30	0.05	5	2	15.94	44
SHAGAN	1978333	sha29nov78	6.07	3.98	3.88	0.05	15.73	15.64	0.03	7	1	16.08	101
SHAGAN	1979032	sha01feb79	5.38	2.74	2.73	0.03	14.97	14.67	0.05	2	7		18
SHAGAN	1979174	sha23jun79	6.22	3.97	3.82	0.10	15.78	15.68	0.05	7	3	16.25	149
SHAGAN	1979188	sha07jul79	5.83	4.25	3.77	0.21	16.16	15.67	0.19	3	7	15.88	97
SHAGAN	1979216	sha04aug79	6.16	4.11	4.02	0.05	15.92	15.84	0.05	11	4	16.27	153
SHAGAN	1979230	sha18aug79	6.12	3.92	3.74	0.08	15.69	15.49	0.12	7	4	16.02	179
SHAGAN	1979301	sha28oct79	5.96	4.12	4.03	0.05	15.88	15.82	0.04	13	3	16.29	139
SHAGAN	1979336	sha02dec79	6.01	4.18	4.11	0.04	16.01	15.94	0.04	8	3	16.14	93
SHAGAN	1979357	sha23dec79	6.18	3.93	3.77	0.06	15.59	15.44	0.06	10	7	15.90	137
SHAGAN	1980164	sha12jun80	5.59	3.35	3.08	0.09	15.17	14.91	0.05	3	8	15.46	37
SHAGAN	1980181	sha29jun80	5.74	3.49	3.35	0.09	15.28	15.11	0.12	5	7	15.61	44
SHAGAN	1980258	sha14sep80	6.21	4.06	3.94	0.07	15.80	15.71	0.06	9	3	16.33	196
SHAGAN	1980286	sha12oct80	5.90	4.03	3.94	0.06	15.89	15.80	0.03	9	5	16.19	102
SHAGAN	1980349	sha14dec80	5.95	3.90	3.85	0.06	15.65	15.61	0.09	10	4	16.10	101
SHAGAN	1980362	sha27dec80	5.88	3.94	3.58	0.13	15.78	15.35	0.11	4	8	15.45	100
SHAGAN	1981088	sha29mar81	5.61	3.55	3.37	0.04	15.30	15.12	0.06	5	7	15.74	30
SHAGAN	1981112	sha22apr81	6.05	4.08	4.03	0.04	15.88	15.87	0.03	9	1	16.11	92
SHAGAN	1981147	sha27may81	5.46	3.16	2.79	0.07	14.86	14.70	0.04	2	8		20
SHAGAN	1981256	sha13sep81	6.18	4.06	4.01	0.06	15.94	15.90	0.06	5	1	16.26	163
SHAGAN	1981291	sha18oct81	6.11	4.19	4.10	0.04	15.94	15.87	0.05	6	8	16.16	107
SHAGAN	1981361	sha27dec81	6.31	4.31	4.17	0.09	15.96	15.86	0.04	9	6	16.19	156
SHAGAN	1982115	sha25apr82	6.10	4.10	3.96	0.03	15.83	15.68	0.04	7	8	16.17	145
SHAGAN	1982339	sha05dec82	6.10	4.16	4.00	0.07	15.94	15.76	0.04	9	6	16.19	119
SHAGAN	1983163	sha12jun83	6.10	4.39	4.27	0.04	16.15	16.04	0.04	12	7	16.30	138
SHAGAN	1983279	sha06oct83	6.00	4.21	4.15	0.07	16.06	15.95	0.05	11	5	16.27	82
SHAGAN	1983299	sha26oct83	6.10	4.25	4.11	0.06	15.98	15.80	0.06	7	9	16.33	114
SHAGAN	1984050	sha19feb84	5.80	4.21	4.08	0.06	15.91	15.79	0.05	8	6	15.98	49
SHAGAN	1984089	sha29mar84	5.90	4.12	3.88	0.08	15.85	15.71	0.04	6	9	16.04	83
SHAGAN	1984116	sha25apr84	5.90	4.35	4.11	0.05	16.17	15.93	0.09	6	7	16.18	76

SHAGAN	1984196	sha14jul84	6.20	4.33	4.28	0.05	16.13	16.06	0.06	14	2	16.28	135
SHAGAN	1984301	sha27oct84	6.20	4.29	4.14	0.08	15.99	15.84	0.06	13	6	16.24	165
SHAGAN	1984337	sha02dec84	5.80	4.08	3.92	0.12	15.80	15.64	0.08	11	7	16.11	79
SHAGAN	1984351	sha16dec84	6.10	4.28	4.17	0.07	16.11	16.01	0.05	14	6	16.36	137
SHAGAN	1984363	sha28dec84	6.00	4.15	4.06	0.05	15.84	15.76	0.05	11	7	15.90	105
SHAGAN	1985041	sha10feb85	5.90	4.26	4.15	0.06	16.07	15.94	0.05	11	5	16.21	62
SHAGAN	1985115	sha25apr85	5.90	4.04	3.75	0.13	15.85	15.55	0.11	8	10		74
SHAGAN	1985166	sha15jun85	6.00	4.00	3.90	0.06	15.71	15.65	0.03	13	5		114
SHAGAN	1985181	sha30jun85	6.00	4.06	3.92	0.05	15.84	15.68	0.03	8	9		86
SHAGAN	1985201	sha20jul85	5.90	4.06	3.90	0.06	15.84	15.64	0.10	9	5		74
SHAGAN	1987071	sha12mar87	5.40	3.81	3.67	0.05	15.63	15.49	0.05	8	13		11
SHAGAN	1987093	sha03apr87	6.20	4.48	4.39	0.05	16.33	16.15	0.07	23	6		140
SHAGAN	1987107	sha17apr87	6.00	4.05	3.68	0.12	15.81	15.56	0.06	6	19		86
SHAGAN	1987171	sha20jun87	6.10	4.01	3.95	0.04	15.82	15.72	0.05	17	4		107
SHAGAN	1987214	sha02aug87	5.90	3.86	3.78	0.03	15.64	15.52	0.07	17	6		72
SHAGAN	1987319	sha15nov87	6.00	4.45	4.39	0.02	16.21	16.17	0.03	23	3		103
SHAGAN	1987347	sha13dec87	6.10	4.28	4.11	0.07	15.98	15.84	0.05	16	8		137
SHAGAN	1987361	sha27dec87	6.10	4.13	3.99	0.08	15.84	15.72	0.03	13	8		117
SHAGAN	1988044	sha13feb88	6.10	4.19	4.10	0.04	15.96	15.89	0.04	15	9		125
SHAGAN	1988094	sha03apr88	6.00	4.30	4.20	0.03	16.05	15.87	0.08	13	7		135
SHAGAN	1988125	sha04may88	6.10	4.34	3.94	0.06	16.20	15.81	0.06	2	15		132
SHAGAN	1988258	shaganjve	6.10	4.28	4.21	0.03	16.09	16.03	0.03	23	3		108
SHAGAN	1988317	sha12nov88	5.26	2.78	2.70	0.03	15.20	14.91	0.06	2	17		15
SHAGAN	1988352	sha17dec88	5.67	4.11	4.01	0.07	15.98	15.87	0.05	18	6		68
SHAGAN	1989022	sha22jan89	5.95	4.26	4.17	0.02	16.05	15.94	0.05	17	4		118
SHAGAN	1989043	sha12feb89	5.71	4.30	4.19	0.05	16.12	15.87	0.09	15	4		63
SHAGAN	1989189	sha08jul89	5.48	3.87	3.78	0.04	15.63	15.51	0.04	17	6		22
SHAGAN	1989245	sha02sep89	4.98	3.46	3.31	0.05	15.32	15.17	0.03	10	10		6
SHAGAN	1989292	sha19oct89	5.74	4.28	4.17	0.05	16.17	16.00	0.08	15	4		70
SN_ZEMLYA	1973270	snz27sep73	5.83	3.68	3.69	0.12	15.55	15.44	0.05	5	1		
SN_ZEMLYA	1973300	snz27oct73	6.90	5.55	5.49	0.10	17.03	16.98	0.08	22	3		
SN_ZEMLYA	1974306	snz02nov74	6.75	5.27	5.25	0.06	16.75	16.73	0.08	21	1		
SN_ZEMLYA	1975291	snz18oct75	6.70	4.92	4.87	0.06	16.47	16.42	0.05	27	3		
USA	1967344	gasbuggy	5.10	3.68	3.61	0.05	15.61	15.55	0.10	4	0		29

Table 1

continued

Testsite	Date	Event Name	m_b	M_s Ave	M_s MLE	Stdev	Log M_0' Ave	Log M_0' MLE	Stdev	Nsig	Nnoise	Log M_i	Yield
USA	1968019	faultless	6.30	5.08	4.99	0.17	17.10	16.97	0.10	3	1		
USA	1969253	rulison	5.30	3.50	3.42	0.12	15.52	15.36	0.12	4	1		40
USA	1973137	rioblanco	5.40	3.61	3.51	0.14	15.68	15.60	0.06	5	1		100
USSR	1970346	sov12dec70	6.60	4.10	4.07	0.12	15.93	15.88	0.14	6	1		80
USSR	1970357	sov23dec70	6.60	4.20	4.23	0.10	16.00	15.95	0.14	6	0		75
USSR	1971082	sov23mar71	5.90	4.19	3.79	0.19	15.84	15.53	0.14	3	2		45
USSR	1976211	sov29jul76	5.90	3.91	3.85	0.09	15.44	15.33	0.04	6	0		58
YUCCA	1965337	corduroy	5.60	4.13	4.07	0.11	16.03	15.96	0.07	6	1		
YUCCA	1966139	dumont	5.80	4.18	4.05	0.07	16.15	15.98	0.08	9	5		
YUCCA	1967054	agile	5.80	4.29	4.05	0.05	15.89	15.77	0.02	2	1		
YUCCA	1967140	commodore	5.90	4.54	4.47	0.06	16.29	16.19	0.03	10	1	16.35	250
YUCCA	1967270	zaza	5.70	4.46	4.40	0.08	16.30	16.19	0.07	8	0		
YUCCA	1967291	lanpher	5.70	3.96	3.90	0.11	15.91	15.79	0.06	8	2		
YUCCA	1968250	noggin	5.60	4.17	4.02	0.10	16.02	15.90	0.07	10	2		
YUCCA	1969302	calabash	5.70	3.88	3.78	0.08	15.89	15.75	0.04	8	5	15.83	110
YUCCA	1970146	flask	5.60	3.72	3.59	0.10	15.74	15.58	0.06	10	5	15.55	105
YUCCA	1970287	tijeras	5.50	4.20	4.11	0.08	16.07	15.97	0.06	13	2		
YUCCA	1970351	carpetbag	5.70	4.36	4.13	0.06	16.10	15.93	0.06	6	7	16.13	220
YUCCA	1972265	oscuro	5.70	4.17	4.11	0.10	16.12	16.03	0.03	14	2		
YUCCA	1973116	starwort	5.60	3.87	3.78	0.06	15.78	15.65	0.08	6	1	15.62	90
YUCCA	1974191	escabosa	5.70	4.39	4.33	0.11	16.23	16.11	0.05	10	3		
YUCCA	1974242	portmanteau	5.80	3.95	3.78	0.07	16.06	15.75	0.08	4	3		
YUCCA	1975059	topgallant	5.70	3.76	3.65	0.08	15.99	15.71	0.07	4	6		
YUCCA	1975154	mizzen	5.70	4.07	3.97	0.11	16.05	15.96	0.05	12	6		
YUCCA	1975354	chiberta	5.70	4.15	4.09	0.06	16.11	16.03	0.08	3	0		
YUCCA	1976035	esrom	5.70	3.97	3.72	0.19	15.85	15.70	0.19	4	5		
YUCCA	1976035	keelson	5.80	4.12	4.01	0.08	16.15	16.02	0.05	5	4		
YUCCA	1976077	strait	5.80	4.41	4.30	0.07	16.23	16.11	0.07	9	2		
YUCCA	1977095	marsilly	5.60	3.76	3.59	0.18	15.77	15.55	0.05	7	8		
YUCCA	1977145	crewline	5.30	3.63	3.50	0.25	16.11	15.96	0.04	2	2		
YUCCA	1977231	scantling	5.60	4.17	4.04	0.07	16.07	15.96	0.04	12	6		

YUCCA	1977313	sandreef	5.70	4.41	4.28	0.08	16.26	16.16	0.06	17	2		
YUCCA	1977348	farallones	5.70	4.17	4.04	0.05	16.02	15.81	0.07	7	5		
YUCCA	1978054	reblochon	5.60	3.87	3.88	0.05	15.77	15.73	0.16	4	0		
YUCCA	1978082	iceberg	5.60	4.01	3.98	0.10	16.07	15.98	0.06	13	2		
YUCCA	1978193	lowball	5.50	4.02	3.95	0.08	15.84	15.74	0.04	16	3		
YUCCA	1978270	rummy	5.70	4.28	4.20	0.06	16.18	16.07	0.08	16	3		
YUCCA	1978322	quargel	5.10	4.07	3.74	0.08	15.60	15.20	0.11	2	4		
YUCCA	1979039	quinella	5.50	4.06	3.92	0.07	15.80	15.71	0.04	10	4		
YUCCA	1979249	hearts	5.80	4.26	4.19	0.07	16.13	16.02	0.05	16	4	16.16	140
YUCCA	1980107	pyramid	5.30	3.88	3.80	0.09	15.84	15.63	0.13	6	4		
YUCCA	1987169	brie		2.81	2.88	0.06	15.17	15.33	0.04	2	15		
YUCCA	1987197	midland	4.80	3.54	3.27	0.18	15.33	15.04	0.07	4	10		
YUCCA	1987197	midlandA		2.22	2.08	0.01	14.47	14.34	0.01	2	10		
YUCCA	1987225	tahoka	5.90	4.26	4.18	0.07	16.08	15.84	0.10	12	6		

Table 2

Surface-wave Magnitude/Yield Relations

Testsite	Number	M_s Ave	M_s MLE	Log M_0' Ave	Log M_0' MLE	Log M_i
ALL	98	2.10 ± 0.26	1.98 ± 0.27	13.91 ± 0.25	13.77 ± 0.25	14.05 ± 0.21 (43)
NTS	12	2.11 ± 0.21	2.02 ± 0.24	13.96 ± 0.17	13.82 ± 0.18	13.86 ± 0.24 (7)
AHAGGAR	2	2.55 ± 0.01	2.48 ± 0.06	14.41 ± 0.08	14.28 ± 0.02	
AMCHITKA	3	2.02 ± 0.03	1.92 ± 0.07	13.63 ± 0.08	13.53 ± 0.04	
CLIMAX	1	2.20	2.20	14.18	14.07	
PAHUTE	5	2.28 ± 0.11	2.21 ± 0.12	14.07 ± 0.14	13.92 ± 0.14	14.27 (1)
SHAGAN	74	2.09 ± 0.26	1.95 ± 0.27	13.90 ± 0.25	13.74 ± 0.25	14.08 ± 0.21 (36)
USA PNE	3	1.91 ± 0.30	1.83 ± 0.32	13.91 ± 24	13.81 ± 0.25	
USSR PNE	4	2.30 ± 0.17	2.18 ± 0.12	14.01 ± 0.23	13.87 ± 0.22	
YUCCA	6	1.95 ± 0.17	1.84 ± 0.19	13.84 ± 0.10	13.70 ± 0.12	13.79 ± 0.18 (6)

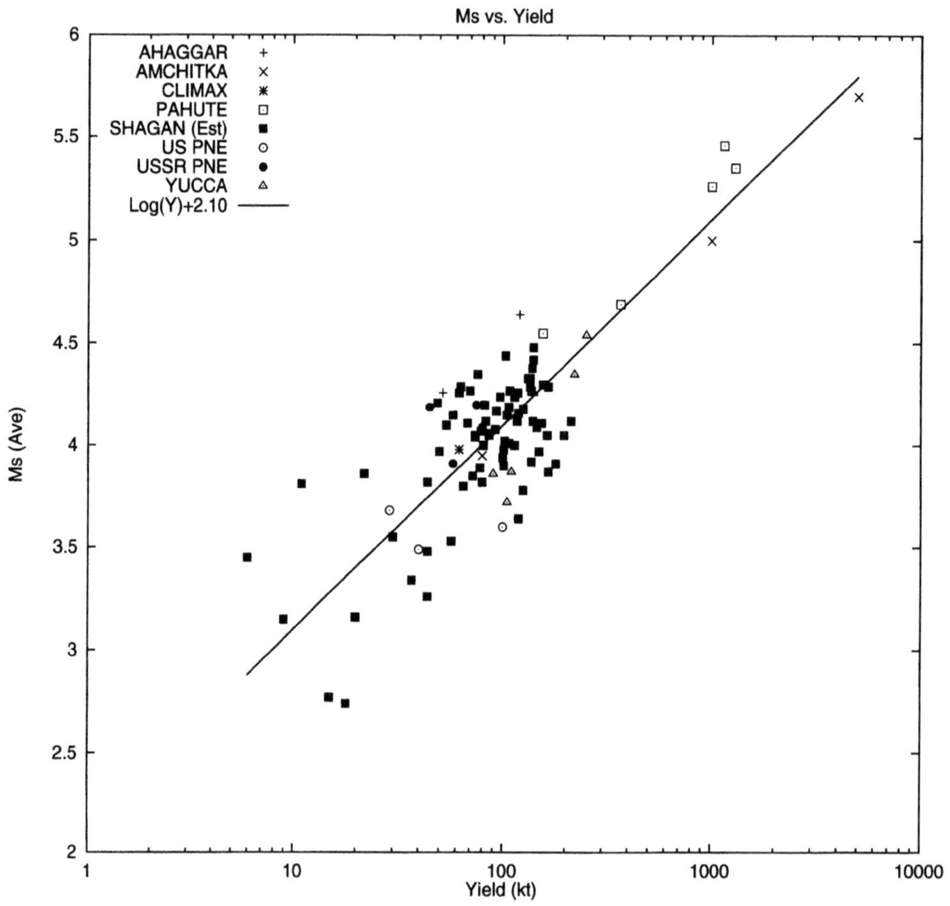

Figure 3
Network average M_s plotted vs. yield for the explosion data set of Table 1.

performed for a fixed depth and moment scales linearly with yield. These relations are compared to the scalar moments (path-corrected spectral magnitudes) estimates, and isotropic moments in Figures 9 and 10. The data have roughly the same magnitude and scatter as the different models, however the moment estimates are in general somewhat lower than the model predictions for hard rock.

Discussion and Conclusions

Table 2 and Figures 3–7 show magnitude/yield relations for five distinct surface-wave magnitude measures. Although there are small differences in scatter between the different magnitudes, all show a standard deviation in the observations of about 0.25 and none is dramatically better than any of the others. The magnitude yield relations for NTS and Shagan River are very close for time domain M_s and path-

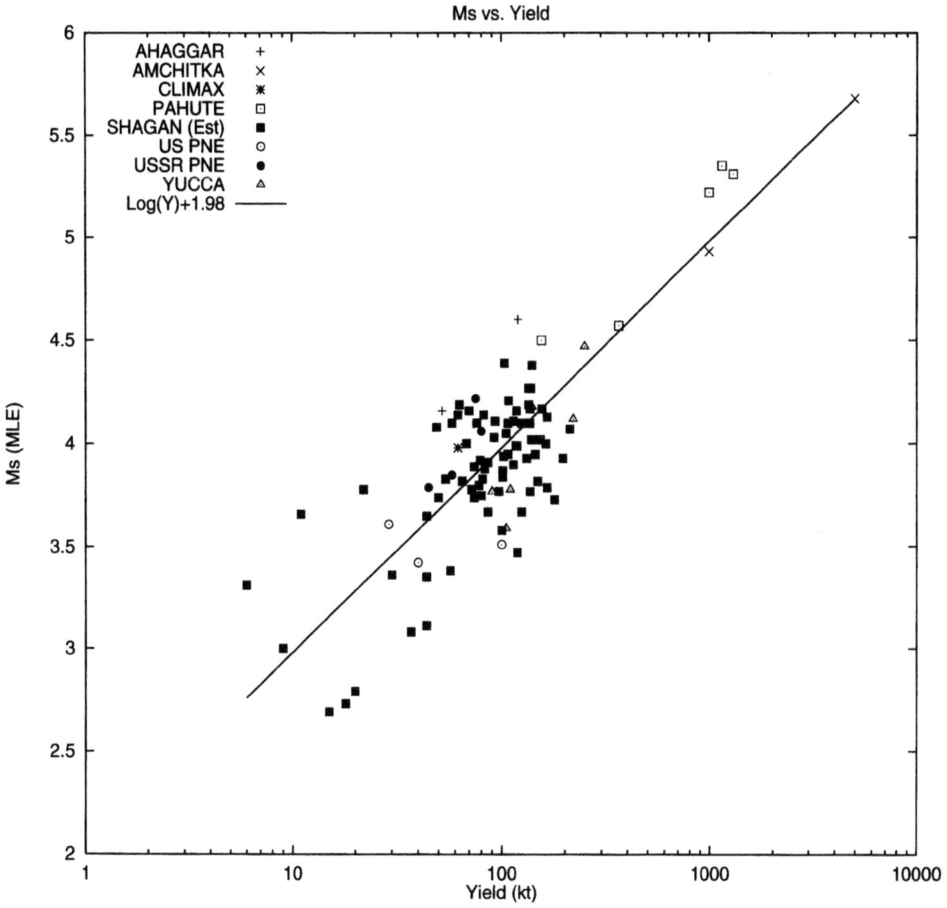

Figure 4
Maximum likelihood M_s plotted vs. yield for the explosion data set of Table 1.

corrected spectral magnitude M_0', however they differ by about 0.2 magnitude units for M_i as determined by moment-tensor inversion. This difference is a direct consequence of the assumption of different tectonic release mechanisms for the two test sites. It could be correct if the material properties for the two sites were very different, however as discussed by EKSTRÖM and RICHARDS (1994) it is likely that the assumption of pure thrust mechanism at Shagan River overcorrects the isotropic moment. A reasonable conclusion that can be drawn from this analysis is that yield estimates accurate to within a standard deviation of 0.25 in log yield can be estimated from surface-wave magnitudes or from moment-tensor inversion. However, the variability in surface waves due to tectonic release and other factors, and the inability to uniquely remove these effects is large enough that it is not possible to reliably make more accurate yield estimates using surface waves. Because the variability in surface-wave amplitude measurements is small, yield estimates can be made to this

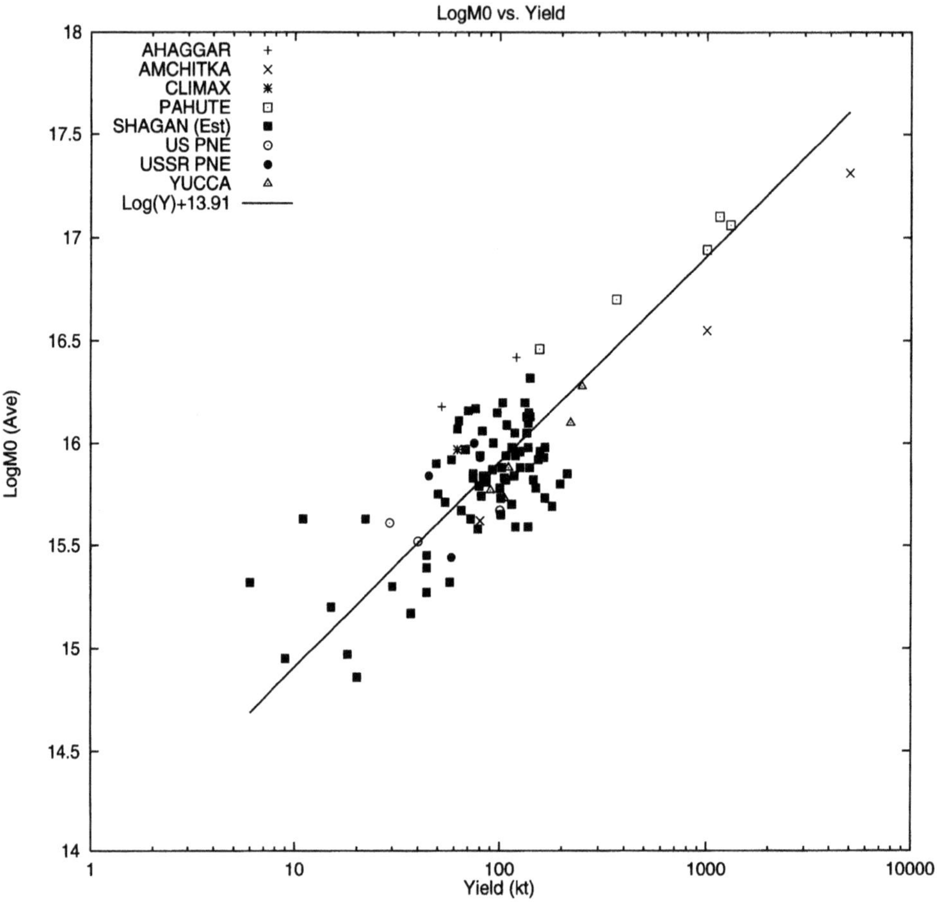

Figure 5
Network average path-corrected spectral magnitude Log M_0' plotted vs. yield for the explosion data set of Table 1.

accuracy from time domain or spectral magnitudes with a limited number of measurements. Moment-tensor inversion requires more measurements in order to solve for three independent quantities and because of uncertainties about the source mechanism does not significantly improve the accuracy of yield estimates.

Appendix A

Surface Wave Generation by a Moment Tensor Source

Surface waves are observed some distance from the source, often after having passed through a series of different earth structures. STEVENS and MCLAUGHLIN

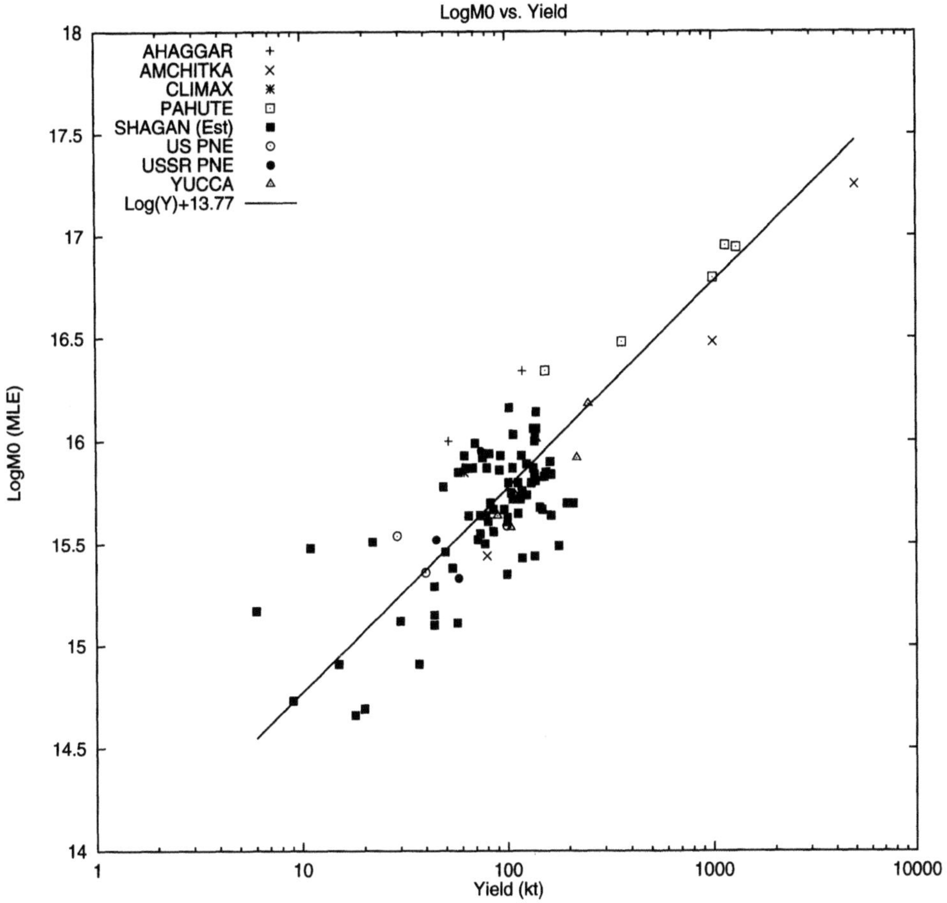

Figure 6
Maximum likelihood path corrected spectral magnitude Log M_0' plotted vs. yield for the explosion data set of Table 1.

(1999) show that the vertical displacement component of the Rayleigh wave, defined with vertical up, for a general source can be written in the form (notation follows HARKRIDER *et al.*, 1994):

$$u_z(\omega, r, \varphi) = \frac{1}{\sqrt{a_e \sin(r/a_e)}} \sqrt{\frac{2A_{R_1}}{\pi\omega c_1^2}} \sqrt{c_2 A_{R_2}} \times \exp\left[i\left(\pi/4 - \omega r/c_p - \gamma_p r\right)\right] F_s(\omega, \varphi, h) \tag{A1}$$

where ω is angular frequency, r is source to receiver distance, a_e is the radius of the earth, φ is azimuth, A_R is the Rayleigh wave excitation function, c is phase velocity, γ is the attenuation coefficient, and the subscripts 1, 2, and p refer to parameters

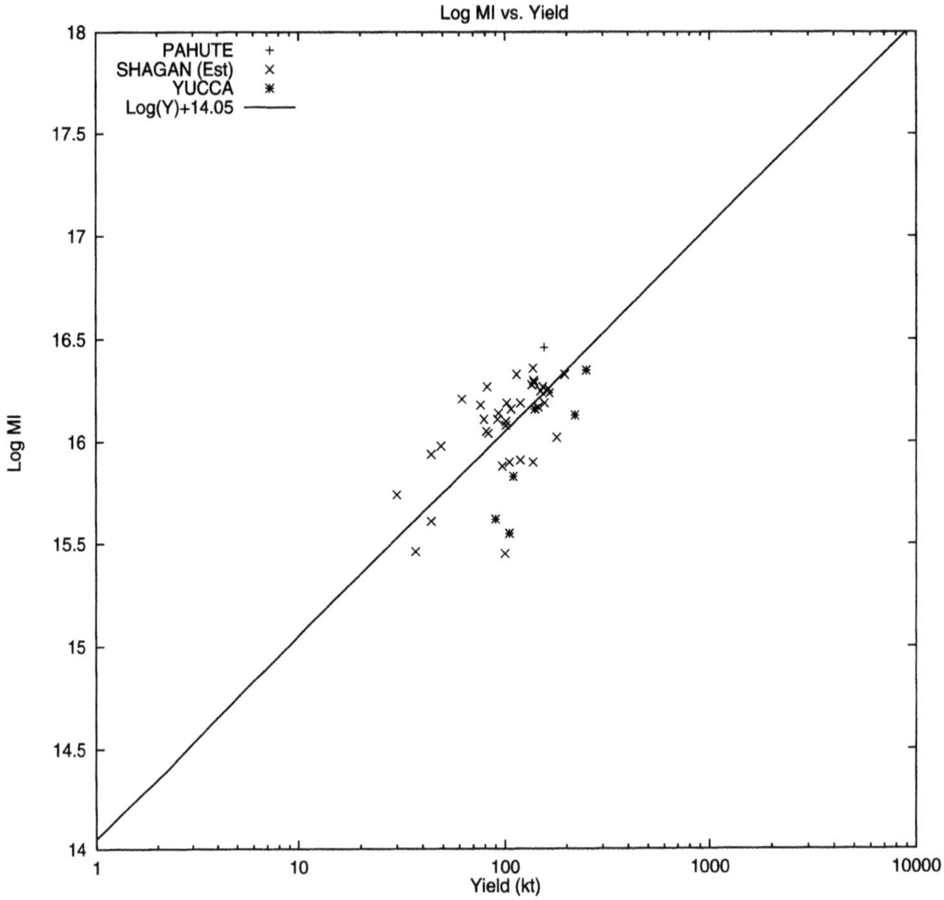

Figure 7
Isotropic moment M_i plotted vs. yield for the explosion data set of Table 1.

derived from the source-region structure, parameters derived from the receiver-region structure, and parameters which are defined by path averages, respectively. All source properties are contained in the function F_s which can be given in the form of moment-tensor components by:

$$F_s(\omega, \varphi, h) = \left[D\frac{\beta^2}{\alpha^2}\frac{M_{11}+M_{22}+M_{33}}{3} - B\frac{M_{11}+M_{22}-2M_{33}}{12}\right.$$
$$\left. + \frac{iC}{2}[M_{13}\cos\varphi + M_{23}\sin\varphi] - \frac{A}{4}[(M_{11}-M_{22})\cos 2\varphi + 2M_{12}\sin 2\varphi]\right] \tag{A2}$$

where

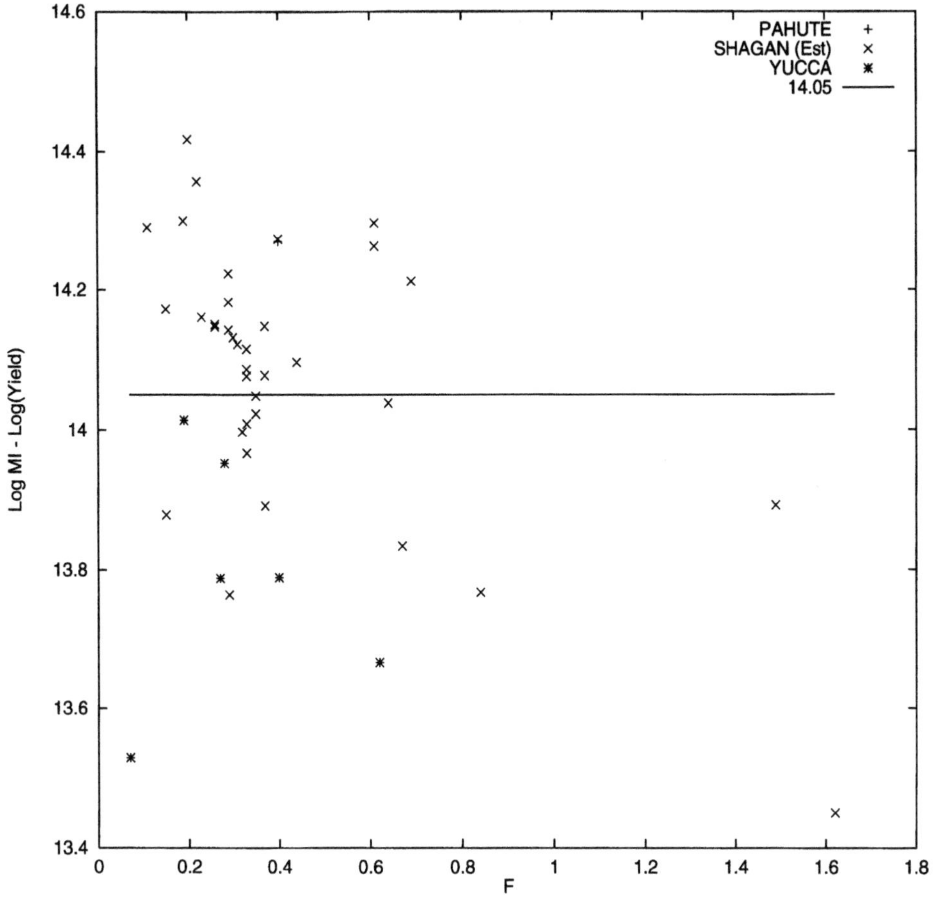

Figure 8
Difference between log moment and log yield shown as a function of the tectonic to isotropic moment ratio *F*.

$$A = -y_3(h)$$

$$B = -\left\{\left(3 - 4\frac{\beta^2}{\alpha^2}\right)y_3(h) + \frac{2}{\rho\alpha^2 k}y_2(h)\right\}$$

$$C = \frac{1}{\mu k}y_4(h)$$

$$D = y_3(h) - \frac{1}{2\mu k}y_2(h)$$

where y_1, y_2, y_3, y_4 are the vertical displacement, normal stress, horizontal displacement, and shear-stress eigenfunctions, respectively (TAKEUCHI and SAITO, 1972) and h is the source depth. Since the stress eigenfunctions both vanish for a free surface, for a shallow source, the coefficients reduce approximately to:

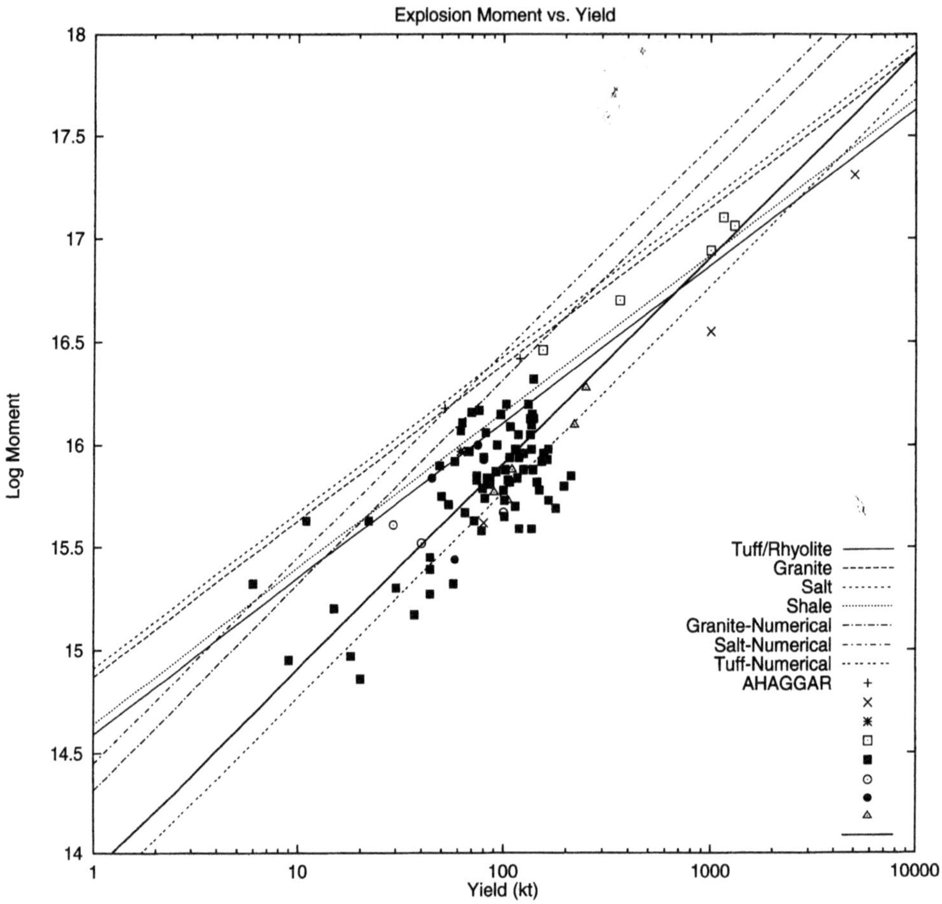

Figure 9
Log M_0' vs. yield curves for different material properties. Four of the curves are derived from the Mueller Murphy source models for tuff/rhyolite, granite, salt, and shale, and the other three from numerical models of explosions in tuff, granite and salt (STEVENS and DAY, 1985).

$$B \approx -\left(3 - 4\frac{\beta^2}{\alpha^2}\right) y_3(h)$$

$$C \approx 0$$

$$D \approx y_3(h) \ .$$

Because C vanishes at the free surface, M_{13} and M_{23} do not contribute to surface-wave excitation for shallow sources at low frequencies, and conversely these moment-tensor components cannot be determined from surface-wave observations. The equations for Love waves can be written in a similar form to equation (A1). Love waves are also independent of M_{13} and M_{23} for shallow events, and are determined only by M_{12} and M_{11}–M_{22}. Moment-tensor inversion for a shallow source therefore

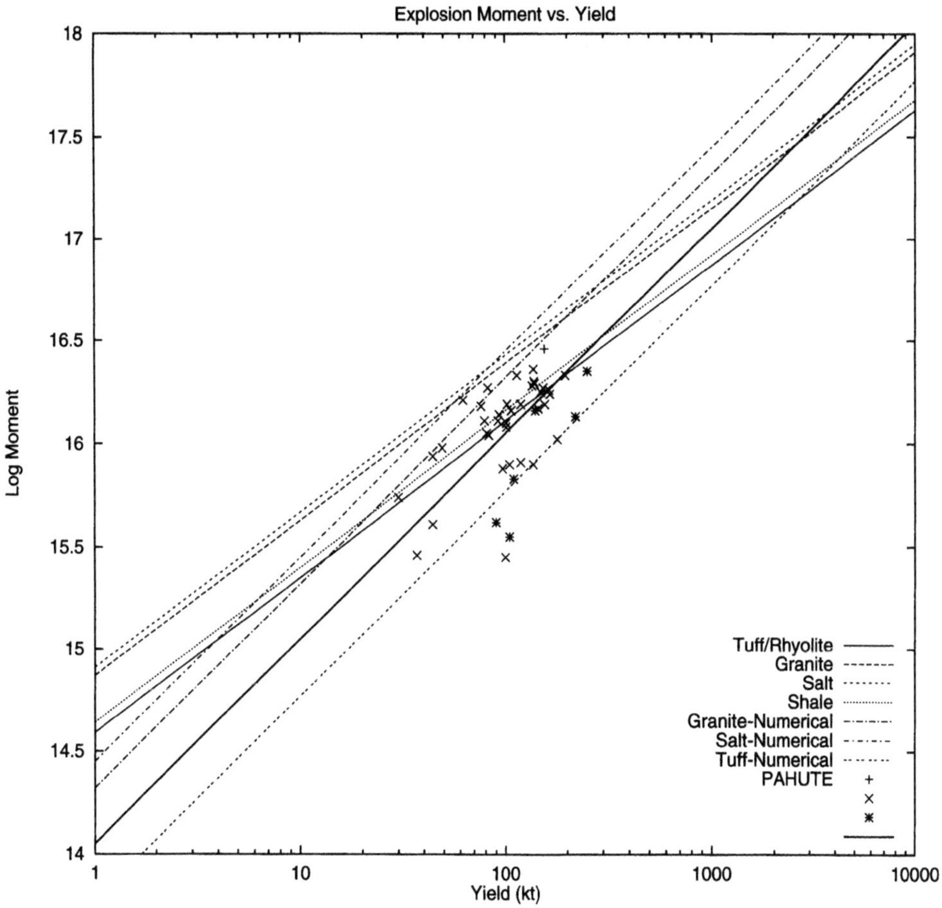

Figure 10

Log M_i vs. yield for different material properties. Models are the same as in Figure 6.

requires solving equation (A1) and the corresponding Love wave equation for the moment-tensor components M_{11}, M_{22}, M_{33}, and M_{12} using observations of Rayleigh and Love waves at multiple locations. Only three of the four parameters are independent, so moment-tensor inversion requires one additional constraint.

The isotropic moment is defined by $M_i = (M_{11} + M_{22} + M_{33})/3$. For an isotropic explosion source, the Rayleigh wave spectrum can be written:

$$u_z(\omega, h_x, r) = M_0' \frac{S_1^x(\omega, h_x) S_2(\omega) \exp[-\gamma_p(\omega) r + i(\varphi_0 - \omega r / c_p(\omega))]}{\sqrt{a_e \sin(r/a_e)}} \tag{A3}$$

where φ_o is the initial phase equal to $-3\pi/4$,

$$S_1^{\alpha}(\omega, h_x) = \sqrt{\frac{2A_{R_1}}{9\pi\omega c_1^2}}\left(\frac{1}{2\mu k}y_2(h_x) - y_3(h_x)\right) ,$$

$S_2(\omega) = \sqrt{c_2 A_{R_2}}$, and $M_0' = 3(\beta^2)/(\alpha^2)M_i$. M_0' is defined this way so that the source region excitation function is not explicitly dependent on local material properties at shallow source depths (STEVENS, 1986).

Acknowledgements

This work was supported by U. S. Air Force contract F19628-95-C-0110 and Defense Threat Reduction Agency contract DSWA01-98-C-0154.

REFERENCES

AKI, K., and TSAI, Y. (1972), *The Mechanism of Love Wave Excitation by Explosive sources*, J. Geophys. Res. *77*, 1452–1475.

ARCHAMBEAU, C. B. (1972), *The Theory of Stress Wave Radiation from Explosions in Prestressed Media*, Geophys. J. R. astr. Soc. *29*, 329–366.

BACHE, T. C. (1982), *Estimating the Yield of Underground Nuclear Explosions*, Bull. Seismol. Soc. Am. *72*, (6), part B, S131–S168.

DAY, S. M., RIMER, N., and CHERRY, J. T. (1983), *Surface Waves from Underground Explosions with Spall: Analysis of Elastic and Nonlinear Source Models*, Bull. Seismol. Soc. Am. *73*, 247–264.

DAY, S. M., CHERRY, J. T., RIMER, N., and STEVENS, J. L. (1987), *Nonlinear Model of Tectonic Release from Underground Explosions*, Bull. Seismol. Soc. Am. *77*, 996–1016.

DAY, S. M., and STEVENS, J. L. (1986), *An Explanation for Apparent Time Delays in Phase-reversed Rayleigh Waves from Underground Nuclear Explosions*, Geophys. Res. Lett. *13*, 1423–1425.

DEPARTMENT OF ENERGY (DOE) (1994), *United States Nuclear Tests, July 1945 through September 1992*, DOE Report DOE/NV-209, December.

EKSTRÖM, G., and RICHARDS, P. G. (1994), *Empirical Measurements of Tectonic Moment Release in Nuclear Explosions from Teleseismic Surface Waves and Body Waves*, Geophys. J. Int. *117*, 120–140.

GIVEN, J. W., and MELLMAN, G. R. (1986), *Estimating Explosion and Tectonic Release Source Parameters of Underground Nuclear Explosions from Rayleigh- and Love-wave Observations*, Sierra Geophysics Final Report to Air Force Geophysics Laboratory, Part 1, AFGL-TR-86-0171, SGI-R-86-126, July.

HARKRIDER, D. G., STEVENS, J. L., and ARCHAMBEAU, C. B. (1994), *Theoretical Rayleigh and Love Waves from an Explosion in Prestressed Source Regions*, Bull. Seismol. Soc. Am. *84* (5), 1410–1442.

HELLE, H. B., and RYGG, E. (1984), *Determination of Tectonic Release from Surface Waves Generated by Nuclear Explosions in Eastern Kazakhstan*, Bull. Seismol. Soc. Am. *74*, 1883–1898.

HERRIN, E., and GOFORTH, T. (1986), *Phase Analysis of Rayleigh Waves from the Shagan River Test Site in the USSR*, Bull. Seismol. Soc. Am. *76*, 1739–1754.

MARSHALL, P. D., SPRINGER, D. L., and RODEAN, H. C. (1979), *Magnitude Corrections for Attenuation in the Upper Mantle*, Geophys. J. *57*, 609–638.

MASSE, R. P. (1981), *Review of Seismic Source Models for Underground Nuclear Explosions*, Bull. Seismol. Soc. Am. *71*, 1249–1269.

MIKHAILOV, V. N. (1996), *USSR Nuclear Weapons Tests and Peaceful Nuclear Explosions*, Russian Federation for Atomic Energy report, ISBN 5-85165-062-1.

MUELLER, R. A., and MURPHY, J. R. (1971), *Seismic Characteristics of Underground Nuclear Detonations, I: Seismic Spectrum Scaling*, Bull. Seism. Soc. Am., *61*, 1675–1692.

MURPHY, J. R., and MUELLER, R. A. (1971), *Seismic Characteristics of Underground Nuclear Detonations, II: Elastic Energy and Magnitude Determinations*, Bull. Seism. Soc. Am., *61*, 693–1704.

MURPHY, J. R. (1977), *Seismic Source Functions and Magnitude Determinations for Underground Nuclear Detonations*, Bull. Seismol. Soc. Am. 135–158.

MURPHY, J. R. (1993), *Development of a Comprehensive Seismic Yield Estimation System for Underground Nuclear Explosions*, S-Cubed Final Report to Phillips Laboratory, SSS-DTR-93-13975, PL-TR-93-2157, May.

REZAPOUR, M., and PEARCE, R. G. (1998), *Bias in Surface-wave Magnitude M_s due to Inadequate Distance Corrections*, Bull. Seismol. Soc. Am. *88*, 43–61.

RINGDAL, F., MARSHALL, P. D., and ALEWINE, R. W. (1992), *Seismic Yield Determination of Soviet Underground Nuclear Explosions at the Shagan River Test Site*, Geophys. J. Int. *109*, 65–77.

RYGG, E. (1979), *Anomalous surface waves from underground explosions*, Bull. Seismol. Soc. Am. *69*, 1995–2002.

SALVADO, C., and MINSTER, J. B. (1980), *Slipping Interfaces: A Possible Source of S Radiation from Explosive Sources*, Bull. Seismol. Soc. Am. *70*, 659–670.

SATO, R. (1967), *Attenuation of Seismic Waves*, J. Phys. Earth *15*, 32–61.

STEVENS, J. L. (1982), *A Model for Tectonic Strain Release from Explosions in Complex Prestress Fields Applied to Seismic Waves from NTS and Eastern Kazakh Explosions*, Systems, Science and Software Technical Report, submitted to VELA Seismological Center, SSS-R-82-5358, January.

STEVENS, J. L., MCLAUGHLIN, K. L., SHKOLLER, B., and DAY, S. M. (1993), *2-D Axisymmetric Calculations of Surface Waves Generated by an Explosion in an Island, Mountain, and Sedimentary Basin*, Geophys. J. Internat. *114*, 548–560.

STEVENS, J. L., BARKER, T. G., DAY, S. M., MCLAUGHLIN, K. L., RIMER, N., and SHKOLLER, B. (1991), *Simulation of teleseismic body waves, regional seismograms, and Rayleigh wave phase shifts using two-dimensional nonlinear models of explosion sources*, In AGU Geophysical Monograph 65: *Explosion Source Phenomenology* (S. Taylor, H. Patton, P. Richards, eds.) ISBN 0-87590-031-3, pp. 239–252.

STEVENS, J. L. (1986), *Estimation of Scalar Moments from Explosion-generated Surface Waves*, Bull. Seismol. Soc. Am. *76*, 123–151.

STEVENS, J. L., and DAY, S. M. (1985), *The Physical Basis of the m_b:M_s and Variable Frequency Magnitude Methods for Earthquake/Explosion Discrimination*, J. Geophys. Res. *90*, 3009–3020.

STEVENS, J. L., and MCLAUGHLIN, K. L. (2001), *Optimization of Surface-wave Identification and Measurement*, Pure Appl. Geophys. *158*, 1547–1582.

TAKEUCHI, H., and SAITO, M. *Seismic surface waves*, In Methods of Computational Physics (B. A. Bolt, ed.), v. 11 (Academic Press, New York, 1972) pp. 217–295.

TOKSÖZ, N., and KEHRER, H. (1972), *Tectonic Strain Release by Explosions and its Effect on Discrimination*, Geophys. J. R. astr. Soc. *31*, 141–161.

VIECELLI, J. S. (1973), *Spallation and the Generation of Surface Waves by an Underground Explosion*, J. Geophys. Res. *78*, 2475–2487.

(Received August 6, 1999, revised July 28, 2000, accepted August 2, 2000)

Pure appl. geophys. 158 (2001) 2253–2274
0033–4553/01/112253–22 $ 1.50 + 0.20/0

Pure and Applied Geophysics

Teleseismic *Lg* of Semipalatinsk and Novaya Zemlya Nuclear Explosions Recorded by the GRF (Gräfenberg) Array: Comparison with Regional *Lg* (BRV) and their Potential for Accurate Yield Estimation

JÖRG SCHLITTENHARDT[1]

Abstract—A comparison of regional and teleseismic log rms (root-mean-square) *Lg* amplitude measurements have been made for 14 underground nuclear explosions from the East Kazakh test site recorded both by the BRV (Borovoye) station in Kazakhstan and the GRF (Gräfenberg) array in Germany. The log rms *Lg* amplitudes observed at the BRV regional station at a distance of 690 km and at the teleseismic GRF array at a distance exceeding 4700 km show very similar relative values (standard deviation 0.048 magnitude units) for underground explosions of different sizes at the Shagan River test site. This result as well as the comparison of BRV rms *Lg* magnitudes (which were calculated from the log rms amplitudes using an appropriate calibration) with magnitude determinations for *P* waves of global seismic networks (standard deviation 0.054 magnitude units) point to a high precision in estimating the relative source sizes of explosions from *Lg*-based single station data. Similar results were also obtained by other investigators (PATTON, 1988; RINGDAL *et al.*, 1992) using *Lg* data from different stations at different distances.

Additionally, GRF log rms *Lg* and *P*-coda amplitude measurements were made for a larger data set from Novaya Zemlya and East Kazakh explosions, which were supplemented with $m_b(Lg)$ amplitude measurements using a modified version of NUTTLI's (1973, 1986a) method. From this test of the relative performance of the three different magnitude scales, it was found that the *Lg* and *P*-coda based magnitudes performed equally well, whereas the modified Nuttli $m_b(Lg)$ magnitudes show greater scatter when compared to the worldwide m_b reference magnitudes. Whether this result indicates that the rms amplitude measurements are superior to the zero-to-peak amplitude measurement of a single cycle used for the modified Nuttli method, however, cannot be finally assessed, since the calculated $m_b(Lg)$ magnitudes are only preliminary until appropriate attenuation corrections are available for the specific path to GRF.

Key words: Seismic yield estimation, nuclear explosions, regional waves, rms coda magnitudes.

Introduction

The concept of using *Lg* signal amplitudes for measuring the strength of earthquakes (and later also explosions) was developed by NUTTLI (1973) more than two decades ago for eastern North America. Since data on underground nuclear

[1] Bundesanstalt für Geowissenschaften und Rohstoffe (BGR), Stilleweg 2, 30655 Hannover, Germany. E-mail: Schlittenhardt@sdac.hannover.bgr.de

explosions became available from modern digital seismic stations installed within the former Soviet Union and China, it became evident that *Lg* signals for undisturbed continental ray paths recorded at regional distances (less than 2000 km) provide an excellent basis not only for detecting seismic events but also for estimating the yield of underground nuclear explosions (RINGDAL, 1983; NUTTLI, 1986a,b; PATTON, 1988; HANSEN *et al.*, 1990). Selected stations at regional distances in the USSR and China (e.g., BRV, Kazakhstan – RICHARDS *et al.*, 1992 and WMQ, China – HANSEN *et al.*, 1990) show much higher signal-to-noise ratio of the *Lg* phase for events at the Eastern Kazakhstan test site, as compared to teleseismic Gräfenberg (GRF) or NORSAR array data. As early as October 1991, shortly after a visit of German seismologists at Borovoye, I received a magnetic tape sent through the post in exchange for recordings of the GRF array. This tape contained digital recordings made at BRV station for eleven underground nuclear explosions detonated at the Semipalatinsk test site in the former Soviet Union (SCHLITTENHARDT, 1993). The data set was complemented with further BRV recordings obtained from Lamont-Doherty Observatory of Columbia University, N.Y., U.S.A. (Richards and Kim, pers. comm.), by exchange.

This paper presents a comparison of regional BRV (distance 690 km) and teleseismic (distance > 4700 km) GRF rms *Lg* magnitudes (m_{Lg}) of Semipalatinsk explosions (see Fig. 1 for locations of the test sites and of the stations used). In addition, teleseismic GRF rms *Lg* and *P*-coda amplitude measurements are presented for a larger data set from Novaya Zemlya and East Kazakh nuclear explosions. These results are supplemented with $m_b(Lg)$ amplitude measurements for Semipalatinsk explosions using a modified version of NUTTLI's (1973, 1986a) method. The relative performance of the three different magnitude scales based on teleseismic *Lg* and *P* coda is then evaluated by a comparison with a reference data set. The calculated regional BRV m_{Lg} magnitudes show a high degree of consistency (standard deviation 0.054 magnitude units) with worldwide m_b. This points to a high precision in relative yield estimation from single station data. Moreover, rms *Lg* amplitudes observed at the GRF array in Germany and at the BRV station in Kazakhstan indicate very similar relative magnitudes for underground nuclear explosions of different sizes at the Shagan River site within the Semipalatinsk test site. From the performance test it is found that the m_{Lg} and the GRF rms *P*-coda magnitudes m_{Pc} perform equally well. Compared to these two magnitude scales, the modified Nuttli $m_b(Lg)$ magnitude reflects greater scatter when plotted against the worldwide m_b reference magnitudes.

Comparison of rms Lg Wave Amplitudes at Regional and Teleseismic Distances: Methodology

The rms *Lg* magnitude is determined from the logarithm of the root of the mean-squared amplitudes in a group velocity window centered around the group velocity

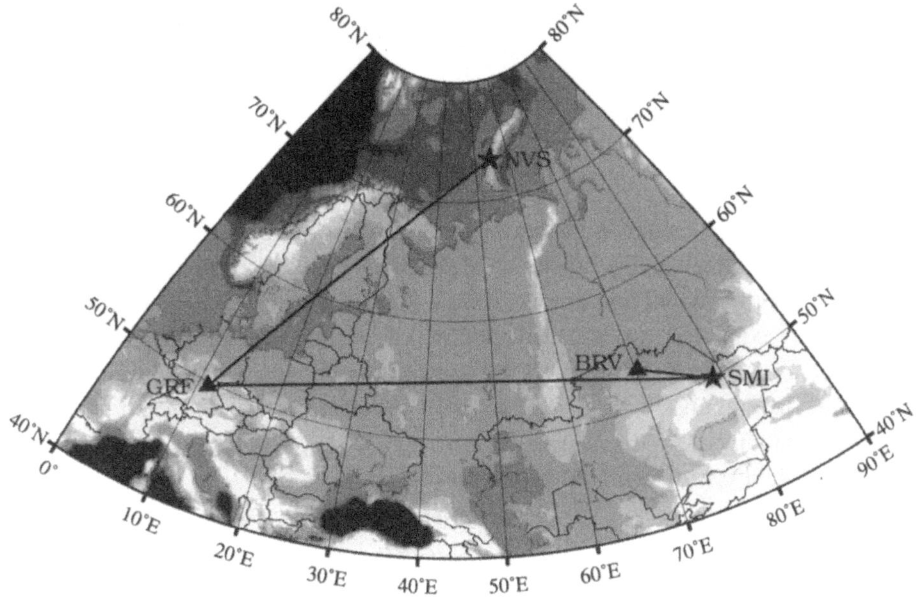

Figure 1
Map of the western part of Eurasia showing the location of the Novaya Zemlya (NVS) and Semipalatinsk (SMI) test sites (stars) and the GRF array in Germany and the BRV station in Kazakhstan (triangles). The black lines delineate the ray paths from the test sites to the GRF array and BRV station. Elevation is shown as shades of gray, with lowest elevations almost black. The ray path from Semipalatinsk to GRF crosses the southern foothills of the Ural mountains, which are visible as an elongated north-south trending feature. Note the large difference in the distances of the GRF and BRV stations from SMI.

of the *Lg* wave. This concept of rms *Lg* magnitude determination is based on work by RINGDAL (1983). In the following it is assumed that the recorded seismogram in the *Lg* window is a superposition of the *Lg* wave and the noise. It is further assumed that the noise is stationary over a certain period of time. It can then be determined in a time window before the *P* onset and can be subtracted from the signal recorded in the *Lg* time window (noise correction). In this model, the recorded signal $x_1(t)$ is a superposition of the stationary noise $x_2(t)$ and the true *Lg* signal $x_3(t)$:

$$x_1(t) = x_2(t) + x_3(t). \tag{1}$$

The mean square amplitudes of x_1 and x_2 in the corresponding time windows are:

$$X_i = \frac{1}{M}\sum_{j=1}^{M} x_i^2(t_j) \quad i = 1,2 \ , \tag{2}$$

where M is the number of samples. The logarithm of the mean-squared amplitudes X_i can be averaged over the J seismogram traces of the GRF array ($J \leq 13$):

$$\bar{X} = \frac{1}{J}\sum_{j=1}^{J}\log_{10}(X_j). \tag{3}$$

The rms Lg magnitude m_{Lg} is then defined as

$$m_{Lg} = \log_{10}\sqrt{10^{\bar{X}_1} - 10^{\bar{X}_2}} + c, \tag{4}$$

where indexes 1 and 2 refer to the corresponding time windows of the above Eq. (1). m_{Lg} is thus determined as the difference of the array averages of the logarithm of the root of the mean-squared amplitudes in the Lg and noise windows. This method can likewise be expanded to the calculation of P-coda magnitudes, simply by interchanging the Lg window with an appropriate P-coda window. It should be noted that the magnitude name m_{Lg} has been chosen to avoid confusion with other magnitude namings.

The constant c is determined by the constraint:

$$\sum_{k=1}^{N}(m_{Lg})_k - m_b(ISC)_k = 0 \tag{5}$$

where N is the number of events, and $m_b(ISC)_k$ is the reference magnitude to which m_{Lg} is calibrated, in this case the ISC body-wave magnitude m_b for event k.

Solving for c yields:

$$c = -\frac{1}{N}\sum_{k=1}^{N}\log_{10}\sqrt{10^{\bar{X}_{1k}} - 10^{\bar{X}_{2k}}} - m_b(ISC)_k \ . \tag{6}$$

In the following analysis only events with a signal-to-noise ratio R above a certain threshold are investigated. R is defined here as the ratio of the mean-squared Lg values and the mean-squared noise-values (Eq. 1):

$$R = \frac{\frac{1}{J}\sum_{j=1}^{J} X_{j,1}}{\frac{1}{J}\sum_{j=1}^{J} X_{j,2}} , \tag{7}$$

where index 1 represents the recorded signal in the Lg window and index 2 represents the noise.

Analysis of Shagan River Explosions Recorded at both GRF and BRV

Table 1 gives a listing of the available recordings of Borovoye station, Kazakhstan and Figure 2 shows a seismogram example recorded by the 'KSM' system with a sampling interval of 0.026 s for the nuclear explosion at the Shagan River test site on Nov. 15, 1987 (for details on seismograph and recording system see ADUSHKIN and AN, 1990; RICHARDS *et al.*, 1992; or KIM and EKSTRÖM, 1996). The

Table 1

Date, origin time and coordinates of fourteen nuclear explosions from the Shagan River test site for which recording from both GRF (Germany) and BRV (Kazakhstan) are available. Ref. Mag. refers to worldwide network-averaged body-wave magnitudes of the ISC and RINGDAL et al. *(1992), used for comparison with the calculated m_{Lg} magnitudes from GRF and BRV data. The last two columns give the signal-to-noise ratio according to equation (7) for both GRF and BRV data. Note the considerably larger signal-to-noise ratios for the BRV regional station*

No.	Date YYMMDD	Origin Time GMT	Coordinates		Ref. Mag.		m_{Lg}		Signal/Noise Ratio	
			Latitude	Longitude	ISC	Ring.	GRF	BRV	GRF	BRV
1	870312	01:57:17.2	49.897 N	78.831 E	5.60	5.31	–	5.329	–	0.916 E + 04
2	870403	01:17:08.1	49.874 N	78.812 E	6.20	6.12	6.212	6.009	9.03	0.585 E + 07
3	870417	01:30:04.9	49.830 N	78.671 E	6.00	5.92	6.006	6.024	6.18	0.936 E + 05
4	870506	04:02:05.8	49.803 N	78.110 E	5.60	–	–	5.584	–	0.753 E + 04
5	870620	00:53:05.0	49.889 N	78.729 E	6.10	6.03	6.018	6.069	8.82	0.122 E + 06
6	870717	01:17:07.1	49.769 N	78.100 E	5.80	–	5.764	5.801	3.21	0.395 E + 04
7	870802	00:58:07.0	49.849 N	78.914 E	5.90	5.83	–	5.995	–	0.590 E + 04
8	871115	03:31:06.8	49.839 N	78.765 E	6.10	5.98	6.071	6.142	5.03	0.399 E + 05
9	871213	03:21:04.9	49.930 N	78.822 E	6.10	6.06	6.135	6.223	7.23	0.315 E + 05
10	880213	30:05:05.9	49.954 N	78.910 E	6.10	5.97	6.116	6.104	4.52	0.119 E + 05
11	880403	01:33:05.8	49.917 N	78.945 E	6.00	5.99	6.130	6.161	9.95	0.903 E + 05
12	880504	00:57:06.8	49.928 N	78.769 E	6.10	6.09	6.146	6.062	10.00	0.130 E + 05
13	880914	03:59:57.4	49.833 N	78.808 E	6.10	6.03	6.049	6.099	4.36	0.104 E + 06
14	881217	04:18:06.9	49.886 N	78.926 E	5.90	5.83	5.795	5.908	2.13	0.113 E + 05

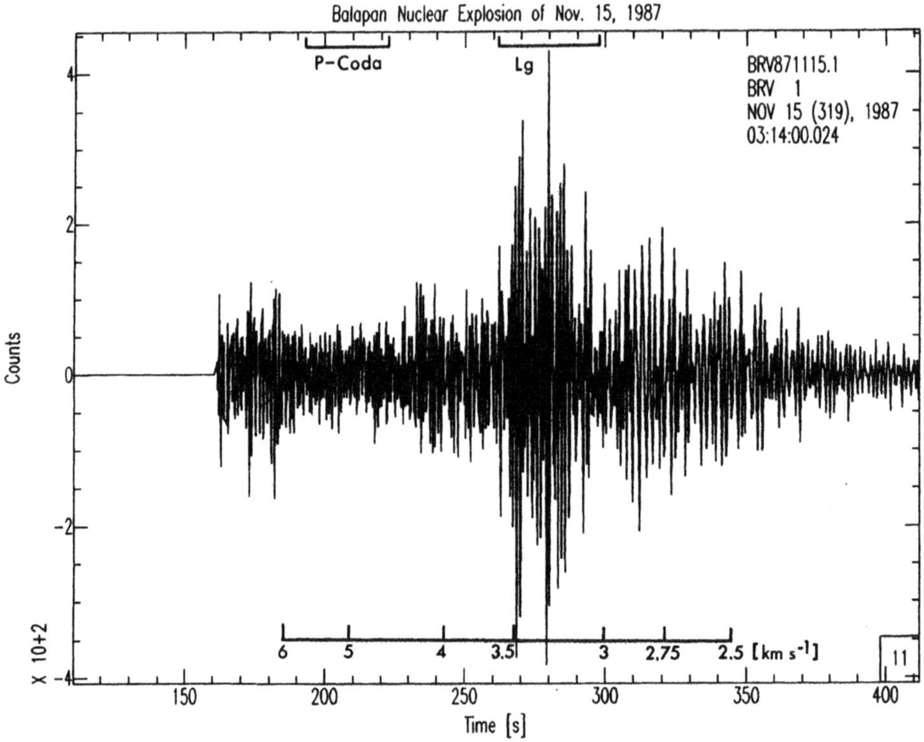

Figure 2
Seismogram (vertical component) of the nuclear explosion of November 15, 1987 (No. 8 in Table 1) at the Shagan River test site, recorded by BRV station in Kazakhstan (distance 690 km). The time windows for the *Lg* and *P*-coda magnitude determination are marked on top of the five-minute time segment. A group velocity axis between 6.0 km/s and 2.5 km/s is plotted at the bottom.

Lg phase is clearly the strongest arrival of the whole seismogram. In Figure 2 the *Lg* window is fixed 100 s after the *P* onset with a duration of 35 s. This time window, which corresponds to a group velocity window between 3.00 and 3.54 km/s, is used for rms amplitude calculation. It should be noted that it is not centered around the often observed *Lg* group velocity of 3.5 km/s (as in the case of GRF data) because the *Lg* amplitude maximum would not lie completely within its boundaries but at its right margin (see group velocity axis in Fig. 2). Figure 3 shows the GRF array recording of the same nuclear explosion of Nov. 15, 1987 (see Fig. 2) with the *Lg* window indicated. It corresponds to a group velocity window between 3.25 and 3.75 km/s. For the 11 common GRF and BRV recordings of nuclear explosions m_{Lg} was calculated from vertical component seismograms according to Eq. (4). Of course, the array averaging could only be applied for GRF data, for the BRV three-component data the vertical component was used. The results are given in Table 1 and the m_{Lg} (GRF) and m_{Lg} (BRV) magnitudes are compared in Figure 4. In this

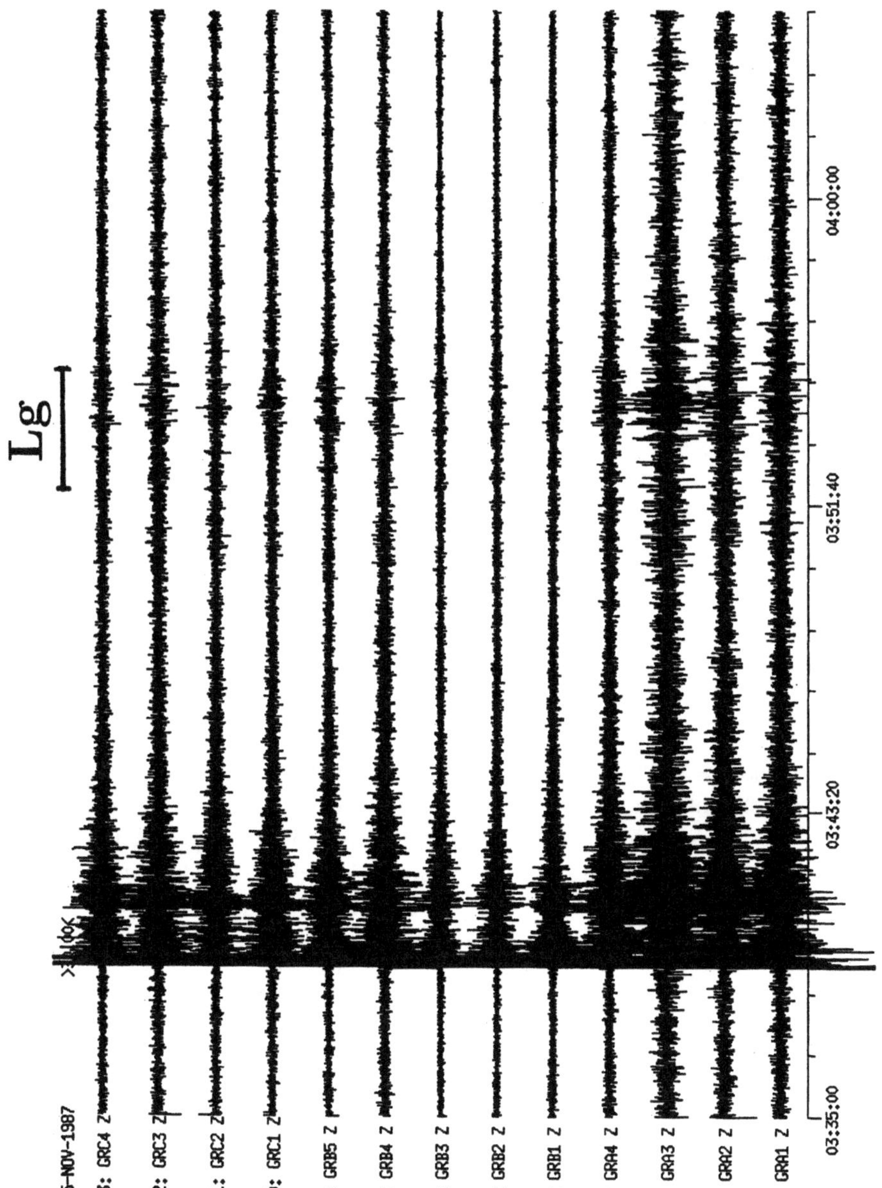

Figure 3

Seismograms (vertical components) of the same nuclear explosion of November 15, 1987 shown in Figure 2, recorded by the GRF array in Germany (distance > 4700 km). The 13 seismograms are simulations of the broadband data according to the WWNSS-SP system transfer function. The group velocity window for the rms L_g amplitude measurement (3.5 km/s ± 0.25 km/s) is marked with a horizontal bar (top). P onsets have been clipped for plotting.

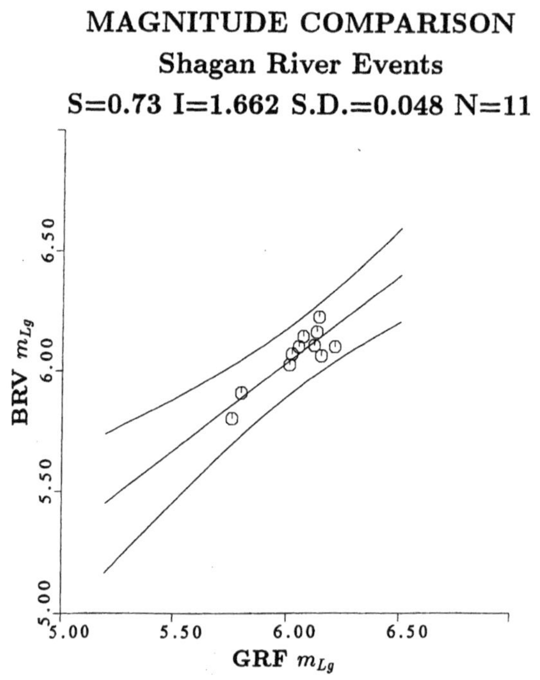

Figure 4

Plot of GRF m_{Lg} magnitudes versus BRV m_{Lg} magnitudes for $N = 11$ underground nuclear explosions from the Shagan River test site. The straight line in the middle is the result of linear regression of the data and the curved lines above and below mark the 95% confidence interval. S, I and $S.D.$ are the slope, intercept and (orthogonal) standard deviation of the linear regression analysis, respectively.

figure, and in Figures 5 and 6 where an additional comparison between m_{Lg}(BRV) and the network magnitudes of the ISC and of RINGDAL *et al.* (1992) is presented, S, I, $S.D.$ and N are the slope, intercept, (orthogonal) standard deviation and number of magnitude pairs of the linear regression analysis, respectively. It should be mentioned that the data samples used in this analysis are rather small and have limited magnitude ranges. In this case, reference to slopes and intercepts can be misleading, and especially intercept estimates may not be reliable. Therefore, the standard deviation $S.D.$ and, to a lesser degree the slope S is used for assessing the reliability of the Lg magnitude scale relative to that of a reference magnitude set. It should be emphasized that for these comparisons the specific choice of the reference magnitude set is unimportant, because the m_{Lg} follow from the log rms amplitudes simply by adding a single constant (Eq. 6), which does not change S or $S.D.$. The m_{Lg} are given priority for plotting over the unusually large negative log rms values. The small standard deviation (0.048 magnitude units) in the regression of m_{Lg} (GRF) and m_{Lg} (BRV) (Fig. 4) is interpreted as an indication that the scaling of root-mean-square Lg-wave amplitudes between explosions of different sizes recorded at the same single station site in the regional distance range appears to be consistent with that at

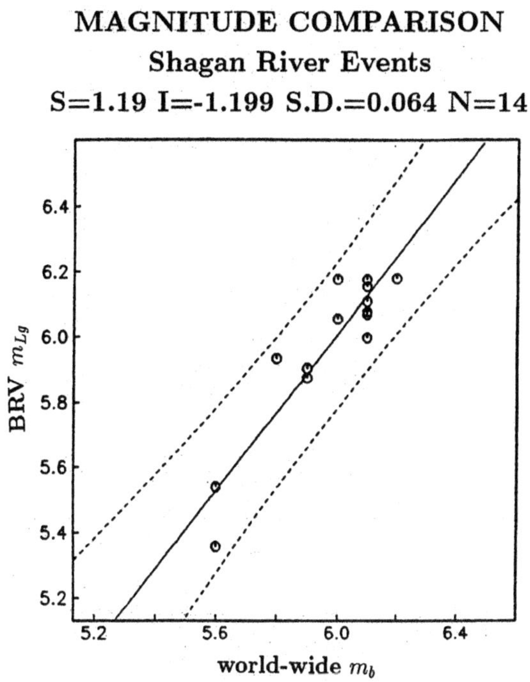

Figure 5

Plot of worldwide m_b magnitudes of the PDE/ISC versus BRV m_{Lg} magnitudes for $N = 14$ underground nuclear explosions from the Shagan River test site. The straight (solid) line is the result of linear regression of the data and the curved (dashed) lines above and below mark the 95% confidence interval. *S*, *I* and *S.D.* are the slope, intercept and (orthogonal) standard deviation of the linear regression analysis, respectively.

another single site in the teleseismic distance range. In other words, rms *Lg* amplitudes observed at the GRF array in Germany and at the station BRV in Kazakhstan show very similar relative magnitudes for underground nuclear explosions of different sizes at the Semipalatinsk test site. Perhaps more important are the small standard deviations (0.064 and 0.054 magnitude units) in the regression of m_{Lg} (BRV) against worldwide network-averaged body-wave magnitudes in Figures 5 and 6. The larger standard deviation for the ISC magnitudes (Fig. 5) compared to the RINGDAL *et al.* (1992) m_b values (Fig. 6) presumably is a result of the higher accuracy of the maximum likelihood magnitudes, but is probably also influenced by the strongly different ISC magnitude for event number one (Table 1), which with $m_b = 5.6$ lies almost 0.3 magnitude units above the available reference magnitudes. The relative magnitudes in Figure 6, which were determined using completely different procedures and seismic wave types, show a high degree of consistency (standard deviation 0.054 magnitude units) with worldwide network-averaged body-wave magnitudes. This interesting result in turn points to a high precision in seismic yield estimation from *Lg*-based single station data.

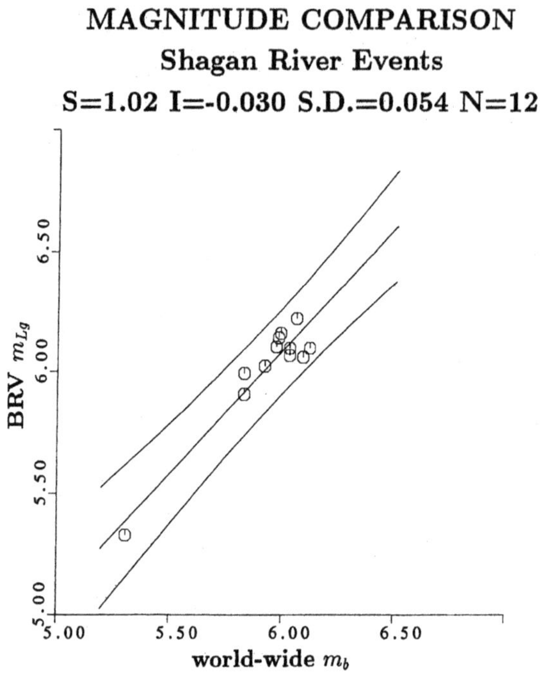

Figure 6
Plot of worldwide maximum-likelihood m_b magnitudes of RINGDAL *et al.* (1992) versus BRV m_{Lg} magnitudes for $N = 12$ underground nuclear explosions from the Shagan River test site. The straight line in the middle is the result of linear regression of the data and the curved lines above and below mark the 95% confidence interval. *S*, *I* and *S.D.* are the slope, intercept and (orthogonal) standard deviation of the linear regression analysis, respectively.

GRF Lg and P-coda Amplitude Measurements for Semipalatinsk and Novaya Zemlya Explosions

Based on the methodology presented above, GRF rms *Lg* and *P*-coda amplitude measurements were made for a larger data set of Novaya Zemlya (Fig. 7 shows, as an example, the GRF array recording of the nuclear explosion on Novaya Zemlya on December 4, 1988) and Semipalatinsk explosions. These results are supplemented with $m_b(Lg)$ amplitude measurements for Semipalatinsk explosions using a modified version of NUTTLI's (1973, 1986a) method. The calculated magnitudes for this large data set of over 60 explosions are then compared to a reference data set using the standard deviation and the slope of the linear regression of the data sets as criteria to evaluate the relative performance of the three different magnitude scales.

The $m_b(Lg)$ magnitude is based on the amplitude and period of *Lg*-waves measured from a short-period World Wide Standard Seismograph Network (WWSSN-SP) instrument in the *Lg* group velocity window. According to NUTTLI (1973) an earthquake in Central North America with a teleseismic *P*-wave magnitude

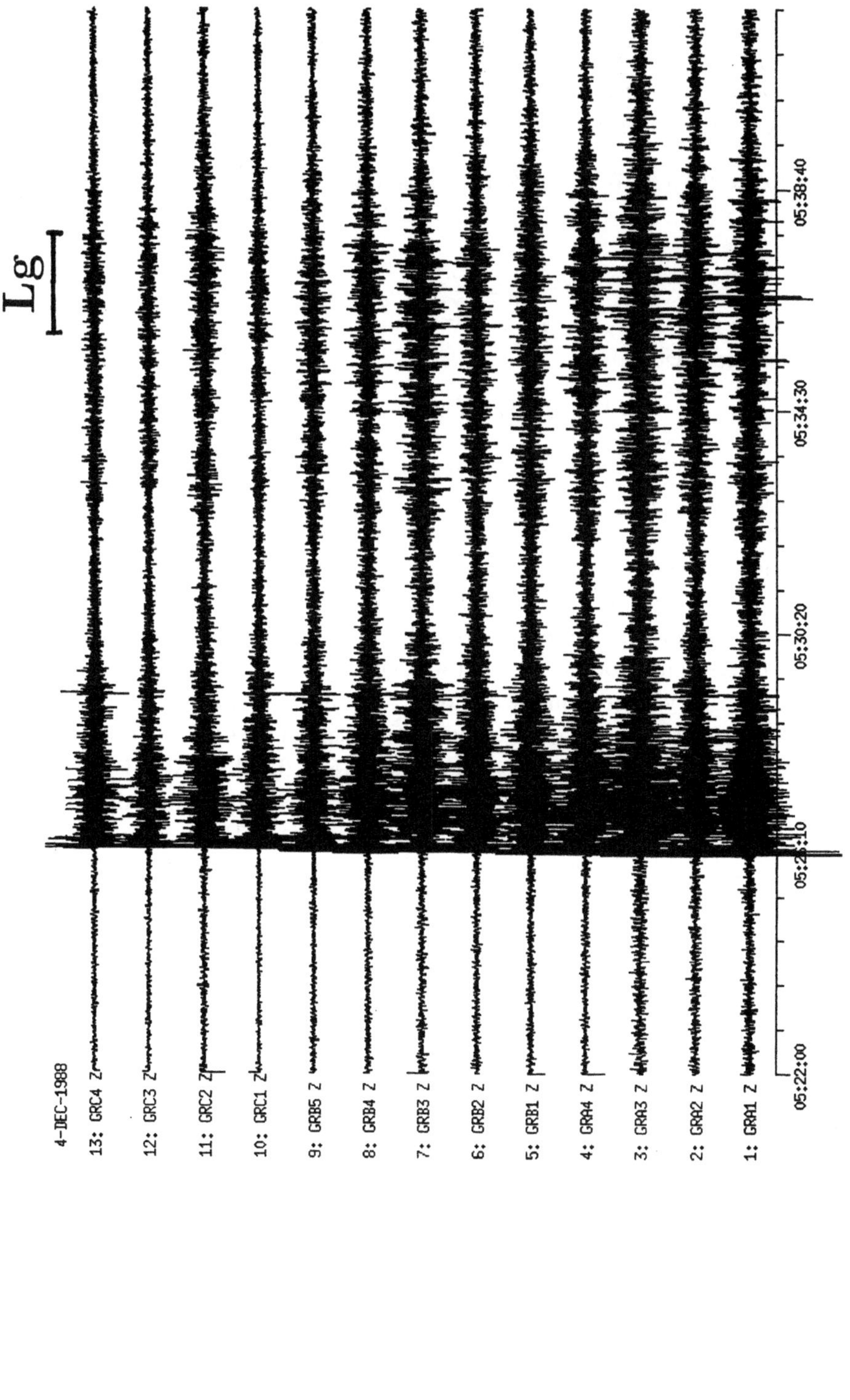

Figure 7

Seismograms (vertical components) of the nuclear explosion of December 4, 1988 on Novaya Zemlya (Table 4), recorded by the GRF array in Germany. The 13 seismograms are simulations of the broadband data according to the WWNSS-SP system transfer function. The group velocity window for the rms L_g amplitude measurement (3.12–3.51 km/s) is marked with a horizontal bar (top). *P* onsets have been clipped for plotting.

m_b equal to 5.0 produces a 1 s Lg amplitude of 110 μm at a distance of 10 km. $m_b(Lg)$ is thus defined as

$$m_b(Lg) = 5.0 + \log(A(10)/110). \tag{8}$$

where $A(10)$ is the Lg-wave amplitude (in μm) at an epicentral distance of 10 km. To calculate $m_b(Lg)$ from the above equation, the measured amplitude $A(\Delta)$ for a Lg-wave with period T at distance Δ must be extrapolated back to a distance of 10 km from the source using the formula

$$A(10) = A(\Delta)(\Delta/10)^{1/3}[\sin(\Delta/111.1)/\sin(10/111.1)]^{1/2}\, e^{\gamma(\Delta-10)} \tag{9}$$

where

$$\gamma(f) = \frac{\pi f}{UQ(f)} \tag{10}$$

is the attenuation coefficient and

$$Q(f) = Q_0 f^{\zeta} \tag{11}$$

is the frequency-dependent quality factor and U and $f = 1/T$ are the group velocity and frequency of the Lg wave, respectively. Q_0, the quality factor at the reference frequency of 1 Hz, and the exponent ζ are path-specific constants which can be determined from the seismic coda (see e.g., HERMANN, 1980). Contrary to the original instruction (NUTTLI, 1986a) to measure the Lg amplitude from the third largest peak of the Lg wavetrain, we measured the Lg amplitude and period from the maximum Lg amplitude in a three-minute-long time window beginning 13 minutes after P, which corresponds to a group velocity window between 3.25 and 3.70 km/s. All amplitude and period readings were taken from the vertical component of the GRF station A1 (if available) after filtering the original broadband data using the WWSSN-SP system transfer function (SEIDL, 1980). Of course, because of the large epicentral distance, accurate γ values are needed to obtain accurate $m_b(Lg)$ estimates. The resulting values are given in Table 2 together with the respective $m_b(Lg)$ values, which have been calculated from amplitude and period using tentatively Q_0 and ζ values of 700 and 0.4, respectively, to correct for attenuation. These Q_0 and ζ values are valid for paths from the Semipalatinsk test site to KON station (Kronsberg, Norway), the closest WWSSN station for which the specific attenuation coefficients were derived by NUTTLI (1986b). It must be mentioned, that contrary to the m_{Lg} reported above which were calibrated against teleseismic m_b, the $m_b(Lg)$ are calculated using an absolute scale which is set such that 110 μm of displacement at a distance of 10 km corresponds to a $m_b = 5.0$. As a consequence, choosing different Q_0 and ζ values would have nonlinear effects on $m_b(Lg)$ through the explicit exponential frequency dependence in Equation (9). The judgement of the performance of this magnitude scale using the standard deviation and the slope of the regression against a reference magnitude is therefore only preliminary until appropriate Q_0 and ζ are available for the specific path to GRF.

Table 2

Date, origin time, coordinates and m_b (source: PDE/ISC) of 71 explosions from the Semipalatinsk test site for which magnitudes were determined using GRF data. m_{Lg} and m_{Pc} are rms amplitude-based Lg and P-coda magnitudes as described in the text. R is the signal-to-noise ratio of the Lg phase according to equation (7) and #C is the number of channels used for array averaging. $m_b(Lg)$ (last column) is a modified version of NUTTLI's (1973) $m_b(L_g)$ magnitude. Amp and Per are the maximum amplitude and period, respectively, measured from WWSSN-SP simulated seismograms from GRF station A1 (when available) in a three minute long group velocity window between 3.25 and 3.70 km/s. No $m_b(Lg)$ was calculated for events with Per greater than 3.0 s

Date ddmmyy	Origin Time (GMT)	Lat. (°N)	Lon. (°E)	m_b	m_{Lg}	m_{Pc}	R	#C	Amp (nm)	Per (s)	$m_b(Lg)$
040776	02:56:57.5	49.909	78.911	5.8	5.808	5.837	10.64	3	42.9	1.8	5.643
280876	02:56:57.6	49.969	78.930	5.8	5.691	5.742	2.74	2	42.3	1.9	5.578
231176	05:02:57.5	50.008	78.963	5.8	5.900	5.828	2.64	3	–	–	–
071276	04:56:57.5	49.922	78.846	5.9	–	–	–	–	419.1	3.5	–
290577	02:56:57.5	49.937	78.770	5.8	–	–	–	–	51.4	2.1	5.559
050977	03:02:57.8	50.035	78.921	5.8	5.843	5.756	2.83	3	94.2	2.1	5.822
301177	04:06:57.6	49.958	78.885	6.0	5.785	5.925	2.62	3	51.3	1.7	5.785
260378	03:56:57.7	49.768	78.044	5.6	–	–	–	–	112.0	2.3	5.808
280778	02:46:57.8	49.756	78.140	5.7	5.620	5.741	2.21	4	41.2	1.9	5.567
290878	02:37:06.4	50.000	78.978	5.9	5.913	5.988	1.54	4	–	–	–
150978	02:36:57.5	49.916	78.879	6.0	5.971	5.945	4.88	3	144.2	2.3	5.918
041178	05:05:57.5	50.034	78.943	5.6	5.739	5.673	2.59	4	73.2	2.2	5.666
230679	02:56:59.0	49.903	78.855	6.2	6.182	6.135	21.03	3	221.3	2.1	6.193
070779	03:46:57.5	50.026	78.991	5.8	6.005	5.835	11.10	4	94.6	1.9	5.928
040879	03:56:57.3	49.894	78.904	6.1	6.171	6.146	22.23	4	200.0	2.3	6.060
180879	02:51:57.3	49.943	78.938	6.1	6.211	6.299	19.09	3	144.7	1.9	6.112
021279	04:36:57.6	49.891	78.796	6.0	6.012	6.017	4.44	4	180.3	2.2	6.058
231279	04:56:57.6	49.916	78.755	6.2	6.123	6.203	5.44	4	146.3	1.9	6.117
250480	03:56:57.4	49.973	78.755	5.5	–	–	–	–	32.6	1.8	5.524
220580	03:56:57.6	49.784	78.082	5.5	–	–	–	–	66.0	2.3	5.578
120680	03:26:57.5	49.980	79.001	5.6	5.608	5.488	2.58	4	34.9	1.8	5.554
290680	02:32:57.7	49.939	78.815	5.7	5.783	5.692	3.92	3	28.0	1.4	5.758
121080	03:34:14.0	49.961	79.028	5.9	6.018	5.837	3.02	4	–	–	–
141280	03:47:06.5	49.899	78.938	5.9	6.023	5.992	3.09	2	121.2	2.0	5.981
271280	04:09:08.2	50.057	78.981	5.9	5.921	5.794	2.36	3	74.4	1.8	5.882
290381	04:03:19.8	50.007	78.982	5.6	–	–	–	–	64.8	2.0	5.709

Table 2

Continued

Date ddmmyy	Origin Time (GMT)	Lat. (°N)	Lon. (°E)	m_b	m_{Lg}	m_{Pc}	R	$\#C$	Amp (nm)	Per (s)	$m_b(Lg)$
220481	01:17:11.4	49.885	78.810	6.0	6.035	5.978	7.62	3	118.3	2.0	5.971
130981	02:17:18.1	49.910	78.915	6.1	6.150	6.148	15.10	4	127.2	1.8	6.115
181081	03:57:02.7	49.923	78.859	6.1	6.020	6.004	6.91	4	147.8	2.2	5.971
291181	03:35:08.6	49.887	78.860	5.7	–	–	–	–	71.0	1.9	5.803
271281	03:43:14.1	49.903	78.876	6.2	6.153	6.211	4.14	4	146.3	2.1	6.013
250482	03:23:05.4	49.903	78.913	6.1	6.122	6.121	9.80	4	157.1	2.0	6.094
051282	03:37:12.6	49.919	78.813	6.1	6.067	6.092	4.19	4	138.9	2.2	5.944
261282	03:35:14.3	50.071	78.988	5.7	5.690	5.625	1.90	4	100.2	2.3	5.759
300583	03:33:44.8	49.735	78.189	5.5	5.355	5.411	1.52	4	–	–	–
061083	01:47:06.6	49.916	78.764	6.0	5.934	5.981	3.16	4	144.0	2.2	5.960
261083	01:55:04.8	49.901	78.828	6.1	6.111	6.036	4.14	4	275.2	2.2	6.241
261283	04:29:06.8	49.810	78.146	5.6	–	–	–	–	497.7	4.0	–
190284	03:57:03.5	49.857	78.806	5.9	5.845	5.829	1.78	4	–	–	–
070384	02:39:06.3	50.049	78.954	5.7	5.668	5.648	2.09	4	595.3	4.4	–
290384	05:19:08.1	49.912	78.955	5.9	6.019	5.941	4.47	4	86.4	1.7	6.012
150484	03:17:09.2	49.758	78.141	5.7	5.700	5.714	3.34	4	47.6	2.0	5.575
250484	01:09:03.5	49.929	78.870	6.0	5.877	5.966	6.61	4	53.8	1.7	5.806
260584	03:13:12.3	49.969	79.006	6.1	6.193	6.007	27.77	4	170.3	2.0	6.129
140784	01:09:10.4	49.893	78.884	6.2	6.116	6.116	19.33	4	149.1	2.0	6.071
271084	01:50:10.6	49.920	78.777	6.2	6.219	6.288	13.11	4	173.3	2.0	6.137
021284	03:19:06.2	49.989	79.011	5.9	5.922	5.764	2.92	4	989.9	4.3	–
161284	03:55:02.7	49.926	78.820	6.1	6.136	6.110	8.99	4	107.5	1.9	5.983
281284	03:50:10.5	49.866	78.703	6.0	6.032	6.085	4.95	4	106.2	1.9	5.978
100285	03:27:07.6	49.888	78.781	5.9	5.944	5.932	2.47	3	136.2	2.1	5.982
250485	00:57:06.5	49.914	78.902	5.9	5.895	5.954	2.95	4	83.3	1.9	5.873
150685	00:57:00.7	49.898	78.845	6.1	6.087	5.997	13.46	4	184.7	2.1	6.114
300685	02:39:02.7	49.848	78.658	6.0	6.016	6.051	12.46	4	100.4	2.1	5.850
200785	00:53:14.5	49.936	78.785	6.0	5.931	5.936	5.45	3	379.9	2.5	6.261
120387	01:57:17.2	49.897	78.831	5.6	–	–	–	–	47.0	1.7	5.747
030487	01:17:08.1	49.874	78.812	6.2	6.212	6.176	9.03	4	90.3	1.9	5.908

170487	01:03:04.9	49.830	78.671	6.0	6.006	6.085	6.18	4	95.1	2.1	5.826
200687	00:53:05.0	49.889	78.729	6.1	6.018	6.142	8.82	3	–	–	–
170787	01:17:07.1	49.769	78.100	5.8	5.764	5.887	3.21	4	156.4	2.1	6.042
151187	03:31:06.8	49.839	78.765	6.1	6.071	6.053	5.03	4	157.7	2.0	6.096
131287	03:21:04.9	49.930	78.822	6.1	6.135	6.127	7.23	4	129.1	1.8	6.122
271287	03:05:04.9	49.823	78.734	6.1	6.111	6.137	3.53	4	595.9	2.9	6.327
130288	03:05:05.9	49.922	78.904	6.1	6.116	6.113	4.52	4	155.9	2.0	6.091
030488	01:33:05.7	49.885	78.955	6.0	6.130	6.102	9.95	4	190.0	2.2	6.080
040588	00:57:06.8	49.922	78.762	6.1	6.146	6.187	10.00	4	209.4	2.2	6.123
140988	03:59:57.4	49.821	78.796	6.1	6.049	6.152	4.36	4	139.0	2.5	5.824
171288	04:18:06.8	49.818	78.910	5.9	5.795	5.785	2.13	3	138.4	2.1	5.989
220189	03:57:06.5	49.915	78.857	6.0	5.964	6.028	3.52	4	104.8	2.0	5.918
120289	04:15:06.7	49.895	78.758	5.8	5.806	5.870	2.13	3	515.2	3.6	–
080789	03:46:57.6	49.873	78.815	5.6	5.465	5.560	1.76	3	47.7	2.0	5.576
191089	09:49:57.1	49.920	79.001	5.9	5.836	5.966	2.02	4	110.9	2.0	5.943

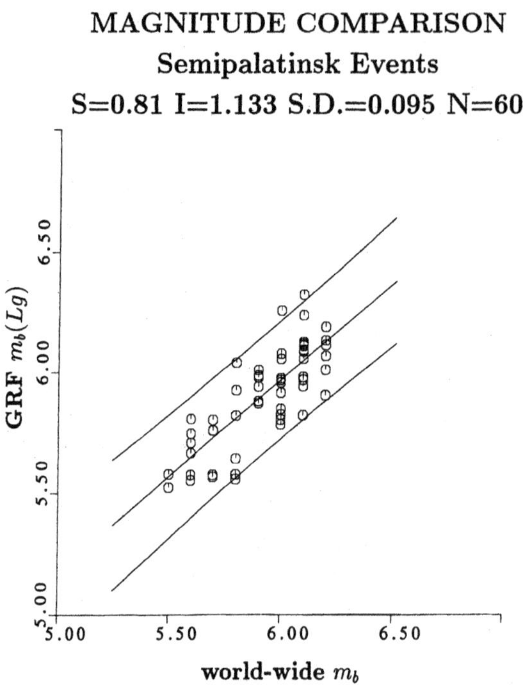

Figure 8

Plot of worldwide m_b magnitudes of the PDE/ISC versus GRF $m_b(Lg)$ magnitudes for $N = 60$ underground nuclear explosions at the Semipalatinsk test site. The straight line in the middle is the result of the linear regression of the data and the lines above and below mark the 95% confidence interval. S, I and $S.D.$ are the slope, intercept and (orthogonal) standard deviation of the linear regression analysis, respectively.

A comparison of $m_b(Lg)$ with worldwide network body-wave magnitudes for 60 explosions is given in Figure 8 (no magnitude was calculated for explosions with Lg periods greater than 3.0 s). It should be noted that the $m_b(Lg)$ values are based on single sensor (A1) data and not on array averages like GRF m_{Lg}. The $m_b(Lg)$ values generally agree well with the body-wave magnitudes but are on average slightly higher (intercept $I = 1.133$), a result that is also found in a comparison (not shown) with the corresponding rms Lg magnitudes. This finding could result from a Q_0 value that is somewhat too low for this specific path, which in turn leads to an overcompensation of the wave attenuation. It should be mentioned that a moderate increase of Q_0 to 750 (keeping $\zeta = 0.4$) gives $I \approx 1.0$, and leaves S and $S.D.$ basically unchanged. However, it must be kept in mind that the intercept values used in this discussion are not very reliable because of the limited range in magnitudes used for the regression.

For the calculation of GRF rms Lg amplitudes for Semipalatinsk explosions from 1976 to 1990 only events with $m_b > 5.5$ were considered. This threshold corresponds roughly to a minimum signal-to-noise ratio $R > 1.5$ of the explosion records studied

(Table 2). For the computations the noise window was fixed to begin 35 s before the *P* onset with a duration of 30 s, whereas the *P*-coda window begins 20 s after the *P* onsets and lasts 30 s as well. The *Lg* group velocity window was kept unchanged between 3.25 and 3.75 km/s, as in the previous section. For a set of altogether 62 underground nuclear explosions (Table 2), m_{Lg} and the *P*-coda magnitude m_{Pc} were determined according to Eq. (4). Both m_{Lg} and m_{Pc} were calibrated (Eq. 6) against m_b as listed in Table 2 to avoid unusually large negative log rms values. Figure 9 shows a comparison of m_{Pc} and the m_b magnitudes of the PDE/ISC. The calculated *P*-coda magnitudes indicate an excellent correlation with the network-averaged body wave magnitudes; the slope of the regression line is 1.05 and the (orthogonal) standard deviation is only 0.048 magnitude units. The resulting slope and standard deviation

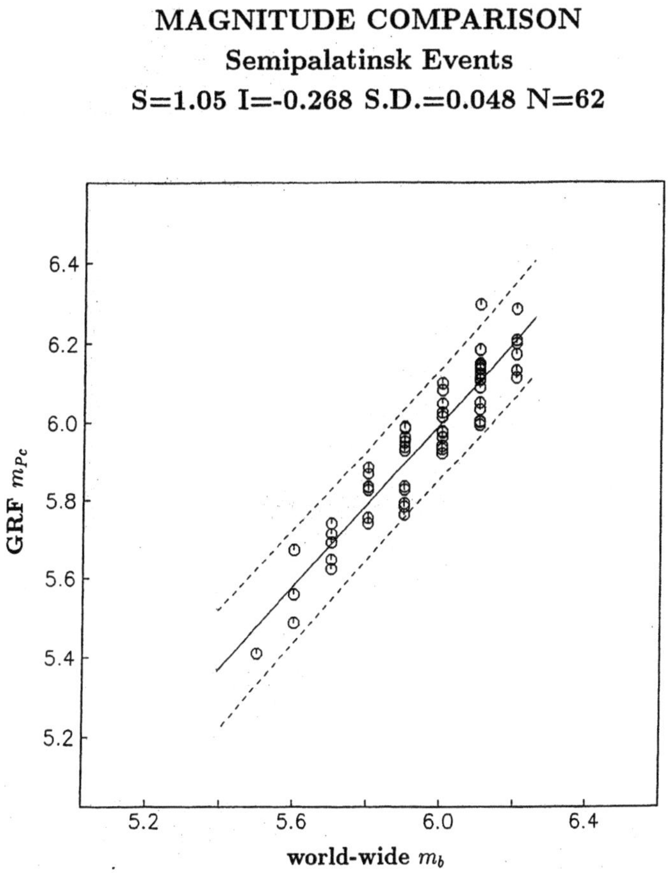

Figure 9

Plot of worldwide m_b magnitudes of the PDE/ISC versus GRF m_{Pc} magnitudes for $N = 62$ underground nuclear explosions at the Semipalatinsk test site. The straight (solid) line is the result of linear regression of the data and the curved (dashed) lines above and below mark the 95% confidence interval. *S*, *I* and *S.D.* are the slope, intercept and (orthogonal) standard deviation of the linear regression analysis, respectively.

S.D. of a corresponding comparison of m_{Lg} for the 62 explosions are given in Figure 10 and in Table 3. For both the Semipalatinsk and Novaya Zemlya data sets in Table 3, slope and *S.D.* values are given for the single-station m_b values for the A1 station of GRF to demonstrate how these results compare with corresponding coda and $m_b(Lg)$ values. For both test sites, the *S.D.* shows that the m_{Pc} and m_{Lg} values correlate slightly better with network m_b values than do the single-station m_b values.

In the case of Novaya Zemlya, 17 explosions from the period 1976 to 1990 with signal-to-noise ratios $R > 1.3$ could be analyzed (Table 4). For these explosions the *Lg* group velocity window was fixed between 3.12 and 3.51 km/s by visual inspection of the records (Fig. 7). All other processing parameters were the same as in the previous analysis. Table 3 summarizes the results of the comparison with PDE/ISC

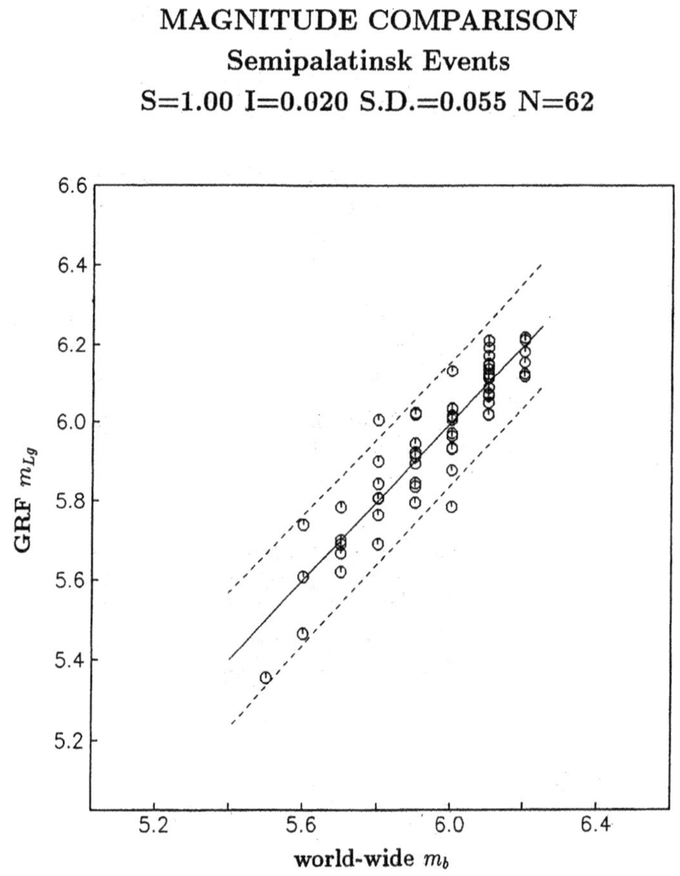

Figure 10

Plot of worldwide m_b magnitudes of the PDE/ISC versus GRF m_{Lg} magnitudes for $N = 62$ underground nuclear explosions from the Semipalatinsk test site. The straight (solid) line is the result of linear regression of the data and the curved (dashed) lines above and below mark the 95% confidence interval. *S*, *I* and *S.D.* are the slope, intercept and (orthogonal) standard deviation of the linear regression analysis, respectively.

Table 3

Slope and orthogonal standard derviation (S.D.) of the regression analysis of GRF m_{Lg}, m_{Pc}, $m_b(Lg)$ and m_b against PDE/ISC body-wave magnitude for explosions at the Semipalatinsk and Novaya Zemlya test sites. N is the number of explosions studied for each magnitude type and test site

	Semipalatinsk				Novaya Zemlya		
	m_{Lg}	m_{Pc}	m_b (Lg)	m_b	m_{Lg}	m_{Pc}	m_b
Slope	1.000	1.050	0.806	1.139	0.970	1.020	1.321
S.D.	0.055	0.048	0.095	0.065	0.078	0.064	0.079
N	62	62	60	60	17	17	17

Table 4

Date, origin time, coordinates and m_b (source: PDE/ISC) of 17 explosions from the Novaya Zemlya test site for which m_{Lg}- and m_{Pc}-coda magnitudes were determined. R is the signal-to-noise ratio of the Lg phase according to equation (7) and #C is the number of channels used for array averaging

Date	Origin Time	Lat.	Lon.	m_b	m_{Lg}	m_{Pc}	R	#C
29.09.1976	02:59:57.7	73.406	54.503	5.8	5.813	5.795	44.34	2
20.10.1976	07:59:57.8	73.402	54.472	5.1	4.955	4.962	1.32	3
01.09.1977	02:59:57.8	73.375	54.411	5.7	5.890	5.768	93.22	3
10.08.1978	07:59:57.7	73.315	54.697	5.9	5.838	5.973	50.11	4
27.09.1978	02:04:58.4	73.383	54.441	5.6	5.686	5.736	16.75	6
24.09.1979	03:29:58.4	73.373	54.578	5.7	5.854	5.867	42.03	7
18.10.1979	07:09:58.5	73.341	54.733	5.8	5.724	5.822	17.62	9
11.10.1980	07:09:57.0	73.353	54.997	5.7	5.766	5.808	13.80	11
01.10.1981	12:14:56.9	73.323	54.554	6.0	5.804	5.880	15.47	7
11.10.1982	07:14:58.4	73.371	54.342	5.6	5.603	5.562	12.83	13
18.08.1983	16:09:58.6	73.377	54.868	5.9	5.852	5.823	65.77	12
25.09.1983	13:09:57.7	73.349	54.377	5.8	5.819	5.744	42.32	12
25.10.1984	06:29:57.6	73.369	54.842	5.8	5.868	5.785	22.37	11
02.08.1987	02:00:00.1	73.346	54.579	5.8	5.835	5.818	76.60	12
07.05.1988	22:49:58.0	73.355	54.428	5.6	5.688	5.608	45.74	12
04.12.1988	05:19:52.8	73.385	54.960	5.9	5.831	5.819	27.95	13
24.10.1990	14:57:58.0	73.360	54.674	5.7	5.580	5.624	4.91	7

magnitudes in terms of the slope and standard deviation of the regression line. Also for this test site there is very good correlation between the rms amplitude values and the network-averaged body wave magnitudes, and, as in the case of Semipalatinsk explosions, the P-coda magnitude performs slightly better than m_{Lg}.

Summary and Conclusions

A comparison of regional and teleseismic log rms Lg amplitude measurements has been made for 14 underground nuclear explosions from the East Kazakh test site recorded at both the BRV station in Kazakhstan and the GRF array in Germany. It was found that, despite the substantial difference in distance, the log rms *Lg* amplitudes observed in the regional and teleseismic range show very similar relative values (standard deviation 0.048 magnitude units) for underground explosions of different sizes at the Shagan River test site. This result as well as the comparison of BRV rms *Lg* magnitudes with magnitude determinations for *P* waves recorded by global seismic networks (standard deviation 0.054 magnitude units) points to a high precision in estimating the relative source sizes of explosions from *Lg*-based single-station data. This high degree of consistency of the scaling of log rms *Lg* amplitudes shows considerable promise in providing precise relative yields of individual explosions. A similar conclusion was reached by RINGDAL *et al.* (1992) in a comparison of teleseismic GRF and NORSAR data. Relative yields can apparently be estimated from observations with high signal-to-noise ratios from a single station or an array with a precision comparable to those from global averages of *P* waves.

In view of the vast distances between the Semipalatinsk and Novaya Zemlya test sites and the GRF array, reliable teleseismic measurements of *Lg* amplitudes can only be made for explosions of approximately $m_b = 5.5$ or greater. It is interesting to estimate to which limit the size of smaller events can be calculated using regional *Lg* amplitude measurements. For this purpose, the *Lg* signal-to-noise ratios of Table 1 can be used for determining how weak an explosion can be and still be detected at the regional distance range. A rough estimate can be achieved by scaling down the signal-to-noise ratio of the Joint Verification Experiment (JVE) explosion (No. 13 in Table 1) conducted at the Soviet test site in Eastern Kazakhstan on 14 September 1988. Although no official yield value has been announced yet for this explosion, it is known that the US-USSR agreement stipulated that the yield of the JVE explosion was to be between 100 and 150 kilotons (kt) TNT equivalent. Indeed, SYKES and EKSTRÖM (1989) give values of 115–122 kt for the Soviet JVE. From the signal-to-noise ratio of 104,000 and an assumed yield of ≈120,000 tons for the JVE, it follows that the *Lg* detection threshold for BRV station is of an order of magnitude of under one hundred tons for explosions on the Shagan River Test Site (using the slope of the linear water-saturated rock calibration curve of NUTTLI (1986a) and assuming that the explosions are fired under comparable conditions). This threshold is far below the anticipated global threshold of 1 kt of the International Monitoring System (IMS), which is being set up to monitor compliance with the CTBT (Comprehensive Nuclear-Test-Ban Treaty). It should be mentioned, however, that the given yield threshold may be too conservative, since LEITH and KLUCHKO (1998) even observed clear signals at BRV from 25 t chemical explosions at the Semipalatinsk test site.

The previous section presented a calculation of GRF m_{Lg} magnitudes for a large data set exceeding 60 explosions, to test the stability of GRF rms *Lg* amplitude measurements against worldwide m_b. As possible alternatives, two other measurements of the radiated seismic energy, rms *P*-coda magnitudes m_{Pc} and a modified version of Nuttli's $m_b(Lg)$ were calculated for the same large data set, and were both also compared to the worldwide m_b to test the relative performance of the three magnitude scales. It results that there is a high degree of consistency of GRF m_{Lg} and worldwide network m_b of the PDE/ISC in terms of slope (close to unity) and small standard deviation (Table 3). From the calculated slope S and standard deviation $S.D.$ values of Table 3, it is found that the m_{Lg} and the GRF rms P-coda magnitudes m_{Pc} perform equally well, the m_{Pc} (smaller $S.D.$) slightly better; this applies for both the Semipalatinsk and Novaya Zemlya explosion data set. Compared to these two magnitude scales, the modified Nuttli $m_b(Lg)$ magnitudes display greater scatter when plotted against the worldwide m_b reference magnitudes (Table 3). However, a final assessment cannot be made regarding whether this result indicates that the rms amplitude measurements are superior to the zero-to-peak amplitude measurement of a single cycle, since the calculated $m_b(Lg)$ magnitudes are only preliminary until appropriate attenuation corrections are available for the specific path to GRF. It is interesting to note in this context that PRIESTLY and PATTON (1997) found no significant difference in scatter using rms and "third peak" (single cycle) amplitude measurements to estimate $m_b(Lg)$ for central Asian earthquakes and explosions. However, despite these qualifications, the results obtained in this study apparently give evidence that the rms measurements are superior to the zero-to-peak amplitude measurements on which the modified NUTTLI (1973) method is based. This seems to be true especially for observations at long distances, for which maximum Lg has periods considerably longer than 1 sec (Table 2) and little is known from previous studies about calibration amplitudes outside the 1 Hz range.

Acknowledgements

This research was funded by the Deutsches Auswärtiges Amt (German Foreign Office) within the scope of work done for the installation of a National Data Center (NDC) at the Federal Institute for Geosciences and Natural Resources, Hannover. I thank the editor and two anonymous reviewers for useful comments, which helped to improve the quality of this paper.

REFERENCES

ADUSHKIN, V. V., and AN, V. A. (1990), *Seismic Observations and Monitoring of Underground Nuclear Explosions at Borovoye Geophysical Observatory*, Izv. Acad. Sci. USSR, Phys. Solid Earth *12*, 1023–1031.

HANSEN, R. A., RINGDAL, F., and RICHARDS, P. G. (1990), *The Stability of RMS Lg Measurements and their Potential for Accurate Estimation of the Yields of Soviet Underground Nuclear Explosions*, Bull. Seismol. Soc. Am. *80*, Part B.

HERMANN, R. B. (1980), *Q Estimates Using the Coda of Local Earthquakes*, Bull. Seismol. Soc. Am. *70*, 447–468.

KIM, W.-Y., and EKSTRÖM, G. (1996), *Instrument Responses of Digital Seismographs at Borovoye, Kazakhstan, by Inversion of Transienst Calibration Pulses*, Bull. Seismol. Soc. Am. *86*, 191–203.

LEITH, W., and KLUCHKO, J. (1998), *Seismic Experiments, Nuclear Dismantlement Go Hand in Hand in Kazakhstan*, EOS, Transact. Am. Geophys. Union *79*, (37), 437–444.

NUTTLI, O. W. (1973), *Seismic wave attenuation and magnitude relations for eastern North America*, J. Geophys. Res. *78*, 876–885.

NUTTLI, O. W. (1986a), *Yield Estimates of Nevada Test Site Explosions Obtained from Seismic L_g Waves*, J. Geophys. Res. *91*, 2137–2151.

NUTTLI, O. W. (1986b), *L_g Magnitudes of Selected East Kazakhstan Underground Explosions*, Bull. Seismol. Soc. Am. *76*, 1241–1251.

PATTON, H. J. (1988), *Application of Nuttli's Method to Estimate Yields of Nevada Test Site Explosions Recorded on Lawrence Livermore National Laboratory's Digital Seismic System*, Bull. Seismol. Soc. Am. *78*, 1759–1772.

PRIESTLY, K. F., and PATTON, H. J. (1997), *Calibration of $m_b(P_n)$, $m_b(Lg)$ Scales and Transportability of the M_0:m_b Discriminant to New Tectonic Regions*, Bull. Seismol. Soc. Am. *87*, 1083–11099.

RICHARDS, P. G., KIM, W.-Y., and EKSTRÖM, G. (1992), *Borovoye Geophysical Observatory, Kazakhstan*, EOS, Transact. Am. Geophys. Union *73*, 201–206.

RINGDAL, F. (1983), *Magnitudes from P Coda and L_g Using NORSAR Data*, NORSAR Semiannual Technical Summary, 1 Oct. 82–31 Mar. 1983, NORSAR Sci. Rep. No. 2–82/83, Kjeller, Norway.

RINGDAL, F., MARSHALL, P. D., and ALEWINE, R. W. (1992), *Seismic Yield Determination of Soviet Underground Nuclear Explosions at the Shagan River Test Site*, Geophys. J. Int. *109*, 65–77.

SCHLITTENHARDT, J. (1993), *Erste Ergebnisse der Datenanalyse der seimischen Station Borovoye (BRV), GUS (in German)*. In *Seismische Verifikation*, Annual reports of Bundesanstalt für Geowissenschaften und Rohstoffe (BGR), Hannover, Germany, 21–35.

SEIDL, D. (1980), *The Simulation Problem for Broadband Seismograms*, J. Geophys. *48*, 84–93.

SYKES, L. R., and EKSTRÖM, G. (1989), *Comparison of Seismic and Hydrodynamic Yield Determinations for the Soviet Joint Verification Experiment of 1988*, Proc. Nat. Acad. Sci., USA, *86*, 3456–3460.

(Received July 15, 1999, revised January 20, 2000, accepted March 8, 2000)

Pure appl. geophys. 158 (2001) 2275–2290
0033–4553/01/112275–16 $ 1.50 + 0.20/0

Pure and Applied Geophysics

Classical and Bayesian Seismic Yield Estimation: The 1998 Indian and Pakistani Tests

ROBERT H. SHUMWAY[1]

Abstract — The nuclear tests in May, 1998, in India and Pakistan have stimulated a renewed interest in yield estimation, based on limited data from uncalibrated test sites. We study here the problem of estimating yields using classical and Bayesian methods developed by SHUMWAY (1992), utilizing calibration data from the Semipalatinsk test site and measured magnitudes for the 1998 Indian and Pakistani tests given by MURPHY (1998). Calibration is done using multivariate classical or Bayesian linear regression, depending on the availability of measured magnitude-yield data and prior information. Confidence intervals for the classical approach are derived applying an extension of Fieller's method suggested by BROWN (1982). In the case where prior information is available, the posterior predictive magnitude densities are inverted to give posterior intervals for yield. Intervals obtained using the joint distribution of magnitudes are comparable to the single-magnitude estimates produced by MURPHY (1998) and reinforce the conclusion that the announced yields of the Indian and Pakistani tests were too high.

Key words: Multivariate regression, calibration, predictive Bayes, magnitude estimation, nuclear explosions, comprehensive test-ban treaty.

1. Introduction

Recent nuclear tests in May of 1998 by India and Pakistan have highlighted the problem of estimating yields from relatively limited seismic data that have been collected while monitoring an uncalibrated test site. In general, the problems involved in monitoring well calibrated test sites such as the U.S. Nevada Test Site (NTS) or the former Soviet Union test site at Semipalatinsk have been studied fairly thoroughly and one is fairly confident that the calibration relations between magnitude measures and yield are quite good. There will still be uncertainty in the estimated yield arising from errors in the magnitude-yield relation so that the confidence intervals obtained from the regression relations may be wider than one would think.

Traditionally, yield has been estimated from seismic magnitudes using empirically calibrated magnitude-yield relations based on the traditional teleseismic m_b and M_s

[1] Division of Statistics, University of California, Davis, CA 95616, USA.
E-mail: shumway@wald.ucdavis.edu

and the regional $M_b(L_g)$ magnitude measures. To illustrate, consider Figure 1 in which plots relating yield to m_b and M_s to yield are shown for a set of 16 explosions from the Semipalatinsk Test Site (STS), as given by BOCHAROV *et al.* (1989) and by VERGINO (1989). The plots also show the fitted straight lines relating the two magnitude measures to log yield and we note that there is a strong linear trend with yield in both magnitude components. Similar plots involving data from NTS show comparable linearity although the coefficients vary; this variability is a concern when magnitude-yield relations are transported to other sites such as those in India and Pakistan. Note also that some experts consider the lowest m_b magnitude in Figure 1 to be an outlier and discard this point from the regression analysis; this leads to a modified intercept and slope of 4.32 and 0.78, respectively. This smaller slope and larger intercept are more consistent with published m_b-yield relation for STS and with other exchange data. In this sense, these values serve as reasonable inputs for a prior distribution in our Bayesian analyses. Evidence accumulated by MURPHY (1998) suggests that the Semipalatinsk experience is more appropriate than the NTS

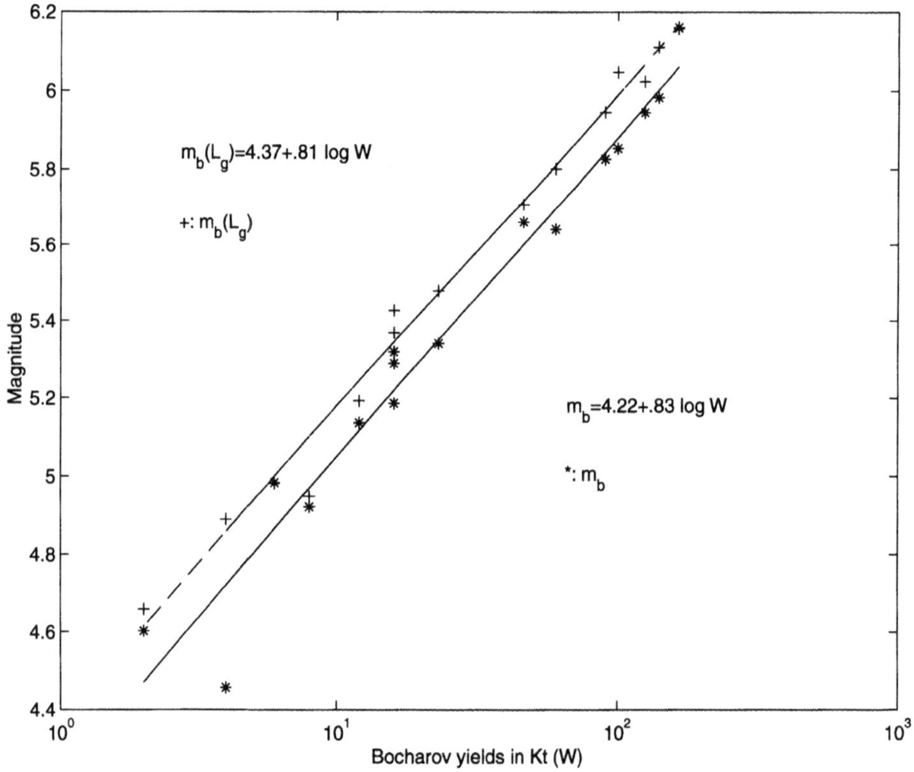

Figure 1

Scatter plots relating log yield to m_b and $m_b(L_g)$ magnitudes and fitted regression lines. Yields are assumed to be exact.

relations for yield estimation at the Indian and Pakistani test sites and we will use STS relations in our yield estimation procedures. There will be prior uncertainty in the magnitude-yield relations that is incorporated via the Bayesian procedure however this uncertainty is insufficient to include the NTS relations and there will still be potential for error that accrues simply from the transport assumption.

For yield estimation, the linearity assumption and the regression relations implied by Figure 1 are used to estimate yield by inverting the linear regression equations shown in the figure. That is, the magnitude measurements available from a presumed nuclear explosion are combined to produce and estimator for yield, regarded as a fixed unknown quantity. In the case of the Indian and Pakistani events, Table 1 shows an extracted set of m_b measures, as compiled by MURPHY (1998), as compiled by Murphy from values published by the Prototype Internationl Data Center (PIDC). The regional magnitudes, $m_b(L_g)$, reported by Murphy, are for only three recording stations; PIDC surface-wave magnitudes are also derived from only three stations. The absence of reliable joint calibration results for common events, even at the Semipalatinsk site, suggests two possible courses of action for deriving yield estimators. One may develop single magnitude yield estimators by inverting the regression relation for each of the magnitudes separately, as in MURPHY (1998). Applying this method to the m_b and $m_b(L_g)$ measures in Table 1, for example, will give yield estimates of 20 kt for the m_b and 8, 5 and 2 kt for the $m_b(L_g)$ measure.

Alternately, we may attempt to combine observed data and prior evidence in terms of prior distributions that express realistically our uncertainty about the magnitude-yield relation and our uncertainty about the accuracy of the measurements associated with the event in question. We can depend on the work of MURPHY (1998) and the calibrated m_b and $M_b(L_g)$ relations that can be obtained from Figure 1 to set sensible prior distributions for the error variances and the regression relations. There will then be an optimal method for combining the estimators into an overall estimate of yield. In the case of the Indian and Pakistani events, MURPHY (1998) has argued that the far-field *P*-wave spectra (see also MURPHY *et al.*, 1989) from the Indian and Pakistani tests are more similar to Semipalatinsk than NTS. Therefore, we assume here that the scatter exemplified by the STS data is indicative also of the

Table 1

Magnitudes recorded by the PIDC for the Indian and Pakistani events (MURPHY, 1998). Number of recording stations in parentheses

Event	m_b	$m_b(L_g)$	M_s	Announced yield (kt)
India (5/11/98)	5.00 (51)	4.87 (3)	3.26 (3)	55
Pakistan 1 (5/28/98)	4.87 (49)		3.64 (3)	40–45
Pakistan 2 (5/30/98)	4.63 (45)			15–18

combined uncertainty of the transferal of the STS curves and the scatter that would be experienced by a large set of tests in India and Pakistan, if such data were available. Hence, it should be possible to combine the calibration data from Figure 1 into our determination of overall yield.

We see the statistical problem of yield estimation, under the uncertainties pointed out above, as being one of calibration with a limited amount of data, subject to uncertainty relative to the slopes and intercepts of the magnitude-yield regression relation as well as in the errors of the magnitude measurements of the event in question. If there are a number of magnitude-yield pairs available for calibration, it is possible to give exact confidence intervals for an unknown future yield using a multivariate version of FIELLER'S (1954) method and we give this solution, due to BROWN (1982). It is also natural to quantify the initial uncertainties in terms of prior distributions and then to provide a unified or combined yield estimator that does not depend on any particular calibration data. We do this by inverting the posterior predictive distribution for magnitude and solving for magnitude. We derive solutions assuming (A) only data are available, (B) only prior information is available and (C) both data and prior distributions are available.

2. *Modeling Data and Prior Information*

Figure 1 implies that a linear model relating log-yield, w, to either of the magnitudes is reasonable and evidence from other research indicates that other kinds of magnitude measures behave in a similar manner. Then, the linear model relating the vector of p magnitudes from the i-th event, $\mathbf{m}_i = (m_{i1}, \dots, m_{ip})', i = 1 \dots, N$, to the log-yields, $w_i, i = 1 \dots, N$, can be written in the form

$$\mathbf{m}_i = \mathbf{a} + \mathbf{b}w_i + \mathbf{e}_i \quad , \tag{1}$$

where $\mathbf{a}$ is the $p \times 1$ vector of unknown intercepts and $\mathbf{b}$ is the $p \times 1$ vector of slopes. We use $'$ to denote the transpose. The error vectors $\mathbf{e}_i$ are assumed to be independent zero-mean multivariate normal random variables with common covariance matrix Σ. The unknown parameters are the slope and intercept vector and the unknown covariance matrix. These will be considered as fixed and unknown in the classical analysis and subject to prior distributions in the Bayesian approach. The yields are assumed to be fixed in our analysis, as opposed to approaches in BROWN (1982) or PICARD and BRYSON (1992), who assign uniform prior distributions to yield.

It is convenient to collect the sample of observed vector magnitudes in the overall $N \times p$ matrix $Y = (\mathbf{m}_1, \dots, \mathbf{m}_N)'$ and to collect the error vectors in a similar matrix, denoted here by $E = (\mathbf{e}_1, \dots, \mathbf{e}_N)'$. Then, defining the matrices

$$X = \begin{pmatrix} 1 & 1 & \cdots & 1 \\ w_1 & w_2 & \cdots & w_N \end{pmatrix}' \tag{2}$$

and

$$B = (\mathbf{a}, \mathbf{b})' \tag{3}$$

leads to writing (1) as the overall multivariate linear model

$$Y = XB + E \tag{4}$$

as is given for example, in ANDERSON (1984, Chapter 8). One can consider several options for estimating B and Σ in this model, depending on whether or not any weight is to be given to prior information. The safest approach is to ask for a reasonably large set of calibration events and then simply let the observed data determine estimators of B and Σ . If $(N-2) > p$, then the usual estimators are

$$\hat{B} = (X'X)^{-1}X'Y \tag{5}$$

and

$$\hat{\Sigma} = (N-2)^{-1}(Y - X\hat{B})'(Y - X\hat{B}) \ . \tag{6}$$

For example, the model implied by Figure 1 for the bivariate vector of magnitudes would have an estimated regression matrix of the form

$$\hat{B} = \begin{pmatrix} 4.22 & 4.37 \\ 0.83 & 0.81 \end{pmatrix} ,$$

where the intercepts appear in the matrix. For later discussion, it is convenient to define the 2×2 matrix

$$C = (XX')^{-1} \ . \tag{7}$$

Noting the partitioned form of B in (3), it is easy to show that

$$\hat{\mathbf{b}} = \frac{\Sigma(w_i - \bar{w})(\mathbf{m}_i - \bar{\mathbf{m}})}{\Sigma(w_i - \bar{w})^2} \tag{8}$$

and

$$\hat{\mathbf{a}} = \bar{\mathbf{m}} - \hat{\mathbf{b}}\bar{w} \ , \tag{9}$$

with $\bar{\mathbf{m}}$ the sample mean of the magnitude vectors, are alternate expressions for the estimated slope and intercept that take the more familiar form.

If there are a limited number of data points, it may be important to bring in prior information about the intercept and slope components of B and the error covariance matrix Σ . Suppose, for example, that the uncertainty in the covariance matrix Σ can be expressed in terms of the inverted-Wishart distribution (see ANDERSON, 1984, Section 7.7.1) with parameters m and $\Psi = m\Sigma_0$, where Σ_0 is the prior covariance matrix corresponding to the assumed prior value of Σ and m is a parameter expressing the initial uncertainty about the value of the error covariance matrix. The value of m corresponds roughly to the degrees of freedom or equivalent sample size achieved by the accumulated prior information. In the case $p = 1$, the distribution reduces to an inverted chi-squared distribution, i.e., $m\sigma_0^2/\sigma^2$ has an ordinary chi-squared distribution.

The slope and intercept components in $B = (\mathbf{a}, \mathbf{b})'$ are assumed to be (conditional on Σ) normally distributed. The stacked vector consisting of the columns of B is assumed to be normally distributed with mean components $(\mathbf{a}_0, \mathbf{b}_0)'$ and covariance matrix with elements of the form

$$\mathrm{cov}\left\{\begin{pmatrix} a_i \\ b_i \end{pmatrix}, \begin{pmatrix} a_j \\ b_j \end{pmatrix}\right\} = \sigma_{ij} D \ , \tag{10}$$

where the matrix D is chosen to reflect the joint uncertainty structure of $\mathbf{a}$ and $\mathbf{b}$. It corresponds with the joint uncertainty structure of the maximum likelihood estimator, $\hat{B}$, given above in (5), for which the comparable stacked vector has covariance matrix $\sigma_{ij} C$, where C is defined in (7).

In order to illustrate some of the considerations involved in specifying the joint prior distributions described above, consider the case in which we record only one magnitude ($p = 1$) and presume that the distribution of the initial variance is inverted chi-square with $m = 10$ or $m = 30$ and $\sigma_0^2 = 0.05^2$.

Figure 2 shows the probability densities of the standard deviation σ corresponding to the cases $m = 10$ and $m = 30$, with the underlying standard deviation

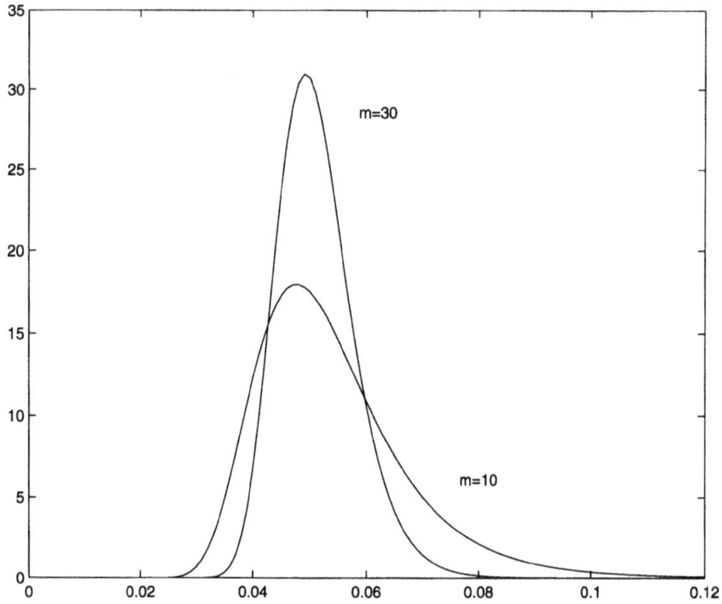

Figure 2
Prior uncertainty of the sample standard deviation for several values of m, the degrees of freedom.

$\sigma_0 = 0.05$. The underlying value is chosen to reflect the uncertainty in the error corresponding to a measured magnitude and will likely be larger than the error in the measured magnitude of the calibrating array. It is seen that the choice $m = 10$ corresponds to a wider uncertainty interval (roughly 0.03–0.12) whereas the choice $m = 30$ narrows the interval (to roughly 0.04–0.07). One must also specify the uncertainty of the intercept and slope components.

In Figure 3, we see a contour plot representing our initial uncertainty about the intercept a_1 and slope b_1 parameters of the magnitude-yield regression involving m_b. In this case, the mean values were assumed to have expectations $a_0 = 4.32$ and $b_0 = 0.78$ and standard deviations equal to 0.05. The values of the prior parameters correspond to those obtained using least-squares regression on the data in Figure 1, with the outlier deleted. The values are comparable to those that might be suggested by various experts. The correlation was taken as -0.9, corresponding to the typical magnitude-yield plot, where the increase in slope must necessarily lead to a decrease in the intercept. We notice from the plot that the initial slope uncertainty puts it in the range roughly between 0.65 and 0.90 about 90/intercept ranging from 4.10 to 4.55. We would also expect, in general, that expert input from panels and the various magnitude-yield calibration studies would be used to

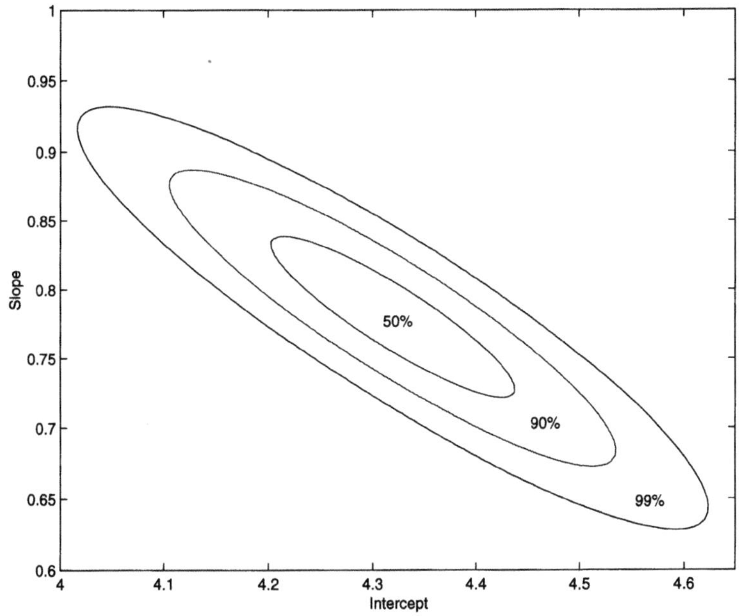

Figure 3
Typical contour plot expressing joint uncertainty of slope and intercept in the magnitude-yield relation. Contours represent 0.90, 0.95 and 0.99 prior probability regions, conditional on a fixed error variance $\sigma^2 = 0.0025$.

establish the prior uncertainty regions for the intercept and slope priors for the parameters **a**, **b** and Σ.

Now, given that the above prior information can be specified, it is natural that the estimates for the slopes, intercepts and error covariance may change from those specified by the maximum likelihood estimators (5) and (6), respectively. To do this, simply take the joint density of Y, B and Σ and derive the joint (conditional on Y) marginal densities of B and Σ. Taking the posterior means of these densities leads to Bayesian estimators proportional to

$$\tilde{B} = (C^{-1} + D^{-1})^{-1}(C^{-1}\hat{B} + D^{-1}B_0) \ , \tag{11}$$

with $\tilde{\mathbf{B}} = (\tilde{\mathbf{a}}, \tilde{\mathbf{b}})$ and

$$\tilde{\Sigma} = \frac{1}{n}\left[m\Sigma_0 + (N-2)\hat{\Sigma} + (\hat{B} - B_0)(C + D)^{-1}(\hat{B} - B_0)'\right] \ , \tag{12}$$

where $n = m + N + 1 - p$ are the pooled degrees of freedom. These estimators are seen to combine the prior panel information and the data to form a combined estimator for the intercepts, slopes and error covariance matrices.

It is clear that there are two routes that one may take towards estimating the slope from an uncalibrated test site. The first is to insist on fully weighting the available observations and use the ordinary multivariate regression estimators given in (5) and (6). If a reasonable number of calibration events are furnished and if one believes the yields to be known for these events, it is possible that the slope and intercept estimators will be reasonably good; the variances of the estimators can be read from the estimated covariance matrix $\hat{\sigma}_{ij}C$. However, there may be compelling reasons for incorporating prior information into the estimation procedure. The first is that there may be considerable geophysical information and expert opinion that can be used to narrow down the possible values for the slope, intercept and error variance. The second reason is that the observed data may have been deliberately modified to contribute to a distorted magnitude-yield picture. The Bayesian solution has the advantage of being able to weight optimally the two components of information contributed by the expert opinion (prior distribution) and the data.

This completes the specification of the method for estimating the slope and intercept vectors along with the error covariance matrix using classical and Bayes regression procedures. In the next section, we consider the problems involved in estimating a future unknown yield using the new magnitude vector and the calibrated regressions derived here.

3. Classical and Predictive Bayes Yield Estimation

We will use the approach described in the previous section and some plausible assumptions for prior distributions to compare the classical and Bayes estimators for the data given shown in Figure 1 in a later section. However, it is necessary first to discuss the estimation of the yield corresponding to a newly measured magnitude vector. That is, given the calibrated regression with values for $(\hat{\mathbf{a}}, \hat{\mathbf{b}})$ and the estimated error covariance $\hat{\Sigma}$, how can we use the new magnitude measurement $\mathbf{m}$ to infer a value and an uncertainty interval for the new unknown yield w?

One approach might be to simply regard the slopes, intercepts and error covariance matrices as known. Then, using the multivariate normal form for the new magnitude vector, $\mathbf{m}$, leads to the maximum likelihood estimator

$$\hat{w} = \frac{\mathbf{b}'\Sigma^{-1}(\mathbf{m} - \mathbf{a})}{\mathbf{b}'\Sigma^{-1}\mathbf{b}} \tag{13}$$

which is unbiased and has variance

$$\sigma^2_{\hat{w}} = \frac{1}{\mathbf{b}'\Sigma^{-1}\mathbf{b}} \tag{14}$$

under the assumption that all parameters are fixed and known. Normal theory then gives a well-defined confidence interval as well as a test of the hypothesis that the estimated yield came from a particular specified threshold against the alternative that it is significantly greater than the threshold. Of course, in (13) and (14) the values of $\mathbf{a}, \mathbf{b}$ and Σ are unknown. Replacing them with the estimators $\hat{\mathbf{a}}, \hat{\mathbf{b}}$ and $\hat{\Sigma}$ leads to intractable distribution theory for the yield estimator, but resampling via the bootstrap could be used to attain a confidence interval from the frequent viewpoint.

Here, we choose to incorporate the prior constraints in the form of joint distributions on the slope, intercept and covariances. We consider the three separate cases mentioned earlier, namely, (A) the *data-only* approach, (B) the *data-free* approach, and (C) the combined *data-prior information* approach. We concentrate on deriving confidence intervals for the new yield considered as a true fixed future observation in case (A) and on posterior predictive probability intervals in cases (B) and (C).

(A) Classical Calibration Intervals (Data-only)

We develop a classical interval by noting that for a new magnitude vector $\mathbf{m}$ and the new unknown yield w, the distribution of the residual

$$\mathbf{m} - \hat{\mathbf{a}} - \hat{\mathbf{b}}w$$

is multivariate normal with mean $\mathbf{0}$ and covariance matrix $(1 + q(\mathbf{x}))\Sigma$ where $\mathbf{x} = (1, w)'$ and

$$q(\mathbf{x}) = \mathbf{x}'C\mathbf{x} \ , \tag{15}$$

where C is given in (7). It is also the case that the residual is independent of $(N-2)\hat{\Sigma}$ in (6), which has the Wishart distribution with $N-2$ degrees of freedom. Then the quadratic form in the residuals with the inverse of the covariance matrix $(1+q(\mathbf{x}))\hat{\Sigma}$ in the middle has Hotelling's T^2 distribution as in ANDERSON (1984, p. 163) or BROWN (1982). Hence, the resulting inequality involving the quadratic form in the residuals can be inverted, using

$$q(\mathbf{x}) = (c_{11} + 2c_{12}w + c_{22}w^2) \ ,$$

where c_{ij} denotes an element of C. We obtain the $(1-\alpha)$ level confidence interval as

$$\frac{d}{c} \pm \frac{\sqrt{d^2 - ce}}{c} \tag{16}$$

where

$$c = \hat{\mathbf{b}}'\hat{\Sigma}^{-1}\hat{\mathbf{b}} - c_{22}K_{p,N}(\alpha) \ , \tag{17}$$

$$d = \hat{\mathbf{b}}'\hat{\Sigma}^{-1}(\mathbf{m} - \hat{\mathbf{a}}) + c_{12}K_{p,N}(\alpha) \ , \tag{18}$$

and

$$e = (\mathbf{m} - \hat{\mathbf{a}})'\hat{\Sigma}^{-1}(\mathbf{m} - \hat{\mathbf{a}}) - (1 + c_{11})K_{p,N}(\alpha) \ , \tag{19}$$

where

$$K_{p,N}(\alpha) = \frac{p(N-2)}{(N-p-1)} F_{p,N-p-1}(\alpha) \ , \tag{20}$$

with $F_{p,N-p-1}(\alpha)$ denoting the $(1-\alpha)$ critical point on the F distribution with p numerator and $N-p-1$ denominator degrees of freedom. The estimated regression coefficients are read from $\hat{B} = (\hat{\mathbf{a}}, \hat{\mathbf{b}})'$ in (5), with $\hat{\Sigma}$ computed as in (6). The elements of $C = \{c_{ij}\}$ are read from (7).

These confidence intervals, derived by BROWN (1982), are the exact analogues of the univariate intervals first given by FIELLER (1954). As in the univariate case, the intervals may be open, although this only occurs when some of the components of the vector vary widely from their regression lines. It should be noted that the interval in (15) and (16) is for log-yield; the interval on yield would be obtained by inverting the limits. OMAN (1988) has proposed an alternative procedure based on geometric arguments in the nonlinear case.

(B) Bayes Calibration Intervals (Prior only)

We may also look at deriving posterior predictive intervals under the assumption that only prior information is available in the form of the joint normal inverted-Wishart priors on B and Σ in Section 2. Now, denoting the prior inverted-Wishart distribution on Σ as $\pi_1(\Sigma)$ and the conditional density of the intercept and slope matrix B by $\pi_2(B|\Sigma)$, we may write the posterior predictive density of $\mathbf{m}$ as

$$P(\mathbf{m}) = \int p(\mathbf{m}|B,\Sigma)\pi_2(B|\Sigma)\pi_1(\Sigma)\, dB\, d\Sigma \ , \tag{21}$$

where $p(\mathbf{m}|B,\Sigma)$ is the probability density of the magnitude observation, conditional on a given slope-intercept combination and the covariance matrix Σ. The result of integrating first over B and then over Σ leads to a multivariate t distribution for $\mathbf{m}$ (see, for example, ANDERSON, 1984, p. 283). This implies that the quadratic form in the multivariate t distribution (involving the residuals $(\mathbf{m} - \mathbf{a}_0 - \mathbf{b}_0 w)$ and the inverse of a particular covariance matrix) has the F distribution. The resulting interval is exactly in the form of (16) with

$$c = \mathbf{b}_0'\Sigma_1^{-1}\mathbf{b}_0 - d_{22}K_{p,m}(\alpha) \ , \tag{22}$$

$$d = \mathbf{b}_0'\Sigma_1^{-1}(\mathbf{m} - \mathbf{a}_0) + d_{12}K_{p,m}(\alpha) \tag{23}$$

and

$$e = (\mathbf{m} - \mathbf{a}_0)'\Sigma_1^{-1}(\mathbf{m} - \mathbf{a}_0) - (1 + d_{11})K_{p,m}(\alpha) \ , \tag{24}$$

where d_{ij} is the i–jth element of D,

$$\Sigma_1 = \frac{m\Sigma_0}{(m - p + 1)} \tag{25}$$

and

$$K_{p,m}(\alpha) = pF_{p,m-p+1}(\alpha) \ . \tag{26}$$

Here, the computations involve the best guesses for the prior values of B and Σ, say $B_0 = (\mathbf{a} - 0, \mathbf{b}_0)'$ and Σ_0 and the degrees of freedom m. The elements of $D = \{d_{ij}\}$ result from substituting Σ_0 for Σ in (10), assuming that the left-hand side is some desired covariance.

(C) Bayesian Intervals (Data and Prior)

Finally, we look at the case in which both data and prior information are available. This means that we introduce the joint likelihood of the calibration data Y in the equation for the posterior predictive density. This leads to

$$P(\mathbf{m}|Y) = \int p(\mathbf{m}|B,\Sigma)L(Y|B,\Sigma)\pi_2(B|\Sigma)\pi_1(\Sigma)\, dB\, d\Sigma \ ,$$

where $L(Y|B,\Sigma)$ denotes the likelihood conditional density of the data Y, given the parameters B and Σ. Note that the conditional distribution of the rows of Y follows from $\mathbf{m}_i$ in (1), which will be multivariate normal with mean vector

$\mathbf{a} + \mathbf{b}w_i$ and covariance matrix Σ. Performing the above integration leads to the posterior t distribution as before and the interval (15) is obtained again with the values

$$c = \tilde{\mathbf{b}}'\tilde{\Sigma}^{-1}\tilde{\mathbf{b}} - a_{22}K_{p,m,N}(\alpha) \ , \tag{27}$$

$$d = \tilde{\mathbf{b}}'\tilde{\Sigma}^{-1}(\mathbf{m} - \tilde{\mathbf{a}}) + a_{12}K_{p,m,N}(\alpha) \tag{28}$$

and

$$e = (\mathbf{m} - \tilde{\mathbf{a}})'\tilde{\Sigma}^{-1}(\mathbf{m} - \tilde{\mathbf{a}}) - (1 + a_{11})K_{p,m,N}(\alpha) \ , \tag{29}$$

where $\tilde{B} = (\tilde{\mathbf{a}}, \tilde{\mathbf{b}})'$ so that $\tilde{\mathbf{a}}$ and $\tilde{\mathbf{b}}$ can be picked out of the Bayesian regression matrix computed in (10) in Section 2 and $\tilde{\Sigma}$ is defined in (11). In these expressions a_{ij} denotes the i–jth element of

$$A = (C^{-1} + D^{-1})^{-1} \tag{30}$$

and the constant is

$$K_{p,m,N}(\alpha) = pF_{p,m+N+1-p}(\alpha) \ . \tag{31}$$

It is interesting to note that confidence intervals from the classical point of view and the posterior Bayesian probability intervals for the logarithm of the yield, $\log w$, all have the same general form (16). While the classical intervals use values of c, d and e computed from (17)–(19), the Bayesian intervals use either (22)–(24) in the pure Bayes case or (27)–(29) when data is combined with prior information. For practical interpretation, it is conventional to convert the transformed log yields to ordinary yield. The estimated yield, in kt, is then

$$\hat{y} = 10^{d/c} \ . \tag{32}$$

The estimated lower and upper limits are $L = \hat{y}/k$ and $U = k\hat{y}$, where

$$k = \frac{\sqrt{d^2 - ce}}{c} \ . \tag{33}$$

The constant k that divides and multiplies corresponds to the *F numbers* used by ALEWINE *et al.* (1988) in their discussion of the issues raised by compliance with a proposed threshold test-ban treaty (TTBT).

4. *Analysis of the Indian and Pakistani Tests*

For the Indian and Pakistani tests, MURPHY (1998) has given a number of estimators for magnitudes represented by the ranges reported in Table 1. These ranges represent different values determined from PIDC data and additional data from the U.S. Geological Survey Earthquake Data Report (EDR). Table 2

Table 2

Single-magnitude estimators for the Indian and Pakistani yields (MURPHY, 1998) based on three separate magnitudes

Event	m_b	$m_b(L_g)$	M_s	Announced yield (kt)
India (5/11/98)	10–14	4–13	18–32	55
Pakistan 1 (5/28/98)	3–9		44	40–45
Pakistan 2 (5/30/98)	2–4			15–18

summarizes the estimators of yield based on the three primary magnitude values $m_b, m_b(L_g)$ and M_s, when those values are available. Note that the yield estimators in Table 2 are substantially lower than the announced yields shown in Table 1.

As we have previously noted, the prior uncertainty implied by the separate estimators from Table 1 can be combined with data to produce various kinds of classical and Bayes estimators, depending on which magnitude measures and which of the methods (A), (B) or (C) we are able to employ.

The magnitude measure m_b are comprised of averages of relatively large numbers of stations (see Table 1) and we focus on these values for estimating yield. The regression of magnitude on yield in Figure 1 produces

$$m_b = 4.22 + 0.83 \log w \ ,$$

using (5) as the basic equation with the estimated variance (6) $\hat{\sigma}^2 = 0.0098$. The estimated yield and 95% confidence interval are computed from (32) and (33) using (17)–(20). The values are comparable to those computed by MURPHY (1998), with the intervals being wider since Murphy computed only point estimators. It is possible that the m_b-yield relations implied by the Bocharov yields are quite far off and therefore the pure Bayesian strategy, described in (B) may be preferable. The second line in Table 3 assumes a magnitude-yield relation that is closer to the result of a regression that ignores the outlier in Figure 1, which implies the equation

$$m_b = 4.32 + 0.78 \log w \ .$$

This form is also more consistent with expert opinion and the results of other studies. The variance of the error pertaining to the regression line was chosen from the distribution plotted in Figure 1 with $\sigma_0 = 0.05$ and $m = 10$, so that we accept the possibility that this standard deviation might be as large as 0.09. Conditionally on the error variance, the joint distribution of the slope and intercept was taken as shown in Figure 3, so that relatively large range of values would be possible for these two parameters, for a fixed σ^2. Specifically, we took the standard deviations of the slope to be 0.05 with a larger standard deviation of 0.10 magnitude units for the intercept; the correlation between the slope and intercept was assumed to be –0.90. It is obvious that these assumptions allow for quite a broad range of errors in the

Table 3

Combined-magnitude estimators for the Indian and Pakistani yields. Table gives the estimated yield (32), with the lower and upper 95% uncertainty bounds in parentheses. The first three columns indicate the basis for the computation (see text)

Method	Data	Prior	India (5/11/98)	Pakistan 1	Pakistan 2
(A)	m_b	–	8.4 (4.5–15.7)	5.9 (3.1–11.0)	3.0 (1.6–5.8)
(B)	–	m_b	7.1 (4.2–12.2)	4.8 (2.7–8.5)	2.3 (1.2–4.4)
(C)	m_b	m_b	8.4 (5.5–12.9)	5.8 (3.8–9.0)	3.0 (1.9–4.7)
(B)	–	M_s		43.0 (22.7–81.2)	
(A)	$m_b, m_b(L_g)$	–	4.9 (3.7–6.5)	–	–
(B)	–	$m_b, m_b(L_g)$	5.3 (3.4–8.4)	–	–
(C)	$m_b, m_b(L_g)$	$m_b, m_b(L_g)$	5.1 (4.6–5.7)	–	–
(B)	–	m_b, M_s		9.0 (Imag.)	
(B)	–	$m_b, m_b(L_g), M_s$	6.4 (4.6–8.9)	–	–

extrapolating of the regression parameters to the uncalibrated Indian and Pakistani test sites. Again, we use (32) and (33) for estimating yield, with the 95% posterior probability intervals determined from (22)–(26). For the interval that uses both measured m_b in Figure 1 and the prior assumptions above, we estimate the 95% posterior probability intervals from (27)–(31). All methods give relatively consistent results that are consistent with those computed by Murphy in Table 2.

For the Bayes estimator, using only the M_s value, it is clear that the estimator does not agree with the yields computed using m_b, although it is more consistent with the announced yield in Table 2. This estimator uses the regression relation

$$M_s = 2.00 + 1.00 \log w$$

suggested by MURPHY (1998), with a larger standard deviation $\sigma_0 = 0.10$ assigned to the error. The estimated yield is much larger than before, with a relatively wide confidence interval. MURPHY (1998) has suggested that the larger M_s value may have been biased by tectonic release effects.

The results of employing the two magnitudes m_b and $m_b(L_g)$ jointly are also shown in Table 3. The regression equations are the same as those given above except that

$$m_b(L_g) = 4.36 + 0.81 \log w$$

is added for L_g magnitude. For methods (B) and (C) employing priors, we used the initial estimators

$$B_0 = \begin{pmatrix} 4.32 & 4.36 & 2.00 \\ 0.78 & 0.81 & 1.00 \end{pmatrix}$$

for the prior regression matrix, as given above. For the prior covariance matrix of the magnitude errors, we chose the diagonal matrix

$$\Sigma_0 = \text{diag}(0.0025, 0.0025, 0.0100) \ ,$$

since an initial bivariate regression with the m_b and $m_b(L_g)$ pairs had a small correlation coefficient ($\hat{\rho} = 0.0219$). The initial covariance matrix of the intercepts and slopes was chosen in accordance with the values mentioned above, with slopes and standard deviations of 0.10 and 0.05 assumed for the $m_b(L_g)$ coefficients. Again, all three methods (A), (B) and (C) give relatively consistent estimators for yield, although the combined estimators are slightly lower than those using m_b alone. The 95% confidence and posterior probability intervals are smaller, showing the effect of using the combined-magnitude approach. The result of using all three magnitudes in the Bayes approach, given in the last line of Table 3, narrows the uncertainty even further. The addition of an M_s magnitude here that is relatively consistent with expectations does not distort the result yield estimators for the Indian test. Attempting to incorporate the inconsistent M_s value for the second Pakistani test gives imaginary posterior probability interval.

5. Discussion

There are many plausible approaches to calibration and estimation in the multivariate case. One must not only be prepared to work with calibration data when it is available and appropriate but must be prepared to develop yield estimators when no calibration data is available and when observations on yield are sketchy or poorly known. In such an environment, the use of prior information provided by experts and by seismic studies of different regions of the earth becomes paramount.

In this paper, we have discussed the application of both data-analytic and Bayes approaches to yield estimation for the recent Indian and Pakistani tests. The results, which indicate with high confidence that the yields are considerably lower than those claimed, confirm the results of MURPHY (1998), who made single-magnitude calculations under various configurations for inputs. It should be noted that much of the analysis given here depends either explicitly or implicitly on the assumption that the Indian and Pakistani have more in common with the former Soviet site at Palatinsk (STS) than with the Nevada Test Site (NTS). The confirmation of this hypothesis awaits the results of more detailed seismological studies of the attenuation characteristics of the Indian and Pakistani sites.

Acknowledgements

The original research reported here was supported, in part, by the Air Force Technical Applications Commission (AFTAC) under Contract No. F19268-91-K-0033, issued by Phillips Laboratory. I am indebted to J.R. Murphy of Maxwell Technologies for providing the technical presentation and the transparencies that suggested applying the unified yield estimation methodology. Reviewers H.L. Gray and K.L. McLaughlin contributed substantive comments that improved the presentation. Any remaining errors are solely my responsibility.

References

ALEWINE, R. W. III., GRAY, H. L., MCCARTOR, G. D., and WILSON, G. L. (1988), *Seismic Monitoring of a Threshold Test-ban Treaty (TTBT) Following Calibration of the Test Site with CORRTEX Experiments*, AFGL-TR-88-0055, Air Force Geophysics Laboratory, Hanscom AFB, Massachusetts 01731.

ANDERSON, T. W., *Multivariate Analysis*, Second Edition (New York: John Wiley, 1984).

BOCHAROV, V. S., ZELENTSOV, S. A., and MIKHAILOV, V. I. (1989), *Characteristics of 96 Underground Nuclear Explosions at the Semipalatinsk Test Facility*, Atomic Energy *67*(3), 210–214 (in Russian).

BROWN, P. J. (1982), *Multivariate Calibration*, J. R. Statist. Soc. *B44*, 287–221.

FIELLER, E. C. (1954), *Some Problems in Interval Estimation*, J. R. Statist. Soc. *B16*, 175–185.

MURPHY, J. R. (1998), *Seismic yield estimation for the Indian and Pakistani nuclear tests of May, 1998*. In *20th Annual Seismic Research Symposium on Monitoring a CTBT*, Sante Fe, NM, September 23, Maxwell Technology Systems Division, Reston Geophysics Office, 11800 Sunrise Valley Drive, Suite 1212, Reston, VA 20191-5309.

MURPHY, J. R., BARKER, B. W., and O'DONNELL, A. (1989), *Network-averaged Teleseismic P-wave Spectra for Undergound Explosions. Part 1. Definitions and Examples*, Bull. Seismol. Soc. Am. *79*, 141–155.

OMAN, S. D. (1988), *Confidence Regions in Multivariate Calibration*, Ann. Statist. *16*, 174–187.

PICARD, R., and BRYSON, M. (1992), *Calibrated Seismic Verification of the Threshold Test-ban Treaty*, J. Am. Statist. Assoc. *87*, 293–299.

SHUMWAY, R. H. (1992), *Multivariate Calibration and Yield Estimation*, Technical Report No. 245, Division of Statistics, University of California, Davis, 95616.

VERGINO, E. S. (1989), *Soviet Test Yields*, EOS *70*, 1511.

(Received June 19, 1999, revised December 23, 1999, accepted January 7, 2000)

GPSR Compliance

The European Union's (EU) General Product Safety Regulation (GPSR) is a set of rules that requires consumer products to be safe and our obligations to ensure this.

If you have any concerns about our products, you can contact us on ProductSafety@springernature.com

In case Publisher is established outside the EU, the EU authorized representative is:

Springer Nature Customer Service Center GmbH
Europaplatz 3
69115 Heidelberg, Germany

Batch number: 10726322

Printed by Printforce, the Netherlands